Kenneth E. Maly
Synthesis of Aromatic Compounds

Also of interest

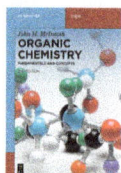

Organic Chemistry.
Fundamentals and Concepts
McIntosh, 2022
ISBN 978-3-11-077820-5, e-ISBN 978-3-11-077831-1

Supramolecular Chemistry.
From Concepts to Applications
Kubik, 2020
ISBN 978-3-11-059560-4, e-ISBN 978-3-11-059561-1

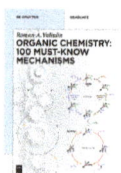

Organic Chemistry: 100 Must-Know Mechanisms
Valiulin, 2020
ISBN 978-3-11-060830-4, e-ISBN 978-3-11-060837-3

High Pressure Organic Synthesis
Margetic, 2019
ISBN 978-3-11-068964-8, e-ISBN 978-3-11-068966-2

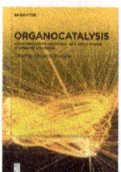

Organocatalysis.
Stereoselective Reactions and Applications in Organic Synthesis
Benaglia (Ed.), 2021
ISBN 978-3-11-058803-3, e-ISBN 978-3-11-059005-0

Kenneth E. Maly

Synthesis of Aromatic Compounds

—

DE GRUYTER

Author
Prof. Dr. Kenneth E. Maly
Department of Chemistry and Biochemistry
Wilfrid Laurier University
University Avenue West
Waterloo
ON
Canada N2L 3C5

ISBN 978-3-11-056267-5
e-ISBN (PDF) 978-3-11-056268-2
e-ISBN (EPUB) 978-3-11-056368-9

Library of Congress Control Number: 2022935630

Bibliographic information published by the Deutsche Nationalbibliothek
The Deutsche Nationalbibliothek lists this publication in the Deutsche Nationalbibliografie;
detailed bibliographic data are available on the Internet at http://dnb.dnb.de.

© 2022 Walter de Gruyter GmbH, Berlin/Boston
Cover image: geopaul/iStock/Getty Images Plus
Typesetting: Integra Software Services Pvt. Ltd.
Printing and binding: CPI books GmbH, Leck

www.degruyter.com

Preface

Aromatic compounds are present everywhere, in the foods we eat, the medicines we take, the clothes we wear, and materials we use on a daily basis. To organic chemists, aromatic compounds feature prominently in natural products, synthetic pharmaceuticals, and materials such as dyes and displays. As such, understanding the reactivity of aromatic compounds and applying this understanding to the synthesis of aromatic compounds is an important endeavor.

This book was developed from a specialized graduate course on the chemistry of aromatic compounds and is targeted at an advanced undergraduate or introductory graduate student level. The hope is that this book will also serve as a useful resource for practicing organic chemists and a source of inspiration for those with an interest in the synthetic organic chemistry. The book begins with some topics that should be review for most students of organic chemistry and builds on this foundation to introduce new reactions, concepts, and structurally interesting targets. An attempt is made to highlight the applications of reactions in the multistep synthesis of complex targets, including natural products, dyes, pharmaceuticals, materials, and structurally challenging aromatic compounds. As we progress through the chapters, the number of examples increases as the number of tools at our disposal is developed.

Chapter 1 begins with a general discussion on the importance of aromatic compounds, their history, and some of the features of aromatic compounds. The question of "what is aromaticity" is an ongoing debate and is not resolved in these pages. Instead, some of the properties and features of aromatic compounds are discussed. Chapter 2 focuses on electrophilic aromatic substitution – a topic that should be mostly review from introductory courses, but which hopefully provides some more depth of coverage. Chapter 3 turns to nucleophilic aromatic substitution and the chemistry of aryldiazonium salts – while some of the topics are likely review, several sections focus on aspects of nucleophilic aromatic substitution that are not usually covered in introductory courses. Chapters 4 and 5 focus on the chemistry of aryllithiums (including directed *ortho* metalation) and benzynes, respectively. While these are sometimes introduced in early organic chemistry courses, these chapters present new reactivity for many readers. Chapters 6 and 7 focus on transition metal-mediated reactions of aromatic compounds – reactions that have become indispensable tools for organic chemists and have modernized the chemistry of aromatic compounds. Chapter 8 explores approaches for constructing aromatic and heteroaromatic rings from nonaromatic precursors. This concept then sets the stage for Chapters 9 and 10, which focus on the reactivity and synthesis of polycyclic aromatic hydrocarbons. The final chapter turns to the synthesis of nonplanar polycyclic aromatic hydrocarbons. The last two chapters highlight some of the fascinating and challenging structures that can be achieved using the reactivity described in earlier chapters.

https://doi.org/10.1515/9783110562682-202

My hope is that readers will not only further their understanding of aromatic compounds and synthetic methods but also appreciate the importance and complexity of aromatic structures as well as their beauty.

Contents

Preface —— V

List of abbreviations —— XIII

1	Introduction – aromatic compounds and aromaticity —— 1	
1.1	Importance of aromatic compounds —— 1	
1.1.1	Synthetic dyes —— 1	
1.1.2	Biologically active compounds —— 3	
1.1.3	Aromatic compounds in materials science —— 4	
1.2	Aromaticity —— 5	
1.2.1	Stability of aromatic compounds —— 6	
1.2.2	Structure and reactivity of aromatic compounds —— 8	
1.2.3	Magnetic properties of aromatic compounds —— 8	
1.3	In this book —— 10	
	References —— 11	

2	Electrophilic aromatic substitution —— 13	
2.1	Introduction —— 13	
2.2	General mechanistic principles —— 13	
2.3	Scope of electrophilic aromatic substitution —— 13	
2.3.1	Nitration —— 14	
2.3.2	Sulfonation —— 15	
2.3.3	Halogenation —— 16	
2.3.4	Carbon electrophiles: Friedel–Crafts alkylation and acylation —— 17	
2.4	Substituent effects in electrophilic aromatic substitution —— 20	
2.4.1	Reactivity —— 20	
2.4.2	Regioselectivity of electrophilic substitution of aromatic compounds —— 21	
2.4.3	Regioselectivity in polysubstituted benzenes —— 23	
2.5	Functional group interconversion to control reactivity and regioselectivity —— 26	
2.5.1	Reduction of nitro groups —— 26	
2.5.2	Reduction of acyl groups —— 27	
2.5.3	Reactions at benzylic positions —— 28	
2.5.4	Acylation of phenols and anilines —— 29	
2.6	Reversibility of electrophilic aromatic substitution —— 30	
2.6.1	Protiodesilylation —— 31	
2.6.2	Halodesilylation —— 31	
2.7	Electrophilic rearrangement – the Fries rearrangement —— 32	

2.8   Electrophilic aromatic substitution of polycyclic aromatic
           compounds —— **33**
2.9   Electrophilic aromatic substitution of heteroaromatic
           compounds —— **34**
2.9.1   Five-membered heterocycles —— **34**
2.9.2   Six-membered heterocycles —— **36**
2.10   Further reading —— **36**
           References —— **36**

3   Nucleophilic aromatic substitution reactions —— 39
3.1   Introduction —— **39**
3.2   Addition–elimination of nucleophiles (S$_N$Ar) —— **39**
3.2.1   Leaving groups in S$_N$Ar —— **41**
3.2.2   Regioselectivity in S$_N$Ar reactions —— **41**
3.2.3   Examples of the utility of S$_N$Ar reactions —— **43**
3.2.4   Carbon nucleophiles in nucleophilic aromatic substitution —— **45**
3.2.5   Reversibility of S$_N$Ar reactions —— **47**
3.2.6   The Smiles rearrangement: an intramolecular S$_N$Ar reaction —— **48**
3.3   Substitution via benzyne formation —— **51**
3.4   Concerted nucleophilic aromatic substitutions —— **52**
3.5   Nucleophilic substitution of heteroaromatic systems —— **55**
3.6   Substitution of aryl diazonium salts —— **58**
3.6.1   Diazonium salt formation —— **58**
3.6.2   Reactions of diazonium salts with nucleophilic species —— **58**
3.6.3   Other reactions of aryl diazonium salts —— **63**
3.6.4   Formation of biaryls: the Gomberg–Bachmann reaction
           and Pschorr cyclization —— **65**
           References —— **67**

4   Reactions of aryllithium species —— 73
4.1   Introduction —— **73**
4.2   Deprotonation of arenes: directed *ortho* metalation —— **74**
4.2.1   Introduction to directed *ortho* metalation —— **74**
4.2.2   Examples of directed *ortho* metalation in synthesis —— **77**
4.2.3   Anionic *ortho*-Fries rearrangement —— **79**
4.3   Directed remote metalation —— **82**
4.4   Lithium–halogen exchange —— **85**
4.5   Rearrangements of haloarenes: halogen dance reactions —— **86**
           References —— **92**

5   Benzynes —— 95
5.1   Introduction to benzynes —— 95
5.2   Formation of benzynes —— 95
5.2.1   Aryne formation by elimination of H–X —— 96
5.2.2   Aryne formation from dihalobenzenes —— 100
5.2.3   Aryne formation by elimination of small molecules —— 101
5.2.4   Fluoride-induced aryne formation —— 102
5.3   Reactions of arynes —— 105
5.3.1   Cycloaddition reactions —— 106
5.3.1.1   [4 + 2] Cycloaddition reactions —— 106
5.3.1.2   Intramolecular Diels–Alder reactions —— 108
5.3.1.3   [2+2] Cycloadditions —— 109
5.3.1.4   1,3-Dipolar cycloadditions —— 111
5.3.2   Reactions of arynes with nucleophiles —— 113
5.3.2.1   Nucleophilic addition reactions —— 113
5.3.2.2   Multicomponent reactions of arynes —— 115
5.3.2.3   Aryne insertion reactions —— 116
5.3.3   Transition metal-mediated reactions of arynes —— 121
5.4   *Meta*- and *para*-didehydrobenzenes (*m*- and *p*-arynes) —— 125
          References —— 126

6   Transition-metal-mediated C–C bond forming reactions of aromatic compounds —— 131
6.1   Introduction —— 131
6.2   Preparation of biaryls via homocoupling —— 132
6.2.1   Copper-mediated coupling: the Ullmann reaction —— 132
6.2.2   Nickel-mediated homocoupling reactions —— 135
6.2.3   Oxidative coupling reactions —— 138
6.3   Preparation of biaryls via cross-coupling reactions —— 139
6.3.1   Metal-catalyzed cross-coupling —— 139
6.3.1.1   Ar–X cross-coupling precursors —— 142
6.3.1.2   Suzuki–Miyaura cross-coupling (aryl boronic acids or boronate esters) —— 143
6.3.1.3   Negishi cross-coupling (organozinc) —— 145
6.3.1.4   Stille cross–coupling (organotin) —— 146
6.3.1.5   The Hiyama cross-coupling (organosilanes) —— 148
6.3.1.6   Kumada–Corriu cross-coupling (Grignard reagents) —— 149
6.4   Other coupling reactions —— 150
6.4.1   The Heck reaction —— 150
6.4.2   The Sonogashira reaction —— 153
6.4.3   Transition metal-mediated cyanation —— 159
6.5   Direct arylation (C–H functionalization) —— 160

6.5.1 Introduction to direct arylation —— 160
6.5.2 Direct arylation with aryl halides —— 161
6.5.2.1 Intramolecular direct arylation reactions with aryl halides —— 162
6.5.2.2 Direct arylation using directing groups —— 163
6.5.2.3 Direct arylation without directing groups —— 166
6.5.2.4 Direct arylation of heteroaromatic compounds —— 167
6.5.3 Direct arylation using organometallic species —— 170
6.5.4 Direct arylation with arenes: dehydrogenative coupling —— 171
6.6 Axially chiral biaryls: synthetic approaches —— 174
6.6.1 Atropisomerism and axial chirality —— 174
6.6.2 Approaches to racemic atropisomeric biaryls —— 176
6.6.3 Stereoselective approaches to axially chiral biaryls —— 177
References —— 182

7 Other transition-metal-mediated reactions of aromatic compounds —— 191
7.1 Introduction —— 191
7.2 Aromatic C–N and C–O bond formation —— 191
7.2.1 Copper-catalyzed aryl C–N and C–O bond formation —— 191
7.2.2 Palladium-catalyzed aryl amination —— 195
7.3 Transition-metal-catalyzed borylation reactions —— 200
7.3.1 Palladium-catalyzed borylation of aryl halides —— 201
7.3.2 Iridium-catalyzed direct borylation —— 204
References —— 211

8 Constructing aromatic rings —— 215
8.1 Introduction —— 215
8.2 Preparing benzene rings from non-aromatic precursors —— 215
8.2.1 Cycloaddition reactions —— 215
8.2.2 Transition metal-catalyzed alkyne cyclotrimerization —— 218
8.2.3 Olefin metathesis —— 224
8.3 Preparing aromatic heterocycles —— 226
8.3.1 Five-membered ring heterocycles —— 227
8.3.2 Six-membered ring heterocycles —— 230
8.3.3 Fused heteroaromatics —— 234
8.3.3.1 Synthesis of indoles —— 234
8.3.3.2 Synthesis of quinolines —— 239
8.3.3.3 Synthesis of isoquinolines —— 242
References —— 244

9   Fused aromatic rings – polycyclic aromatic hydrocarbons —— 247
9.1   Introduction to polycyclic aromatic hydrocarbons —— 247
9.1.1   Classification and nomenclature of PAHs —— 247
9.2   Stability and reactivity of PAHs —— 250
9.2.1   The aromatic sextet —— 250
9.2.2   General reactivity of PAHs —— 253
9.3   Synthetic approaches to PAHs —— 256
9.3.1   Haworth synthesis —— 256
9.3.2   The Pschorr synthesis —— 260
9.3.3   The Elbs reaction —— 260
9.3.4   Diels–Alder cycloadditions —— 261
9.3.5   The Wittig reaction —— 266
9.3.6   Photocyclization (the Mallory–Katz reaction) —— 267
9.3.7   Cyclodehydrogenation reactions —— 268
9.3.8   Cyclodehydrohalogenation reactions —— 275
9.3.9   Aryne cyclotrimerization and related reactions —— 276
9.3.10   Ring-closing metathesis —— 277
9.3.11   Alkyne benzannulation —— 279
9.3.12   Asao–Yamamoto benzannualation —— 282
             References —— 284

10   Synthesis of select classes of polycyclic aromatic hydrocarbons —— 289
10.1   Introduction —— 289
10.2   Synthesis of linear fused aromatic structures – the acenes —— 289
10.2.1   Synthetic approaches to pentacenes —— 290
10.2.2   Synthesis of larger acenes —— 293
10.3   Synthesis of phenacenes —— 295
10.4   Perylenes and rylenes —— 299
10.5   Synthesis of triphenylenes and related compounds —— 301
10.5.1   Triphenylenes —— 301
10.5.2   Trinaphthylenes and larger starphenes —— 306
10.6   From coronenes to larger PAHs —— 311
10.7   Graphene nanoribbons —— 317
             References —— 324

11   Nonplanar aromatic compounds —— 329
11.1   Introduction —— 329
11.2   Helicenes and related contorted polycyclic compounds —— 329
11.2.1   Helicenes —— 329
11.2.2   Stereoselective helicene syntheses —— 336
11.2.3   Other benzannulated PAHs —— 340
11.3   Twisted acenes —— 344

11.4	Circulenes and related curved polycyclic aromatic hydrocarbons —— **347**	
11.4.1	[5]-Circulenes – corannulenes —— **347**	
11.4.2	[7]-Circulenes and related systems —— **350**	
11.5	Cyclophanes and cyclophenylenes —— **351**	
11.5.1	Cyclophanes —— **352**	
11.5.2	Cycloparaphenylenes —— **354**	
	References —— **361**	

Index —— **365**

List of abbreviations

Ac$_2$O	Acetic anhydride
AcOH	Acetic acid
AIBN	Azobis(isobutyronitrile)
API	Active pharmaceutical ingredient
Ar	Aryl
B	Base
bpy	2,2′-Bipyridine
BHT	2,6-Di-*tert*-butyl-4-methylphenol
BINAP	(2,2′-Bis(diphenylphosphino)-1,1′-binaphthyl)
BINOL	1,1′-Bi-2-naphthol
Bn	Benzyl
Boc	*tert*-Butoxycarbonyl
B$_2$pin$_2$	Bis(pinacolato)diboron
BTMSA	Bis(trimethylsilyl)acetylene
Bu	Butyl
CDHC	Cyclodehydrochlorination
CIPE	Complex-induced proximity effect
cod	Cyclooctadiene
Cp	Cyclopentadienyl
CPP	Cycloparaphenylene
CuTC	Copper(I) thiophene-2-carboxylate
Cy	Cyclohexyl
dba	Dibenzylideneacetone
DBU	1,8-Diazabicyclo[5.4.0]undec-7-ene
DDQ	Dichlorodicyanoquinone
DMA	Dimethylacetamide
DMAP	4-(Dimethylamino)pyridine
DME	Dimethoxyethane
DMF	*N,N*-Dimethylformamide
DMG	Directed metalation group
DMSO	Dimethyl sulfoxide
DoM	Directed *ortho* metalation
DreM	Directed remote metalation
dppf	1,1′-Bis(diphenylphosphino)ferrocene
Et	Ethyl
equiv	Equivalents
FVP	Flash vacuum pyrolysis
GNR	Graphene nanoribbon
Het	Heterocycle
HMDS	Hexamethyldisilazane
HMPA	Hexamethyl phosphoric triamide
HOMA	Harmonic oscillator model of aromaticity
HOMO	Highest occupied molecular orbital
HPLC	High-performance liquid chromatography
i-Pr	Isopropyl
kcal	Kilocalorie
L	Ligand

https://doi.org/10.1515/9783110562682-204

LCD	Liquid crystal display
LDA	Lithium diisopropylamide
LiTMP	Lithium tetramethylpiperide
LUMO	Lowest unoccupied molecular orbital
M	Metal
mCPBA	*meta*-Chloroperoxybenzoic acid
Me	Methyl
MeOH	Methanol
mol	Mole
MOM	Methoxymethyl
NBS	*N*-Bromosuccinimide
NCS	*N*-Chlorosuccinimide
NHC	*N*-Heterocyclic carbene
NICS	Nucleus-independent chemical shift
NMO	*N*-Morpholine oxide
NMP	*N*-Methylpyrrolidine
NMR	Nuclear magnetic resonance
Nu	Nucleophile
OAc	Acetate
OLED	Organic light-emitting diode
OPiv	Pivalate
OTf	Triflate
o-Tol	*ortho*-Tolyl
PAH	Polycyclic aromatic hydrocarbon
Ph	Phenyl
Pin	Pinacol
PIDA	Phenyliodine(III) diacetate
PIFA	(Bis(trifluoroacetoxy)iodo)benzene
Piv	Pivyl
PMMA	Poly(methyl methacrylate)
py	Pyridyl
rt	Room temperature
RCM	Ring-closing metathesis
s-Bu	*sec*-Butyl
S_EAr	Electrophilic aromatic substitution
S_NAr	Nucleophilic aromatic substitution
TADF	Thermally activated delayed fluorescence
TAPA	2-(2,4,5,7-Tetranitro-9-fluorenylidene-aminooxy)propionic acid
TASF	Tris(dimethylamino)sulfonium difluorotrimethylsilicate
TBAF	Tetrabutylammonium fluoride
TBDMS	*tert*-Butyldimethylsilyl
t-Bu	*tert*-Butyl
TES	Triethylsilyl
TFA	Trifluoroacetic acid
TFAA	Trifluoroacetic anhydride
Tf	Trifluoromethanesulfonyl
Tf_2O	Triflic anhydride
THF	Tetrahydrofuran
TIPS	Triisopropylsilyl

TMEDA	Tetramethylethylenediamine
TMS	Trimethylsilyl
TMSCl	Trimethylsilyl chloride
Ts	Tosyl
UV	Ultraviolet
VID	Valence isomerization/dehydration
X	Halide

1 Introduction – aromatic compounds and aromaticity

Since it was first isolated by Faraday in 1825 [1], benzene has captured the interest of chemists. At first, the question focused on the structure of benzene, until Kekulé's famously proposed the cyclic structure that corresponds to what we understand today [2]. For an account of Kekulé's story of the structure of benzene, see Japp's memorial lecture [3]. This story is not without controversy [4], but remains an enduring story in the history of organic chemistry. After this, attention turned to some of the unexpected properties of benzene. For example, benzene has a propensity to undergo electrophilic substitution instead of electrophilic addition. Furthermore, if the double bonds are localized in the benzene ring, we would expect to see two different isomers of an *ortho*-disubstituted benzene, but only one is observed. At the time, these were important questions, and especially challenging since a detailed understanding of chemical bonding was in its infancy. It was also in the nineteenth century that the term "aromatic" was used at first to describe compounds with a pleasant aroma, but soon this term became associated with benzene and related structures and was used to describe a class of structurally related compounds.

1.1 Importance of aromatic compounds

Since the discovery and structural elucidation of benzene, our understanding of the reactivity, synthesis, and properties of aromatic compounds has evolved considerably. Moreover, aromatic compounds have become important parts of our everyday lives, being found widely in compounds ranging from pharmaceuticals to dyes and materials. As such, the synthesis of aromatic compounds has become an essential tool for the preparation of dyes, active pharmaceutical ingredients, as well as compounds for applications in modern electronic devices.

1.1.1 Synthetic dyes

Among the first practical uses of aromatic compounds was in the preparation of commercial dyes. The color of dyes stems from their absorption of visible light, which is a consequence of the compounds having a low highest occupied molecular orbital (HOMO)– lowest unoccupied molecular orbital (LUMO) gap. The low HOMO–LUMO gap is the result of extended π-conjugated structures which often incorporate aromatic rings. The structures of some representative dyes are shown in Figure 1.1.

Dyes have been used for millennia, and were, until the late nineteenth century, derived from natural sources. However, in the 1850s, during an attempted

https://doi.org/10.1515/9783110562682-001

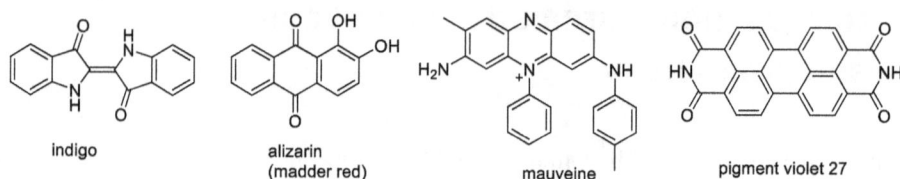

Figure 1.1: Examples of dye structures.

synthesis of quinine (a natural product that was used to treat malaria), Perkin serendipitously prepared compounds that were able to stain cloth a deep purple color. This purple dye, which was called mauveine, was the result of a reaction of toluidine and aniline in the presence of potassium dichromate as an oxidizing agent (Figure 1.2) [5,6].

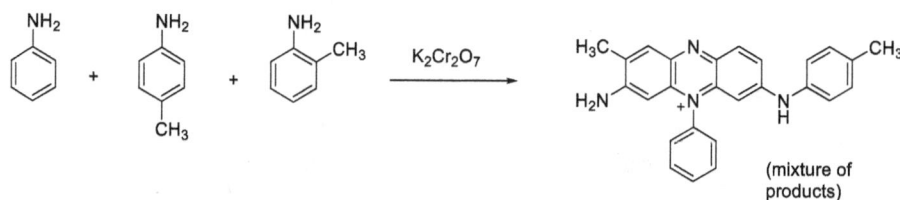

Figure 1.2: Perkin's synthesis of mauveine.

Around the same time, synthetic routes to naturally occurring dyes were developed. For example, alizarin (also known as madder red) is a natural dye from the red madder plant (*Rubia tinctorum*). In the 1860s, a synthetic route from anthracene was devised. Anthracene was oxidized to the corresponding anthraquinone. Sulfonation of anthraquinone gave the sulfonic acid, which reacted with potassium hydroxide to give alizarin (Figure 1.3) [5].

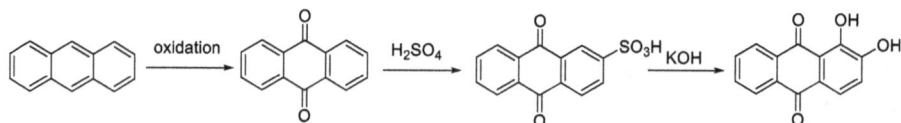

Figure 1.3: Synthesis of alizarin.

As another example, indigo is a natural dye that is familiar to us as the dye in denim jeans. It was historically extracted from the plant *Indigofera tinctoria*. However, in the late nineteenth century, von Baeyer reported the synthesis and structure of indigo. Following this, the commercial synthetic route to indigo was developed by Heumann in 1890 (Figure 1.4). The synthesis began by an alkylation of anthranilic acid using chloroacetic acid. This was followed by a base-mediated cyclization and decarboxylation to give 3-hydroxyindole (indoxyl), which was converted to indigo by oxidation in air. Since then, an alternative approach was developed where anthranilic acid was replaced by the much less expensive aniline [5].

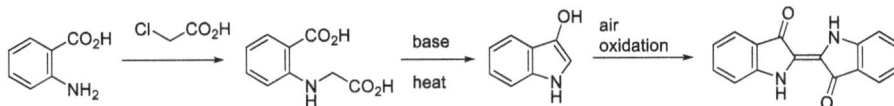

Figure 1.4: First commercial synthesis of indigo.

The development of synthetic routes to known and new dye molecules arguably served as a starting point for the development of the commercial dye industry and led to the acceleration of the development of synthetic methods for aromatic compounds.

1.1.2 Biologically active compounds

Aromatic compounds are important from a biological perspective and consequently many aromatic compounds have biological activity. Naturally occurring aromatic compounds include simple structures such as the amino acids phenylalanine, tyrosine, and tryptophan, but also include more complex structures, some of which have therapeutic use or potential use as pharmaceuticals (Figure 1.5).

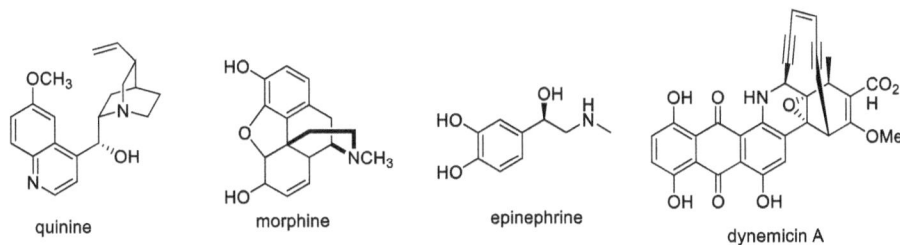

quinine   morphine   epinephrine   dynemicin A

Figure 1.5: Examples of aromatic natural products with biological activity.

Aromatic rings are also ubiquitous structures in synthetic small-molecule pharmaceutical drugs. Indeed, a 2011 survey of drug candidates from a Pfizer database showed that 99% of the compounds contained an aromatic or heteroaromatic ring [7]. As another example of the importance of aromatic compounds in pharmaceuticals, 19 of the top 20 selling brand name drugs in 2008 contained at least one aromatic ring [8]. These examples highlight the prevalence of aromatic rings in active pharmaceutical ingredients and the need for synthetic methods to prepare these compounds. Figure 1.6 shows a few examples of small molecule pharmaceutical drugs.

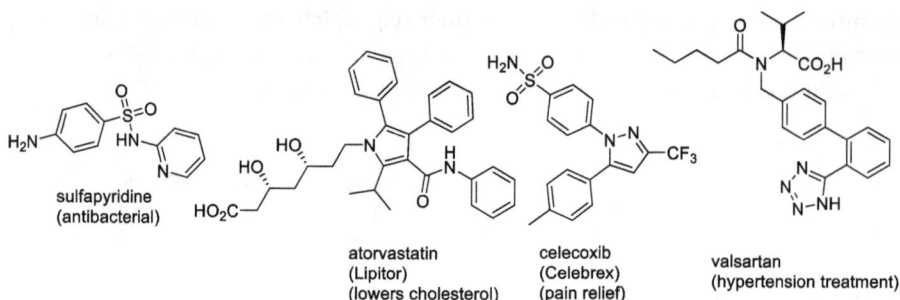

Figure 1.6: A selection of small molecule active pharmaceutical ingredients containing aromatic rings.

1.1.3 Aromatic compounds in materials science

Aromatic compounds have also found use in electronic devices and show potential in new electronic materials. For example, liquid crystalline compounds used in liquid crystal display applications are composed of aromatic compounds such cyanobiphenyls (Figure 1.7). Similarly, discotic liquid crystals that are generally composed of disk-shaped polycyclic aromatic hydrocarbons are used as optical compensation films in these displays.

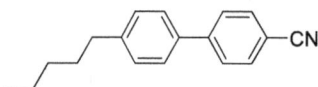

Figure 1.7: A nematic liquid crystalline compound of the type used in liquid crystal displays.

Aromatic compounds have properties that make them candidates for applications in organic electronics, light-emitting displays, and solar cells. In particular, the relatively low HOMO–LUMO gap in some aromatic compounds – particularly conjugated polycyclic aromatic compounds, make them suitable candidates as semiconducting materials. This same low HOMO–LUMO gap makes them capable of absorbing visible light, which contributes not only to their use as dyes, but also as potential chromophores in light harvesting materials for solar energy conversion. In addition, the relatively rigid structure of aromatic compounds means that they are often luminescent, making them potential components of light emitting materials for display applications (e.g., organic light-emitting diodes – OLEDs). Aromatic units are also incorporated into functional polymeric materials, ranging from conducting polymers, to macromolecular frameworks (such as covalent organic frameworks) with potential applications in separations, gas storage, and catalysis.

Overall, these established and emerging applications for aromatic compounds continue to drive the development of new synthetic approaches for the preparation of aromatic compounds and the exploration of new aromatic structures.

1.2 Aromaticity

Readers are likely already familiar with the criteria for aromaticity based on Hückel: cyclic, planar, fully conjugated polyenes with $4n+2$ π-electrons are aromatic, while comparable systems with $4n$ electrons are antiaromatic. This definition or some version of it are found in most organic chemistry texts. Hückel's $4n+2$ rule stems from calculations of the π-molecular orbitals. He showed that the orbital energy diagram for cyclic conjugated systems featured a lowest energy orbital, followed by pairs of double degenerate orbitals [9]. In cases where an even number of p-orbitals are present (resulting from an even number of number of atoms in the ring), there is also a non-degenerate highest energy orbital. For two selected examples, a sketch of the molecular orbital energy levels for benzene and cyclobutadiene is shown in Figure 1.8. Based on the orbital energy levels, any cyclic conjugated system with $4n+2$ π-electrons will form a closed-shell stable aromatic ring, while any system with $4n$ π-electrons will have an open-shell configuration.

Figure 1.8: Molecular orbital energy levels and electron occupancy for benzene and cyclobutadiene.

While this definition does work in many cases, it does have its limitations. For example, it does not formally apply to polycyclic systems, does not account for systems that deviate from planarity, and does not provide insight into differences in aromatic stabilization energies. Consequently, the definition of aromaticity is the subject of continued discussion and several experimental and computational approaches have been developed to describe aromaticity. Indeed, there are multiple measures of aromaticity, and which has led to questions of what aromaticity is and whether there can be a single unified definition [10]. The discussion of aromaticity has been referred to as *an exercise in futility* [11], and as recently as 2009, Stanger has posed the question *Aromaticity . . . What is it?* and has questioned whether a unified definition of aromaticity can be achieved [12].

Here, we will certainly not resolve this ongoing question. Instead, we will consider some of the general criteria for aromaticity, focusing on the stability of aromatic compounds, as well as the structural and the magnetic aspects of aromaticity. In other words, we can describe some of the important features of aromatic compounds without actually defining aromaticity.

1.2.1 Stability of aromatic compounds

One of the defining characteristics of benzene and related aromatic systems is their unusual stability. A common illustration of the stability of benzene involves a comparison of the heats of hydrogenation of benzene as compared with cyclic alkenes [13]. The hydrogenation of alkenes is exothermic, and as shown in Figure 1.9, depends on the number of π-bonds present. For example, the enthalpy of hydrogenation of 1,4-cyclohexadiene ($\Delta H = -57.2$ kcal/mol) is essentially double the enthalpy of hydrogenation of cyclohexene ($\Delta H = -28.6$ kcal/mol), suggesting that each π-bond releases 28.6 kcal/mol upon hydrogenation. Based on this, a hypothetical cyclohexatriene is expected to have an enthalpy of hydrogenation of 3x(−28.6 kcal/mol) = −85.5 kcal/mol. However, benzene has an enthalpy of hydrogenation of only −49.7 kcal/mol, suggesting that it is significantly more stable than expected. The difference between the actual heat of hydrogenation and the expected heat of hydrogenation is 36.1 kcal/mol, which is referred to as the aromatic stabilization energy. The comparison shown in Figure 1.9 does neglect two potential confounding factors: stabilization that may result from conjugation (as opposed to aromaticity), as well as changes that may result from ring strain. However, even if one accounts for these, an aromatic stabilization energy is over 30 kcal/mol, which still shows a substantial stabilization.

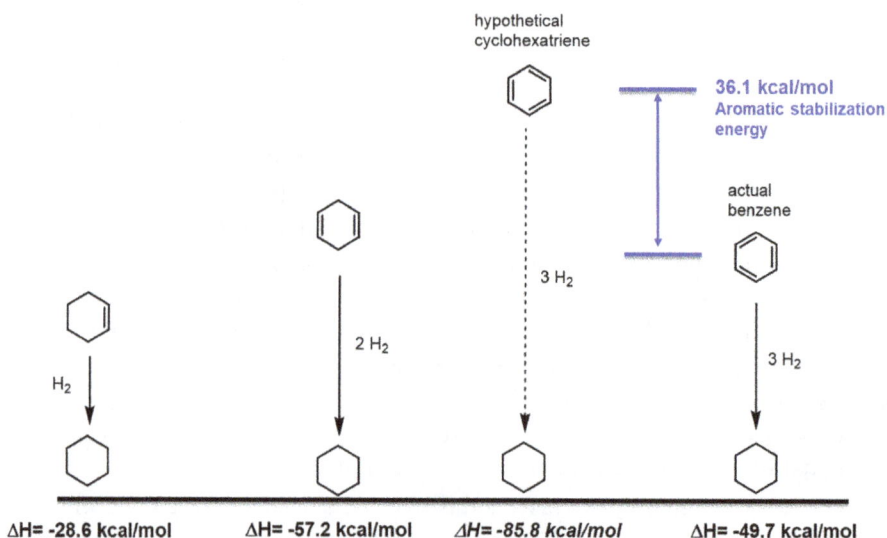

hypothetical cyclohexatriene

36.1 kcal/mol Aromatic stabilization energy

actual benzene

3 H$_2$

2 H$_2$

3 H$_2$

H$_2$

ΔH= -28.6 kcal/mol ΔH= -57.2 kcal/mol   *ΔH= -85.8 kcal/mol* ΔH= -49.7 kcal/mol

Figure 1.9: Heats of hydrogenation of benzene and cyclic alkenes. Adapted from D. E. Lewis, *Advanced Organic Chemistry* [14].

Heats of hydrogenation can be used to determine the aromatic stabilization energies of other aromatic compounds and thereby provide a comparison of the relative stabilization associated with the aromatic rings. Figure 1.10 shows the aromatic stabilization energies of some other common aromatic compounds [14]. These heats of hydrogenation can provide clues about the relative stabilities of aromatic compounds and shows that aromatic stabilization energies can vary significantly. In some cases, these differences have implications for reactivity, as we will see in this book.

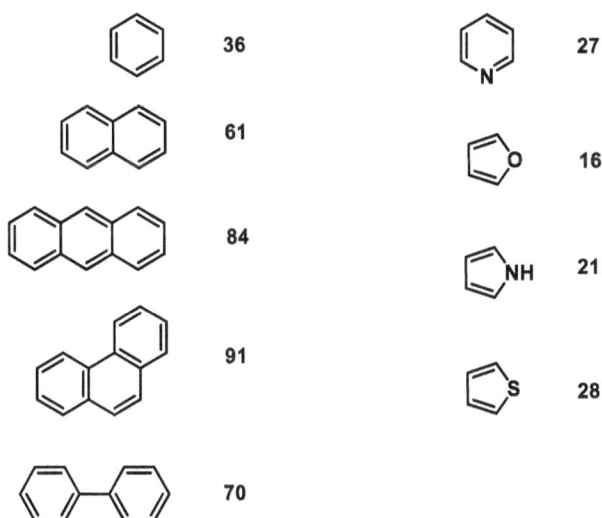

Figure 1.10: Aromatic stabilization energies of representative aromatic compounds (in kcal/mol). Adapted from D. E. Lewis, *Advanced Organic Chemistry* [14].

Another approach for determining the stabilization associated with aromaticity is by using isodesmic reactions. An isodesmic reaction is a hypothetical chemical process where the number of bonds of each type remains the same on each side of the equation. Standard enthalpies of formation of each of the components can be used to determine the enthalpy change of the reaction. If one considers the hypothetical conversion of three equivalents of cyclohexene to one equivalent of benzene and two cyclohexane, one can see that the number and type of bond remains constant in this reaction (Figure 1.11) [15]. Since the only significant difference is the formation of an aromatic ring, the reaction is exothermic and the enthalpy change is consistent with the aromatic stabilization energy of benzene. This approach still does not take into account changes in ring strain – however in this case these are not large contributors. In contrast, if one considers a similar isodesmic reaction for the formation of cyclobutadiene, which is antiaromatic, the reaction is highly endothermic, suggesting that cyclobutadiene is much less stable.

3 ⬡   ⟶   ⬡ $+ 2$ ⬡   ΔH = -35.5 kcal/mol

ΔH$_f$ = -1.2 kcal/mol   ΔH$_f$ = +19.7 kcal/mol   ΔH$_f$ = -29.4 kcal/mol

2 ⊡   ⟶   □   ⊞   ΔH = +45.7 kcal/mol

ΔH$_f$ = +37.5 kcal/mol   ΔH$_f$ = +6.7 kcal/mol   ΔH$_f$ = +114 kcal/mol

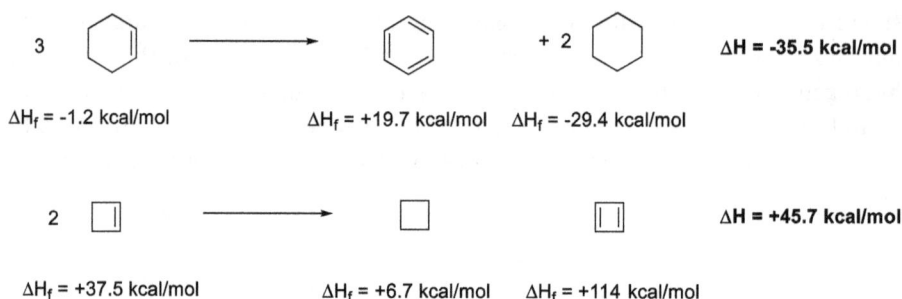

Figure 1.11: Isodesmic reactions for the formation of benzene and cyclobutadiene. Adapted from Anslyn and Dougherty, *Modern Physical Organic Chemistry* [15].

1.2.2 Structure and reactivity of aromatic compounds

Another defining feature of aromatic rings is their bonding geometry. A manifestation of the delocalization of the π-bonds is that all of the C–C bond lengths in benzene are equal (ca. 1.40 Å) and intermediate between a single bond and a double bond. The variation in bond length has been used as a metric for aromaticity. For example, the harmonic oscillator model of aromaticity (HOMA) is computational tool that looks at deviation of bond lengths from optimal, creating a scale that allows the assessment of aromaticity [16].

One of the properties of aromatic compounds is that their reactivity is distinctly different from other unsaturated compounds such as alkenes. For example, in the presence of electrophiles, they usually undergo substitution (with loss of a hydrogen) rather than the electrophilic addition that is typical of alkenes. This ties back to the thermodynamic stability of aromatic compounds described above and influences how we approach the synthesis of aromatic compounds. Indeed, the reactivity of aromatic compounds and application of this reactivity for the preparation of aromatic compounds will be the major focus of this book.

1.2.3 Magnetic properties of aromatic compounds

A key observable property of aromatic compounds is their behavior in a magnetic field. In the presence of an external magnetic field, a ring current in aromatic rings is induced, leading to a local magnetic field in the vicinity of the ring (Figure 1.12) [17]. This phenomenon is known as diamagnetic anisotropy.

A consequence of this is that the local magnetic field reinforces the magnetic field in nuclear magnetic resonance (NMR) for protons attached to the aromatic ring, leading to deshielding of those protons. For nuclei centered above an aromatic ring (or, in the middle of the aromatic ring), the magnetic field opposes the main

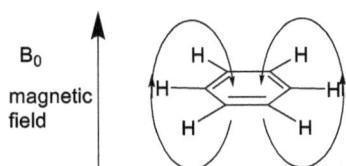

Figure 1.12: Induced magnetic field by ring current.

magnetic field and leads to shielding. Figure 1.13 shows some representative examples of the NMR chemical shifts of some aromatic compounds [18]. The chemical shift of the protons in benzene is 7.27 ppm, which is significantly deshielded compared to cyclohexene (5.6 ppm). The other examples illustrate that the protons on the periphery of the ring are deshielded, while protons situated above or below the ring (as with the CH_2 and CH_3 groups in the second and third examples) are significantly shielded because the induced magnetic field opposes the main magnetic field. The last example of [18] annulene is quite striking: at low temperature the outer protons are deshielded at 9.28 ppm, while the inner protons are highly shielded at – 3.0 ppm.

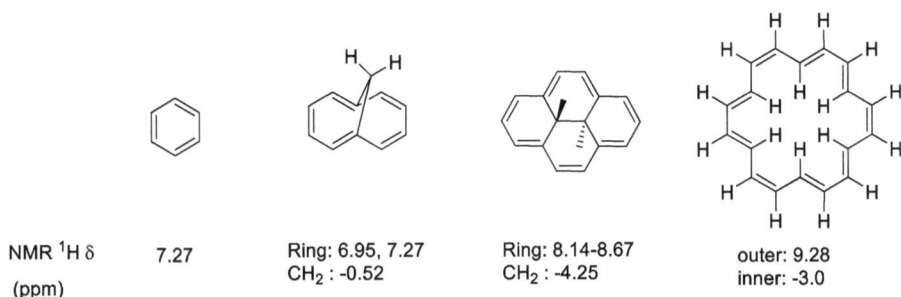

| NMR ^{1}H δ (ppm) | 7.27 | Ring: 6.95, 7.27 CH_2 : -0.52 | Ring: 8.14-8.67 CH_2 : -4.25 | outer: 9.28 inner: -3.0 |

Figure 1.13: Chemical shifts (δ, ppm) of some representative conjugated cyclic structures [18].

In contrast to aromatic systems, antiaromatic systems display a paratropic ring current and resulting magnetic field in the opposite direction. Consequently, protons on the outer periphery of the ring are unusually shielded, while any protons above or below the ring are deshielded. Hess et al. showed this trend for a series of aromatic and antiaromatic dehydro[n]annulenes [19]. As another example, when [18] annulene described above undergoes two-electron reduction using potassium, the resulting dianion, which is now antiaromatic, shows outer protons at – 1.13 ppm and inner protons at ca. 28–30 ppm [20].

As an extension of the experimental evidence by NMR, computational probes of aromaticity based on diamagnetic anisotropy and ring currents have been developed. Among the best known computational measures of aromaticity is known as nucleus-independent chemical shift (NICS) [21–23]. This computational technique consists of placing a nucleus directly above the center of ring and calculating the chemical shift using established computational methods. A negative NICS value suggests aromatic character, while a positive NICS value indicates aromatic character (Figure 1.14). This tool allows individual rings in a polycyclic system to be probed and provides a scale of relative aromaticity because NICS values correlate with aromatic stabilization energies [21]. In other words, the more negative the NICS value, the "more aromatic" the ring is.

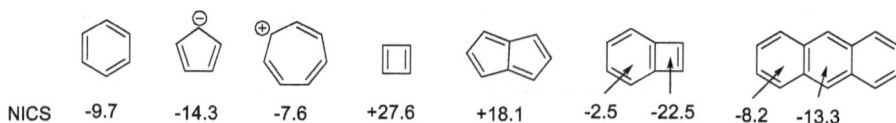

| NICS | -9.7 | -14.3 | -7.6 | +27.6 | +18.1 | -2.5 | -22.5 | -8.2 | -13.3 |

Figure 1.14: Representative NICS values for some conjugated cyclic systems [21].

Other computational methods consider the calculation of ring-current densities as a measure of aromaticity. Among the best known of these methods is the anisotropy of induced current density method, which serves as a tool for quantifying and visualizing the delocalization of electrons [24].

1.3 In this book

While a unified definition of aromaticity may elude us, the compounds that we generally consider to be aromatic are important for our modern society. Furthermore, aromatic compounds display an incredible diversity of structures and chemical reactivity. As such, an understanding of the synthesis of aromatic compounds is important from both a practical perspective for the preparation of useful compounds and because it allows us to explore new and exciting chemical structures with as yet undiscovered properties.

In this book, we will explore the reactivity of aromatic compounds with an attention to reaction mechanisms. We will also see how this reactivity can be exploited in synthesis. Throughout the chapters, examples of how the reactions discussed are used in the synthesis of a variety of compounds, ranging from biologically active compounds such as natural products and active pharmaceutical ingredients, to compounds with applications in materials science. We will also explore the synthesis of complex structures that are designed to probe the nature of aromaticity and advance the creative limits of synthetic chemistry.

References

[1] Faraday M. On new compounds of carbon and hydrogen, and on certain other products obtained from the decomposition of oil by heat. Phil Trans Roy Soc London. 1825;115:440–66.

[2] Kekulé A. Sur la constitution des substances aromatiques. Bull Soc Chim. 1865;3:98–110.

[3] Japp FR. Kekule memorial lecture. J Chem Soc Trans. 1898;73:97–138.

[4] Roth HD. 150 years later: A look at Loschmidt's contributions to organic chemistry. J Phys Org Chem. 2013;26:755–63.

[5] Zollinger H. Color Chemistry. Syntheses, Properties, and Applications of Organic Dyes and Pigments. Weiheim: VCH; 1987.

[6] Nicolaou KC, Montagnon T. Molecules That Changed the World. Darmstadt: Wiley-VCH; 2008.

[7] Roughley SD, Jordan AM. The medicinal chemist's toolbox: An analysis of reactions used in the pursuit of drug candidates. J Med Chem. 2011;54:3451–79.

[8] McGrath NA, Brichacek M, Njardarson JT. A graphical journey of innovative organic architectures that have improved our lives. J Chem Educ. 2010;87:1348–49.

[9] Hückel E. Quantentheoretische beiträge zum benzolproblem – I. Die Elektronenkonfiguration des benzols und verwandter verbindungen. Z Phys. 1931;70:204–86.

[10] Kramer K. The search for the grand unification of aromaticity. Chem World. 2021; https://www.chemistryworld.com/features/the-search.

[11] Binsch G. Aromaticity – An exercise in chemical futility? Naturwissenschaften. 1973;60 (8):369–74.

[12] Stanger A. What is aromaticity: A critique of the concept of aromaticity – can it really be defined? Chem Commun. 2009;1939–47.

[13] Slayden SW, Liebman JF. The energetics of aromatic hydrocarbons: An experimental thermochemical perspective. Chem Rev. 2001;101(5):1541–66.

[14] Lewis DE. Advanced Organic Chemistry. New York: Oxford University Press; 2016.

[15] Anslyn EV, Dougherty DA. Modern Physical Organic Chemistry. Sausalito: University Science Books; 2006.

[16] Krygowski TM, Cyrański MK. Structural aspects of aromaticity. Chem Rev. 2001;101 (5):1385–419.

[17] Gomes JANF, Mallion RB. Aromaticity and ring currents. Chem Rev. 2001;101(5):1349–83.

[18] Mitchell RH. Measuring aromaticity by NMR. Chem Rev. 2001;101(5):1301–15.

[19] Hess BA, Schaad LJ, Nakagawa M, Linear A. Relation between nuclear magnetic resonance chemical shifts of tetra- tert-butyldehydro[n]annulenes and resonance energies per pi electron. J Org Chem. 1977;42:1669–70.

[20] Oth JFM, Woo EP, Sondheimer F. The dianion of [181Annulene. J Am Chem Soc. 1973;95:7337–45.

[21] Schleyer PVR, Maerker C, Dransfeld A, Jiao H, van Eikema Hommes NJR. Nucleus-independent chemical shifts : A simple and efficient aromaticity probe. J Am Chem Soc. 1996;118:6317–18.

[22] Chen Z, Wannere CS, Corminboeuf C, Puchta R, von Ragué Schleyer P. Nucleus-independent chemical shifts (NICS) as an aromaticity criterion. Chem Rev. 2005;105(10):3842–88.

[23] Jiao H, Schleyer PVR, Mo Y, McAllister MA, Tidwell TT. Magnetic evidence for the aromaticity and antiaromaticity of charged fluorenyl, indenyl, and cyclopentadienyl systems. J Am Chem Soc. 1997;119:7075–83.

[24] Geuenich D, Hess K, Köhler F, Herges R. Anisotropy of the induced current density (ACID), a general method to quantify and visualize electronic delocalization. Chem Rev. 2005;105:3758–72.

2 Electrophilic aromatic substitution

2.1 Introduction

By virtue of their π-bonds, aromatic compounds undergo reactions with electrophiles, often in the presence of a Lewis acid catalyst. However, unlike alkenes, which undergo electrophilic addition reactions, aromatic compounds generally undergo electrophilic *substitution*, where initial electrophilic addition is followed by a deprotonation to re-aromatize. In this chapter, we will review the basic principles of electrophilic aromatic substitution, the scope of electrophiles that can be introduced onto an aromatic ring, reactivity and regioselectivity considerations, and applications in the synthesis of substituted aromatic compounds.

2.2 General mechanistic principles

A generic electrophilic aromatic substitution (also referred to as an S_EAr mechanism) proceeds via an arenium ion mechanism, where the electrophile is attacked by the π-electrons of the aromatic ring to form a cationic intermediate, which is re-aromatized upon deprotonation to give the substituted aromatic ring (Figure 2.1).

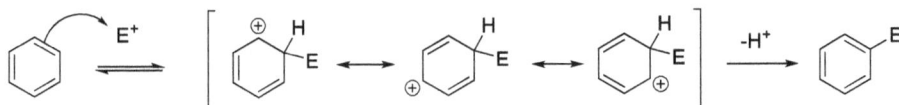

Figure 2.1: Generalized mechanism for electrophilic aromatic substitution.

It should be noted that the first step is generally the rate-determining step because aromatic stabilization is broken when the electrophile adds to the ring. The intermediate, referred to as an arenium ion or a Wheland intermediate, does not possess aromatic stabilization but is stabilized by resonance. Deprotonation to form the final product occurs rapidly because aromaticity is restored when the final product is formed. As we will see, the reactivity and regiochemistry of electrophilic aromatic substitution can be explained by this mechanism and in particular by the nature of the arenium ion intermediate.

2.3 Scope of electrophilic aromatic substitution

Electrophilic aromatic substitution can be used to introduce a variety of functional groups onto an aromatic ring, including halogens, nitro groups, sulfonic acids, and

https://doi.org/10.1515/9783110562682-002

alkyl, acyl and formyl groups (Figure 2.2). In each of these cases, the active electrophile is generated in situ, usually with the help of a Lewis or Brønsted acid.

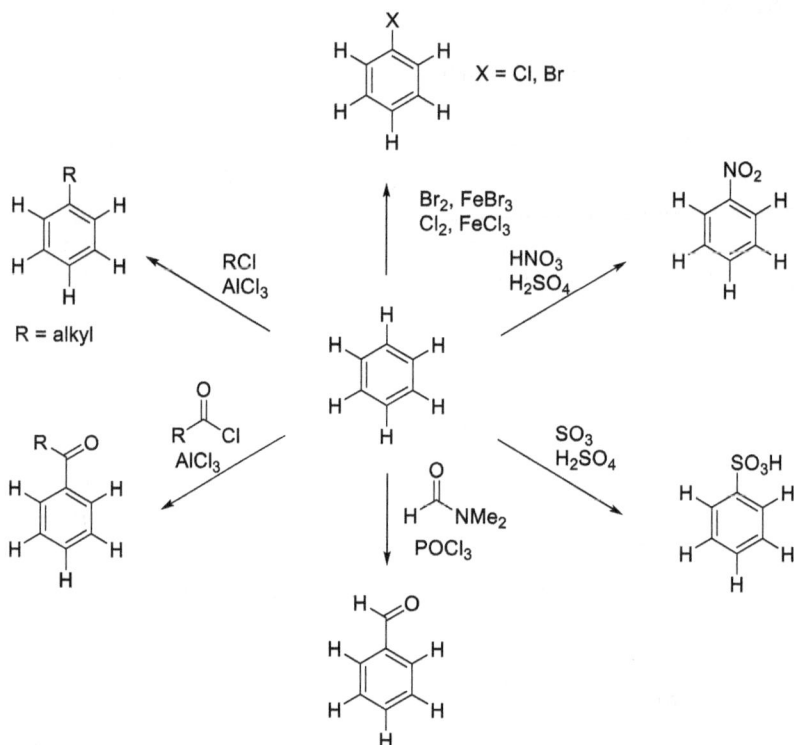

Figure 2.2: Examples of common electrophilic aromatic substitution reactions.

As mentioned above, the active electrophile is often formed in situ and the mechanism of electrophile formation depends on the electrophile being used.

2.3.1 Nitration

Nitration is one of the most important electrophilic substitution reactions because nitro groups can be reduced to the corresponding amines, which, as we will see in Chapter 3, can be converted into versatile diazonium salts. Electrophilic nitration is typically conducted with nitric acid in the presence of sulfuric acid, although it can be conducted in nitric acid alone or in combination with water or acetic acid. The active electrophile is NO_2^+, which is generated by protonation and dehydration of nitric acid (Figure 2.3) [1].

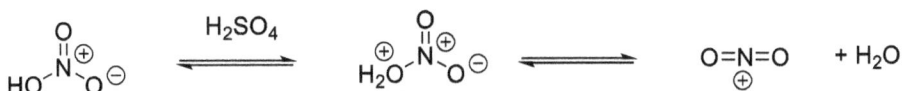

Figure 2.3: Formation of the nitronium ion electrophile.

These nitration conditions are harsh, which has prompted the development of milder nitrating agents, such as nitronium tetrafluoroborate ($NO_2^+BF_4^-$) or $AgNO_3/BF_3$ [2, 3]. Lanthanide salts can also be used to catalyze nitration. For example, nitration can be carried out with aqueous nitric acid with $Yb(O_3SCF_3)_3$, or using $Sc(O_3SCF_3)_3$ with $LiNO_3$ or $AlNO_3$ [4, 5]. In these reactions, the lanthanide salt likely catalyzes the formation of the nitronium ion in solution.

2.3.2 Sulfonation

The introduction of sulfonic acid groups via electrophilic aromatic substitution can be achieved using sulfuric acid, or using SO_3 in sulfuric acid (also called fuming sulfuric acid or oleum) [6]. The electrophilic species may vary with the reagent, but is often considered to be SO_3 or SO_3H^+. The latter can be generated from sulfuric acid or by protonation of SO_3 (see Figure 2.4). A closely related reaction is chloro-sulfonation using chlorosulfuric acid. The resulting aryl sulfonyl chlorides can be used for making sulfonate esters and sulfonamides.

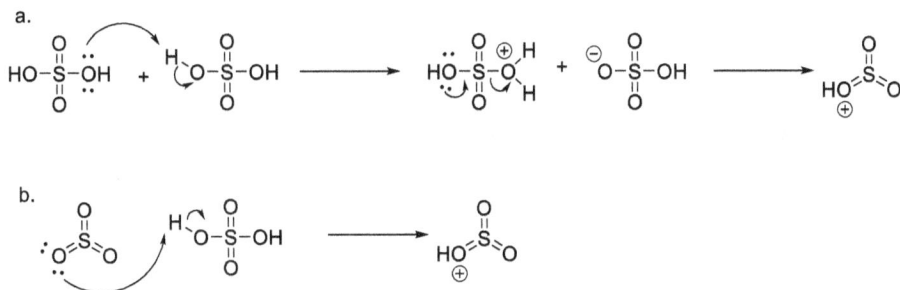

Figure 2.4: Generation of electrophiles for sulfonation.

An important feature of sulfonation is that it is reversible, so the sulfonic acid group can be removed under dilute aqueous acid. This reversibility means that sul-fonic acid groups can be used as protective groups, a strategy that will be explored in Section 2.6.

2.3.3 Halogenation

Electrophilic halogenation, especially to introduce bromo substituents is an important transformation because aryl halides are widely used in transition metal-catalyzed coupling reactions (which we will see in Chapters 6 and 7). Bromination and chlorination are typically carried out using Br_2 or Cl_2 in the presence of a Lewis acid such as $FeCl_3$ [1]. Brominations are often carried out in the presence of a catalytic amount of iron, which generates $FeBr_3$ in situ. The Lewis acid activates the halogen toward electrophilic attack, and the electrophilic species is either the free halogen cation (Br^+ or Cl^+) or a polarized molecular halogen after complexation with the Lewis acid.

For activated aromatic compounds, bromination and chlorination can often be carried out without a Lewis acid. Indeed, highly activated aromatic compounds such as anilines and phenols can undergo bromination at all of the activated positions, giving polybrominated products. To mitigate this "overreaction," less reactive brominating agents such as tetraalkylammonium tribromides can lead to more controlled reactions [7–9]. Similarly, for activated substrates, N-bromosuccinimide (NBS) or N-chlorosuccinimide can be used as the electrophilic reagent [10, 11].

In contrast to bromine and chlorine, iodine is relatively unreactive as an electrophile. Electrophilic iodination often requires an oxidant such as iodic or periodic acid to oxidize iodine to generate I^+ as the active electrophile [12–14]. Other electrophilic iodination reagents include iodine monochloride [15].

Fluorine is highly reactive and is not typically introduced via direct electrophilic substitution. An alternative to fluorine as an electrophile is acetyl hypofluorite, which is generated from fluorine gas and sodium acetate [16]. However, this reagent still requires the use and manipulation of fluorine, which is highly reactive and toxic. Some electrophilic fluorinating agents that are easier to handle include N-fluoro-bis(trifluoromethansulfonyl)amine, which reacts with activated aromatics, as well as N-fluoropyridinium sulfonate and N,N'-difluoro-1,4-diazoniabicyclo[2.2.2] octane salts (Figure 2.5) [17–19].

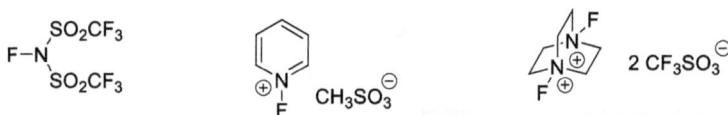

Figure 2.5: Electrophilic fluorination reagents.

2.3.4 Carbon electrophiles: Friedel–Crafts alkylation and acylation

The reaction of an aromatic compound with an alkyl halide in the presence of a Lewis acid to form the corresponding alkyl-substituted aromatic is referred to as Friedel–Crafts alkylation [20]. In this reaction, the electrophilic species is thought to be the alkyl carbocation, which is produced by reaction of the alkyl halide with the Lewis acid (Figure 2.6).

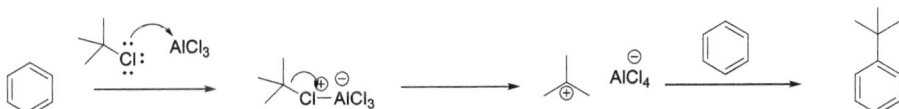

Figure 2.6: Formation of carbocation electrophile for Friedel–Crafts alkylation.

One challenge associated with Friedel–Crafts alkylations is the instability of the carbocation electrophile. For example, Friedel–Crafts reactions of linear alkyl halides often lead to rearrangement results from hydride shifts (Figure 2.7) [21].

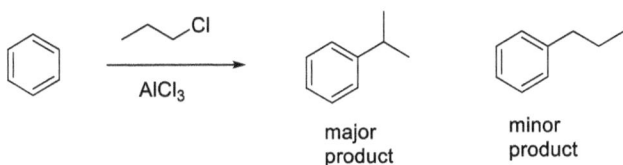

Figure 2.7: Example of a rearrangement in a Friedel–Crafts alkylation.

Another challenge with Friedel–Crafts alkylation is that the alkyl-substituted aromatic compounds is more activated than the starting material. Consequently, it can undergo further electrophilic substitution reactions and lead to a mixture of products. These problems can be circumvented by using a Friedel–Crafts acylation (see below), or by using more modern metal-mediated coupling approaches (See Chapter 6).

Variations on the Friedel–Crafts alkylation include the use of alcohols in the presence of Brønsted or Lewis acids to generate the carbocation electrophile, or protonation of alkenes (Figure 2.8).

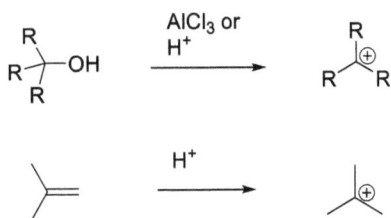

Figure 2.8: Formation of carbocation electrophiles from alcohols and alkenes.

Aromatic compounds can be acylated using an acid halide in the presence of a Lewis acid (referred to as Friedel–Crafts acylation). Similar to alkylation, the Lewis acid reacts with the acid halide to form a cationic acylium ion, which is the active electrophilic species (Figure 2.9). One important distinction between Friedel–Crafts alkylation and acylation that we will explore in the next section is that while alkyl groups are activating and *ortho/para*-directing, acyl groups are deactivating and *meta*-directing. This difference has implications for subsequent transformations once these groups are introduced.

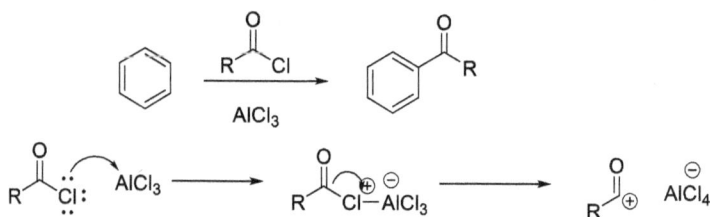

Figure 2.9: Friedel–Crafts acylation.

It is important to note two limitations of Friedel–Crafts alkylation and acylation. The first is that the reaction does not work for highly deactivated substrates such as nitro- and cyano-substituted benzenes. The second limitation is that they do not work for compounds such as anilines that can serve as Lewis bases. Lewis acid-base reactions with the Lewis acid will effectively compete with electrophile formation.

Bromomethyl and chloromethyl groups can be introduced directly onto activated aromatic rings by reaction with formaldehyde in the presence of HCl or HBr (Figure 2.10) [22, 23].

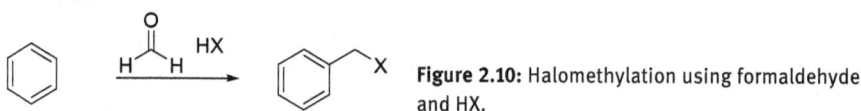

Figure 2.10: Halomethylation using formaldehyde and HX.

The reaction is thought to involve the formation of a chloromethylium or bromomethylium ion by reaction of formaldehyde with the acid (Figure 2.11). The halomethylium ion then serves as the electrophile in a standard electrophilic aromatic substitution.

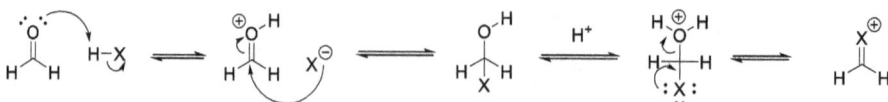

Figure 2.11: Formation of the halomethylium ion electrophile.

There are several approaches for the direct electrophilic introduction of formyl groups onto aromatic rings [24]. One of the best known approaches for introducing formyl groups uses a formamide (such as DMF) in the presence of POCl$_3$, followed by hydrolysis [25]. This reaction is referred to as a Vilsmeier or Vilsmeier–Haack reaction. The active electrophilic species in this reaction is the chloroiminium ion shown (Figure 2.12), which is produced by reaction of the formamide with POCl$_3$ [26]. The initial product of electrophilic aromatic substitution is the formyl iminium species, which is readily hydrolyzed to the corresponding aldehyde under aqueous conditions.

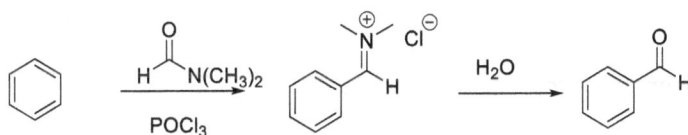

Electrophile formation:

Figure 2.12: Vilsmeier–Haack formylation.

Other methods for introducing formyl groups via electrophilic aromatic substitution have also been reported. For example, the Rieche formylation uses dichloromethyl methyl ether in the presence of TiCl$_4$, followed by aqueous acid work-up (Figure 2.13) [27]. The reaction bears a resemblance to the Vilsmeier reaction in terms of the active electrophile, and the fact that the initial electrophilic aromatic substitution product requires hydrolysis to give the formyl product. This reaction is most effective with activated aromatic substrates.

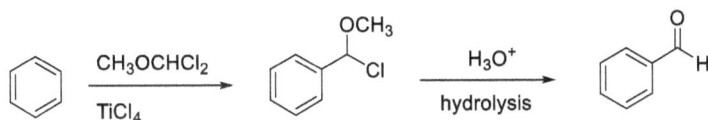

Electrophile formation:

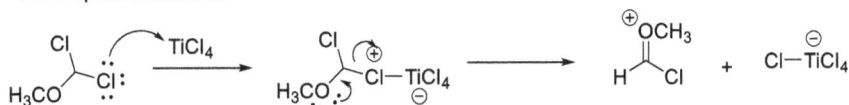

Figure 2.13: Rieche formylation and formation of the active electrophilic species.

2.4 Substituent effects in electrophilic aromatic substitution

2.4.1 Reactivity

When a substituted benzene derivative undergoes electrophilic aromatic substitution, the substituents will influence the overall reactivity of the compound as well as the regiochemistry of substitution. In general, electron-donating groups increase the rate of electrophilic substitution, while electron-withdrawing substituents decrease the rate of electrophilic substitution. As such, electron-donating substituents are often referred to as activating groups, while electron-withdrawing substituents are said to be deactivating. This trend can be rationalized based on the idea that electron-donating groups increase the electron density on the aromatic ring and therefore make them more nucleophilic, while electron-withdrawing groups do the opposite. Another way to rationalize this observation is based on the effect of substituents on the stability of the cationic intermediate (as well as the transition state leading to that intermediate). To illustrate the effects of substituents on the rate of electrophilic aromatic substitution, one can compare the relative rates of nitration as compared to benzene. Toluene reacts 25 times faster, while phenol reacts 1,000 times faster than benzene. In contrast, chlorobenzene reacts approximately 30 times slower than benzene, while nitrobenzene reacts approximately seven orders of magnitude slower [28, 29].

Based on the relative rates of reaction, substituents can be classified as activating (more reactive than benzene) or deactivating (less reactive than benzene). They can also be compared in terms of their relative activating and deactivating effects (Figure 2.14). The substituent effects for electrophilic aromatic substitution show a reasonable correlation to Hammett σ^+ values, which accounts for direct resonance stabilization of cationic intermediates [30]. An examination of Figure 2.14 shows that many of the strongly deactivating groups are able to withdraw electron density by resonance, while the weaker deactivating groups are inductively electron withdrawing. Similarly, the strongly activating substituents are able to donate electrons by resonance, while the weakly activating groups are those that are inductively electron donating.

Compounds such as phenols, aryl ethers, and anilines are highly activated by virtue of the strongly electron-donating groups. Unlike many electrophilic substitution reactions, which require a Lewis acid, these compounds will undergo electrophilic substitution readily in the absence of a Lewis acid. For example, phenol undergoes electrophilic bromination without a Lewis acid at relatively low temperatures (Figure 2.15). Aniline is an even more extreme example: it is reactive enough that controlling the electrophilic substitution becomes challenging. It will overreact to yield di- and tribromoanilines.

Strongly **deactivating** groups

Weakly **deactivating** groups

—NO₂   —CN   —SO₃H   (structure: C(=O)R)   —CF₃   —F   —Cl   —Br   —I

Strongly **activating** groups

Weakly **activating** groups

—NR₂   —OH   —OR   (structure: —N(H)—C(=O)R)   —CH₃

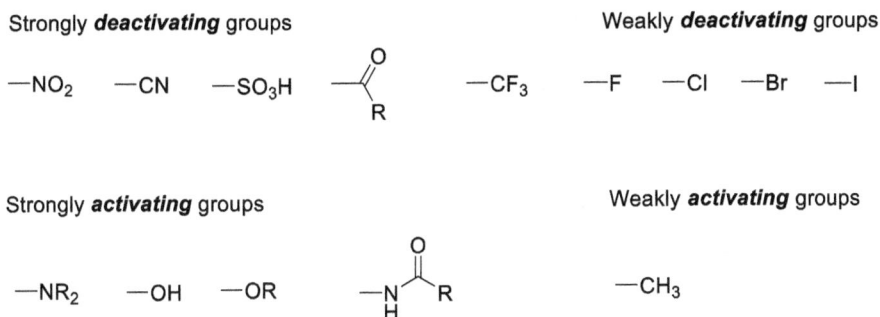

Figure 2.14: Activating and deactivating substituents in electrophilic aromatic substitution.

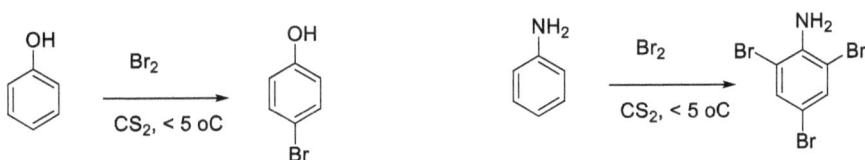

Figure 2.15: Reactions of phenol and aniline in electrophilic aromatic substitution.

In order to control the electrophilic aromatic substitution in these highly activated systems, the compounds can be acetylated using acetic anhydride or acetyl chloride. The corresponding acetates and acetamides are still activated, but their reactivity is somewhat attenuated, making it easier to control the substitution, as well as influencing the regioselectivity through steric interactions (see below).

2.4.2 Regioselectivity of electrophilic substitution of aromatic compounds

In terms of regiochemistry of substitution, electron-donating groups generally favor substitution at the *ortho* and *para* positions, while electron-withdrawing groups favor substitution at the *meta* position. As a demonstration of the regiochemical outcome of electrophilic aromatic substitution, consider the nitration of a series of substituted benzenes shown in Table 2.1. For electron-withdrawing groups such as nitro, cyano, carboxy, and trifluoromethyl, the *meta*-substituted product is predominant, while for electron-donating and activating substituents, a mixture of *ortho* and *para* isomers are formed. The halogens are an apparent exception: while they are electron-withdrawing and deactivating, they yield a mixture of *ortho* and *para* isomers as the major products.

To understand the regiochemistry, it is helpful to consider the resonance structures of the intermediate for *ortho*, *meta*, or *para* substitution (Figure 2.16), with a particular focus for the resonance structures where the positive charge is on the

Table 2.1: Distribution of product isomers for nitration of some monosubstituted benzenes.

Substituent (X)	% Ortho	% Meta	% Para
NO_2	5–8	91–93	0–2
CN	15–17	81–83	~2
CO_2H	15–20	75–85	~1
CF_3	6	91	3
F	9–13	0–1	86–91
Cl	30–35	~1	64–70
Br	36–43	1	56–62
I	38–45	1–2	54–60
CH_3	56–63	2–4	34–41
CH_2CH_3	46–59	2–4	46–51
OCH_3	30–40	0–2	60–70

Selected data from Carey and Sundberg, *Advanced Organic Chemistry, Part A: Structure and Mechanisms*, 5th ed. [31].

carbon bearing the substituent. In those resonance structures, the positive charge is *stabilized* by electron-donating substituents, and *destabilized* by electron-withdrawing groups. Based on this, substitution at the *ortho* and *para* positions leads to structures that are stabilized by electron-donating groups and there for the *ortho* and *para* products are favored. In contrast, the same structures are *destabilized* by electron-withdrawing substituents, and consequently *meta* substitution is preferred because of the absence of a significant destabilization.

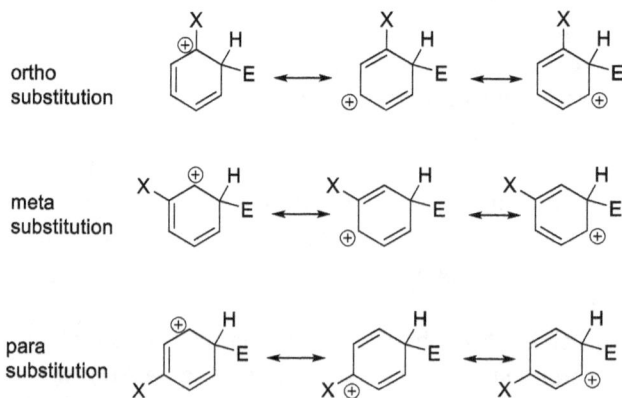

Figure 2.16: Resonance structures for cationic intermediates in electrophilic substitution.

A closer examination of the directing effects of substituents reveals some nota-ble exceptions to the general trend that electron-donating (activating) substituents are *ortho/para* directors and electron-withdrawing (deactivating) substituents are *meta* directors. Specifically, the halogen, which are weakly *deactivating* by virtue of their electronegativity, are also *ortho/para* directors. This apparent contradiction can be rationalized by the fact that the halogens have available nonbonding elec-tron pairs that can serve to stabilize the cationic intermediates by resonance when substitution takes place in the *ortho* or *para* positions.

When carrying out electrophilic aromatic substitution using monosubstituted ben-zene derivatives with *ortho/para*-directing groups, how does one predict or control the ratio of *ortho* and *para* products? Based purely on probability of reaction, one might expect a 2:1 ratio of *ortho:para* substitution because there are two *ortho* sites available to react and only one *para* site. However, other factors, such as sterics, can influence the ratio of *ortho* and *para* substitution products. If the substituent attached to the ring is bulky, it is possible to favor *para* substitution over *ortho* substitution.

One example of a reaction where *ortho* substitution is observed almost exclu-sively is in the Kolbe–Schmitt process for the preparation of acetyl salicylic acid (ASA, known commonly under the trade name Asparin) shown in Figure 2.17. In this industrial process, sodium phenoxide, which is highly activated toward elec-trophilic substitution, reacts with carbon dioxide as the electrophile to form the *ortho*-substituted product only. This high selectivity is attributed to a chelation of the sodium ion with the phenoxide oxygen and the developing negative charge on the carbon dioxide as the electrophilic attack proceeds [32]. As such, the sodium phenoxide is directing the electrophilic attack to take place in proximity to the ex-isting group. Similar effects to this will be discussed in Chapter 4 in the context of lithiation (deprotonation) of aromatic rings.

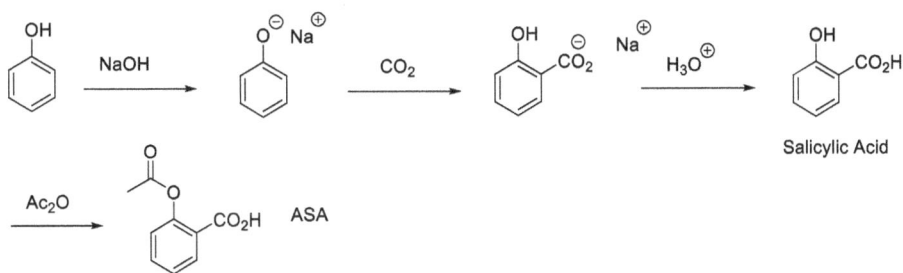

Figure 2.17: Kolbe–Schmitt process for the preparation of ASA.

2.4.3 Regioselectivity in polysubstituted benzenes

When two or more substituents are attached to the benzene ring, both can contribute to directing the regiochemistry of electrophilic aromatic substitution. Two scenarios can be envisioned: one where the substituents work cooperatively and direct to the same position(s), and the second where substituents compete and direct to different positions. In the former case, the outcome of the electrophilic substitution is relatively easy to predict. Consider the bromination of 4-hydoxybenzaldehyde, which is part of a synthesis of the herbicide bromoxynil (Figure 2.18). The hydroxyl group is an *ortho/para* director, and therefore directs substitution to the *ortho* position, since the *para* position is occupied. The aldehyde is a *meta* director, and thus directs to the same positions. Treatment with two equivalents of bromine leads to dibromination as shown.

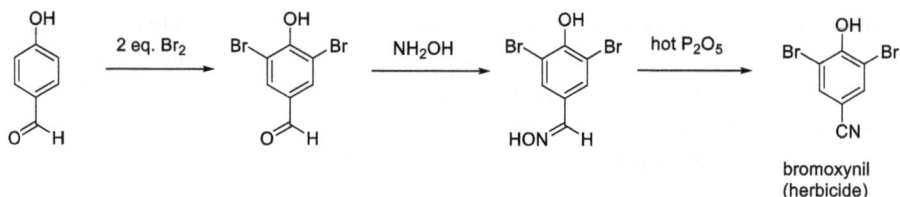

Figure 2.18: Synthesis of the herbicide bromoxynil.

In contrast, there are many situations where substituents compete and direct to different positions, whether it be two activating groups that compete, or an activating group and a deactivating group. In these cases, the activating group (or more strongly activating group in the case of two competing activating groups) generally determines the regiochemistry of substitution. For example, consider the reaction below to make butylated hydroxytoluene, commonly known as BHT, which is an antioxidant used as a stabilizer in fuels, solvents, and even as a food additive (Figure 2.19). The *t*-butyl groups both add *ortho* to the hydroxyl and not the methyl group despite the fact that both OH and CH_3 are activating groups and *ortho/para* directors. Since OH is a much stronger activating group, it controls the regiochemistry of substitution.

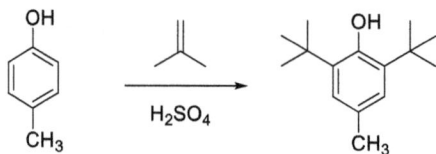

Figure 2.19: Preparation of BHT.

Several other examples of electrophilic substitution of polysubstituted benzenes are shown in Figure 2.20. Entries 1–3 in Figure 2.20 illustrate examples where the substituents on the ring work cooperatively in terms of their directing effects. Even when substituents work cooperatively, entries 1 and 2 show that a mixture of

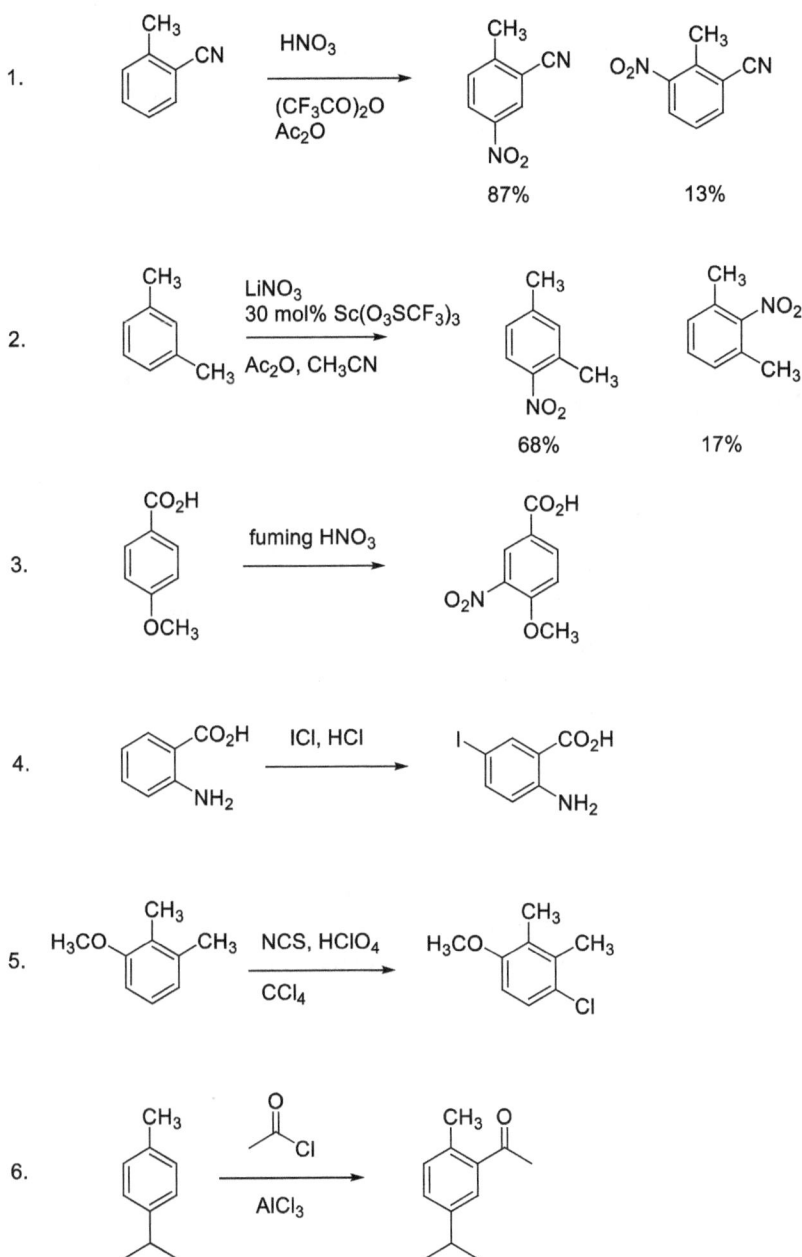

Figure 2.20: Representative examples of electrophilic aromatic substitution in polysubstituted benzenes.

products can still result [5, 33]. In these two examples, the 1,2,4-trisubstituted products are favored over the 1,2,3-trisubstituted products, presumably because of steric effects. In entry 3, only the expected product is observed [34]. Entries 4 and 5 have substituents working competitively in terms of directing effects [15, 35]. In both cases, the stronger activating group (the NH_2 and the OCH_3, respectively) determine the outcome of substitution. In the last entry, the two substituents direct to different positions, but are comparable in terms of their electronic directing effects [36]. Here, the outcome of the reaction is presumably determined by sterics.

2.5 Functional group interconversion to control reactivity and regioselectivity

Electrophilic aromatic substitution has a broad scope for the synthesis of polysubstituted benzenes and generally well-defined regiochemistry, depending on the substituents in place. But what if the desired product is not amenable to preparation because the directing effects of the substituents would lead to a different product? Alternatively, what if the desired reaction will not work because of functional group incompatibility? To circumvent some of these potential problems, there are some key functional group interconversion reactions that are important for the synthesis of aromatic compounds that are worth reviewing.

2.5.1 Reduction of nitro groups

Aromatic nitro groups can readily be reduced to the corresponding anilines. This is typically carried out by hydrogenation using palladium on carbon, or using stannous chloride (Figure 2.21). Stannous chloride can either be used directly in solvents such as ethanol, or it can be generated in situ from tin and aqueous HCl.

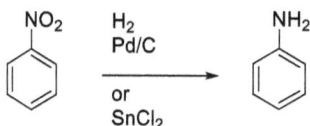

Figure 2.21: Reduction of aromatic nitro groups.

This reaction is significant because it converts a deactivating *meta*-directing group into and activating *ortho/para*-directing group. As such, it has implications for planning syntheses involving multiple aromatic substitution reactions. This reaction is also important because, as we will see in Chapter 2, anilines can be converted into their diazonium salts, which allows a variety of transformations to take place.

2.5.2 Reduction of acyl groups

We have seen that aromatic ketones can be prepared by Friedel–Crafts acylation (Section 2.3.4). The acyl groups can be reduced to the corresponding alkyl groups, usually by a Wolff–Kishner reduction or a Clemmensen reduction (Figure 2.22). The Wolff–Kishner reduction takes place using hydrazine under basic conditions, while the Clemmensen reduction uses zinc under strongly acidic conditions. Since one reaction takes place under basic conditions and the other under acidic conditions, the choice of whether to use a Wolff–Kishner or Clemmensen reduction will sometimes be determined by the compatibility of other functional groups with acidic or basic conditions.

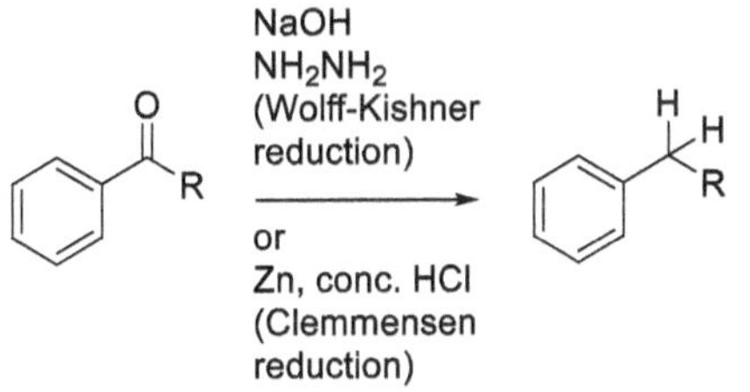

Figure 2.22: Reduction of aromatic ketones.

The Wolff–Kishner reduction proceeds by formation of the hydrazone using hydrazine and ultimately involved expulsion of nitrogen gas. The mechanism of the reaction is outlined in Figure 2.23

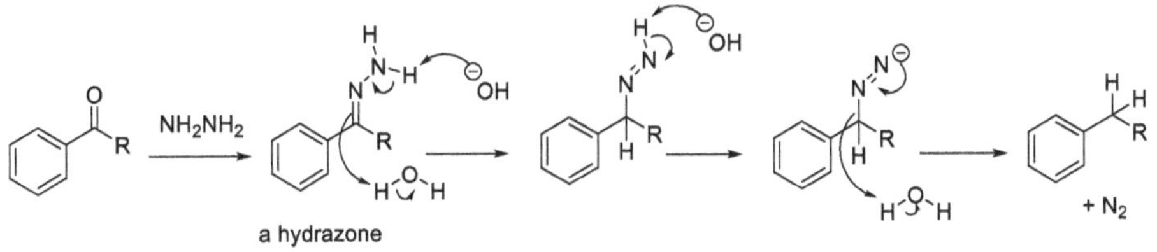

a hydrazone

Figure 2.23: Mechanism of Wolff–Kishner reduction.

In contrast to the Wolff–Kishner reduction, the mechanism of the Clemmensen reduction is not clearly established. One proposed mechanism proceeds via a radical anion, while another involves a zinc–carbenoid mechanism.

In Section 2.3.4, we saw that Friedel–Crafts alkylation is problematic for two reasons: (1) The product of Friedel–Crafts alkylation is more activated than the starting material, and as such multiple alkylations are favored, and (2) Friedel–Crafts alkylation does not work effectively for the introduction of linear alkyl groups because rearrangements occur. To avoid both of these problems, an effective alternative, a two-step sequence of Friedel–Crafts acylation, is followed by reduction (Figure 2.24). The product of Friedel–Crafts acylation is less reactive than the starting material, so

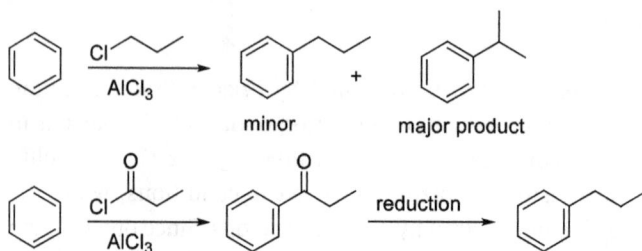

Figure 2.24: Comparison of Friedel–Crafts alkylation and Friedel–Crafts acylation and reduction.

reaction at multiple sites is prevented, because the initial reaction is an acylation, there is no risk of rearrangement.

2.5.3 Reactions at benzylic positions

There are several other reactions at benzylic carbons that are important for the elaboration of aromatic compounds. For example, radical bromination of benzylic methyl groups is important for further functionalization of aromatic compounds (Figure 2.25). This reaction is usually carried out using either bromine with irradiation, or with NBS either photochemically or with a radical initiator such as benzoyl peroxide or azobis(isobutyronitrile).

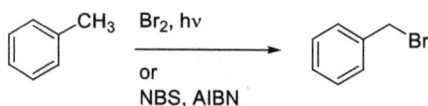

Figure 2.25: Benzylic bromination.

Mechanistically, these reactions are radical chain reactions. Taking photochemical bromination using bromine as an example, the initiation step involves homolytic cleavage of bromine to give bromine atoms (Figure 2.26). In the propagation steps, a bromine radical abstracts a hydrogen from the benzylic position to create a carbon–centered radical, which then reacts with molecular bromine to form the carbon–bromine bond and generate another bromine radical. Termination consists of recombination of two radicals – shown as the recombination of the benzyl radical and bromine radical here. When the reaction is carried out with NBS and a radical initiator, the NBS serves to form Br_2 in solution in a low and steady-state concentration [37, 38].

Oxidation of the benzylic positions is another important transformation for aromatic compounds. Benzyl alkyl groups can be oxidized to the corresponding carboxylic acid (Figure 2.27). This reaction is carried out using reagents such as $KMnO_4$, acidic dichromate, or nitric acid. The reaction is most commonly used for aromatic methyl groups, but alkyl side chains bearing at least one benzylic hydrogen can all be oxidized to the corresponding carboxylic acid.

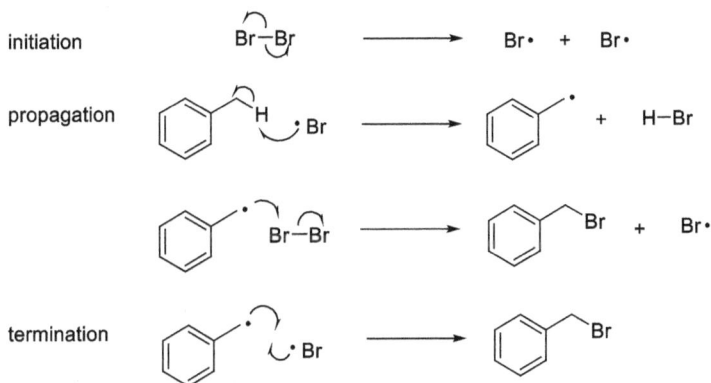

initiation

propagation

termination

Figure 2.26: Mechanism of benzylic bromination.

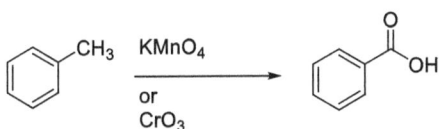

Figure 2.27: Benzoic acids via benzylic oxidation.

2.5.4 Acylation of phenols and anilines

Phenols and anilines can be readily acylated using acetic anhydride (Figure 2.28). Acylation attenuates the reactivity of the highly activated phenols and anilines and makes the regiochemistry of electrophilic aromatic substitution reactions easier to control. These groups can be readily cleaved by acid hydrolysis.

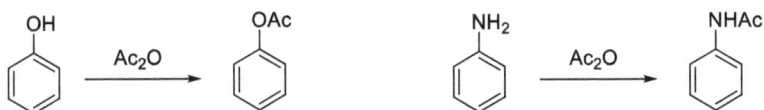

Figure 2.28: Acetylation of phenols and anilines with acetic anhydride.

This reaction can be particularly important for anilines, which are very activated toward electrophilic aromatic substitution. As we already saw, bromination of aniline is difficult to control, often leading to bromination at all of the activated positions (Figure 2.29) even at low temperature. In contrast, if aniline is acetylated, it can be brominated at room temperature and give predominantly the *para*-substituted product. The better control of reactivity is mostly due to the electronic deactivation of the aniline, but the preference for the *para* isomer may also be influenced by steric hindrance from the acetyl group.

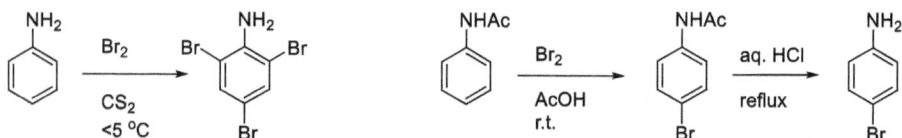

Figure 2.29: Comparison of reactivity of aniline and *N*-acetylaniline toward bromination.

2.6 Reversibility of electrophilic aromatic substitution

Some electrophilic aromatic substitution reactions are reversible. Specific examples include sulfonation and the introduction of *t*-butyl groups (Figure 2.30).

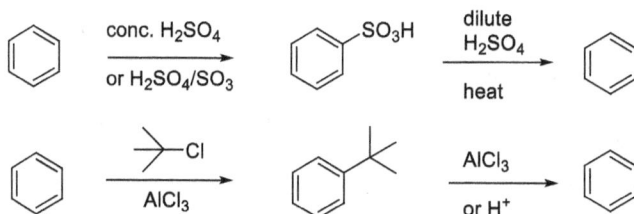

Figure 2.30: Reversible sulfonation or Friedel–Crafts alkylation.

The reversibility of these reactions can be exploited in the synthesis of substituted aromatic compounds in order to circumvent the normal directing effects of substituents on an aromatic ring. For example, the preparation of o-bromophenol is challenging because the bromination of phenol leads to p-bromophenol as the major product. However, but first sulfonating, the *para* and one *ortho* position can be blocked, forcing bromination to occur *ortho* to the phenolic OH (Figure 2.31) [39]. Finally, removal of the sulfonyl groups (by distillation of the product from the reaction mixture) leads to o-bromophenol as the major product.

 The reversibility of electrophilic aromatic substitution as illustrated by sulfonation shows that it is in principle possible to remove functional groups and either replace them with a hydrogen or another functional group. Here, we will consider a few examples where removal or replacement of a functional group has practical applications in synthesis or implications for synthesis. The utility of these reactions is that functional groups can either serve as a blocking group to prevent reaction at a given position of an aromatic ring as described above, or to act as a latent reactive site that is revealed upon electrophilic substitution.

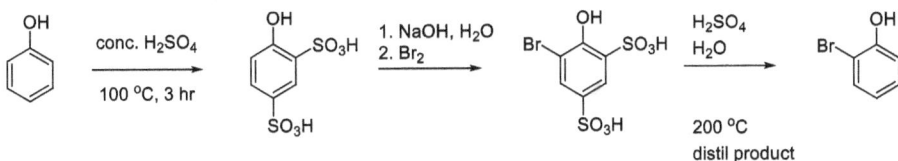

Figure 2.31: Use of sulfonation as blocking groups to control regiochemistry of electrophilic substitution.

2.6.1 Protiodesilylation

Trialkylsilyl groups can be removed under acidic conditions to produce the corresponding benzene derivative [26]. The reaction typically involves HCl or HBr and consists of *ipso*-protonation, followed by loss of the silyl group to regenerate the aromatic ring (Figure 2.32). In the context of synthesis, a silyl group such as trimethylsilyl (TMS) can serve as a protecting group, blocking positions that would otherwise undergo other types of reactions.

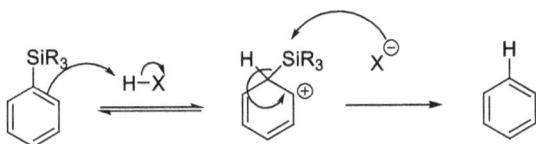

Figure 2.32: Protiodesilylation of arylsilanes.

2.6.2 Halodesilylation

Similarly, silyl groups can be removed by halogens such as Br_2 or ICl to replace the silyl group by a halogen (Figure 2.33) [26]. The reaction is typically carried out in CCl_4, $CHCl_3$, or acetic acid as the solvent.

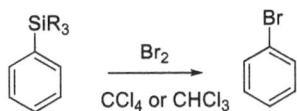

Figure 2.33: A halodesilylation reaction.

Mechanistically, one could imagine a mechanism that resembles protiodesilylation. However, in nonpolar solvents such as CCl_4, the reaction is second order in bromine and takes place with inversion of configuration at silicon. These observations and others have led to the proposal of a concerted reaction with a cyclic transition state (Figure 2.34) [40].

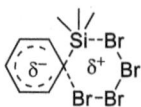

Figure 2.34: Proposed transition state for a concerted bromodesilylation reaction.

In the context of synthetic chemistry, the silyl group can serve as a masked aryl halide, which can be used for a variety of important synthetic transformations, in particular as substrates for transition metal-catalyzed reactions that we will see in Chapter 6.

2.7 Electrophilic rearrangement – the Fries rearrangement

In the presence of Lewis acids, aryl esters can undergo a rearrangement where the acyl group migrates from the phenolic oxygen to the aromatic ring (Figure 2.35) [41]. This reaction is known as the Fries rearrangement, named after Karl Fries who reported the reaction in 1908 [42].

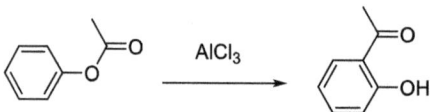

Figure 2.35: The Fries rearrangement.

Mechanistically, the reaction is an electrophilic rearrangement (Figure 2.36). The Lewis acid first coordinates to one of the ester oxygens, which is followed by fragmentation to produce an acyl cation and a phenoxide coordinated to the Lewis acid. The acyl cation then serves as an electrophile, undergoing aromatic substitution at either *ortho* or *para* to the phenoxide to give *o*- and *p*-acyl phenols.

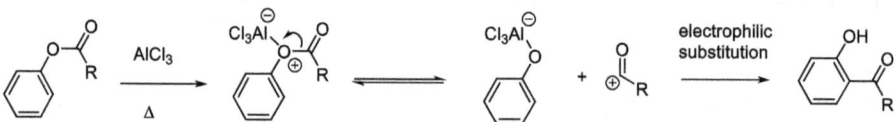

Figure 2.36: Mechanism of the cationic Fries rearrangement.

This reaction has also been observed photochemically. Referred to as the photo-Fries rearrangement, this reaction is thought to proceed *via* a radical mechanism, involving homolytic cleavage of the ester C–O bond, to form an acyl radical and a phenoxy radical, which can recombine to form the acyl phenol. The photo-Fries rearrangement has the advantage that it does not require high temperatures and does not require a Lewis acid. While these reactions have not been used widely in the synthesis of aromatic compounds, a similar reaction under anionic conditions, referred to as the anionic-*ortho*-Fries reaction, has been explored extensively for the preparation of substituted aromatic compounds. This reaction will be discussed in more detail in Chapter 4.

2.8 Electrophilic aromatic substitution of polycyclic aromatic compounds

Polycyclic aromatic compounds also undergo electrophilic aromatic substitution. In addition to the reactivity, the regiochemistry of substitution should be considered. Naphthalene, for example, typically undergoes electrophilic substitution at the 1-position (Figure 2.37). The observed regiochemistry is explained by considering the stability of the cationic intermediate. In the case of substitution at the 1-position, the carbocation intermediate has greater resonance stabilization than when substitution occurs at the 2-position. This argument applies when the substitution is carried out under kinetic control, which is typical for many electrophilic aromatic substitution reactions. If the reaction is carried out at higher temperatures in conditions where the reaction is reversible, then substitution at the 2-position to yield the thermodynamic product is preferred.

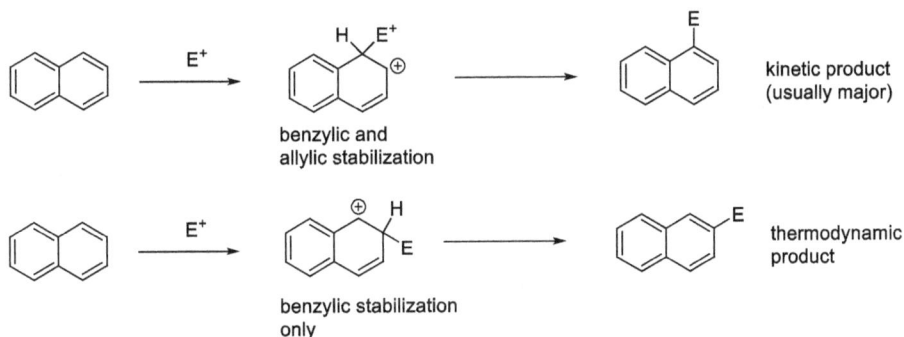

Figure 2.37: Electrophilic aromatic substitution of naphthalene.

Anthracene and phenanthrene typically react at the central ring (Figure 2.38). This reactivity is explained by the fact that the cationic intermediate for substitution at the central ring can maintain two intact aromatic rings.

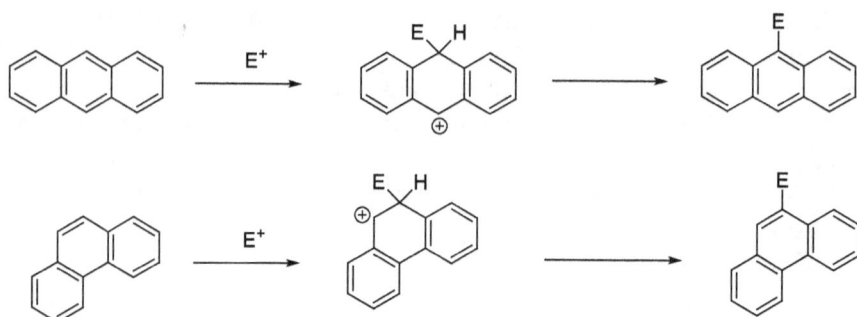

Figure 2.38: Electrophilic aromatic substitution of anthracene and phenanthrene.

However, because the cationic intermediates maintain aromatic stabilization, the driving force for the last step (deprotonation to reform the π-bond) is significantly less. As such, both phenanthrenes and anthracenes are prone to competing electrophilic addition reactions. For example, reaction of phenanthrene with bromine yields the dibrominated addition product (Figure 2.39).

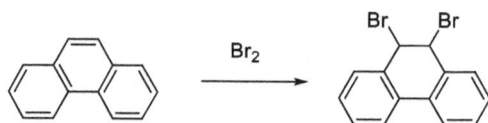

Figure 2.39: Reaction of phenanthrene with bromine.

2.9 Electrophilic aromatic substitution of heteroaromatic compounds

Heteroaromatic compounds also undergo electrophilic aromatic substitution. However, the heteroatom can have a dramatic effect on the rates of reaction and also brings its own regiochemical considerations, with the position of electrophilic substitution being largely determined by the heteroatom.

2.9.1 Five-membered heterocycles

Five-membered aromatic heterocycles such as pyrrole, furan, and thiophene can undergo electrophilic aromatic substitution, and the reactivity and regiochemistry is determined by the heteroatom. In terms of overall reactivity toward electrophiles, the five-membered heterocycles are *more* reactive than benzene. The five-membered heterocycles react readily with electrophiles without the need for a Lewis acid. Indeed, controlling reactivity can be challenging. For example, reaction of pyrrole with bromine at 0 °C gives a mixture of polybrominated pyrroles including tetra-bromopyrrole (Figure 2.40).

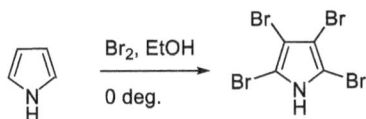

Figure 2.40: Electrophilic bromination of pyrrole.

Pyrrole is so reactive that it readily polymerizes under acidic conditions (Figure 2.41). This is why pyrrole is often requires purification by distillation immediately prior to use in reactions.

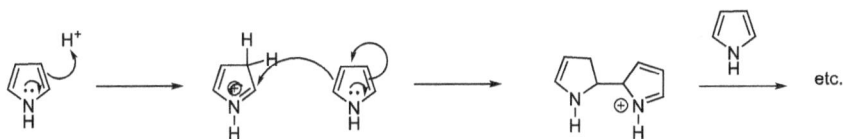

Figure 2.41: Mechanism of electrophilic pyrrole polymerization under acidic conditions.

Despite the high reactivity of pyrrole that can make controlling electrophilic substi-tution challenging, there are some examples where reactivity can be controlled. For example, the Vilsmeier reaction occurs readily to introduce formyl groups onto the 2-position of pyrrole, adjacent to the heteroatom (Figure 2.42).

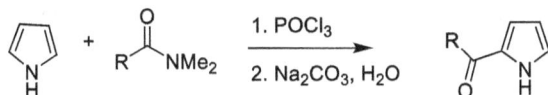

Figure 2.42: Vilsmeier–Haack reaction of pyrrole.

Furans and thiophenes are less reactive than pyrroles, but are still highly activated toward electrophilic substitution. As in the case of pyrroles, substitution occurs pref-erentially at the 2- and 5-positions – adjacent to the heteroatoms (Figure 2.43).

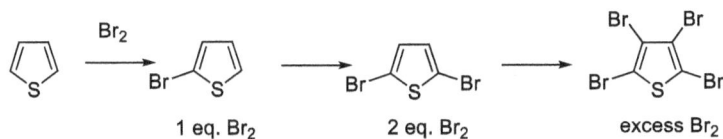

Figure 2.43: Bromination of thiophene.

2.9.2 Six-membered heterocycles

The most common six-membered heterocycles include pyridine (with one nitrogen atom), pyrimidine, pyrazine, and pyridazine (with two nitrogen atoms), and the tri-azines (with three nitrogen atoms). Unlike the five-membered heteroaromatics, pyr-idine and related heteroaromatics are relatively unreactive toward electrophilic substitution. We will see later on that these compounds undergo nucleophilic aro-matic substitution more readily (Section 3.5). The low reactivity of pyridine and de-rivatives can be rationalized by the presence of the nitrogen atom which lowers the energy of the highest occupied molecular orbital.

While pyridine is unreactive toward electrophiles, it can be oxidized to the cor-responding pyridine-*N*-oxide, which is reactive toward electrophiles (Figure 2.44). Typically, electrophilic substitution of pyridine *N*-oxides occurs at the 4-position. The *N*-oxide can then be reduced using trivalent phosphorus reagents such as PCl_3 or $P(OMe)_3$.

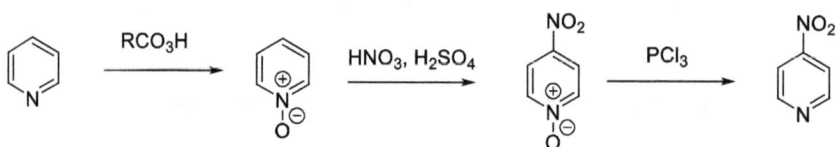

Figure 2.44: Electrophilic aromatic substitution of pyridine via the pyridine *N*-oxide.

2.10 Further reading

The topic of electrophilic aromatic substitution is covered in most organic chemistry texts. For some detailed discussion of the topic, see F. A. Carey and R. J. Sundberg, *Advanced Organic Chemistry Part A, Structure and Mechanisms* (Chapter 9) and *Part B, Reactions and Synthesis* (Chapter 11) 5th ed. Springer, 2007; M. B. Smith and J. March, *March's Advanced Organic Chemistry*, 6th ed. John Wiley & Sons, 2007. For resources focused on the topic, see: R. Taylor, *Electrophilic Aromatic Substitution*, John Wiley & Sons, 1990. For more on heteroaromatic compounds, T. L. Gilchrist, *Heterocyclic Chemistry*, 3rd ed. Addison Wesley Longman Ltd., 1997.

References

[1]   Mare DL, David PB. Aromatic Substitution; Nitration and Halogenation. London: Butterworths;1959.
[2]   Kuhn SJ, Olah GA. Aromatic substitution. VII. Friedel-Crafts type nitration of aromatics. J Am Chem Soc. 1961;83(22):4564–71.

[3] Olah GA, Kuhn SJ. Aromatic substitution XII.1 steric effects in nitronium salt nitrations of alkylbenzenes and halobenzenes. J Am Chem Soc. 1962;84(19):3684–87.

[4] Waller FJ, Barrett AGM, Braddock DC, Ramprasad D. Lanthanide(III) triflates as recyclable catalysts for atom economic aromatic nitration. Chem Commun. 1997;613–14.

[5] Kawada A, Takeda S, Yamashita K, Abe H, Harayama T. Scandium(III) trifluoromethanesulfonate catalyzed aromatic nitration with inorganic nitrates and acetic anhydride. Chem Pharm Bull. 2002;50(8):1060–65.

[6] Gilbert EE. Sulfonation and Related Reactions. New York: Wiley; 1965.

[7] Kajigaeshi S, Kakinami T, Yamisaki H, Fujisaki S, Okamoto T. Halogenation using quaternary ammonium polyhalides xi. bromination of acetanilides by use of tetraalkylammonium polyhalides. Bull Chem Soc Jpn. 1988;61:2681–83.

[8] Berthelot J, Guette C, Desbène P-L, Basselier -J-J, Chaquin P, Masure D. Bromation régiosélective en série aromatique. I: Monobromation en position para de phénols et d'aminés aromatiques par le tribromure de tétrabutylammonium. Can J Chem. 1989;67 (12):2061–66.

[9] Gervat S, Léonel E, Barraud JY, Ratovelomanana V. High regioselectivity of bromination of anilines by tetraethylammonium chloride / methanol system as cocatalysts under mild conditions. Tetrahedron Lett. 1993;34(13):2115–18.

[10] Carreno MC, Garcia Ruano JL, Sanz G, Toledo MA, Urbano A. N-Bromosuccinimide in acetonitrile: A mild and regiospecifc nuclear brominating reagent for methoxybenzenes and naphthalenes. J Org Chem. 1995;60(16):5328–31.

[11] Zysman-Colman E, Arias K, Siegel JS. Synthesis of arylbromides from arenes and Nbromosuccinimide bromosuccinimide (NBS) in acetonitrile – A convenient method for aromatic bromination. Can J Chem. 2009;87(2):440–47.

[12] Butler AR. Mechanism of aromatic iodination. J Chem Educ. 1971;48(8):508.

[13] Stavber S, Jereb M, Zupan M. Electrophilic iodination of organic compounds using elemental iodine or iodides. Synthesis (Stuttg). 2008;(10):1487–513.

[14] Brazdil LC, Cutler CJ. Selective production of diiodobenzene and iodobenzene from benzene. J Org Chem. 1996;61(26):9621–22.

[15] Wallingford VH. A KP. 5-Iodoanthranilic acid. Org Syn Coll. 1943;2:349.

[16] Purrington ST, Kagen BS, Patrick TB. The application of elemental fluorine in organic synthesis. Chem Rev. 1986;86:997–1018.

[17] Umemoto T, Nagayoshi MN. N′-Difluoro-1,4-diazoniabicyclo[2.2.2]octane salts, highly reactive and easy-to-handle electrophilic fluorinating agents. Bull Chem Soc Jpn. 1996;69 (8):2287–95.

[18] Singh S, DesMarteau DD, Zuberi SS, Witz M, Huang H-N. N-Fluoroperfluoroalkylsulfonimides. Remarkable new fluorination reagents. J Am Chem Soc. 1987;109(23):7194–96.

[19] Umemoto T, Fukami S, Tomizawa G, Harasawa K, Kawada K, Tomita K. Power and structure-variable fluorinating agents. The N-fluoropyridinium salt system. J Am Chem Soc. 1990;112 (23):8563–75.

[20] Olah GA editor. Friedel Crafts and Related Reactions. New York: Wiley InterScience; 1963.

[21] Sharman SH. Alkylaromatics. Part I. Friedel-crafts alkylation of benzene and alkyl-substituted benzenes with n-alkyl bromides. J Am Chem Soc. 1962;84:2945–51.

[22] Olah GA, Yu SH. Protonated chloromethyl alcohol and chloromethyl ethers. Proof for the intermediacy of the elusive chloromethyl alcohol. J Am Chem Soc. 1975;97(8):2293–95.

[23] Belen'kii LI, Vol'kenshtein YB, Karmanova IB. New data on the chloromethylation of aromatic and heteroaromatic compounds. Russ Chem Rev. 1977;46(9):891–903.

[24] Olah GA, Ohannesian L, Arvanaghi M. Formylating agents. Chem Rev. 1987;87(4):671–85.

[25] Haack A AV. Über die Einwirkung von Halogenphosphor auf Alkyl-formanilide. Eine neue methode zur Darstellung sekundärer und tertiärer p-Alkylamino-benzaldehyde. Ber Dtsch Chem Ges. 1927;60:119–22.

[26] Taylor R. Electrophilic Aromatic Substitution. Toronto: J. Wiley;1990.

[27] Rieche A, Gross H, Hoft E. Über α-Halogenäther, IV. Synthesen aromatischer Aldehyde mit Dichlormethyl-alkyläthern. Chem Ber. 1960;93(1):88–94.

[28] Hughes ED, Ingold CK, Reed RI. Kinetics and mechanism of aromatic nitration. Part II. Nitration by the nitronium ion, NOz+, derived from nitric acid. J Chem Soc. 1950;2400–40.

[29] Anslyn EV, Dougherty DA. Modern Physical Organic Chemistry. Sausalito: University Science Books; 2006.

[30] Brown HC, Okamoto Y. Electrophilic substituent constants. J Am Chem Soc. 1958;80:4979–87.

[31] Carey FA, Sundberg RJ. Advanced Organic Chemistry; Part A: Structure and Mechanism. 5th ed. New York: Springer; 2007.

[32] Hales JL, Jones JI, Lindsey AS. Mechanism of the Kolbe-Schmitt reaction. Part I. Infra-red studies. J Chem Soc. 1954;3145–51.

[33] Smith K, Gibbins T, Millar RW, Claridge RP. A novel method for the nitration of deactivated aromatic compounds. J Chem Soc Perkin Trans. 2000;1:2753–58.

[34] Dinescu L, Maly KE, Lemieux RP. Design of photonic liquid crystal materials: Synthesis and evaluation of new chiral thioindigo dopants designed to photomodulate the spontaneous polarization of ferroelectric liquid crystals. J Mater Chem. 1999;9(8):1679–86.

[35] Goldberg Y, Alper H. Biphasic electrophilic halogenation of activated aromatics and heteroaromatics with N-halosuccinimides catalyzed by perchloric acid. J Org Chem. 1993;58 (11):3072–75.

[36] Allen CFH. Aceto-p-Cymene. Org Syn Coll. 1943;2:3.

[37] Russell GA, DeBoer C, Desmond KM. Mechanisms of benzylic bromination. J Am Chem Soc. 1963;85(3):365–66.

[38] Pearson RE, Martin JC. The identity of the chain-carrying species in brominations with n-bromosuccinimide: Selectivity of substituted n-bromosuccinimides toward substituted toluenes. J Am Chem Soc. 1963;85(20):3142–46.

[39] Huston RC, Ballard MM. o-Bromophenol. Org Syn Coll. 1943;2:97.

[40] Eaborn C, Steward OW. Organosilicon compounds. Part XXX. The stereochemistry of the clavage of a silicon-aryl bond by bromine. J Chem Soc. 1965;521–27.

[41] Blatt AH. The fries reaction. Chem Rev. 1940;27:413–36.

[42] Fries K, Finck G. Über Homologe des Cumaranons und ihre Abkömmlinge. Ber Dtsch Chem Ges. 1908;41:4271–84.

3 Nucleophilic aromatic substitution reactions

3.1 Introduction

In the previous chapter, we saw that aromatic rings themselves are usually nucleophiles, reacting with electrophiles in electrophilic aromatic substitution reactions. Furthermore, S_N2 reactions cannot take place at the sp^2 hybridized carbon atoms of an aromatic ring. Nonetheless, as we will see in this chapter, nucleophilic displacements on aromatic rings can occur. We will explore the mechanistic scenarios for nucleophilic aromatic substitution with particular attention to the addition–elimination mechanism. We will also consider the chemistry of aryl diazonium salts, which bear superficial resemblance to nucleophilic aromatic substitution.

3.2 Addition–elimination of nucleophiles (S_NAr)

The most common type of nucleophilic aromatic substitution at an aromatic ring proceeds first by nucleophilic addition to the carbon bearing the leaving group to form a delocalized carbanion intermediate, followed by expulsion of the leaving group and rearomatization [1]. Because the reaction proceeds by an anionic intermediate, it usually involves electron-deficient aromatic rings. The rate-determining step is typically nucleophilic addition and the reaction usually requires electron withdrawing groups *ortho* or *para* to the site of nucleophilic attack (Figure 3.1).

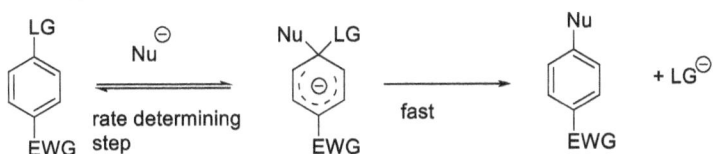

Figure 3.1: Generalized S_NAr mechanism.

As shown in Figure 3.1, initial nucleophilic attack at the carbon bearing the leaving group results in a delocalized carbanion intermediate, referred to as a Meisenheimer complex. The reaction often requires electron-withdrawing groups *ortho* or *para* to the leaving group in order to stabilize the negative charge of the intermediate. The most common electron-withdrawing groups used to activate compounds toward nucleophilic aromatic substitution are nitro groups, but other frequently used groups include cyano, carbonyl, and sulfonate groups. All of these groups can stabilize the negative charge by resonance. This direct resonance stabilization explains the importance of having electron-withdrawing at the *ortho-* and/or *para*-positions – groups in the *meta*-positions can only stabilize the negative charge inductively. As strong

https://doi.org/10.1515/9783110562682-003

support for this mechanism, several Meisenheimer intermediates have been isolated and characterized (Figure 3.2) [2].

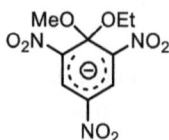

Figure 3.2: Structure of an isolated Meisenheimer complex.

Having multiple electron-withdrawing groups serve to increase the electrophilicity of the aromatic ring. For example, 2,4-dinitro-1-chlorobenzene, which is prepared by dinitration of chlorobenzene, reacts readily with hydrazine to form 2,4-dinitrophenylhydrazine (Figure3.3). 2,4-Dinitrophenylhydrazine was commonly used as a qualitative test for the presence of aldehydes and ketones through the formation of the corresponding phenylhydrazone derivative.

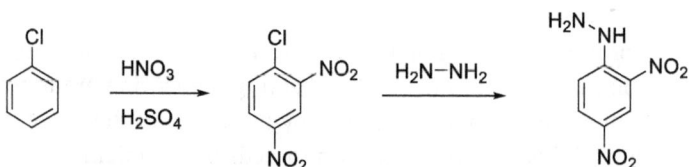

Figure 3.3: Synthesis of 2,4-dinitrophenylhydrazine via nucleophilic aromatic substitution.

Again, the electron withdrawing groups increase reactivity toward nucleophiles, provided they are in the *ortho* and/or *para* positions relative to the site of nucleophilic attack. This behavior can be rationalized by the nature of the anionic intermediate, where the negative charge is only delocalized to the *ortho* and *para* positions. As an illustration of the importance of the positions of the electron withdrawing groups, consider the following reaction, where there are two possible sites for nucleophilic attack. As shown in Figure 3.4, only the chloride *ortho* to the nitro group undergoes substitution, while the chloro group that is *meta* to the nitro group is unreactive.

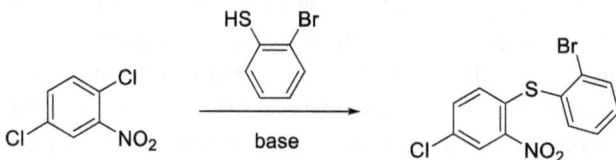

Figure 3.4: Regiochemistry of nucleophilic aromatic substitution.

3.2.1 Leaving groups in S_NAr

Unlike aliphatic nucleophilic aromatic substitution reactions, where bromide and io-dide are generally the best leaving groups, the trend in leaving group ability in S_NAr is distinctly different, with fluoride being the best leaving group (Figure 3.5) [3–5].

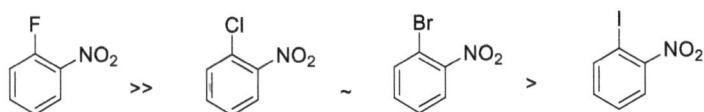

Figure 3.5: Trends in leaving group ability in S_NAr reactions.

This observation can be explained by the rate-determining nucleophilic addition step of the reaction. This step is accelerated when the carbon bearing the leaving group is more electrophilic. Since fluoride is the most electronegative of the halo-gens, it withdraws electron density most effectively from the site of attack, making it more electrophilic.

3.2.2 Regioselectivity in S_NAr reactions

While the position of substitution in many S_NAr reactions is determined by the posi-tion of the leaving group, for compounds bearing multiple potential leaving groups, regioselectivity of substitution becomes an issue. We already saw that the electron-withdrawing group activates *ortho* and *para* positions toward nucleophilic attack and that the *meta* position is not effectively activated. This result can be explained by considering the resonance structures of the Meisenheimer intermediate. But what do we expect for nucleophilic aromatic substitution where there are leaving groups *ortho* and *para* to the activating group? In many cases, nucleophilic substi-tution occurs preferentially *para* to the activating group. For example, for nucleo-philic substitution on fluoro-substituted esters and nitriles, selective substitution *para* to the ester or nitrile was observed provided the reaction was carried out at low temperature (Figure 3.6) [6]. Substitution at the *ortho*-position was readily achieved at higher temperature, allowing a controlled and sequential set of nucleo-philic aromatic substitutions. If the reaction was carried out at higher temperature, the regioselectivity was significantly less. It should be emphasized that the posi-tions *meta* to the electron-withdrawing group do not undergo substitution because they are not effectively activated. As an interesting counter example, in the case of 2,3,4-trifluorobenzonitrile, substitution takes place *ortho* to the nitrile, suggesting that the 3-fluoro group is playing some kind of directing effect (Figure 3.6).

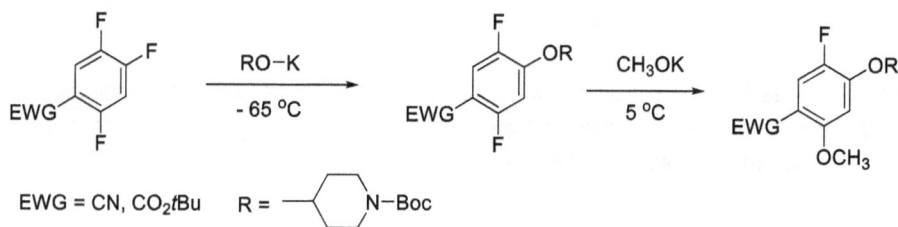

EWG = CN, CO$_2$tBu   R = [N-Boc]

Figure 3.6: Regioselective S$_N$Ar reactions [6].

As further demonstration of the tendency for S$_N$Ar to take place *para* to the activating group is highlighted in the reactions of fluorinated benzophenones (Figure 3.7) [7]. In these compounds, substitution takes place first *para* to the carbonyl groups using a variety of nucleophiles. The products are then able to undergo substitution *ortho* to the carbonyls. In this case, the latter nucleophiles like potassium hydroxide, sodium sulfide, and primary amines undergo twofold nucleophilic aromatic substitution, resulting in cyclization to give the corresponding xanthones, thioxanthones, and acridones.

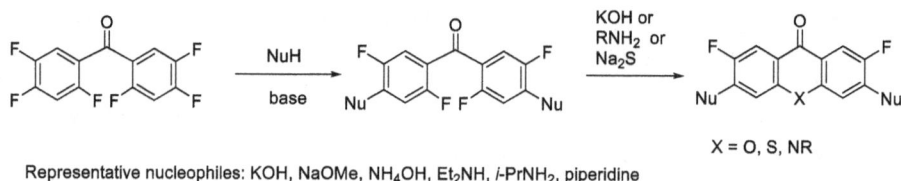

X = O, S, NR

Representative nucleophiles: KOH, NaOMe, NH$_4$OH, Et$_2$NH, *i*-PrNH$_2$, piperidine

Figure 3.7: Regioselective S$_N$Ar reactions to prepare xanthones, thioxanthones, and acridones [7].

In contrast to the previous examples where *para*-substitution is generally preferred, when polyhalogenated anilines are subjected to nucleophilic aromatic substitution with xanthate nucleophiles, substitution takes place exclusively at the *ortho*-position [8]. This reaction is followed by subsequent cyclization (Figure 3.8). On the one hand, the reactivity is surprising because the electron-rich anilines are deactivating with respect to S$_N$Ar. However, the interesting feature is the selectivity for *ortho*-substitution. This high regioselectivity is thought to be due to precoordination of the nucleophile to the aniline via hydrogen bonding. Support for this hypothesis is provided by the fact that 3,4-difluoroaniline, which has leaving groups in only the *meta* and *para* positions, shows no reaction under these conditions.

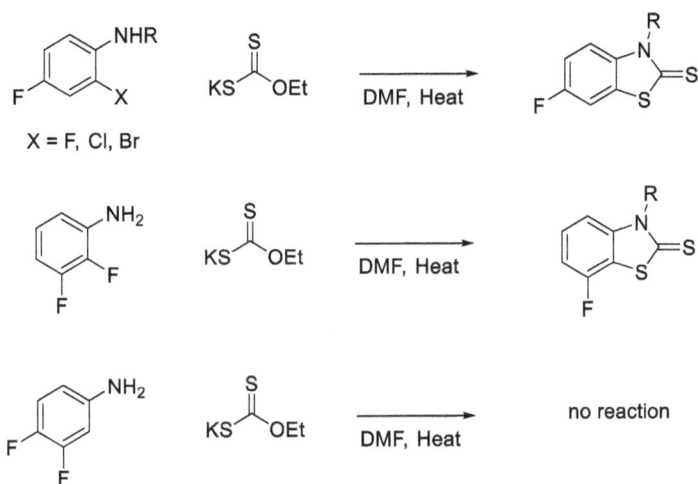

Figure 3.8: *Ortho*-selective S$_N$Ar reactions [8].

3.2.3 Examples of the utility of S$_N$Ar reactions

Due to the prevalence of aromatic compounds in pharmaceuticals, there are numerous examples of the use of nucleophile aromatic substitution reactions in the synthesis of active pharmaceutical ingredients. S$_N$Ar substitution reactions are useful for introduction of O, N, and S linkages onto the aromatic ring.

For example, a key step in the synthesis of duloxetine, which is used as an antidepressant and for treatment of fibromyalgia, features a nucleophilic aromatic substitution where an alkoxide nucleophile displaces the fluoride on 1-fluoronaphthalene (Figure 3.9) [9]. A similar reaction is used for one of the reported asymmetric syntheses of fluoxetine, commonly known as Prozac™ (Figure 3.9) [10].

As another example of the utility of nucleophilic aromatic substitution, consider the synthesis of ofloxacin (an antibiotic used to treat bacterial infections) from the compound shown (Figure 3.10) [11].

The first two steps in this synthetic sequence are intramolecular nucleophilic aromatic substitutions involving first the amine and then the alcohol as the nucleophiles (Figure 3.11) [11]. Following this, an intermolecular S$_N$Ar takes place using *N*-methylpiperazine as the nucleophile. The final step is ester hydrolysis to produce ofloxacin.

Nucleophilic aromatic substitution reactions have seen utility beyond the synthesis of active pharmaceutical ingredients. This methodology has been exploited in materials chemistry for the preparation of luminescent compounds, as well as polymeric materials and frameworks. For example, carbazole anions generated by deprotonation of carbazole with sodium hydride were used as nucleophiles

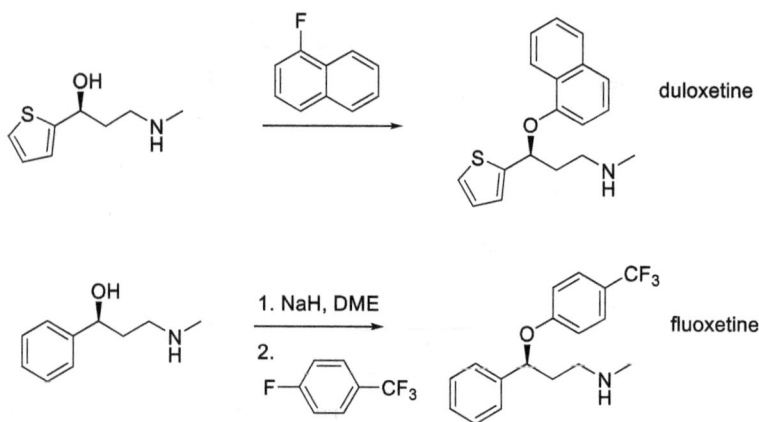

Figure 3.9: Synthesis of antidepressant pharmaceuticals via nucleophilic aromatic substitution [9, 10].

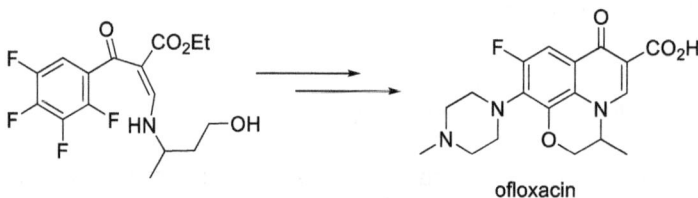

Figure 3.10: Synthetic approach to ofloxacin.

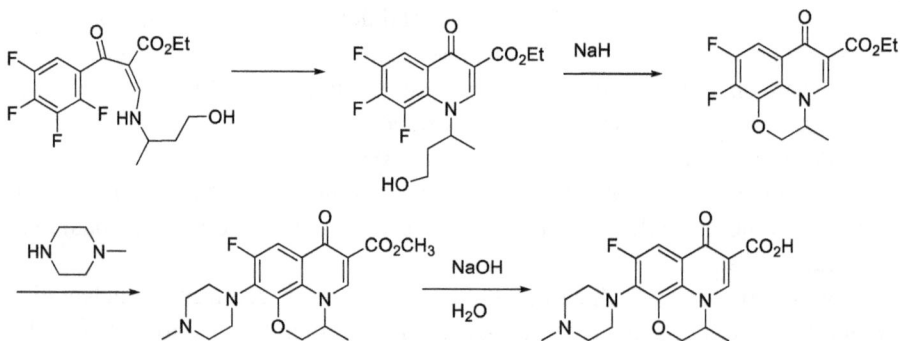

Figure 3.11: Synthesis of ofloxacin via a sequence of nucleophilic aromatic substitution reactions [11].

on electron-deficient polyfluorinated cyanoarenes (Figure 3.12) [12]. These compounds are highly luminescent, and because they consist of nearly orthogonal electron-rich components and an electron-deficient ring, they are reported to display thermally activated delayed fluorescence (TADF) [12]. Thermally activated delayed fluorescence has emerged as an important strategy for achieving efficient organic light-emitting diodes.

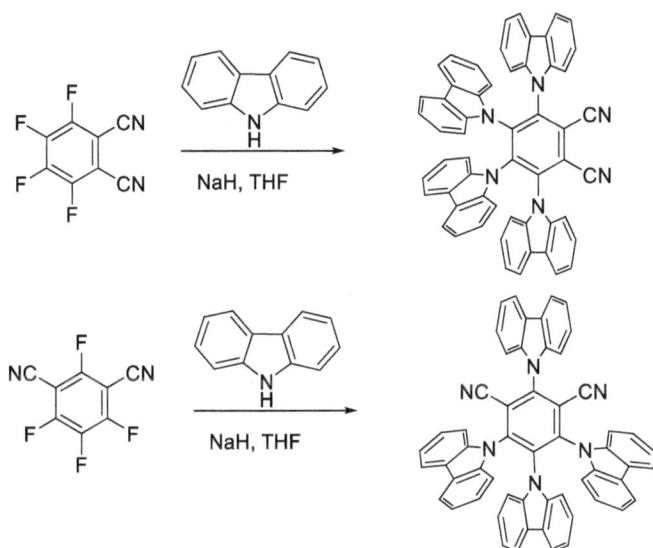

Figure 3.12: Synthesis of TADF emitters via nucleophilic aromatic substitution [12].

Bidentate nucleophiles such as catechols, 2-aminophenols, and 2-aminothio-phenols have also been used in S$_N$Ar reactions with tetrafluoroterephthalonitrile to prepare highly luminescent compounds [13]. Depending on reaction conditions and stoichiometry, the corresponding tricyclic dioxins and phenoxazines, as well as the corresponding pentacyclic heteroacene analogs have been prepared. Select examples are shown in Figure 3.13. Because the S$_N$Ar reaction can be controlled by reaction conditions to produce either tricyclic or pentacyclic compounds, the reaction can be performed in a stepwise manner to prepare dissymmetric systems with varied heteroatom substitution patterns. Not only has this approach been used to prepare luminescent materials, but has also been used to prepare functional polymers and covalent organic frameworks [14–16].

3.2.4 Carbon nucleophiles in nucleophilic aromatic substitution

In contrast to other nucleophiles, carbon-based nucleophiles have not been used as extensively for nucleophilic aromatic substitution. There are examples of nucleophilic aromatic substitution with carbanion nucleophiles generated from relatively acidic carbons, such as malonic esters or acetoacetic esters [1]. As a relatively recent example, some cyanoacetates have been used in nucleophilic aromatic substitutions on activated fluorobenzenes (Figure 3.14) [17].

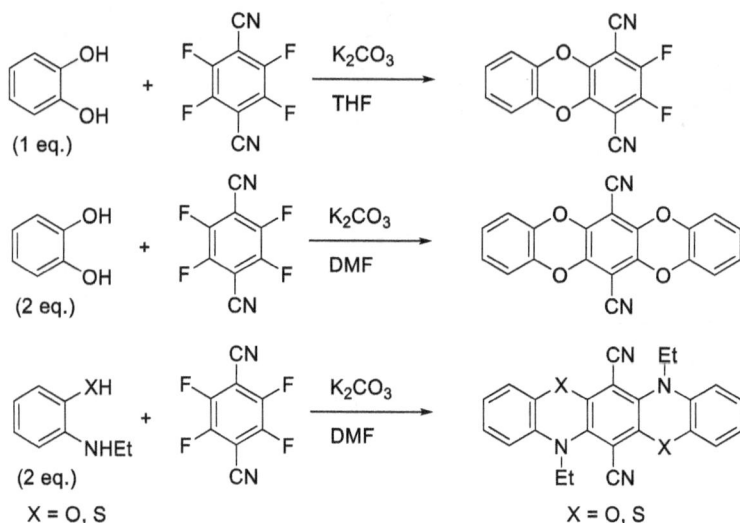

Figure 3.13: Synthesis of polycyclic compounds via S$_N$Ar reactions using bidentate nucleophiles.

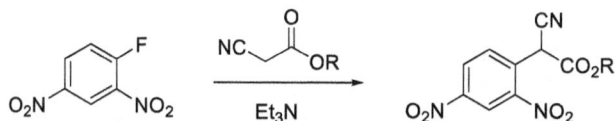

Figure 3.14: Example of nucleophilic aromatic substitution with a carbon-based nucleophile.

An interesting variation of nucleophilic aromatic substitution is the use of *N*-heterocyclic carbenes (NHCs) to catalyze the nucleophilic displacement of aryl fluorides with aryl aldehydes to prepare benzophenones (Figure 3.15) [18, 19].

Figure 3.15: NHC-catalyzed aroylation of aryl fluorides.

The reaction takes place in the presence of sodium hydride and a catalytic amount of an NHC precursor such as 1,3-dimethylimidazolium iodide. The NHC is generated in situ by deprotonation of the imidazolium salt and serves to convert the normally electrophilic aldehyde into a nucleophile – an umpolung that bears resemblance to a benzoin condensation. The proposed mechanism is outlined in Figure 3.16 [18, 19].

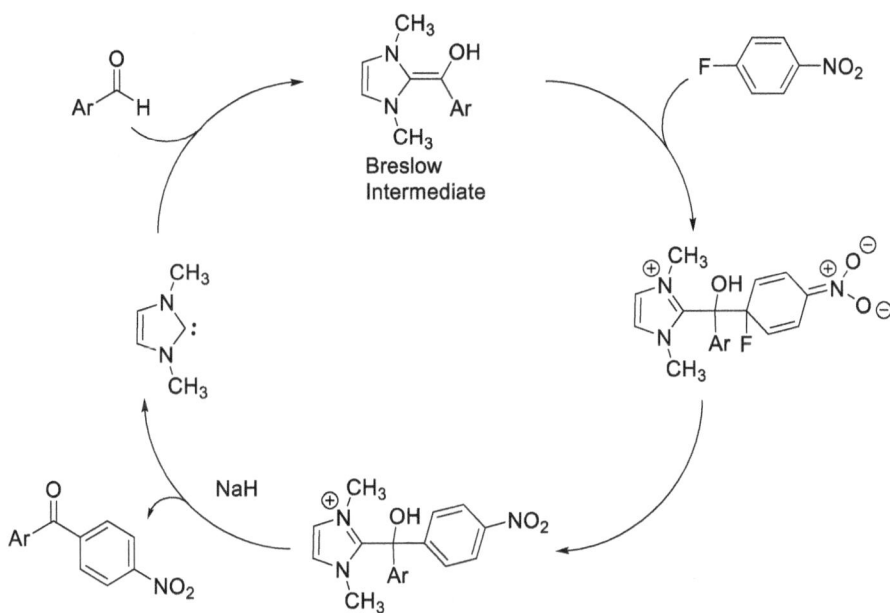

Figure 3.16: Proposed mechanism of NHC-catalyzed aroylation of aryl fluorides [18, 19].

Initially, the NHC reacts with the aldehyde to produce an intermediate (often referred to as a Breslow intermediate), which acts as a nucleophile in the displacement of fluoride. Deprotonation then results aldehyde formation and liberation of the NHC catalyst. This reaction has been applied to the synthesis of several xanthone natural products [20, 21].

3.2.5 Reversibility of S$_N$Ar reactions

Nucleophilic aromatic substitutions are generally thought of as being irreversible and indeed in many cases this is true. However, there are several examples that show evidence for the reversibility of S$_N$Ar reactions. This reversibility means that S$_N$Ar reactions can be dynamic and controlled by thermodynamics – a form of "dynamic covalent chemistry," which is an emerging tool in supramolecular and materials chemistry [22, 23]. For example, dichlorotetrazine undergoes a typical S$_N$Ar reaction with phenols such as cresol to displace the chlorides. The resulting product, when treated with a different phenol, undergoes another nucleophilic aromatic substitution, leading to an exchange of the aryl ethers under thermodynamic control (Figure 3.17) [24]. This exchange can in some cases reach equilibrium in a matter of minutes.

Another example demonstrating the dynamic and reversible nature of the S$_N$Ar reaction was reported by Ong and Swager [25]. They demonstrated that reaction of

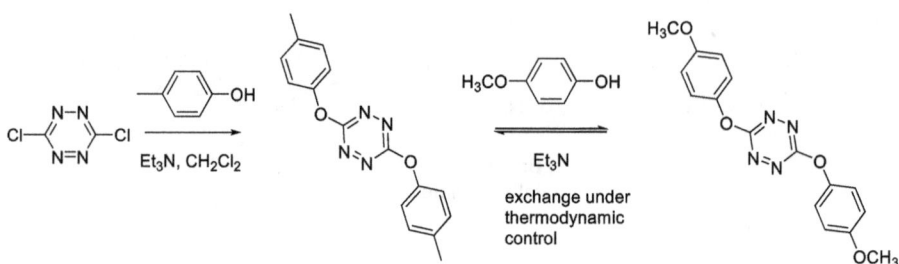

Figure 3.17: Exchange of aryl ethers based on S$_N$Ar of tetrazines [24].

two equivalents of 1,2-benzenedithiol with tetrafluorophthalonitrile results in a four-fold S$_N$Ar reaction to give the corresponding dithianthrene in good yield (Figure 3.18). When one equivalent of benzenedithiol is used, the symmetrical thianthrene is formed. Treatment of this thianthrene with a second equivalent of benzenedithiol gives the same dithianthrene (Figure 3.18). This product can only be the result of a nucleophilic ring opening, which is outlined in Figure 3.19. As a further demonstration of the dynamics of the reaction, when the dithianthrene is treated with substituted benzendithiols under basic conditions, exchange of the dithiols to form a mixture of dithianthrene products is observed.

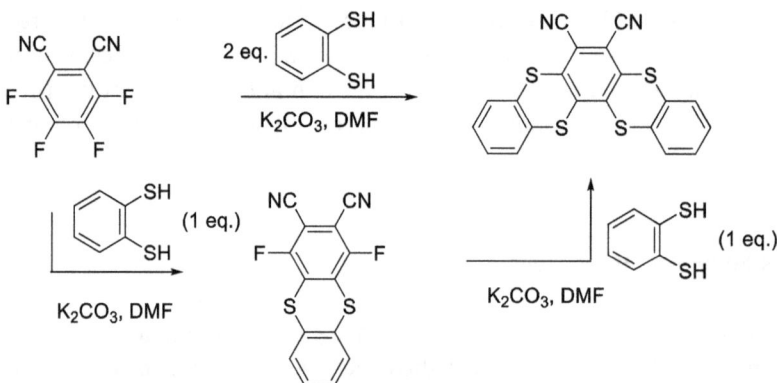

Figure 3.18: Dynamic formation of dithianthrenes via S$_N$Ar [25].

3.2.6 The Smiles rearrangement: an intramolecular S$_N$Ar reaction

The Smiles rearrangement consists of an aryl migration with the generalized reaction shown in Figure 3.20 [26, 27]. Mechanistically, it is an intramolecular S$_N$Ar reaction. The reaction was first reported in the late 1800s, but the scope of the reaction was developed by Smiles in the 1930s [28–32].

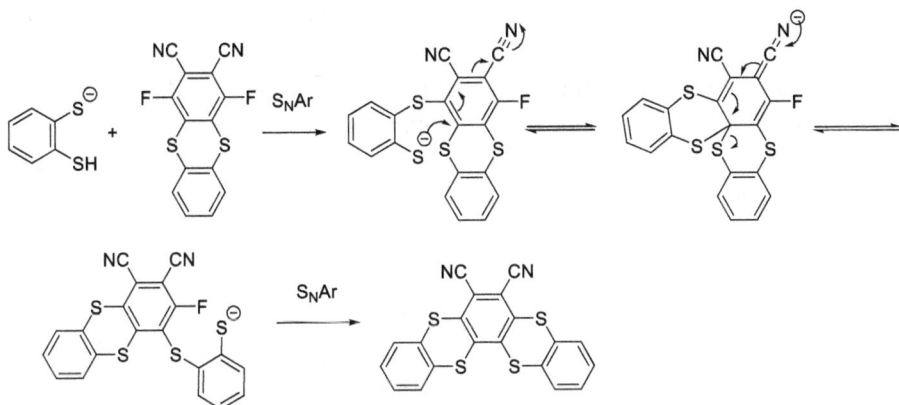

Figure 3.19: Proposed mechanism of the dynamic S$_N$Ar reaction [25].

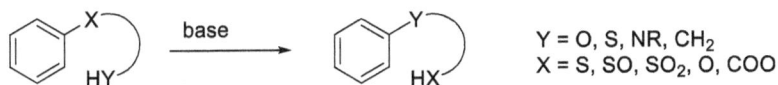

Y = O, S, NR, CH$_2$
X = S, SO, SO$_2$, O, COO

Figure 3.20: The Smiles rearrangement.

The nucleophile is typically an anionic heteroatom (Y = O, S, NR) generated by deprotonation but can also be a carbanion (Y = RCH$_2$). The leaving group is often another heteroatom (X = S, SO, SO$_2$, O, or COO), with the SO$_2$ being common. The linking group is often another aromatic ring. An example is shown below, involving a diaryl sulfone (Figure 3.21) [31, 32]. Since the reaction usually proceeds via an S$_N$Ar mechanism, electron-withdrawing groups such as nitro groups are often needed to stabilize the Meisenheimer intermediate.

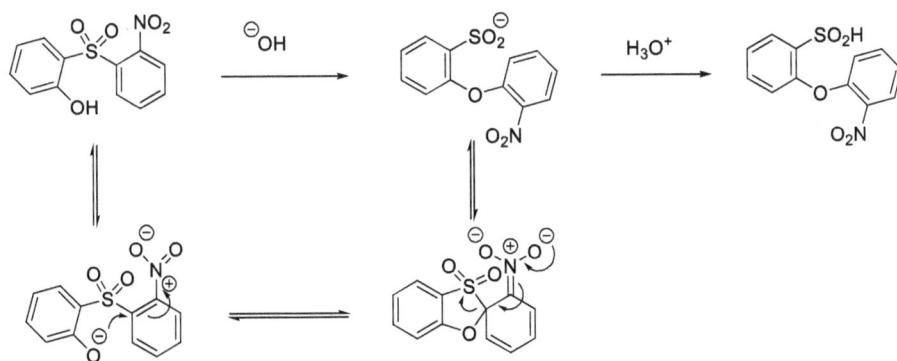

Figure 3.21: Example of the Smiles rearrangement of a diarylsulfone.

Evidence for the S_NAr mechanism is found in the example below, where the spirocyclic Meisenheimer complex is long-lived enough to be characterized by UV/Vis and ^1H NMR spectroscopy (Figure 3.22) [33]. The ability to detect the intermediate allows a more detailed examination of the mechanism and shows that in these systems the rate-limiting step in the reaction is the breakdown of the anionic Meisenheimer complex (the product-forming step). Specifically, the formation of the anionic intermediate is faster with more strongly electron-withdrawing groups, but the rate of product formation is slower.

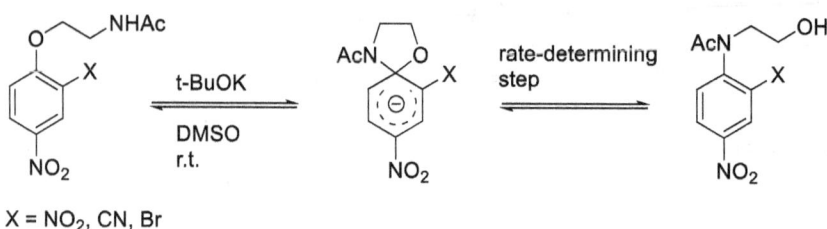

X = NO_2, CN, Br

Figure 3.22: Meisenheimer complex as an intermediate in the Smiles rearrangement [33].

As mentioned above, in some cases, carbanions can serve as the nucleophile [34, 35]. This variation is often referred to as the Truce–Smiles rearrangement. In these cases, since a strongly nucleophilic anion is involved, electron-withdrawing groups are not needed as activating groups. The reaction presents an interesting strategy for carbon–carbon bond formation in arenes. For example, deprotonation at one of the benzylic positions is achieved with n-BuLi, and the resulting anion serves as the nucleophile (Figure 3.23) [34].

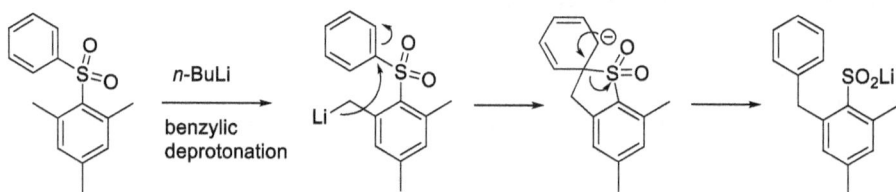

Figure 3.23: An example of the Truce–Smiles rearrangement [34].

The following examples show some of the scope of the Truce–Smiles rearrangement. In the first example, the nucleophilic carbanion is generated by deprotonating the α-carbon of the ester using sodium hydride (Figure 3.24). Following the Truce–Smiles rearrangement, the phenoxide leaving group is able to react with the ester to form the lactone [36]. The second example shows a one-pot intermolecular nucleophilic aromatic substitution followed by a Truce–Smiles reaction (Figure 3.24). In the S_NAr

reaction, the phenolate displaces fluoride from o-fluoronitrobenzene, while the subsequent Truce–Smiles rearrangement involves the α-carbon of the ketone [37].

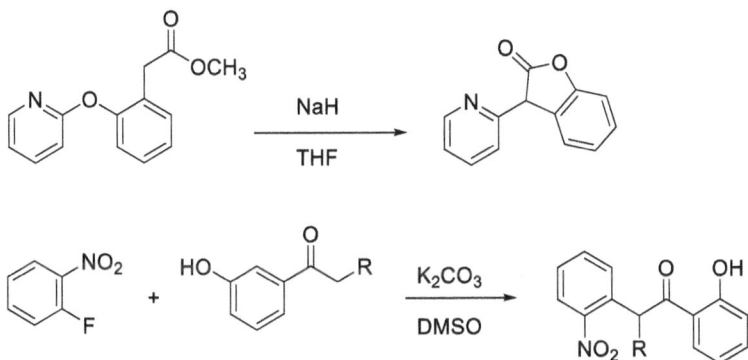

Figure 3.24: More examples of the Truce–Smiles rearrangement.

A variant of the Truce–Smiles rearrangement involves the base-mediated rearrangement of aryl ureas, carbamates, and thiocarbamates (Figure 3.25) [38–40]. This reaction is sometimes referred to as the Clayden rearrangement and takes place in the presence of an alkyllithium, which deprotonates the methylene adjacent to the heteroatom. This carbanion then serves as the nucleophile in a Truce-Smiles rearrangement.

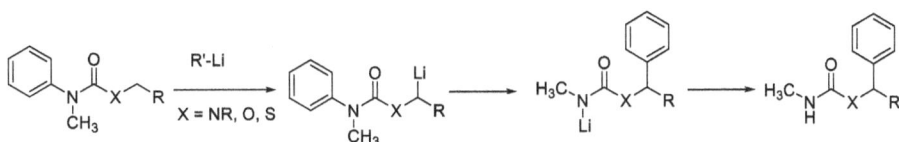

Figure 3.25: Clayden rearrangement of aryl ureas, carbamates, and thiocarbamates.

3.3 Substitution via benzyne formation

While nucleophilic aromatic substitutions of aryl bromides are generally slow, treatment of bromobenzene with sodium amide produces aniline (Figure 3.26) [41, 42]. This reaction proceeds, despite the fact that the aromatic ring is not activated by electron withdrawing groups.

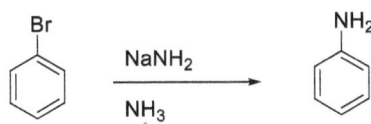

Figure 3.26: Nucleophilic substitution of bromobenzene with sodium amide.

Interestingly, when substituted bromobenzenes are used, an approximately equal mixture of isomers is observed (Figure 3.27) [42].

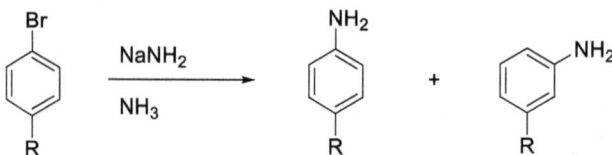

Figure 3.27: Nucleophilic substitution of substituted bromobenzenes with sodium amide.

These observations can be explained by sodium amide acting as a base to eliminate HBr, resulting in the formation of a benzyne (Figure 3.28). Excess amide then acts as a nucleophile, attacking either carbon of the benzyne. This mechanism is sometimes referred to as the elimination–addition mechanism.

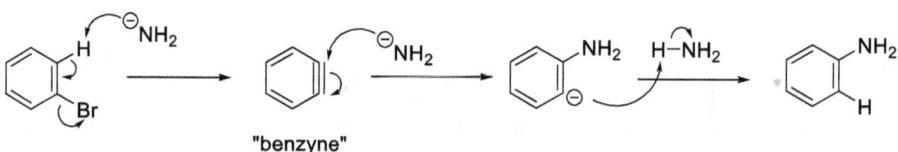

Figure 3.28: Mechanism of nucleophilic aromatic substitution via a benzyne intermediate.

By virtue of the strain of a triple bond constrained to a six-membered ring, these intermediates are highly reactive and have emerged as useful intermediates in synthesis. We will explore the chemistry of benzynes in Chapter 5.

3.4 Concerted nucleophilic aromatic substitutions

While nucleophilic aromatic substitution via the addition–elimination mechanism or benzyne formation are well-established stepwise mechanisms that likely explain most nucleophilic aromatic substitution reactions, there are an increasing number of reactions that are thought to occur via a *concerted* mechanism. The evidence in support of a concerted nucleophilic aromatic substitution mechanism comes from both experimental and computational studies, and suggests that a concerted mechanism has a surprisingly broad scope [43].

An early example of a reaction that is thought to proceed via concerted mechanism is the reaction of aryl halides with potassium hydride, where the hydride nucleophile displaces the halide (Figure 3.29) [44]. Unlike the typical addition–elimination mechanism, this reaction takes place in the absence of electron-withdrawing groups.

Furthermore, the trend in leaving group ability is the opposite of that of normal nucleophilic aromatic substitutions (i.e., Ar–I > Ar–Br > Ar–Cl > Ar–F).

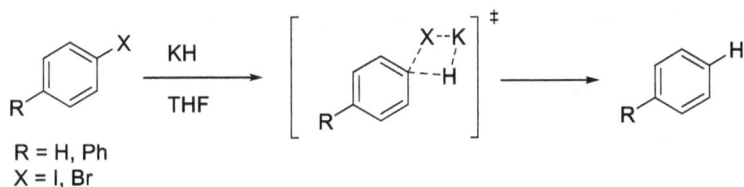

R = H, Ph
X = I, Br

Figure 3.29: Concerted nucleophilic displacement of aryl halides with potassium hydride.

More recently, Ritter and coworkers developed a deoxyfluorination of phenols using PhenoFluor and cesium fluoride (Figure 3.30) [45, 46]. The reaction involves nucleophilic addition of the phenol to the PhenoFluor reagent to form an imidazolium salt, which could be isolated and characterized by X-ray crystallography. Following this, fluoride from CsF adds to the imidazolium intermediate to give a tetrahedral species, which then delivers the fluoride in a concerted displacement of the oxygen to yield the fluorobenzene. Evidence for a concerted process for the displacement stems from $^{16}O/^{18}O$ kinetic isotope effects and computational studies.

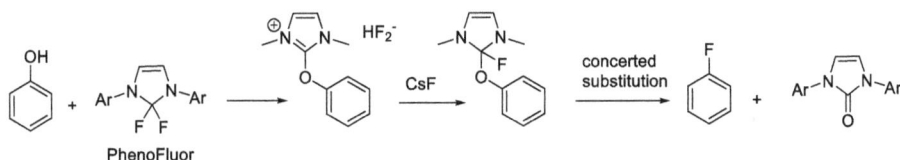

PhenoFluor

Figure 3.30: Concerted deoxyfluorination of phenols using PhenoFluor [45].

Deoxyfluorination of using sulfuryl fluoride and tetramethylammonium fluoride is another nucleophilic displacement that is thought to proceed via a concerted mechanism (Figure 3.31) [47]. The first step of this reaction is the activation of the phenol using sulfuryl fluoride to produce the fluorosulfonate. The second step is a nucleophilic aromatic substitution with tetramethylammonium fluoride to give the corresponding fluoroarene. This has a broad substrate scope, including both electron-poor and electron-rich substrates, and computational studies suggest that the second step is a concerted substitution [47].

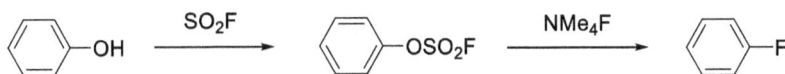

Figure 3.31: Deoxyfluorination of phenols via concerted nucleophilic aromatic substitution [47].

Building on the studies by Ritter and coworkers, Jacobsen and coworkers carried out a comparative study using kinetic isotope effects in combination with computational studies. Their study showed that arenes bearing strongly electron-withdrawing groups do indeed proceed by a stepwise mechanism, substitution on a bromopyridines with fluoride are concerted, and substitution using fluoride on an electron-deficient benzene is borderline (Figure 3.32) [48].

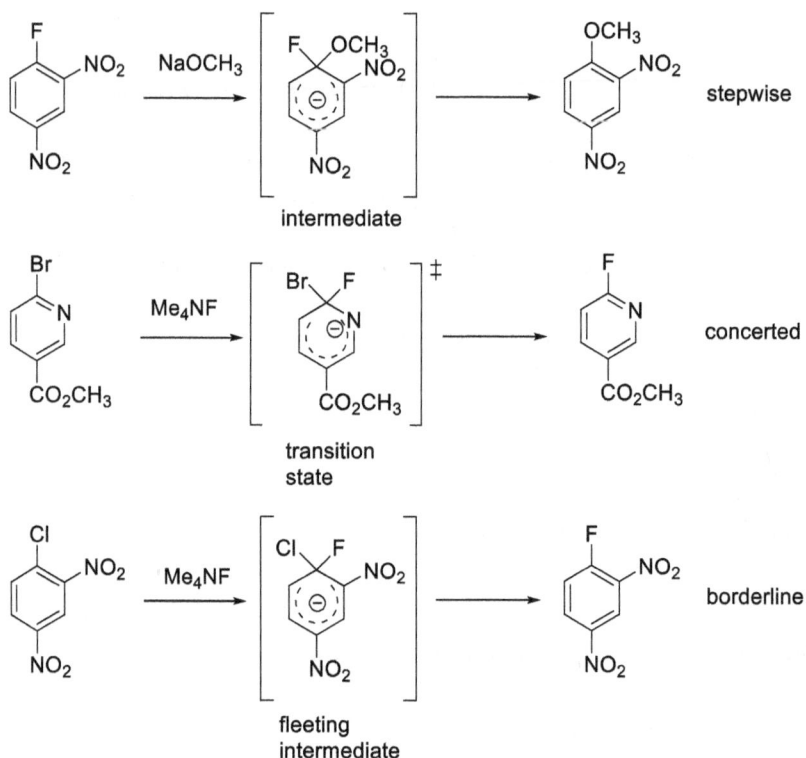

Figure 3.32: Stepwise and concerted nucleophilic aromatic substitutions reported by Jacobsen and coworkers [48].

In many cases, the evidence for concerted nucleophilic aromatic substitution is derived from computational studies, although there is increasingly experimental evidence to support concerted mechanisms. It is possible that some of the reactions we typically assume are stepwise S_NAr reactions may in fact proceed via a concerted mechanism.

3.5 Nucleophilic substitution of heteroaromatic systems

Electron-deficient heteroaromatic compounds such as pyridines are excellent candidates for nucleophilic aromatic substitution, especially when the leaving group is in the 2- or 4-position (Figure 3.33). These reactions are generally thought to occur through an addition–elimination mechanism (S_NAr) and the anionic intermediate is stabilized by placing the charge on the electronegative nitrogen. However, based on the study by Jacobsen and coworkers discussed above, it is possible that many of the nucleophilic aromatic substitution reactions of heteroaromatic compounds are actually concerted [48].

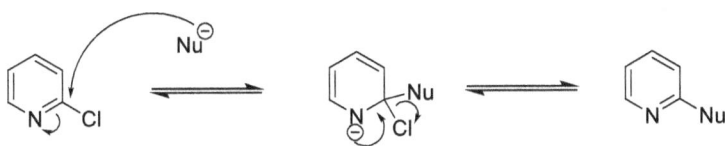

Figure 3.33: Generalize nucleophilic aromatic substitution on a chloropyridine.

Pyridyl chlorides are commonly used substrates for nucleophilic aromatic substitution because they can be readily prepared from the corresponding pyridones using reagents such as $POCl_3$ or PCl_5 (Figure 3.34). Mechanistically, the reaction bears analogies with the Vilsmeier–Haack reaction we saw in Chapter 2. The pyridone oxygen attacks $POCl_3$, liberating chloride that can then perform a nucleophilic addition to the carbon adjacent to the nitrogen. Rearomatization is accompanied by expulsion of the leaving group, giving the chloropyridine upon deprotonation.

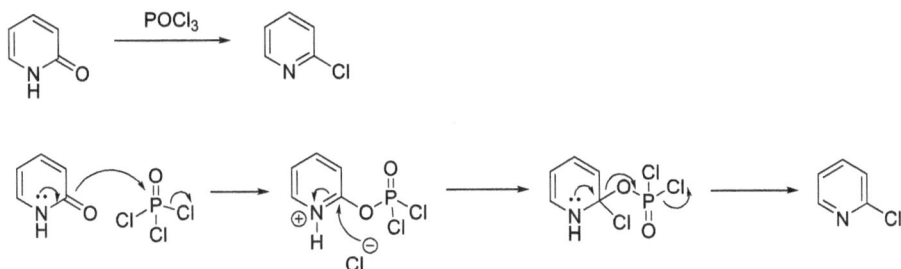

Figure 3.34: Synthesis of 2-chloropyridine from the corresponding 2-pyridone.

Interestingly, pyridyl ethers can also serve as substrates for nucleophilic aromatic substitution, despite the fact that ethers are not usually considered to be good leaving groups. For example, in part of the synthesis of the analgesic flupirtine, the methoxy group of a substituted pyridine is replaced by an amine (Figure 3.35) [49].

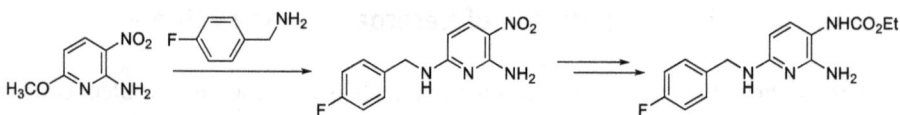

Figure 3.35: Nucleophilic aromatic substitution of a pyridine in the synthesis of flupirtine.

Much like pyridines, other six-membered nitrogen-containing aromatics such as pyrimidines, pyridazines, pyrazines, and triazines also readily undergo nucleophilic substitution. For example, cyanuric chloride (trichlorotriazine) can undergo nucleophilic substitution at all three sites and typically makes use of amine, alcohol, and sulfur-based nucleophiles. As an example, reaction of cyanuric acid with dimethylamine is used to prepare hexamethylmelamine, also known as the anticancer drug altretamine (Figure 3.36) [50].

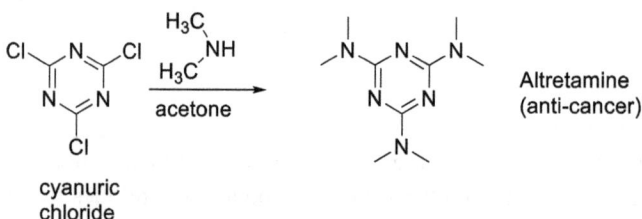

Figure 3.36: Synthesis of altretamine from cyanuric chloride [50].

Furthermore, since the replacement of electron-withdrawing chloro groups with electron-rich substituents, the reactivity with each successive substitution reaction on the triazine ring decreases. An important consequence of this attenuation of reactivity is that the nucleophilic substitution can be controlled to produce dissymmetric triazines. With amine nucleophiles, the first substitution takes place below room temperature, the second takes place with gentle heating above room temperature, while the third substitution takes place only upon heating to higher temperatures (Figure 3.37) [50, 51].

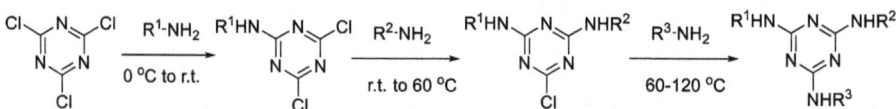

Figure 3.37: Stepwise synthesis of dissymmetric triazines.

As an example, the ability to control substitution of cyanuric chloride can be used to prepare the herbicide atrazine (Figure 3.38) [52].

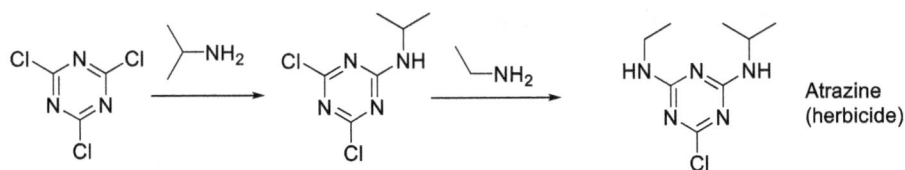

Figure 3.38: Synthesis of atrazine from cyanuric chloride [52].

Cyanuric halides can also react with carbon nucleophiles. For example, the re-action of acetylides with cyanuric fluoride yields the corresponding trialkynyl tria-zines (Figure 3.39) [53]. This direct nucleophilic substitution can be contrasted with transition metal-catalyzed reactions that will be explored in Chapter 6.

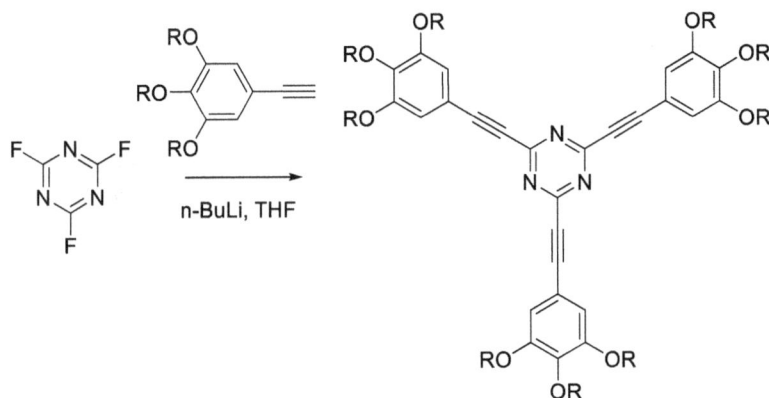

Figure 3.39: Synthesis of tris(phenylethynyl)triazines via nucleophilic aromatic substitution [53].

In contrast to pyridines and related compounds, electron-rich heteroaromatics (such as pyrroles, furans, and thiophenes) rarely undergo nucleophilic aromatic substitu-tion. The reaction typically requires electron-withdrawing groups (e.g., carbonyl, sulfonyl, nitro, or cyano) to activate the substrate toward nucleophilic aromatic substitution. As one example, the synthesis of the painkiller ketorolac involves a nucleophilic aromatic substitution (Figure 3.40) [54].

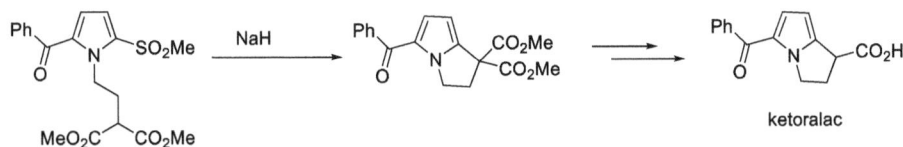

Figure 3.40: Nucleophilic aromatic substitution on a pyrrole in the synthesis of ketorolac [54].

3.6 Substitution of aryl diazonium salts

3.6.1 Diazonium salt formation

Another type of substitution reaction that is widely used for the synthesis of aromatic compounds is the substitution of aryl diazonium salts, which are generated from the corresponding aniline derivatives, usually using sodium nitrite or isoamyl nitrite (Figure 3.41) [55].

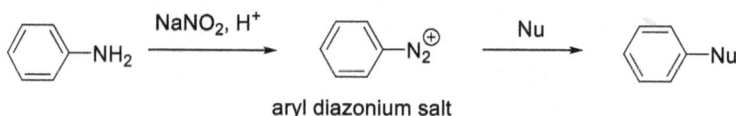

Figure 3.41: Synthesis and reaction of an aryl diazonium salt.

Generally, the aryl diazonium salt is generated in situ, and subsequently treated with the appropriate nucleophile. The mechanism of diazonium salt formation involves protonation of nitrite under the acidic reaction conditions and loss of water to form NO^+ (Figure 3.42). The nucleophilic aniline derivative then reacts with the electrophilic NO^+, forming the N–N bond. A series of proton transfers ultimately results in the loss of water and formation of the N–N triple bond.

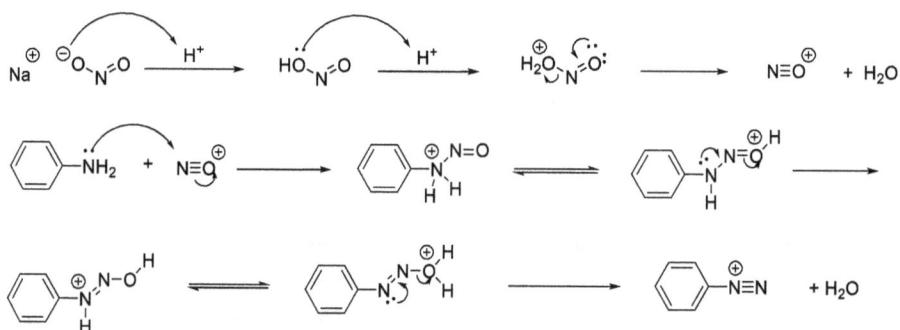

Figure 3.42: Mechanism of diazonium salt formation.

3.6.2 Reactions of diazonium salts with nucleophilic species

The scope of the reaction is broad, allowing the introduction of a number of different functional groups onto the aromatic ring, including halogens, nitriles, phenols, and hydrogen (Figure 3.43).

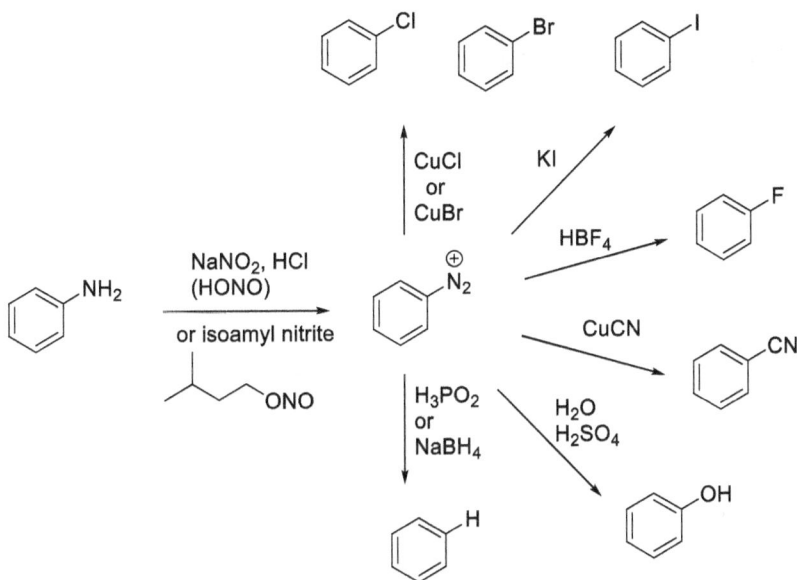

Figure 3.43: Summary of reactions of aryl diazonium salts.

The mechanism for the substitution of the diazonium salt depends on the nucleophile used and reaction conditions. One mechanism for the substitution reaction can be viewed as an S_N1 reaction, where the departure of N_2 results in an unstable aryl cation that is rapidly trapped by the nucleophile. This mechanism is thought to apply for the preparation of phenols, where water is the nucleophile that traps the carbocation (Figure 3.44) [56].

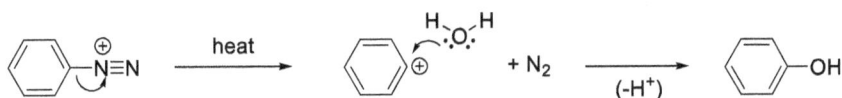

Figure 3.44: Formation of phenols from a diazonium salt.

A second plausible mechanism consists of nucleophilic addition to the diazonium nitrogen, followed by decomposition of the adduct. This mechanism is thought to apply to the reaction of diazonium salts with the azide ion as the nucleophile [57].

In the 1880s, Sandmeyer discovered that copper(I) salts such as CuCl, CuBr, and CuCN react with diazonium salts to prepare the corresponding aryl halide or nitrile [58, 59]. Known as the Sandmeyer reaction, the reaction is thought to be a radical process, where a single electron transfer from the copper(I) salt (Figure 3.45) [60]. This is followed by bond cleavage to yield an aryl radical which reacts with the copper salt, resulting in substitution. Alternatively, the reaction may involve

oxidative addition of the aryl diazonium to copper(I) to give a copper(III) intermediate, which undergoes reductive elimination to form the carbon–halogen bond [61].

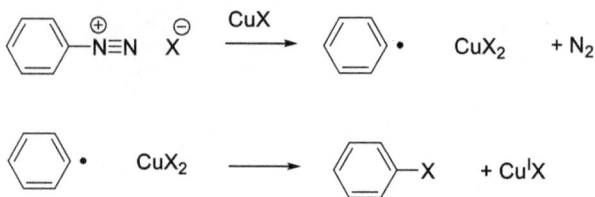

Figure 3.45: Proposed mechanism of the Sandmeyer reaction (X = Cl, Br, CN) [60].

Reductive removal of the diazonium ion to replace it with a hydrogen atom is typically achieved using hypophosphorous acid (H_3PO_2) [62] or sodium borohydride [63].
Alternatively, it can be done in the presence of an alkyl nitrite (such as isoamyl nitrite) in DMF [64]. The use of hypophosphorous acid is the most common approach,
and is thought to proceed via a radical chain reaction (Figure 3.46) [65]. The reaction
is thought to be initiated by a one-electron transfer from the hypophosphorous acid,
which results in homolytic bond cleavage to give an aryl radical and nitrogen. At this
point, the aryl radical abstracts a hydrogen from H_3PO_2, giving the H-substitution on
the benzene and producing an H_2PO_2 radical, which can react with another aryl diazonium to propagate the reaction. The $H_2PO_2^+$ produced reacts with water to form
phosphorous acid (H_3PO_3).

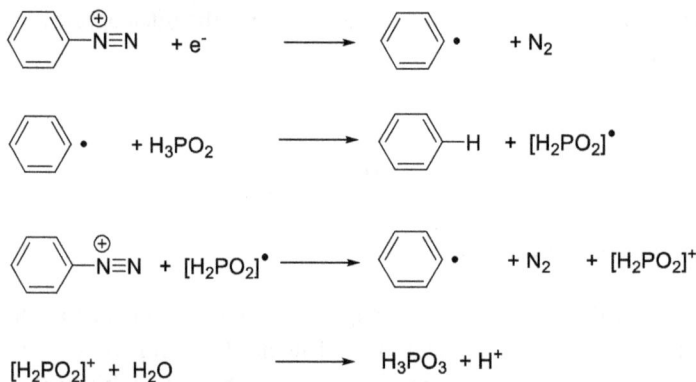

Figure 3.46: Proposed mechanism for reaction of a diazonium salt with hypophosphorous acid.

As mentioned above, aryl diazonium salts are usually prepared in situ and are not
isolated. One exception is the introduction of fluoro substituents. The formation
of the diazonium salt in the presence of tetrafluoroboric acid (HBF_4) produces an
isolable diazonium tetrafluoroborate salt. Subsequent thermal decomposition of

this salt produces the corresponding aryl fluoride and BF_3 (Figure 3.47) [66]. This reaction is known as the Schiemann reaction or Balz–Schiemann reaction and is thought to proceed through the formation of an aryl cation [67]. This reaction presents a convenient approach for the introduction of fluoro groups onto an aromatic ring – a transformation that is not easily achieved using electrophilic aromatic substitution.

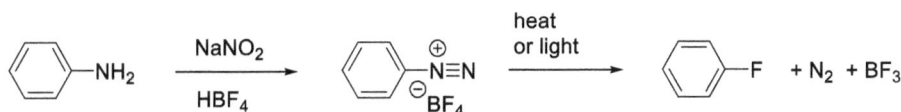

Figure 3.47: Formation of aryl fluorides by decomposition of diazonium tetrafluoroborates.

Substitution reactions of aryl diazonium salts are widely used in the synthesis of aromatic compounds and are frequently used to replace anilines with other functional groups. We will consider a few examples of the application of diazonium substitutions below. More examples of the use of diazonium chemistry will appear in later chapters.

In the first example, we see that diazonium chemistry can be applied to the synthesis of 1,3,5-tribromobenzene (Figure 3.48) [68]. Recall from Chapter 2 that bromo substituents are *ortho/para* directors. As such, the direct bromination of benzene would not give the desired *meta*-substitution pattern. However, by brominating aniline with an excess of bromine, the desired substitution pattern of the bromo substituents is achieved. The amino can then be reductively removed by converting it to the diazonium salt. While H_3PO_2 is the typical reagent used, in this case the removal is achieved with ethanol and sulfuric acid, where presumably the ethanol is serving as a source of hydrogen atoms.

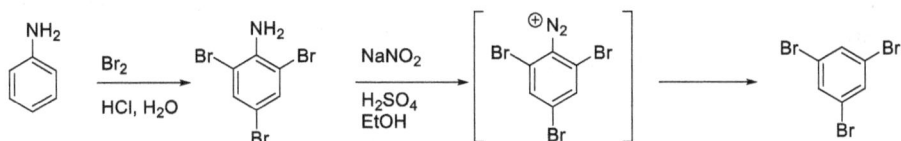

Figure 3.48: Synthesis of 1,3,5-tribromobenzene [68].

For an example of the use of diazonium chemistry in the synthesis of pharmaceutical agents, let us consider the synthesis of thymoxamine, also known as moxisylyte, a vasodilator (Figure 3.49) [69]. The compound is prepared from thymol by nitrosation, followed by reduction. Following this, the aniline is protected as the acetamide to allow alkylation of the phenol. Removal of the protecting group by hydrolysis gave the aniline, which was treated with sodium nitrite under acidic conditions, and then heated in the presence of water to give the phenol. Finally, acetylation of the phenol gave the target thymoxamine.

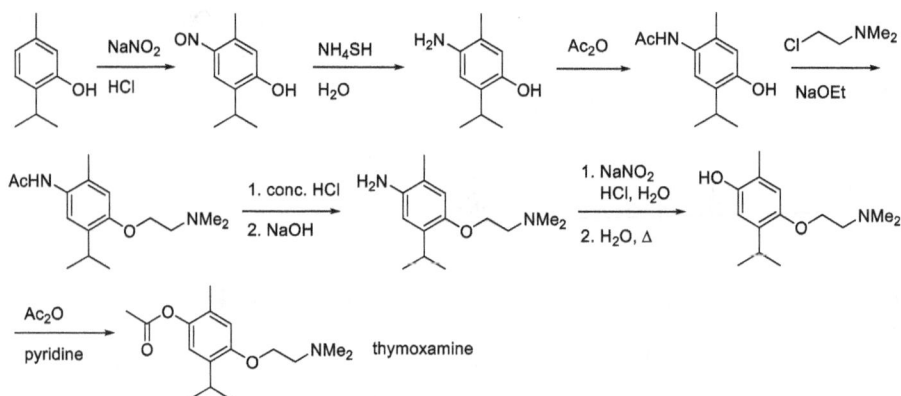

Figure 3.49: Synthesis of thymoxamine.

As another example, we will consider the synthesis of a substituted thioindigo dye (structurally related to indigo), which was investigated as a photoswitchable component for liquid crystals. The synthesis, outlined in Figure 3.50 [70], highlights some of the basic electrophilic aromatic substitution reactions we saw in Chapter 2, but also highlights nucleophilic aromatic substitution reactions and diazonium chemistry. The synthesis began with nitration of 4-methoxybenzoic acid, followed by reduction using hydrogen and palladium on carbon. The amino group was then acetylated, which allowed a subsequent nitration reaction to take place *para* to the NHAc group. Removal of the acetamide protecting group gave the amine, which was treated with sodium nitrite and cuprous chloride to replace the amino group with a chloro substituent. The carboxylic acid was then converted into the methyl ester, and the nitro group was reduced to the amine. A Balz–Schiemann reaction using sodium nitrite and tetrafluoroboric acid, followed by heat, gave the desired fluoro-substituted compound. The methoxy group was replaced with a chiral chain using a sequence of deprotection with BBr$_3$, Mitsunobu reaction, and re-esterification. This compound was then treated with methyl thioglycolate in a nucleophilic aromatic substitution of the thiol on the fluoride, followed by an intramolecular condensation to give the substituted benzothiophene. Hydrolysis and decarboxylation under basic conditions gave the corresponding benzothiophenone, which was oxidized using potassium ferricyanide to give the thioindigo dye.

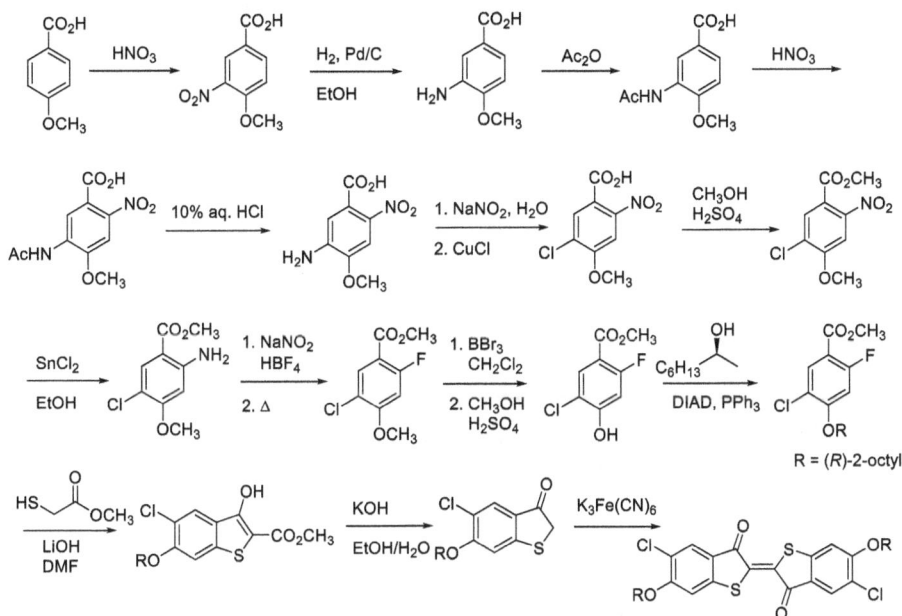

Figure 3.50: Synthesis of a substituted thioindigo dye [70].

3.6.3 Other reactions of aryl diazonium salts

Aryl diazonium salts are not uniquely used for substitution reactions. Diazonium salts can be reduced to form the corresponding phenylhydrazines. For example, the aryl diazonium salt in Figure 3.51 was reduced with $SnCl_2$ to form the corresponding phenylhydrazine, which was used to prepare Celecoxib (Celebrex-TM), a non-steroidal anti-inflammatory drug [71].

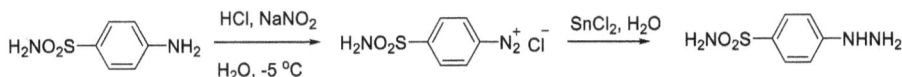

Figure 3.51: Example of the reduction of a diazonium salt to the corresponding phenylhydrazine.

Aryl diazonium salts can also act as electrophiles in electrophilic aromatic substitution, where an activated aromatic compound attacks at the terminal nitrogen to form the corresponding azobenzenes (Figure 3.52), which are a well-known class of dyes.

Aryl diazonium salts also react with primary or secondary amines to form the corresponding triazenes (Figure 3.53) [72]. The reaction is conceptually similar to that of the formation of azobenzenes, where the nucleophilic amine attacks the terminal nitrogen of the diazonium salt.

Figure 3.52: Formation of azobenzenes from diazonium salts.

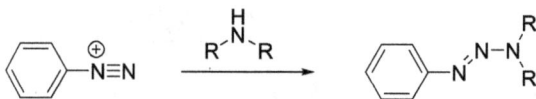

Figure 3.53: Formation of aryl triazenes by reaction of diazonium salts with amines.

Aryltriazenes are of therapeutic interest for the treatment of cancer. 1-Phenyl-3,3-dimethyltirazene has been studied extensively in this context, and there is evidence to suggest that the compound undergoes an enzymatic demethylation and ultimately hydrolysis to produce aniline and a methyldiazonium ion, which can methylate DNA [73, 74]. Dacarbazine (Figure 3.54) has been used for more than 40 years for the treatment of cancers such as melanoma and Hodgkin's disease [74]. It is prepared from 5-aminoimidazole-4-carboxamide by diazotization using sodium nitrite, followed by addition of dimethylamine in methanol (Figure 3.54).

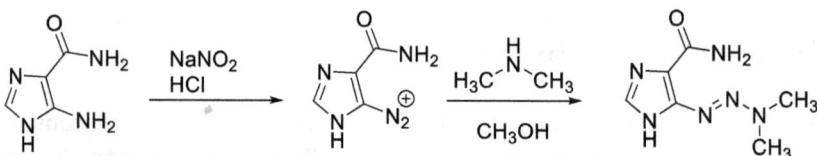

Figure 3.54: Synthesis of the anticancer agent dacarbazine [74].

Triazenes can be converted into the corresponding anilines, so in principle, they could be used as protecting groups for anilines. However, the availability of other amine protecting groups has meant that the use of triazenes in this context is has been limited. On the other hand, aryltriazenes have been extensively used in the context of synthesis as masked aryl iodides. Moore and coworkers reported that aryltriazenes could be converted to the corresponding aryl iodides by heating in excess iodomethane in a sealed tube (Figure 3.55) [75]. Under these conditions, the iodomethane can methylate the terminal nitrogen, which then leads to decomposition to the diazonium salt. In the presence of the iodide liberated from the iodomethane, the diazonium group is displaced by the iodide (Figure 3.55).

Subsequently, Moore and coworkers showed that the additions of iodine lead to improved yields and allowed the reaction to take place at lower temperature, sometimes even in the absence of iodomethane [76]. The reaction is inhibited by oxygen, which suggests a radical reaction. Other methods to convert triazenes into aryl

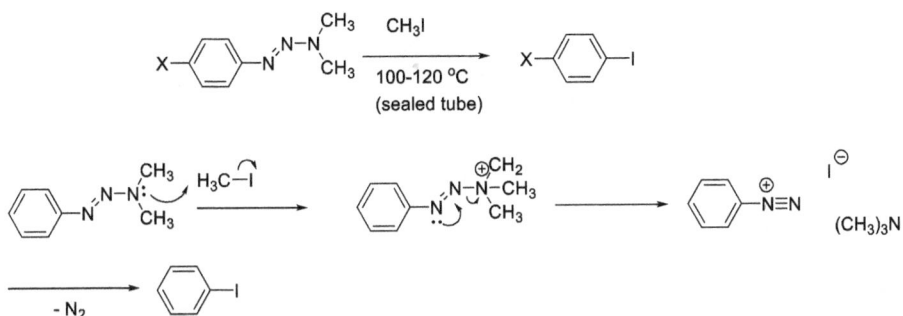

Figure 3.55: Preparation of aryl iodides from the corresponding triazenes.

iodides have been developed, including trimethylsilyl iodide [77], or sodium iodide with an ion exchange resin [78]. However, the method using iodomethane is simple and convenient. This reaction has been used to enable the preparation of phenyl-acetylene oligomers and macrocycles [79–82]. We will see an example of this reaction in Chapter 6 in the context of transition metal-catalyzed reactions.

3.6.4 Formation of biaryls: the Gomberg–Bachmann reaction and Pschorr cyclization

Aryl diazonium salts can also react with other aromatic compounds to form a biphenyl [83]. The synthesis of biaryls from aryl diazonium salts was first reported by Gomberg and Bachmann in 1924 [84]. In the Gomberg–Bachmann reaction, an aryl diazonium salt is reacted with another aromatic compound under basic conditions or in the presence of copper salts (Figure 3.56).

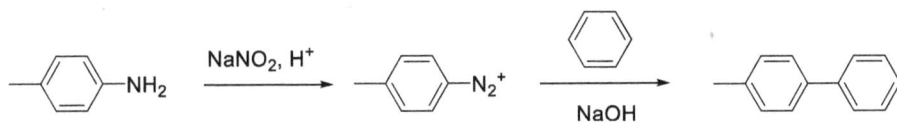

Figure 3.56: The Gomberg–Bachmann reaction.

The reaction is thought to proceed via a radical mechanism. Specifically, under basic conditions, the aryl diazonium reacts with hydroxide to form an azoanhydride (Figure 3.57) [85, 86]. This species undergoes homolytic bond cleavage to form an aryl radical, which is coupled with the other aromatic reactant, forming the aryl-aryl bond and a delocalized radical. Finally, H atom abstraction leads to rearomatization to generate the biaryl product.

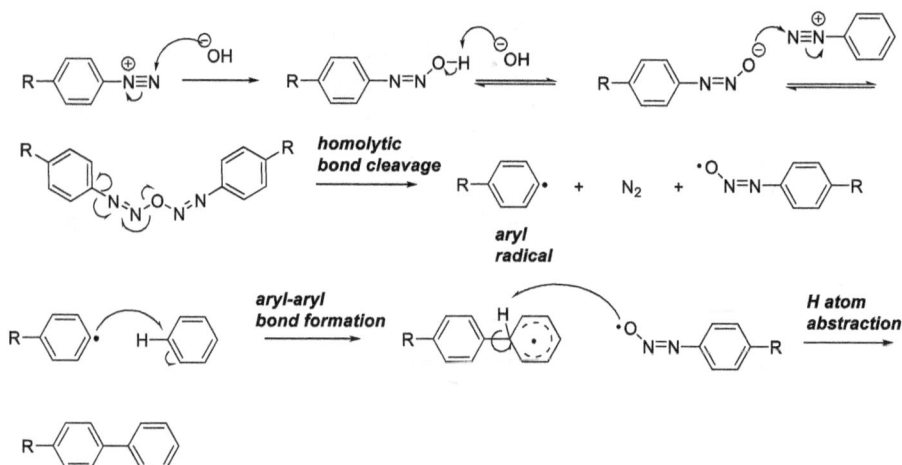

Figure 3.57: Proposed mechanism of the Gomberg–Bachmann reaction.

It should be noted that this is one of the earliest approaches for preparing dissymmetric biaryls. The reaction has the disadvantage that the reaction conditions are harsh and yields tend to be low. Furthermore, if a substituted aromatic is coupled with a diazonium salt, the regiochemistry of the coupling can be difficult to control. Another potential drawback of the reaction that may contribute to lower yields is competing coupling to form the corresponding azobenzenes, as described above. In Chapter 6, we will see modern cross coupling methods for the formation of aryl–aryl bonds that are more commonly used and avoid the drawbacks of the Gomberg–Bachmann reaction.

Despite the fact that the Gomberg–Bachmann reaction is a classical approach that has largely been supplanted by other reactions, the interest in methods for preparing biaryls as biologically active molecules means that an interest in the Gomberg–Bachmann reaction is sustained. For example, there are recent reports of a photoactivated version of the Gomberg–Bachmann reaction [87]. Specifically, the Gomberg reaction can be carried out by reacting diazonium tetrafluoroborates with haloarenes in the presence of pyridine and exposure to visible light, resulting in halogenated biphenyls (Figure 3.58). In this reaction pyridine is thought to complex to the diazonium ion to produce an "electron donor–acceptor" (EDA) complex, which undergoes hemolytic bond cleavage to produce the aryl radical. At that point, the reaction mechanism resemble the classical Gomberg–Bachmann reaction. Unlike the classical conditions, this version of the reaction takes place under mild conditions and in some cases is high yielding. Furthermore, carrying out the reaction with haloarenes leads to halogenated biphenyls, providing a handle for further transformations. Unfortunately, regiochemistry of the reaction is still relatively poor, except for substrates such as 1,4-dichlorobenzene where only one regioisomer is possible.

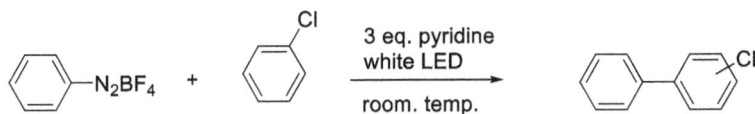

Figure 3.58: Photochemical Gomberg–Bachmann reaction [87].

Aryl-aryl bonds can also be formed intramolecularly via the diazonium salt [88]. This intermolecular reaction is often carried out in the presence of copper and is referred to as the Pschorr cyclization (Figure 3.59) [83, 89].

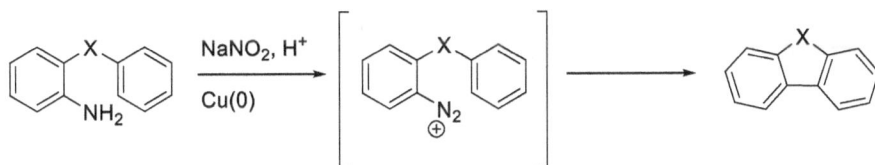

Figure 3.59: The Pschorr cyclization.

Under neutral or basic conditions in the presence of copper, the reaction proceeds by a radical mechanism, where one-electron reduction of the diazonium salt by copper leads to loss of N_2 to generate an aryl radical, which undergoes carbon–carbon bond formation, followed by rearomatization similar to the Gomberg-Bachmann reaction above. Under acidic conditions, the Pschorr reaction is thought to proceed by loss of nitrogen to form an aryl cation, which acts as the electrophile in a substitution at the adjacent benzene ring. In later chapters, we will see examples how the Pschorr reaction of stilbene derivatives can be used for the preparation of polycyclic aromatic hydrocarbons.

References

[1] Bunnett JF, Zahler RE. Aromatic nucleophilic substitution reactions. Chem Rev. 1951;49:273–412.
[2] Strauss MJ. Anionic sigma Complexes. Chem Rev. 1970;70(6):667–711.
[3] Peter BG, Miller J, Liveris M, Lutz PG. The SN mechanism in aromatic compounds. Part VIII J Chem Soc. 1954;1265–66.
[4] Bunnett JF, Garbisch EW, Pruitt KM. The "element effect" as a criterion of mechanism in activated aromatic nucleophilic substitution reactions. J Am Chem Soc. 1957;79(2):385–91.
[5] Bartoli G, Todesco PE. Nucleophilic substitution. Linear free energy relationships between reactivity and physical properties of leaving groups and substrates. Acc Chem Res. 1977;10(4):125–32.
[6] Wells KM, Shi Y-J, Lynch JE, Humphrey GR, Volante RP, Reider PJ. Regioselective nucleophilic substitutions of fluorobenzene derivatives. Tetrahedron Lett. 1996;37(36):6439–42.

[7] Woydziak ZR, Fu L, Peterson BR. Synthesis of fluorinated benzophenones, xanthones, acridones, and thioxanthones by iterative nucleophilic aromatic substitution. J Org Chem. 2012;77(1):473–81.

[8] Zhu L, Zhang M. Ortho-selective nucleophilic aromatic substitution reactions of polyhaloanilines with potassium/sodium O-ethyl xanthate: A convenient access to halogenated 2(3H)-benzothiazolethiones. J Org Chem. 2004;69(21):7371–74.

[9] Suzuki Y, Iwata M, Yazaki R, Kumagai N, Shibasaki M. Concise enantioselective synthesis of duloxetine via direct catalytic asymmetric aldol reaction of thioamide. J Org Chem. 2012;77(9):4496–500.

[10] Robertson DW, Krushinski J, Fuller RW, David Leander J. Absolute configurations and pharmacological activities of the optical isomers of fluoxetine, a selective serotonin-uptake inhibitor. J Med Chem. 1988;31(7):1412–17.

[11] Dinakaran M, Senthilkumar P, Yogeeswari P, China A, Nagaraja V, Sriram D. Novel ofloxacin derivatives: Synthesis, antimycobacterial and toxicological evaluation. Bioorg Med Chem Lett. 2008;18(3):1229–36.

[12] Uoyama H, Goushi K, Shizu K, Nomura H, Adachi C. Highly efficient organic light-emitting diodes from delayed fluorescence. Nature. 2012;492(7428):234–38.

[13] Hiscock LK, Yao C, Skene WG, Dawe LN, Maly KE. Synthesis of emissive heteroacene derivatives via nucleophilic aromatic substitution. J Org Chem. 2019;84:15530–37.

[14] Zhang B, Wei M, Mao H, Pei X, Alshmimri SA, Reimer JA, et al. Crystalline dioxin-linked covalent organic frameworks from irreversible reactions. J Am Chem Soc. 2018;140(40):12715–19.

[15] Short R, Carta M, Bezzu CG, Fritsch D, Kariuki BM, McKeown NB. Hexaphenylbenzene-based polymers of intrinsic microporosity. Chem Commun. 2011;47(24):6822–24.

[16] Taylor RGD, Carta M, Bezzu CG, Walker J, Msayib KJ, Kariuki BM, et al. Triptycene-based organic molecules of intrinsic microporosity. Org Lett. 2014;16(7):1848–51.

[17] Gololobov YG, Gording IR, Terrie F, Petrovskii PV, Lyssenko KA, Garbuzovaa IA. Synthesis and structure of 2,4-dinitrophenylcyanoacetamide derivatives as CH acids and their organic salts. Russ Chem Bull. 2009;58(12):2443–48.

[18] Suzuki Y, Toyota T, Imada F, Sato M, Miyashita A. Nucleophilic acylation of arylfluorides catalyzed by imidazolidenyl carbene. Chem Commun. 2003;1314–15.

[19] Suzuki Y, Ota S, Fukuta Y, Ueda Y, Sato M. N-heterocyclic carbene-catalyzed nucleophilic aroylation of fluorobenzenes in this nucleophilic substitution, the halogen substituents of these compounds are replaced by aroyl groups originating from mol % without a significant decrease in the product yi. J Org Chem. 2008;73(d):2420–23.

[20] Suzuki Y, Fukuta Y, Ota S, Kamiya M, Sato M. Xanthone natural products via N -heterocyclic carbene catalysis: Total synthesis of atroviridin. J Org Chem. 2011;76(10):3960–67.

[21] Ito S, Kitamura T, Arulmozhiraja S, Manabe K, Tokiwa H, Suzuki Y. Total Synthesis of Termicalcicolanone A via Organocatalysis and Regioselective Claisen Rearrangement. Org Lett. 2019;21(8):2777–81.

[22] Rowan SJ, Cantrill SJ, Cousins GRL, Sanders JKM, Stoddart JF. Dynamic covalent chemistry. Angew Chem Int Ed. 2002;Vol. 41:898–952.

[23] Jin Y, Wang Q, Taynton P, Zhang W. Dynamic covalent chemistry approaches toward macrocycles, molecular cages, and polymers. Acc Chem Res. 2014;47(5):1575–86.

[24] Santos T, Rivero DS, Pérez-Pérez Y, Martín-Encinas E, Pasán J, Daranas AH, et al. Dynamic nucleophilic aromatic substitution of tetrazines. Angew Chem Int Ed. 2021;60:18783–91.

[25] Ong WJ, Swager TM. Dynamic self-correcting nucleophilic aromatic substitution. Nat Chem. 2018;10(10):1023–30.

[26] Holden CM, Greaney MF. Modern aspects of the Smiles rearrangement. Chem Eur J. 2017;23 (38):8992–9008.

[27] Snape TJ. A truce on the Smiles rearrangement: Revisiting an old reaction – The Truce-Smiles rearrangement. Chem Soc Rev. 2008;37(11):2452–58.

[28] Warren LA, Smiles S. Dehydro-2-naphtholsulphone. J Chem Soc. 1930;1327–31.

[29] Warren LA, Smiles S. A rearrangement of ortho-amino-sulphones. J Chem Soc. 1932;2774–78.

[30] Levi A, Warren LA, Smiles SA. Rearrangement of o-acetumido-sulphoxides. J Chem Soc. 1933;1490–93.

[31] Galbraith F, Smiles S. The rearrangement of o-hydroxysulphones. Part V J Chem Soc. 1935;1234–38.

[32] Mcclement CS, Smiles S. The rearrangement of ortho-hydroxy-sulphones. Part VI J Chem Soc. 1937;1016–21.

[33] Okada K, Sekiguchi S. Aromatic nucleophilic substitution. 9. Kinetics of the formation and decomposition of anionic complexes in the Smiles rearrangements of N-acetyl-beta-aminoethyl 2-X-4-nitro-1-phenyl or N-acetyl-beta-aminoethyl 5-nitro-2-pyridyl ethers in aqueous dimethyl. J Org Chem. 1978;43(3):441–47.

[34] Truce WE, Ray WJ, Norman OL, Eickemeyer DB. Rearrangements of aryl sulfones. I. The metalation and rearrangement of mesityl phenyl sulfone. J Am Chem Soc. 1958;80:3625–29.

[35] Truce WE, Ray WJ. Rearrangements of aryl sulfones. II. The synthesis and rearrangement of several o-methyldiaryl sulfones to o-benzylbenzenesulfinic acids. J Am Chem Soc. 1959;81:481–84.

[36] Erickson WR, McKennon MJ. Unexpected Truce-Smiles type rearrangement of 2-(2'-pyridyloxy) phenylacetic esters: Synthesis of 3-pyridyl-2-benzofuranones. 2000;41:4541–44.

[37] Mitchell LH, Barvian NC. A homologous enolate Truce-Smiles rearrangement. Tetrahedron Lett. 2004;45(29):5669–71.

[38] Clayden J, Dufour J, Grainger DM, Helliwell M. Substituted diarylmethylamines by stereospecific intramolecular electrophilic arylation of lithiated ureas. J Am Chem Soc. 2007;129(24):7488–89.

[39] Clayden J, Farnaby W, Grainger DM, Hennecke U, Mancinelli M, Tetlow DJ, et al. N to C aryl migration in lithiated carbamates: α-arylation of benzylic alcohols. J Am Chem Soc. 2009;131 (10):3410–11.

[40] MacLellan P, Clayden J. Enantioselective synthesis of tertiary thiols by intramolecular arylation of lithiated thiocarbamates. Chem Commun. 2011;47:3395–97.

[41] Roberts JD, Simmons HE, Carlsmith LA, Vaughan CW. Rearrangement in the reaction of chlorobenzene-C-14 with potassium amide. J Am Chem Soc. 1953;75:3290–91.

[42] Roberts JD, Vaughan CW, Carlsmith LA, Semenow DA. Orientation in animations of substituted halobenzenes. J Am Chem Soc. 1956;78:611–14.

[43] Rohrbach S, Smith AJ, Pang JH, Poole DL, Tuttle T, Chiba S, et al. Concerted nucleophilic aromatic substitution reactions. Angew Chem Int Ed. 2019;58(46):16368–88.

[44] Pasquini MA, Le Goaller R, Pierre JL. Effets de cryptands et activation de bases – VII. Tetrahedron. 1980;36(9):1223–26.

[45] Neumann CN, Hooker JM, Ritter T. Concerted nucleophilic aromatic substitution with 19F- and 18F-. Nature. 2016;534(7607):369–73.

[46] Neumann CN, Ritter T. Facile C-F bond formation through a concerted nucleophilic aromatic substitution mediated by the PhenoFluor reagent. Acc Chem Res. 2017;50(11):2822–33.

[47] Schimler SD, Cismesia MA, Hanley PS, Froese RDJ, Jansma MJ, Bland DC, et al. Nucleophilic deoxyfluorination of phenols via aryl fluorosulfonate intermediates. J Am Chem Soc. 2017;139 (4):1452–55.

[48] Kwan EE, Zeng Y, Besser HA, Jacobsen EN. Concerted nucleophilic aromatic substitutions. Nat Chem. 2018;10:917–23.

[49] Orth W, Engel J, Emig P, Scheffler G, Pohle H. 2-Amino-3-nitro-6-(4-fluorobenzylamino) pyridine and 2-amino-3-carbethoxyamino-6-(4-fluorobenzylamino)pyridine. Ger Offen. 1986;3608762.

[50] Kaiser DW, Thurston JT, Dudley JR, Schaefer FC, Hechenbleikner I, Holm-Hansen D. Cyanuric chloride derivatives. II. Substituted melamines. J Am Chem Soc. 1951;73(7):2984–86.

[51] Pearlman WM, Banks CK. Substituted chlorodiamino-s-triazines. J Am Chem Soc. 1948;70:3726–28.

[52] Barton B, Gouws S, Schaefer MC, Zeelie B. Evaluation and optimisation of the reagent addition sequence during the synthesis of atrazine (6-chloro-N2-ethyl-N4-isopropyl-1,3,5-triazine-2,4-diamine) using reaction calorimetry. Org Proc Res Dev. 2003;7(6):1071–76.

[53] Pieterse K, Lauritsen A, Schenning APHJ, Vekemans JAJM, Meijer EW. Symmetrical electron-deficient materials incorporating azaheterocycles. Chem.-Eur J. 2003;9:5597–604.

[54] Franco F, Greenhouse R, Muchowski JM. Novel syntheses of 5-aroyl-1,2-dihydro-3H-pyrrolo [1,2-a]pyrrole-1 -carboxylic acids. J Org Chem. 1982;47(9):1682–88.

[55] Zollinger H. Reactivity and stability of arenediazonium ions. Acc Chem Res. 1973;6 (10):335–41.

[56] Lewis ES, Hartung LD, Mckay BM. The reaction of diazonium salts with nucleophiles. XIII. Identity of the rate- and product-determining steps. J Am Chem Soc. 1969;91(2):419–25.

[57] Ritchie CD, Virtanen POI. Cation-anion combination reactions. IX. A remarkable correlation of nucleophilic reactions with cations. J Am Chem Soc. 1972;93:4966–71.

[58] Sandmeyer T. Ueber die Ersetzung der Amid-gruppe durch Chlor, Brom und Cyan in den aromatischen Substanzen. Chem Ber. 1884;17:2650–53.

[59] Hodgson HH. The sandmeyer reaction. Chem Rev. 1947;40:251–77.

[60] Kochi JK. The mechanism of the Sandmeyer and Meerwein reactions. J Am Chem Soc. 1957;79 (11):2942–48.

[61] Galli C. Radical reactions of arenediazonium ions: An easy entry into the chemistry of the aryl radical. Chem Rev. 1988;88(5):765–92.

[62] Alexander ER, Burge RE. The hypophosphorous acid deamination of diazonium salts in deuterium oxide. J Am Chem Soc. 1950;72:3100–03.

[63] Hendrickson JB. Reduction of diazonium borofluorides by sodium borohydrides. J Am Chem Soc. 1961;83:1961.

[64] Doyle MP, Dellaria JF, Siegfried B, Bishop SW. Reductive deamination of arylamines by alkyl nitrites in N,N-dimethylformamide. A direct conversion of arylamines to aromatic hydrocarbons. J Org Chem. 1977;42(22):3494–98.

[65] Kornblum N, Cooper GD, Taylor JE. The chemistry of diazo compounds. II. Evidence for a free radical chain mechanism in the reduction of diazonium salts by hypophosphorous acid. J Am Chem Soc. 1950;72:3013–21.

[66] Balz G, Schiemann G. Über aromatische Fluorverbindungen, I.: Ein neues Verfahren zu ihrer Darstellung. Chem Ber. 1927;60:1186–90.

[67] Swain CG, Rogers RJ. Mechanism of formation of aryl fluorides from arenediazonium fluoborates. J Am Chem Soc. 1975;97:799–800.

[68] Coleman GH, Talbot WF. sym.-Tribromobenzene. Org Syn. 1933;13:96.

[69] Florand I. 1-dimethylamino-2-(6-acetoxythymyloxy)ethane. 1962;p. FR 1313297.

[70] Dinescu L, Maly KE, Lemieux RP. Design of photonic liquid crystal materials: Synthesis and evaluation of new chiral thioindigo dopants designed to photomodulate the spontaneous polarization of ferroelectric liquid crystals. J Mater Chem. 1999;9(8):1679–86.

[71] Penning TD, Talley JJ, Bertenshaw SR, Carter JS, Collins PW, Docter S, et al. Synthesis and biological evaluation of the 1, 5-diarylpyrazole class of cyclooxygenase-2 inhibitors: Identification of 4-[5-(4-methylphenyl)-3-(trifluoromethyl)-1 H-pyrazol-1-yl] benzenesulfonamide (SC-58635, celecoxib). J Med Chem. 1997;40(9):1347–65.

[72] Kimball DB, Haley MM. Triazenes: A versatile tool in organic synthesis. Angew Chem Int Ed. 2002;41(18):3338–51.

[73] Kleihues P, Kolar GF, Margison GP. Interaction of the carcinogen 3,3-dimethyl-1-phenyltriazene with nucleic acids of various rat tissues and the effect of a protein-free diet. Cancer Res. 1976;36:2189–93.

[74] Gescher A, Hickman JA, Simmonds RJ, Stevens MFG, Vaughan K. Studies of the mode of action of antitumour triazenes and triazines – II. Investigation of the selective toxicity of 1-aryl-3,3-dimethyltriazenes. Biochem Pharmacol. 1981;30(1):89–93.

[75] Moore JS, Weinstein EJ, Wu Z. A convenient masking group for aryl iodides. Tetrahedron Lett. 1991;32(22):2465–66.

[76] Wu Z, Moore JS. Iodine-promoted decomposition of 1-aryl-3,3-dialkyltriazenes: A mild method for the synthesis of aryl iodides. Tetrahedron Lett. 1994;35(31):5539–42.

[77] Ku H, Barrio JR. Convenient synthesis of aryl halides from arylamines via treatment of 1-ary 1-3,3-dialkyltriazenes with trimethylsilyl halides. J Org Chem. 1981;46(25):5239–41.

[78] Satyamurthy N, Barrio JR. Cation exchange resin (hydrogen form) assisted decomposition of l-aryl-3,3-dialkyltriazenes. A mild and efficient method for the synthesis of aryl iodides. J Org Chem. 1983;48(23):4394–96.

[79] Zhang J, Moore JS, Xu Z, Aguirre RA. Nanoarchitectures. 1. Controlled synthesis of phenylacetylene sequences. J Am Chem Soc. 1992;114(6):2273–74.

[80] Macrocycles P, Zhang J, Pesak DJ, Ludwick JL, Moore JS, Arbor A. Geometrically-controlled and site-specifically-functionalized phenylacetylene macrocycles. J Am Chem Soc. 1994;116 (10):4227–39.

[81] Haley MM, Bell ML, English JJ, Johnson CA, Weakley TJR. Versatile synthetic route to and DSC analysis of dehydrobenzoannulenes: Crystal structure of a heretofore inaccessible [20] annulene derivative received November 25, 1996 The synthesis and chemistry of dehydroannulenes and their benzannelated anal. J Am Chem Soc. 1997;119(12):2956–57.

[82] Pak JJ, Weakley TJR, Haley MM. Stepwise assembly of site specifically functionalized dehydrobenzo [18] annulenes. J Am Chem Soc. 1999;121(36):8182–92.

[83] Felpin FX, Sengupta S. Biaryl synthesis with arenediazonium salts: Cross-coupling, CH-arylation and annulation reactions. Chem Soc Rev. 2019;48(4):1150–93.

[84] Gomberg M, Bachmann WE. The synthesis of biaryl compounds by means of the diazo reaction. J Am Chem Soc. 1924;46(10):2339–43.

[85] Rüchardt C, Merz E. Der mechanismus der Bachmann-Gomberg reaktion. Tetrahedron Lett. 1964;5(36):2431–36.

[86] Eliel EL, Saha JG, Meyerson S. The Gomberg-Bachmann reaction with benzene-d. J Org Chem. 1965;30:2451–52.

[87] Lee J, Hong B, Lee A. Visible-light-promoted, catalyst-free Gomberg-Bachmann reaction: Synthesis of biaryls. J Org Chem. 2019;84(14):9297–306.

[88] Pschorr R. Neue Synthese des Phenanthrens und seiner Derivate. Ber Dtsch Chem Ges. 1896;29(1):496–501.

[89] Leake PH. The Pschorr synthesis. Chem Rev. 1956;56(1):27–48.

4 Reactions of aryllithium species

4.1 Introduction

Organometallic reagents derived from aromatic compounds are nucleophilic and are important for a variety of synthetic transformations. These can include not only aryl Grignard reagents, but also a variety of other organometallic species. Among these organometallic reagents, aryllithium species have found wide-spread use in the preparation of substituted aromatic compounds. Aryllithium species are strong bases and highly nucleophilic, and as such react with a wide variety of electrophiles. Aryllithium reagents are highly reactive and moisture sensitive and are generated in situ. The two main methods for preparing aryllithium reagents are by deprotonation or by lithium–halogen exchange (Figure 4.1).

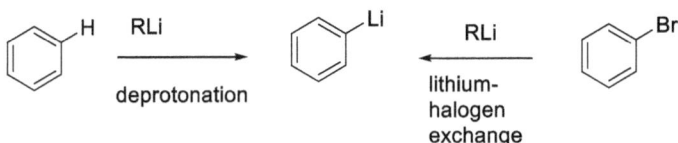

Figure 4.1: Preparation of aryllithium species.

Aryllithium species can undergo reactions with a wide variety of electrophiles, making them useful intermediates for the preparation of a variety of substituted aromatics (Figure 4.2).

The reactions of aryllithium species with a number of different electrophiles outlined in Figure 4.2 is not comprehensive, but provides an indication of the scope of reactivity. These include the formation of aryl–carbon bonds, arylsilanes, aryl halides, as well as other organometallic species such as arylboronic acids and arylstannanes. These last two are useful reagents for transition metal cross-coupling reactions, which will be discussed in Chapter 6.

One obvious limitation for the use of aryllithiums is that the compounds must be compatible with the strong basic conditions required to generate them. A second limitation of aryllithiums is that they are sometimes incompatible with strongly electron-withdrawing groups such as nitro or cyano-substituted aromatics. For these substrates, it is possible that electron transfer from the alkyllithium to the electron-deficient aromatic compound takes place.

In this chapter, we will explore the preparation and reactivity of aryllithium species. The regiochemistry of aryllithium formation will be a particular focus, with a discussion of the role of directing groups in controlling the site of deprotonation. We will also see some select examples of the application of aryllithium species in

https://doi.org/10.1515/9783110562682-004

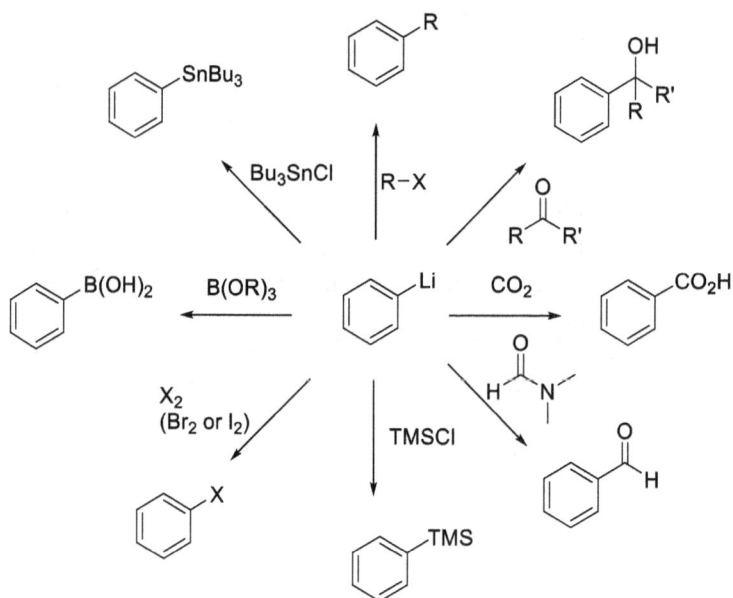

Figure 4.2: Summary of reactions of aryllithium species.

synthesis. More examples will be considered in later chapters, especially in conjunction with transition metal-mediated cross-coupling reactions (Chapter 6).

4.2 Deprotonation of arenes: directed *ortho* metalation

4.2.1 Introduction to directed *ortho* metalation

Benzene has a pK_a of approximately 43 [1], which means that it can be deprotonated using strongly basic alkyllithiums such as *n*-butyllithium (Figure 4.3).

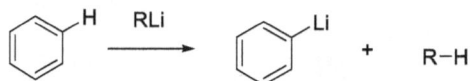

Figure 4.3: Deprotonation of benzene using an alkyllithium base.

For substituted aromatic compounds controlling the regiochemistry of deprotonation presents a challenge. In 1939 and 1940, it was discovered that deprotonation of anisole by *n*-BuLi occurred *ortho* to the methoxy group in two separate reports by Gilman and Bebb and Wittig and Furhman (Figure 4.4) [2, 3].

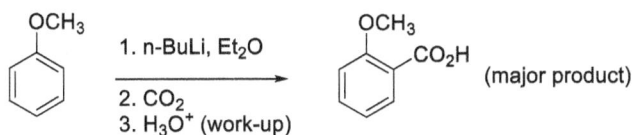

Figure 4.4: Early example of *ortho*-lithiation of anisole [2].

This discovery spawned the development of a broader understanding of the factors that influence the site of deprotonation in substituted aromatics. In general, several different types of substituents, usually containing heteroatoms, direct deprotonation to the *ortho* site [4]. This reaction has become known as directed *ortho* metalation (DoM), and presents an opportunity to introduce electrophiles selectively *ortho* to functional groups collectively known as "directed metalation groups" or DMGs (Figure 4.5) [4, 5].

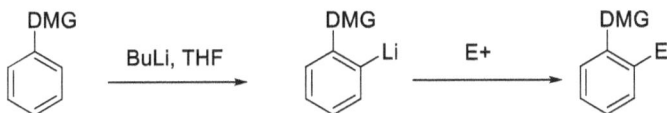

Figure 4.5: General directed *ortho* metalation (DoM) reaction.

DoM typically requires very strong bases, such as alkyllithiums. The reactivity of the alkyllithium is an important determining factor in DoM reactions. Alkyllithium bases are soluble in organic solvents, but are aggregated. For example, *n*-butyllithium and *sec*-butyllithium exist as a hexamers hydrocarbon solvents and as a tetramers in coordinating solvents like THF or ether [4, 6–8]. Similarly, *t*-butyllithium is a tetramer in hydrocarbon solvents and a dimer in THF. To enhance reactivity of the alkyllithiums, bidentate ligands such as tetramethylethylenediamine (TMEDA) cause aggregates to dissociate even further and enhance reactivity [4, 8]. As such, TMEDA is often used to improve reactivity in DoM reactions. Lithium dialkyl amides such as lithium diisopropyl amide (LDA) are not basic enough to quantitatively deprotonate aromatic rings. However, as we will see later, they can be used to deprotonate reversibly under thermodynamic conditions.

Mechanistically, many of these DoM reactions are thought to be a two-step process, where the alkyllithium forms a complex with the directed metalation group, bringing the reactive sites into close proximity prior to deprotonation. This phenomenon is referred to as the complex-induced proximity effect (CIPE) [5, 9]. A schematic representation of the CIPE is shown in Figure 4.6. While there is in many cases evidence for this stepwise pre-complexation before deprotonation, it is also possible that the complexation and deprotonation occur simultaneously [10]. Nevertheless, the CIPE concept remains a useful tool for explaining the directing effects of groups in lithiation reactions such as DoM.

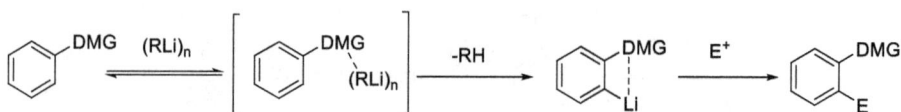

Figure 4.6: Complex-induced proximity effect in directed lithiation [5, 9].

A wide variety of functional groups can serve as directed metalation groups (DMGs) [4]. The common feature of these groups is that they contain heteroatoms that can complex with the organolithium base. That said, some groups are more effective as DMGs than others, leading to a hierarchy of DMGs, as shown in Figure 4.7 [4, 5].

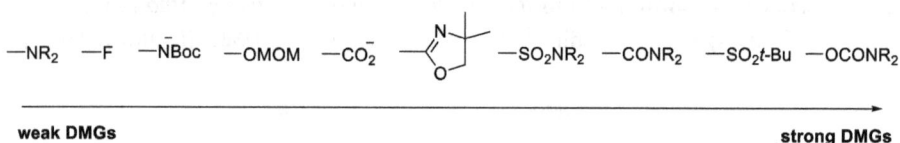

weak DMGs   **strong DMGs**

Figure 4.7: Relative "power" of DMGs.

The scale above highlights some of the common DMGs used and shows them in order of increasing efficiency of the DMG. This scale was established by competition experiments between DMGs (either inter- or intramolecular) [4]. For example, treatment of a mixture of two compounds containing different DMGs with only one equivalent of alkyllithium, followed by trapping of the aryllithium allows a direct comparison of two DMGs. Alternatively, the experiment can be conducted with a disubstituted benzene bearing two DMGs can be treated with one equivalent of alkyllithium, followed by determining the site of deprotonation [11].

Among the most powerful and commonly used DMGs are carbamates, amides, sulfonamides, and sulfones. In the context of synthesis, it is also important to consider how these DMGs can be prepared. DMGs such as carbamates and methoxymethyl ether (MOM) groups can be prepared from the corresponding phenols, while amides, oxazolines, and carboxylates are derived from the corresponding carboxylic acids. Sulfones and sulfonamides can be derived from the corresponding sulfides and sulfonic acids. Examples of some of the general approaches for installing some of these common DMGs are illustrated in Figure 4.8.

Because DoM is selective for substitution at the *ortho* position, it presents an alternative to electrophilic aromatic substitution for the preparation of substituted aromatic compounds. For example, in systems bearing an activating an *ortho–para*-directing group which can also serve as a DMG, the *ortho* product can be favored over a mixture of *ortho-* and *para* isomers. Even more interestingly, deactivating *meta* directors for electrophilic aromatic substitution can serve as DMGs for *ortho* metalation and therefore provide complementary reactivity.

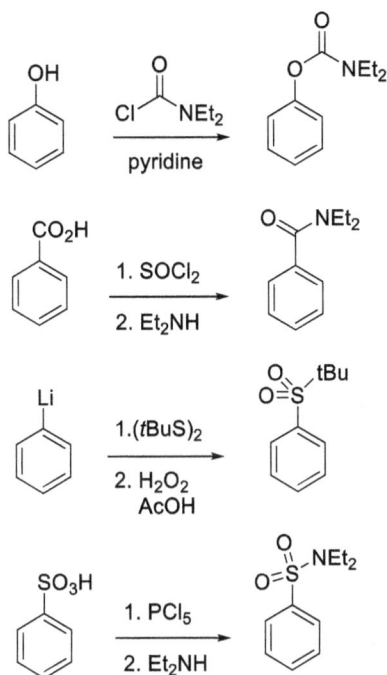

Figure 4.8: Examples of typical methods for installing common DMGs.

Some representative examples of typical DoM reactions are shown in Figure 4.9. These examples are chosen to highlight typical reaction conditions and demonstrate some of the electrophiles that can be trapped by aryllithiums. The first example shows DoM with *N,N*-diethylbenzamide and trapping with trimethylsilyl chloride as the electrophile to install a TMS group [12]. The second example shows that the amide is a stronger DMG than a methoxy group and highlights the use of DMF as the electrophile to install a formyl group [13]. The third example uses a carbamate DMG with methyl iodide as the electrophile [14]. The final example highlights the potential cooperativity of two DMGs. In this case, the strong amide DMG and the weaker methoxymethyl ether (OMOM) DMG work cooperatively to lithiate in between the two groups, despite the potential steric constraints [15]. The aryllithium is trapped with 1,2-diiodoethane, which serves as a source of electrophilic iodine.

4.2.2 Examples of directed *ortho* metalation in synthesis

As an example of the applications of DoM in synthesis, we will consider the synthesis of unsymmetrical anthraquinones (Figure 4.10). The general strategy consists of DoM using a secondary amide as a directing group and an aryl aldehyde as the electrophile. Subsequent cyclization and oxidation gives the substituted anthraquinone. An

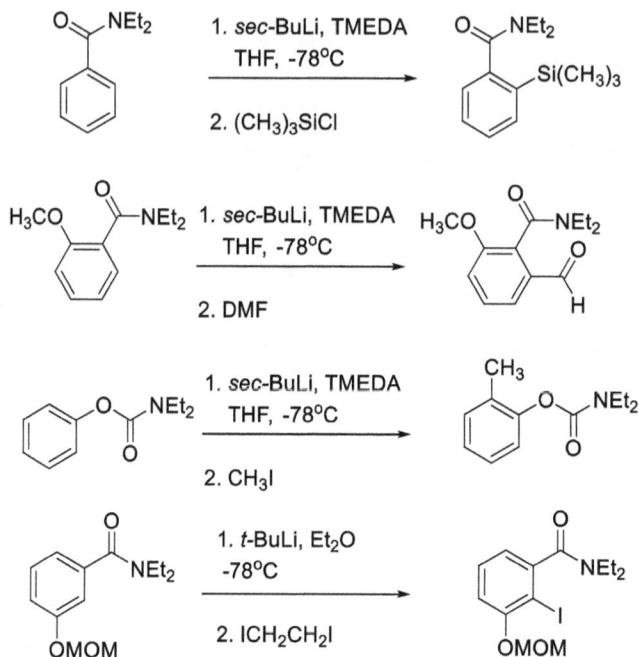

Figure 4.9: Some selected examples of DoM reactions.

example of this strategy is shown for the synthesis of the anthraquinone natural products catenarin and erythroglaucin [16].

Another example of *ortho* lithiation in synthesis is seen in the synthesis of the sesquiterpene cacalol (Figure 4.11) [17]. This synthesis also features three Friedel–Crafts reactions. The first step of the synthesis involves lithiation of 4-methylanisole, where deprotonation occurs *ortho* to the methoxy-directing group. This aryllithium is trapped with 5-iodopent-1-ene to give the alkylated product, which upon treatment with AlCl₃ underwent a Friedel–Crafts cyclization reaction with the alkene. A subsequent electrophilic formylation using dichloromethyl methyl ether and TiCl₄ gave the aldehyde shown as the major product. Baeyer–Villiger oxidation and hydrolysis converted the formyl group into the corresponding phenol, which was alkylated with 2-chloroacetone. This product was then treated with concentrated sulfuric acid, resulting in an electrophilic cyclization reaction at the carbonyl, followed by dehydration to give the benzofuran. Finally, the methoxy group was cleaved using BBr₃ to give cacalol.

During the synthesis of fredericamycin A, Kelly and coworkers used three sequential lithiations to prepare one of the key intermediates in the synthesis (Figure 4.12) [18]. After introducing a MOM group onto the phenolic position of 4-inadanol, treatment with s-BuLi led to lithiation *ortho* to the MOM group, which we saw is an effective DMG. The resulting aryllithium was trapped with N,N-diethyl carbamoyl chloride to introduce the diethyl amide, which is also a strong DMG. Treatment of the amide

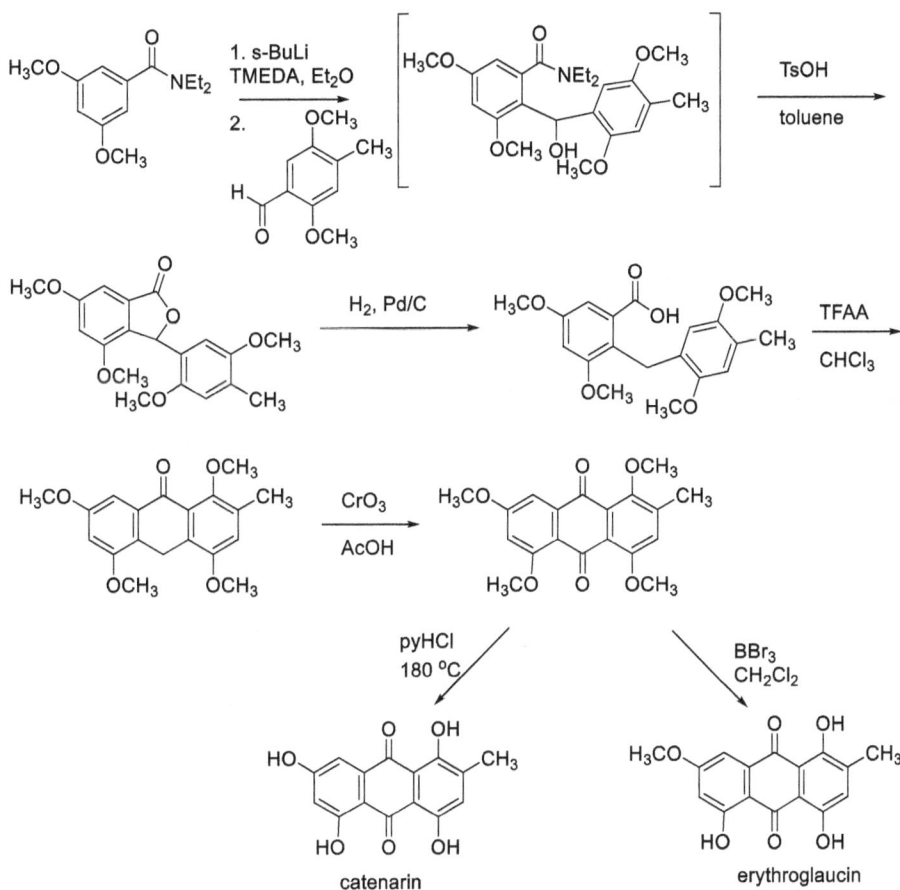

Figure 4.10: General synthetic approach toward unsymmetrical anthraquinones [16].

with *t*-BuLi led to metalation *ortho* to the amide, resulting in an aryllithium that was trapped with methyl iodide. In the next step lithium tetramethylpiperide (LiTMP) is used, which is a base that is similar to LDA, leading to deprotonation under thermo-dynamic control at the benzylic position adjacent to the amide DMG. The benzylic anion was trapped with diethoxyacetonitrile, which cyclized to form the desired pyr-idone. This last lithiation step, where deprotonation occurs not on the aromatic ring, but at a benzylic position, will be explored in more detail in Section 4.3.

4.2.3 Anionic *ortho*-Fries rearrangement

When a carbamate DMG is used for directed metallation and the resulting aryllithium is not treated with an external electrophile, it can undergo rearrangement to the

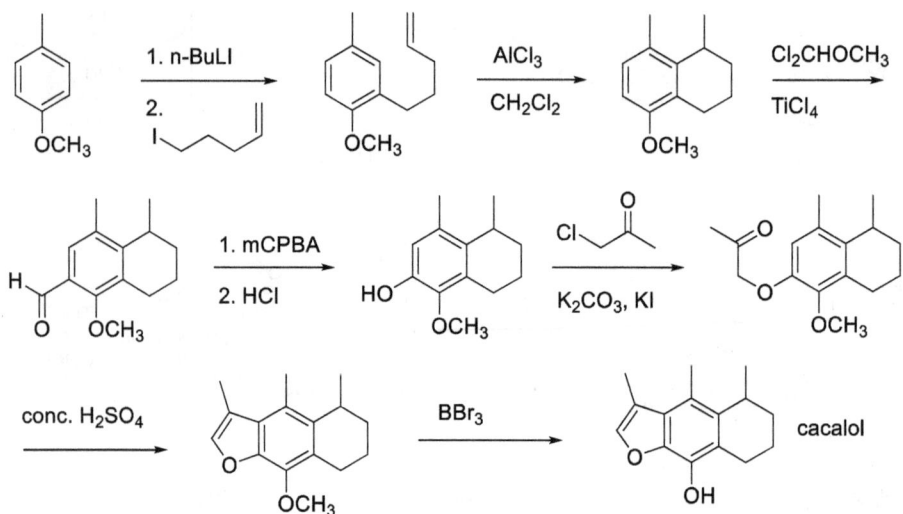

Figure 4.11: Synthesis of cacalol [17].

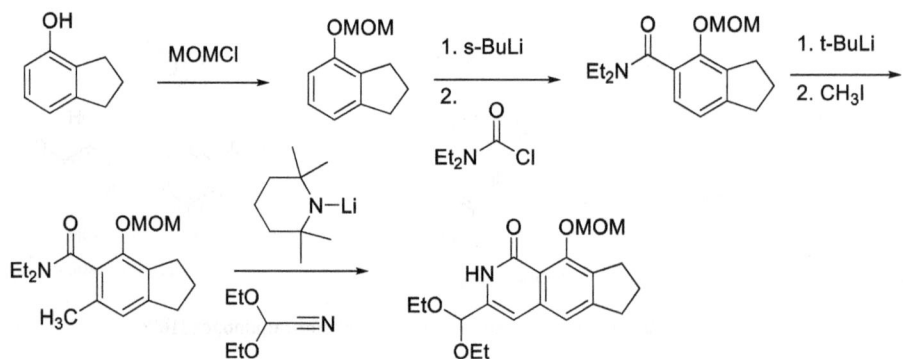

Figure 4.12: Sequential lithiations as part of the synthesis of fredericamycin [18].

corresponding *ortho*-hydroxy amide (Figure 4.13) [14]. The reaction is analogous to the Fries rearrangement we saw in Chapter 2, so this reaction is generally called an anionic *ortho*-Fries rearrangement [19].

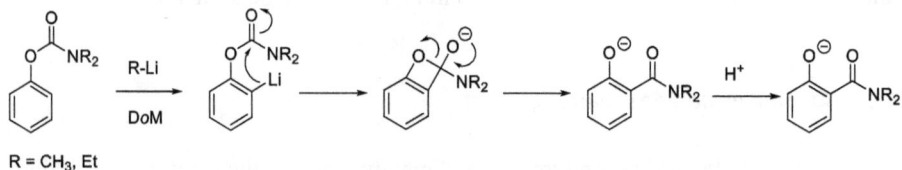

R = CH₃, Et

Figure 4.13: Mechanism of the anionic *ortho*-Fries rearrangement.

The anionic *ortho*-Fries reaction depends on the substrate (i.e., the structure of the carbamate) and is also sensitive to the reaction time for metalation and the temperature of metalation. Consider the following reactions for *O*-aryl carbamates with s-BuLi, followed by trapping with methyl iodide (Figure 4.14) [14]. When *ortho* lithiation is carried out under standard conditions at low temperature, trapped with methyl iodide, and then allowed to warm to room temperature, the major product is the methylation at the *ortho* position. In contrast, if the reaction mixture is allowed to warm to room temperature before quenching, the rearranged product is observed as the major product.

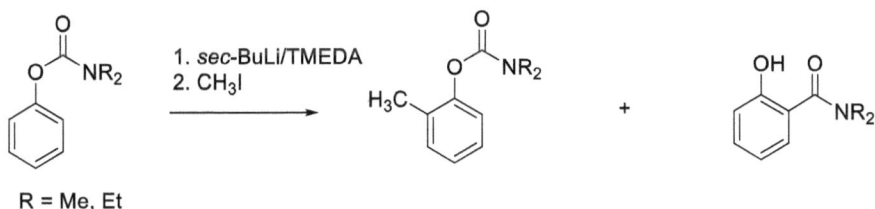

R = Me, Et

Figure 4.14: Standard DoM reaction vs. anionic *ortho*-Fries rearrangement.

The anionic *ortho*-Fries rearrangement can be useful for the synthesis of polysubstituted aromatics because the product of rearrangement installs and amide group on the ring, which can then serve as a DMG for subsequent DoM reactions. As an example of the utility of DoM and the anionic *ortho*-Fries reaction, we will consider the synthesis of ochratoxin B (Figure 4.15) [20]. Starting with the carbamate, *ortho*-lithiation using s-BuLi in the presence of TMEDA gave the aryllithium, which was trapped with *N,N*-diethyl carbamoyl chloride to install the amide. This product was again treated with s-BuLi/TMEDA and allowed to warm to room temperature to effect the Fries rearrangement. The phenoxide was trapped with methyl iodide to form the corresponding methyl ether. At this point, a third DoM reaction was carried out, followed by a transmetallation using MgBr$_2$ and trapping the resulting aryl

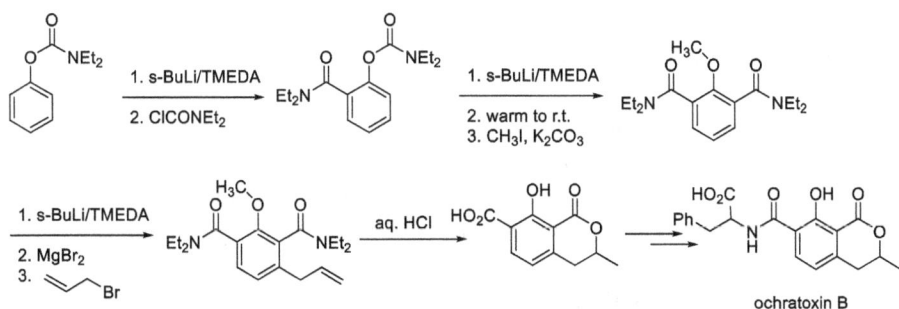

Figure 4.15: Synthesis of ochratoxin B [20].

Grignard species with allyl bromide. This product was treated with 6 M HCl, which resulted in a one-pot amide hydrolysis, acid-catalyzed hydration and lactonization, and ether demethylation to give the isocoumarin carboxylic acid, which could be converted into ochratoxin B using established procedures.

Another example of the utility of the anionic *ortho*-Fries rearrangement is shown in Figure 4.16, where it was used to prepare an *o*-trimethylsilyl aryl triflate [21], which can serve as a precursor to a benzyne – an important reactive species that we will explore in Chapter 5.

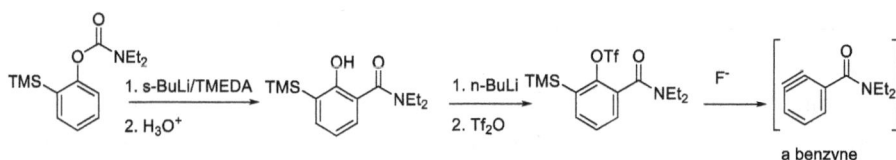

Figure 4.16: Preparation of an *o*-trimethylsilyl aryl triflate via anionic *ortho*-Fries reaction.

The concept of the anionic *ortho*-Fries was extended to an anionic *N*-Fries rearrangement using *N*-carbamoyl diarylamines instead of carbamates (Figure 4.17) [22]. The *ortho*-lithiation reaction takes place in the presence of an alkyllithium such as *t*-BuLi, or with LDA. The reaction shown has a symmetric starting material, so only one product is observed. However, if the two aryl rings are different, deprotonation could happen on either ring, resulting in two isomeric products.

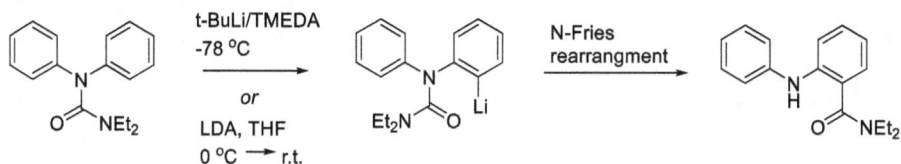

Figure 4.17: Anionic *N*-Fries rearrangement [22].

4.3 Directed remote metalation

In substituted aromatic compounds bearing more than one aromatic ring, the site of deprotonation is still normally governed by the strongest DMG present. However, when the *ortho* site of a DMG is blocked, or when the reaction is carried out under thermodynamic control using a weaker base such as LDA, deprotonation can occur on an adjacent aromatic ring or benzylic position that is in proximity to the DMG (Figure 4.18). This metalation at a "remote" site is referred to as directed remote metalation (DreM) [23].

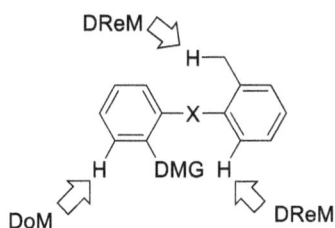

Figure 4.18: Sites of directed *ortho*-metalation (DoM) versus directed remote metalation (DreM).

In these reactions, LDA is typically used as a base instead of an alkyllithium, so the deprotonation step is reversible. Indeed, the reversibility of aryllithium formation is very important for DreM reactions. In cases where an *ortho* site is available for deprotonation, DoM is usually kinetically preferred, while DreM is favored thermodynamically. For this reason, most DoM reactions take place at low temperature using strong alkyllithium bases, whereas DreM reaction are usually carried out at higher temperature with amide bases. Furthermore, in most DreM reactions, the remote aryllithium is usually trapped intramolecularly, resulting in either a remote anionic Fries rearrangement or a cyclization reaction.

Consider the example of the carbamate-substituted biphenyl (Figure 4.19) [24]. When treated with LDA, the compound undergoes *ortho* lithiation, which is then followed by an anionic *ortho*-Fries reaction, as described above. However, in the second example, where the *ortho* site is blocked, deprotonation occurs on the adjacent ring in closest proximity to the DMG. Again, this is followed by an anionic Fries reaction to install the amide on the adjacent ring. Interestingly, the product can then undergo acid-catalyzed cyclization to produce the corresponding lactone.

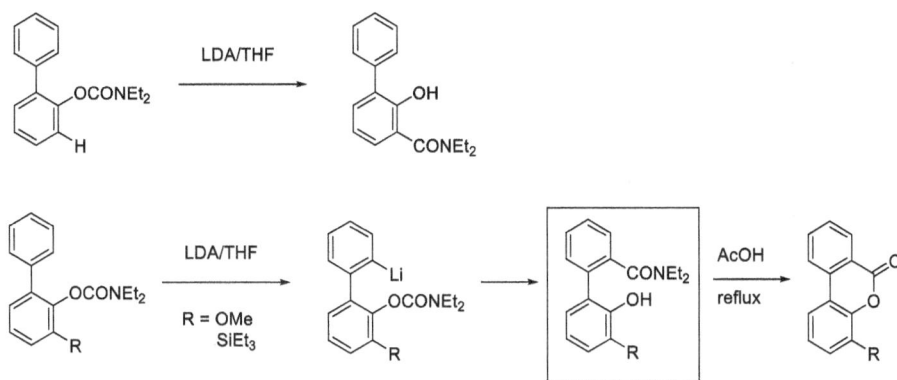

Figure 4.19: DoM versus DreM in an anionic Fries rearrangement.

When an amide DMG is used for DreM, the aryllithium is typically trapped intramolecularly by a cyclization reaction with the amide. In the case of a remote metalation on the adjacent aromatic ring, intramolecular cyclization leads to the formation of fluorenones (Figure 4.20) [25].

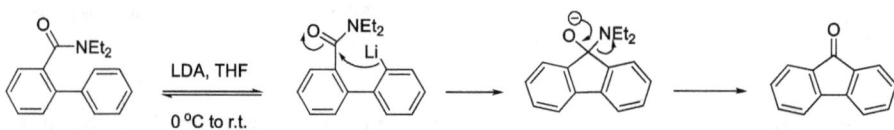

Figure 4.20: Formation of fluorenones via DreM [25].

This approach has been used for the preparation of several fluorenone-containing natural products such as dengibsin and dengibsinin [24, 26]. We will explore the synthesis of dengibsinin in more detail in Chapter 6. This methodology has also been used for the synthesis of chiral fluorenols, which served as dopants for ferroelectric liquid crystals [27].

For biphenyl derivatives with an amide DMG where there are benzylic protons available on the remote ring, deprotonation takes at the benzylic position (sometimes referred to as directed lateral remote metalation). The resulting anion is trapped intramolecularly in a cyclization reaction that ultimately leads to the formation of the phenanthrol (Figure 4.21) [28].

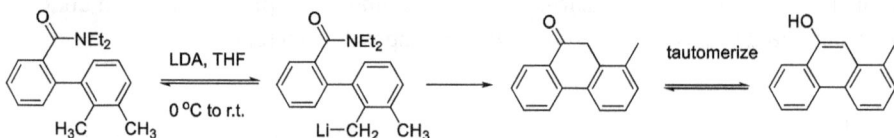

Figure 4.21: Remote benzylic deprotonation to form phenanthrols [28].

DreM at the benzylic position followed by cyclization has been applied for the synthesis of several phenanthrene natural products. For example, this methodology has been used in the synthesis of gymnopusin and eupolauramine (Figure 4.22) [29, 30]. For the synthesis of eupolauramine, the phenyl pyridine starting material shown undergoes DreM at the benzylic position in proximity to one of the amide DMGs, and the resulting anion cyclizes onto the amide to give the substituted phenanthrol, which is ultimately converted to eupolauramine. The synthesis of gymnopusin is slightly more elaborate. First, DreM involving the carbamate DMG leads to a remote anionic Fries rearrangement where the phenolate is trapped with methyl iodide. With the amide group now migrated to the previously "remote" ring, a second DreM leads to benzylic deprotonation and cyclization to give the phenanthrol, which can readily be converted to gymnopusin.

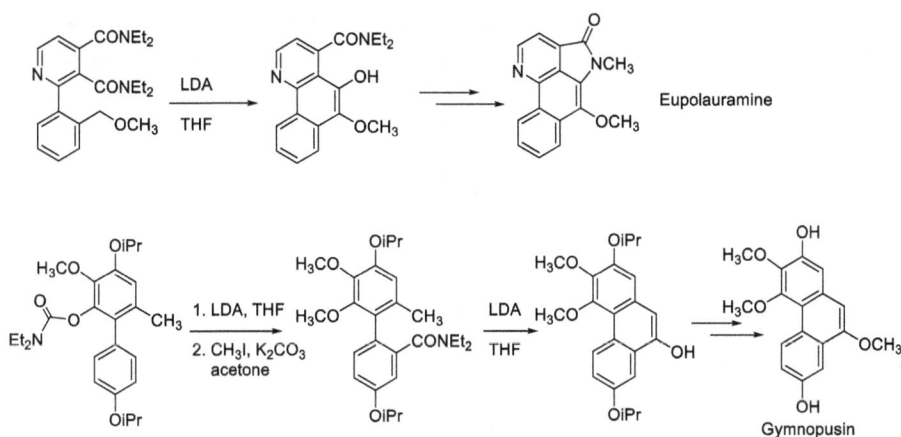

Figure 4.22: DreM cyclizations in the synthesis of phenanthrene natural products [29, 30].

4.4 Lithium–halogen exchange

Aryllithiums can be generated by treating haloarenes with alkyllithiums such as bu-tyllithium (Figure 4.23). Lithium–halogen exchange is typically carried out using bromo- or iodoarenes. First developed by Gilman and Wittig [31, 32], lithium–halogen exchange has become a widely used tool for the preparation of aryllithium species. It is complementary to methods such as DoM or DReM because the site of lithiation is determined by the position of the halogen.

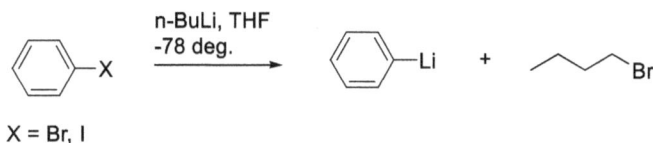

X = Br, I

Figure 4.23: Formation of aryllithiums via lithium–halogen exchange.

Typical lithium–halogen exchange reactions involve an aryl bromide or iodide in THF or ether and take place rapidly (i.e., within seconds) at low temperatures – usually –78 °C. When lithium–halogen exchange is carried out using *t*-BuLi, 2 M equivalents of the alkyllithium are needed because the by-product of the first lith-ium–halogen exchange is *t*-butyl bromide, which reacts readily with the aryl-lithium in an elimination reaction to produce isobutene (Figure 4.24) [33]. To prevent this undesired quenching of the aryllithium by *t*-butyl bromide, a second equivalent of *t*-BuLi is used, which reacts more rapidly with the *t*-butyl bromide, leaving the ar-yllithium intact. For this reason, when *t*-BuLi is used for lithium–halogen exchange, two equivalents of *t*-BuLi are almost always used. However, the assumption that two equivalents are necessary has been questioned by Waldmann and coworkers, who showed that depending on the substrate, a range of 1–1.8 equivalents of *t*-BuLi provided good yields of the desired product [34].

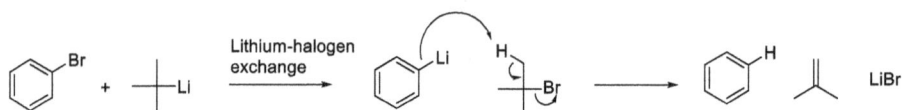

Figure 4.24: Undesired side reaction of an aryllithium with *t*-butyl bromide after initial lithium–halogen exchange.

The mechanism of lithium–halogen exchange reactions has been debated, with several mechanistic scenarios being proposed [35]. Among the mechanisms considered are a concerted exchange proceeding through a four-membered ring transition state, a sequence of single-electron transfer reactions, or a nucleophilic attack of the alkyllithium on the halogen of the aryl halide, resulting in either a concerted displacement of an aryl anion intermediate, or a stepwise formation of an anionic complex. The nucleophilic displacement mechanisms are both considered to be the most plausible (Figure 4.25). Since both of these possibilities are on the same mechanistic continuum, it is possible that the mechanism varies between the two, depending on the reactants and reaction conditions.

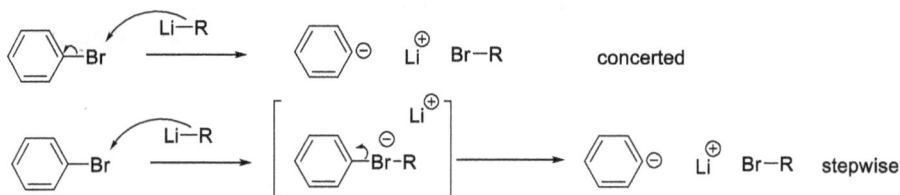

Figure 4.25: Proposed mechanisms of lithium–halogen exchange.

Once generated, the aryllithium can react with a variety of electrophiles as discussed earlier in this chapter (see Figure 4.2). It is a particularly useful tool for preparing organometallic reagents such as aryl boronic acids, aryl stannanes, and aryl zinc species that are used in transition metal-mediated cross-coupling reactions that we will see in Chapter 6.

4.5 Rearrangements of haloarenes: halogen dance reactions

Under strongly basic conditions, some haloarenes have can undergo rearrangements, where the position and number of halogen substituents can be changed. This based-promoted rearrangement is referred to as a halogen dance reaction [36–38]. For example, reaction of 1,2,4-tribromobenzene with potassium anilide (PhNHK) in liquid ammonia produces predominantly the isomerized 1,3,5-tribromobenzene, along with minor amounts of di- and tetra-bromobenzenes (Figure 4.26) [39]. Dibromoanilines are also observed as minor side products.

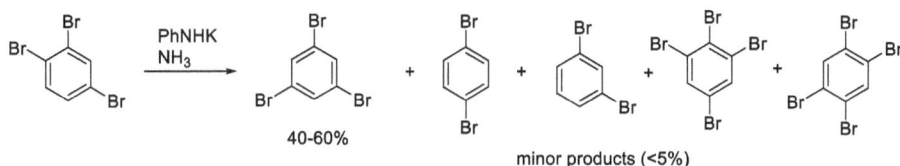

40-60%

minor products (<5%)

Figure 4.26: Rearrangement of aryl halides under strongly basic conditions (halogen dance) [36, 39].

Bunnett and others explored several mechanistic explanations for these observations [36, 39]. One possible mechanism involves benzyne formation (Chapter 5), since many of the observed examples take place under similar conditions to that of benzyne formation and indeed the presence of substituted anilines as minor products are likely due to benzyne formation. Another possible mechanism is deprotonation followed by an intramolecular halogen shift. However, neither of these mechanisms can explain the formation of dibromo- and tetrabromo-products. Bunnett proposed the generally accepted mechanism where the reaction proceeds via intermolecular reactions of aryl anions with halobenzenes to generate a new aryl anion [36, 39]. Specifically, the first step in the reaction is deprotonation of the haloarene by the strong base (often an amide base such as ArNHK, KNH_2, or $NaNH_2$), followed by intermolecular metal–halogen exchange to generate a new haloarene and an aryl anion (Figure 4.27). These compounds can react further in a cascade, ultimately producing a mixture of rearrangement and disproportionation products. Below is a plausible mechanism for the isomerization of 1,2,4-tribromobenzene to produce 1,3,5-tribromobenzene. After the initial deprotonation, a sequence of intermolecular metal–halogen exchange reactions take place, which can be described as nucleophilic attack of the aryl anion on the halogen of the aryl halide. It should be apparent from this reaction that this is one possible sequence, and that several other deprotonation and halogen exchange reactions could occur. This explains why a mixture of products is often observed, and importantly accounts for the formation of dibromo- and tetrabromo-benzenes as minor side products. It should also be pointed out that these reactions typically also show some recovered starting material and rarely go to completion.

Given the numerous possible outcomes of the halogen dance reaction, the major products are determined by the formation of the most thermodynamically stable aryl anion intermediate. Only in cases where there is a thermodynamic preference for one of the aryl anions will there be good conversion. Where there is little energy difference between different aryl anions, a mixture of products will result.

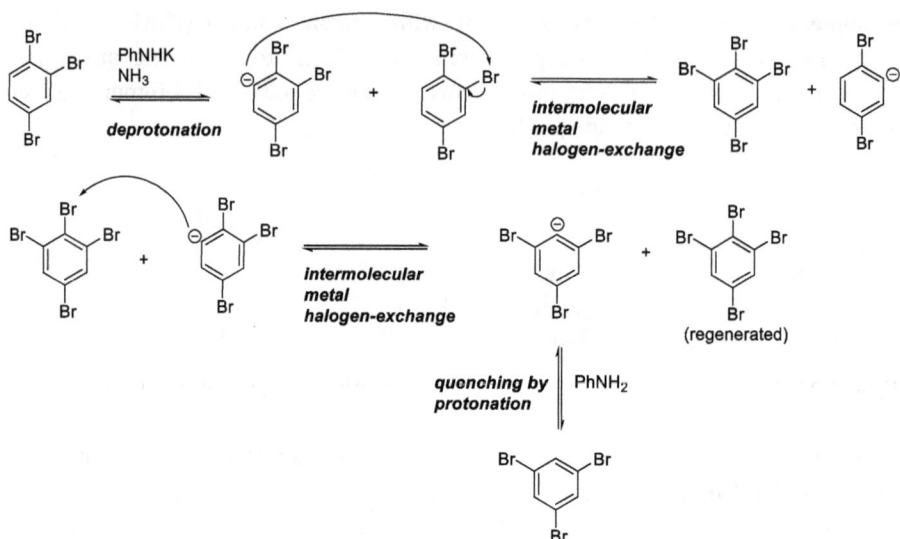

Figure 4.27: Bunnett's proposed mechanism for the halogen dance reaction.

While the first examples of halogen dance reactions involved bases such as NaNH$_2$, KNH$_2$, or PhNHK, it is more common to use bases such as LDA or LiTMP. A key requirement of the halogen dance reaction is the simultaneous presence of both the starting material and initial aryl anion in order to allow the intermolecular reaction. If deprotonation is rapid and quantitative, the halogen dance reaction cannot occur. Similarly, the quenching step by an electrophile (in some cases simply a proton) should be slow in order to provide enough time for the halogen exchange sequence to occur. If the electrophile is present in the reaction, it needs to react slowly. An alternative is to add the electrophile after sufficient time has elapsed to allow the exchange to take place.

To illustrate the conditions under which halogen dance reactions can be promoted or prevented, let us consider this example, where o-bromo-trifluoromethylbenzene is reacted with LTMP (Figure 4.28) [40]. When this reaction is carried out at –100 °C, the aryllithium species is generated by deprotonation and simply trapped with CO$_2$ to generate the corresponding carboxylic acid with no halogen dance occurring. However, upon warming to –75 °C, a halogen dance reaction occurs, generating the more stable aryllithium where the anion is adjacent to the electron-withdrawing CF$_3$ group. Upon quenching, the isomeric benzoic acid derivative is obtained.

Halogen dance reactions can also be initiated by lithium–halogen exchange using alkyllithiums (such as BuLi) instead of deprotonation. For example, 3-bromo-2-chloropyridine can be converted into 4-bromo-2-chloropyridine by treatment with 0.5 equivalents of n-BuLi (Figure 4.29) [41]. In this example, the reaction is quenched with MeOD to produce the deuterated derivative.

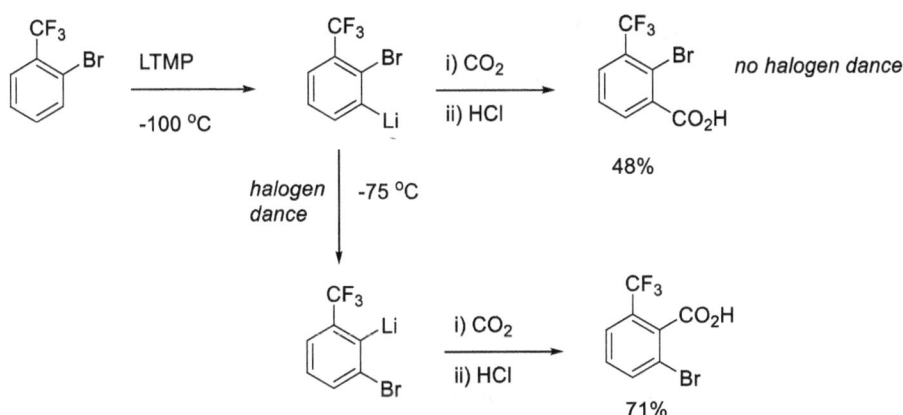

Figure 4.28: Temperature dependence of halogen dance reaction [40].

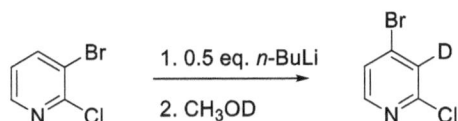

Figure 4.29: Halogen dance reaction after lithium–halogen exchange [41].

The substoichiometric amount of n-BuLi is important to ensure that some starting material is present to react with the anion formed upon lithium–halogen exchange. The cascade of reactions is outlined in Figure 4.30 [41]. It should be noted that starting material is regenerated after the sequence of two intermolecular lithium–halogen exchange reactions, resulting in a chain reaction.

Another example of a halogen dance reaction via lithium–halogen exchange involves bromo-substituted hexaphenylbenzenes (Figure 4.31) [42]. Specifically, reaction of hexakis(4-bromophenyl) benzene with six equivalents of t-BuLi gives the C_3-symmetric product as the major product. Recall that for lithium–halogen exchange using t-BuLi, two equivalents of the alkyllithium are needed, so six equivalents is only able to effect lithium–halogen exchange at three positions. The selectivity of this trilithiation is surprising given the number of products possible.

The mechanism of this reaction was investigated by quenching the reaction mixture with TMSCl at various time intervals. These experiments showed that initial lithium–halogen exchange produces a 1:1 mixture of the hexalithiated species as well as unreacted starting material. However, upon gradual warming, this mixture is converted to the major trilithiated species, consistent with a halogen dance reaction (Figure 4.32). This approach provided access to C_3-symmetric hexaphenylbenzene derivatives, which are not easily prepared selectively using other methods.

While the halogen dance reaction has the potential to yield multiple products, there are several examples where it has been exploited in multistep synthesis. As one

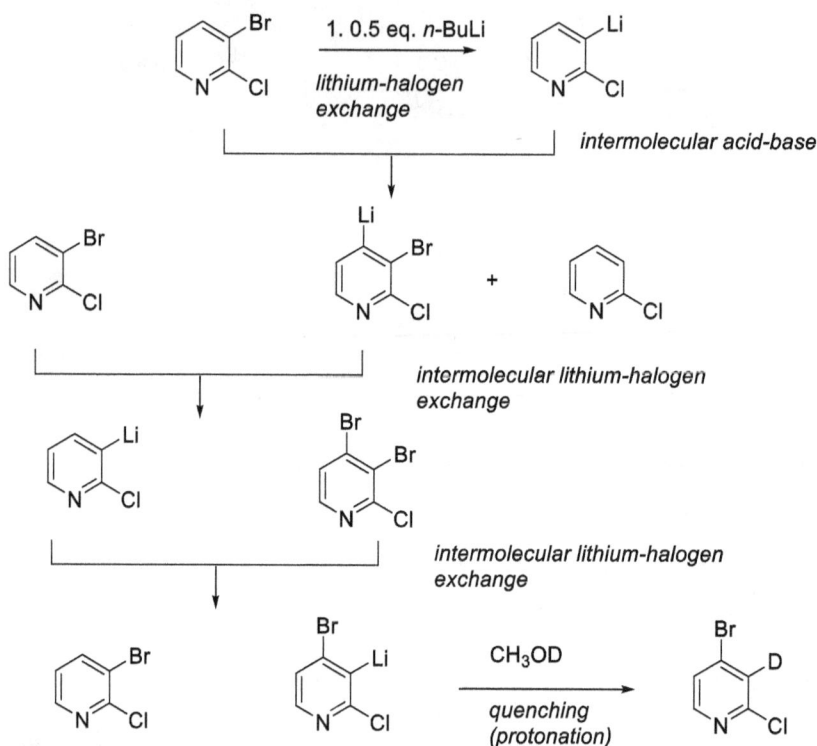

Figure 4.30: A cascade of reactions in a halogen dance process.

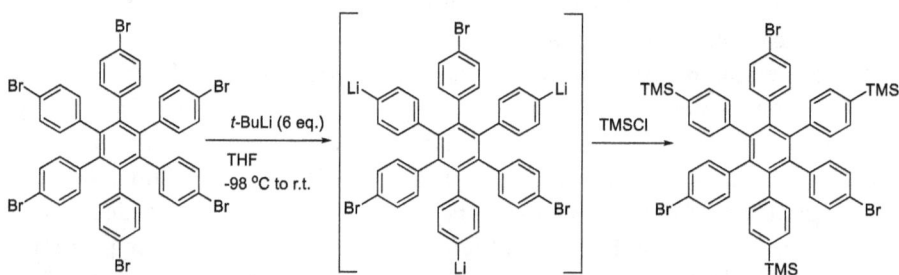

Figure 4.31: Halogen dance reactions in brominated hexaphenylbenzenes [42].

example, Sammakia et al. reported the synthesis of caerulomycin C, which is derived from *Streptomyces caeruleus* and possesses weak antibiotic properties (Figure 4.33) [43]. The synthesis, outlined below, begins with a nucleophilic aromatic substitution (see Chapter 3) reaction of the chloropyridine derivative, followed by a DoM reaction to install the iodo group next to the amide directing group. The iodo-substituted product was then treated with LDA to effect the halogen dance reaction, which resulted in a shift of the iodo group from the 3-position to the 5-position. The iodo

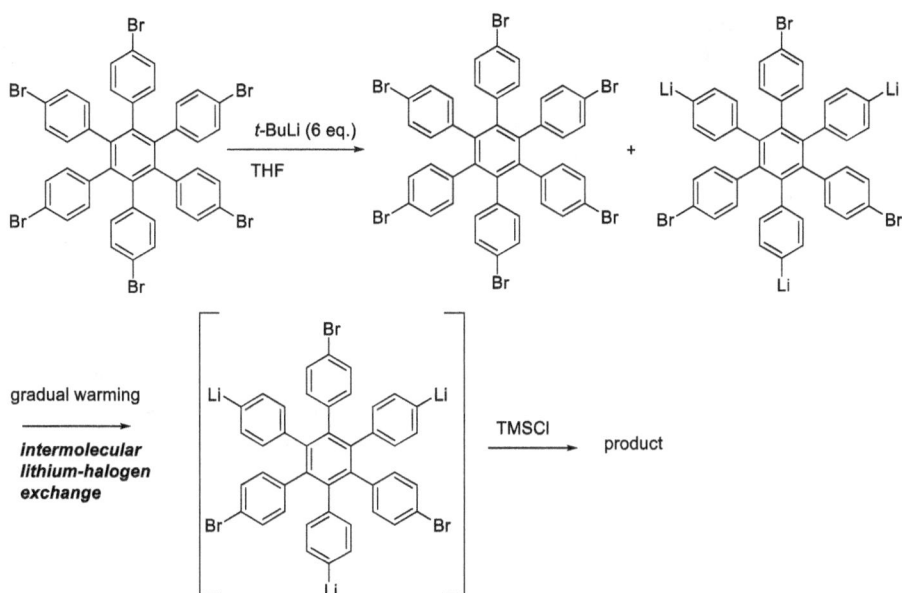

Figure 4.32: Proposed mechanism of halogen dance reaction with brominated hexaphenylbenzene [42].

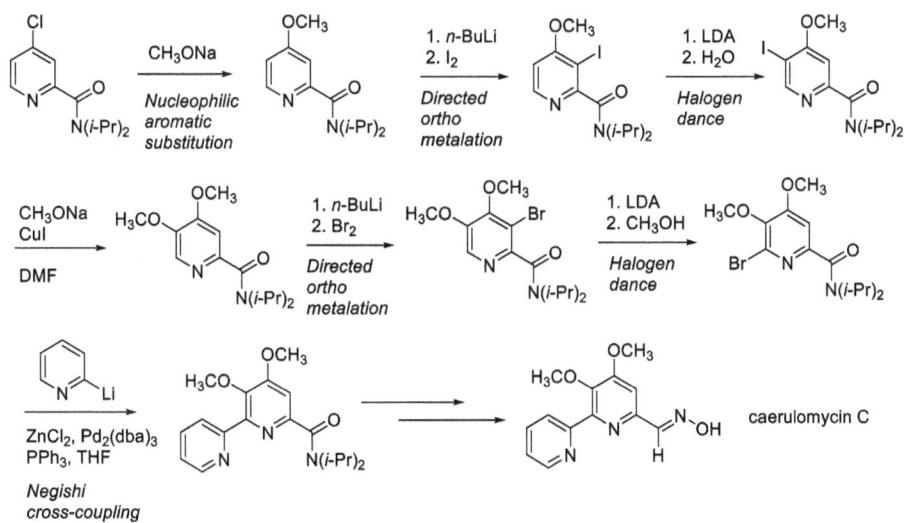

Figure 4.33: Synthesis of caerulomycin C [43].

group was then replaced by a methoxy group in a copper-mediated nucleophilic substitution reaction (see Chapter 7). The product was once again subjected to a DoM reaction to install a bromide, followed by another halogen dance reaction using LDA. This product was then used in a Negishi cross-coupling reaction – a coupling reaction we will see in Chapter 6 – to furnish the bipyridine, which as then converted into caerulomycin C in two steps.

The same group employed a similar strategy for the preparation of WS75624 B a natural product with potential as an antihypertensive reagent. In this case, both fragments were prepared using a halogen dance reaction (Figure 4.34) [44].

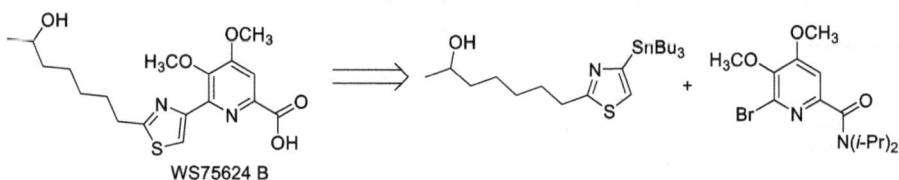

WS75624 B

Figure 4.34: Halogen dance reactions were used to prepare the two precursors to WS75624 B [44].

References

[1]   Streitweiser A, Scannon PJ, Niemeyer HM. Acidity of hydrocarbons. XLIX. Equilibrium ion pair acidities of fluorinated benzenes for cesium salts in cyclohexylamine. Extrapolation to pK of benzene. J Am Chem Soc. 1972;94:7936–37.

[2]   Gilman H, Bebb R. Relative reactivities of organometallic compounds. XX. Metalation. J Am Chem Soc. 1939;61:109–12.

[3]   Wittig G, Fuhrmann G. On the performance of halogenised anisole towards phenyl-lithium. Chem Ber. 1940;73:1197–218.

[4]   Snieckus V. Directed ortho metalation. Tertiary amide and O-carbamate directors in synthetic strategies for polysubstituted aromatics. Chem Rev. 1990;90(6):879–933.

[5]   Whisler MC, MacNeil S, Snieckus V, Beak P. Beyond thermodynamic acidity: A perspective on the complex-induced proximity effect (CIPE) in deprotonation reactions. Angew Chem Int Ed. 2004;43:2206–25.

[6]   Eastham JF, Gibson GW. Solvent effects in organometallic reactions. II. J Am Chem Soc. 1963;85:2171–72.

[7]   West P, Waack R. Colligative property measurements on oxygen- and moisture-sensitive compounds I. Organolithium reagents in donor solvents at 25. J Am Chem Soc. 1967;89(17):4395–99.

[8]   Lewis HL, Brown TL. Association of alkyllithium compounds in hydrocarbon media. Alkyllithium-base interactions. J Am Chem Soc. 1970;92(15):4664–70.

[9]   Beak P, Meyers AI. Stereo- and regiocontrol by complex induced proximity effects: Reactions of organolithium compounds. Acc Chem Res. 1986;19(11):356–63.

[10] van Eikema Hommes NJR, von Ragué Schleyer P "Kinetically enhanced metalation" – How substituents direct ortho lithiation. Angew Chem Int Ed Engle. 1992;31:755–58.

[11]   Beak P, Brown RA. The tertiary amide as an effective director of ortho lithiation. J Org Chem. 1982;47(1):34–46.

[12]   Beak P, Brown RA. The ortho lithiation of tertiary benzamides. J Org Chem. 1977;42 (10):1823–24.

[13]   de Silva SO, Reed JN, Snieckus V. Direct lithiation of N,N-diethylbenzamines. Regiospecific synthesis of contiguously tri- and tetra-substituted alkoxybenzenes. Tetrahedron Lett. 1978;51:5099–102.

[14]   Sibi MP, Snieckus V. The directed ortho lithiation of O-aryl carbamates. An anionic equivalent of the fries rearrangement. J Org Chem. 1983;48(11):1935–37.

[15]   Winkle MR, Ronald RC. Regioselective metalation reactions of some substituted (methoxymethoxy)arenes. J Org Chem. 1982;47(11):2101–08.

[16]   de Silva SO, Watanabe M, Snieckus V. General route to anthraquinone natural products via directed metalation of N,N-diethylbenzamides. J Org Chem. 1979;44(26):4802–08.

[17]   Kedrowski BL, Hoppe RW. A concise synthesis of cacalol. J Org Chem 2008;73(13):5177–79.

[18]   Kelly TR, Ohashi N, Armstrong-Chong RJ, Bell SH. Synthesis of fredericamycin A. J Am Chem Soc. 1986;108(22):7100–01.

[19]   Korb M, Lang H. The anionic Fries rearrangement: A convenient route to: Ortho-functionalized aromatics. Chem Soc Rev. 2019;48:2829–82.

[20]   Sibi MP, Chattopadhyay S, Dankwardt JW, Snieckus V. Combinational O-aryl carbamate and benzamide directed ortho metalation reactions. Synthesis of ochratoxin A and ochratoxin B. J Am Chem Soc. 1985;107(22):6312–15.

[21]   Shankaran K, Snieckus V. Silicon in benzamide directed ortho metalation. Formation and reactions of benzamide benzynes. Tetrahedron Lett. 1984;25(27):2827–30.

[22]   MacNeil SL, Wilson BJ, Snieckus V. Anionic N-fries rearrangement of N-carbamoyl diarylamines to anthranilamides. Methodology and application to acridone and pyranoacridone alkaloids. Org Lett. 2006;8(6):1133–36.

[23]   Tilly D, Magolan J, Mortier J. Directed remote aromatic metalations: Mechanisms and driving forces. Chem-Eur J. 2012;18(13):3804–20.

[24]   Wang W, Snieckus V. Remote directed metalation of biaryl o-carbamates. Ring to ring carbamoyl transfer route to biaryls, dibenzo[6,d]pyranones, and the natural fluorenone dengibsin. J Org Chem. 1992;57(2):424–26.

[25]   Fu J, Zhao B, Sharp MJ, Snieckus V. Remote aromatic metalation. An anionic Friedel-Crafts equivalent for the regioselective synthesis of condensed fluorenones from biaryl and m-teraryl 2-amides. J Org Chem. 1991;56(5):1683–85.

[26]   Fu JM, Zhao BP, Sharp MJ, Snieckus V. Ortho and remote metalation – Cross coupling strategies. Total synthesis of the naturally occurring fluorenone dengibsinin and the azafluoranthene alkaloid imeluteine. Can J Chem. 1994;72(1):227–36.

[27]   McCubbin JA, Tong X, Zhao Y, Snieckus V, Lemieux RP. Directed metalation route to ferroelectric liquid crystals with a chiral fluorenol core: The effect of restricted rotation on polar order. J Am Chem Soc. 2004;126(4):1161–67.

[28]   Cai X, Brown S, Hodson P, Snieckus V. Regiospecific synthesis of alkylphenanthrenes using a combined directed ortho and remote metalation – Suzuki-Miyaura cross coupling strategy. Can J Chem. 2004;82(2):195–205.

[29]   Wang X, Snieckus V. Synthetic strategies basedon aromatic metalation – Cross coupling links. Regiospecific synthesis of the phenanthrene natural product gymnopusin. Tetrahedron Lett. 1991;32(37):4879–82.

[30]   Wang X, Snieckus V. Synthetic strategies based on aromatic metalation – Cross coupling links. A concise formal synthesis of the azaphenanthrene alkaloid eupolauramine. Tetrahedron Lett. 1991;32(37):4883–84.

[31] Gilman H, Jacoby AL. Dibenzothiophene: Orientation and derivative. J Org Chem. 1938;3:108–19.

[32] Wittig G, Pockels U, Droge H. The interchangeability of aromatically linked hydrogen to lithium via phenyl-lithiums. Ber Dtsch Chem Ges. 1938;71:1903–12.

[33] Seebach D, Neumann H. Brom-Lithium-Austausch an Vinyl- und Aryl-bromiden mit tert-Butyllithium Zur Ringerweiterung über Dibromcarbenaddukte. Chem Ber. 1974;107 (3):847–53.

[34] Waldmann C, Schober O, Haufe G, Kopka K. A closer look at the bromine-lithium exchange with tert-butyllithium in an aryl sulfonamide synthesis. Org Lett. 2013;15(12):2954–57.

[35] Bailey WF, Patricia JJ. The mechanism of the lithium – Halogen interchange reaction: A review of the literature. J Organomet Chem. 1988;352(1–2):1–46.

[36] Bunnett JF. The base-catalyzed halogen dance, and other reactions of aryl halides. Acc Chem Res. 1972;5(4):139–47.

[37] Schnürch M, Spina M, Khan AF, Mihovilovic MD, Stanetty P. Halogen dance reactions – A review. Chem Soc Rev. 2007;36(7):1046–57.

[38] Schlosser M. The 2 x 3 toolbox of organometallic methods for regiochemically exhaustive functionalization. Angew Chem Int Ed 2005;44(3):376–93.

[39] Moyer CE, Bunnett JF. Base-catalyzed isomerization of trihalobenzenes. J Am Chem Soc. 1963;85(12):1891–93.

[40] Mongin F, Desponds O, Schlosser M. Reagent-modulated optional site selectivities: The metalation of o-, m-, and p-halobenzotrifluorides. Tetrahedron Lett. 1996;37(16):2767–70.

[41] Mallet M, Quéguiner G. Action du n-butyllithium sur les bromo-3 halogeno-2 pyridines fluoree, chloree et bromee. Principe et etude d'une possibilite reactionnelle nouvelle: L'homotransmetallation. Tetrahedron 1985;41(16):3433–40.

[42] Kojima T, Hiraoka S. Selective alternate derivatization of the hexaphenylbenzene framework through a thermodynamically controlled halogen dance. Org Lett. 2014;16:1024–27.

[43] Sammakia T, Stangeland EL, Whitcomb MC. Total synthesis of caerulomycin C via the halogen dance reaction. Org Lett 2002;4(14):2385–88.

[44] Stangeland EL, Sammakia T. Use of thiazoles in the halogen dance reaction: Application to the total synthesis of WS75624 B. J Org Chem. 2004;69(7):2381–85.

5 Benzynes

5.1 Introduction to benzynes

1,2-Didehydrobenzene (benzyne) was first proposed as a reactive intermediate by Wittig in 1942 [1]. Subsequently, the existence of benzyne was established by Roberts in 1953 by the reaction of ^{14}C-labeled chlorobenzene with sodium amide (Figure 5.1) [2].

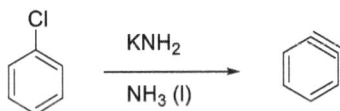

Figure 5.1: Preparation of benzyne by Roberts [2].

Arynes (the generalized term for didehydroaromatic compounds) are highly reactive intermediates and are therefore generated in situ. Arynes are typically represented as a strained alkyne with the second π-bond being orthogonal to the aromatic system. They can also be viewed as a singlet diradical species, although computational and experimental studies suggest that there is more bonding character than radical character [3]. Because the π-bond is perpendicular to the aromatic π-system, the reactivity of the aryne is not affected much by resonance – rather inductive effects of substituents dominate.

Arynes undergo a variety of reactions, including cycloadditions, reactions with nucleophiles, and transition metal-mediated reactions [4, 5]. Indeed, aryne chemistry has developed to such an extent that arynes are often used in the total synthesis of natural products [6, 7], as well as the preparation of other complex aromatic architectures such as polycyclic aromatic hydrocarbons [8]. In this chapter, we will explore the methods for generating arynes, describe their reactivity, and highlight their utility in the organic synthesis.

5.2 Formation of benzynes

The early methods for forming arynes involved either strong bases or high temperatures. For example, benzyne can be generated by reaction of a halobenzene with a strong base such as sodium amide in a beta elimination reaction, elimination of fluoride from an *ortho*-fluoro Grignard or aryllithium species, or by expulsion of CO_2 and N_2 from an *ortho*-aryl diazonium carboxylate. More recently, methods for formation of arynes using milder conditions, such as the fluoride induced elimination of trimethylsilyl aryl triflates have been developed. The most common methods for forming arynes are summarized in Figure 5.2 and will be described in more detail below.

https://doi.org/10.1515/9783110562682-005

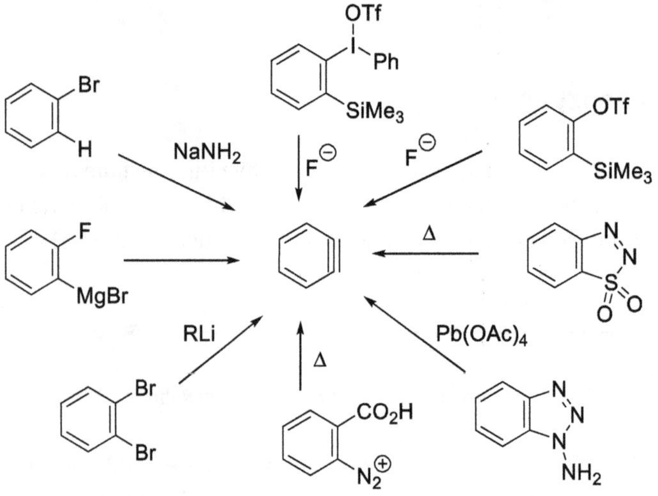

Figure 5.2: Summary of methods to form benzyne.

5.2.1 Aryne formation by elimination of H–X

In Chapter 3, we introduced benzynes in the context preparing anilines from the corresponding aryl halides using sodium amide. Sodium amide acts as a strong base that is able to deprotonate the proton adjacent to the halogen leaving group, resulting in the benzyne. Subsequently, a second equivalent of sodium amide as a nucleophile, adding to the benzyne to generate an aryl anion that is subsequently protonated (Figure 5.3) [2, 9].

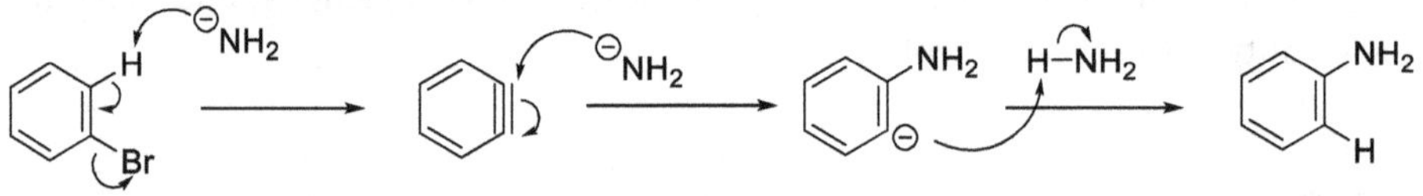

Figure 5.3: Aryne formation by elimination of H–X using sodium amide.

The reaction is not restricted to sodium amide, but can also be carried out in the presence of other strong bases, such as potassium t-butoxide or phenyllithium (Figure 5.4).

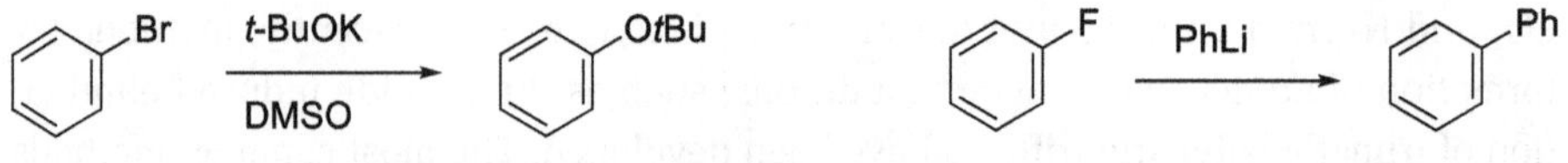

Figure 5.4: Aryne formation by H–X elimination using other strong bases.

The reaction mechanism is often described as an elimination–addition for nucleophilic substitution of arynes because the benzyne formation step is an elimination, which is followed by nucleophilic addition. The reaction can be carried out with all of the halobenzenes, although the relative reactivity depends on the reaction conditions being used. For example, when organolithium bases are used in ethereal solvents, the trend in reactivity is F > Cl > Br > I [10]. This trend is rationalized by the electron-withdrawing effects, where the halogen influences the acidity of the proton being removed. Thus, a fluoro group, being the most electron-withdrawing, makes the adjacent proton more acidic. In contrast, for the reaction with potassium amide (KNH$_2$) in liquid ammonia, the trend in reactivity is Br > I >Cl ≫ F [11]. This unusual reactivity trend is through to be a compromise between the inductive effects described above and the strength of the Ar-X bond.

In the case of the substituted aryl halide shown in Figure 5.5, the benzyne that is formed reacts with more sodium amide to produce the two regioisomeric products in an approximately 1:1 ratio.

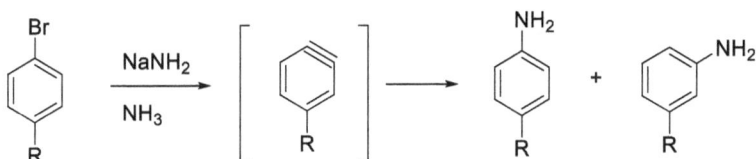

Figure 5.5: Regiochemistry of reactions of benzynes with nucleophiles.

The example highlights one of the challenges of using arynes in synthesis: controlling the regiochemistry of benzyne reactions. The challenge is further amplified if one noticed that the example above could only yield one benzyne. In other cases, one must consider the regioselectivity of benzyne formation as well. The regiochemical considerations for benzynes can be summarized as follows:

- In cases where more than one regioisomeric benzyne can be formed by elimination of HX, the regioselectivity is governed by the relative acidities of the protons being removed.
- Once formed, the nucleophilic addition step onto the benzyne is mostly determined by the stability of the resulting aryl anion, although steric hindrance of nucleophilic attack may also play a role.

Taking into consideration both of these factors, substituents on the benzene ring influence the regiochemistry of the reaction primarily through *inductive* electronic effects, since both the σ-C–H bond and the aryl anion involve sp^2 hybrid orbitals that are orthogonal to the π-system, resonance effects are not important. To demonstrate these principles in action, we will consider the regiochemistry of a series of substituted halobenzenes. Consider the reaction of the series of bromoanisoles with sodium amide

(Figure 5.6) [9]. For o-bromoanisole, only one benzyne is possible, which could lead to two regioisomers upon nucleophilic addition (o-anisidine and m-anisidine). In this reaction, the *meta*-substituted product is formed exclusively. This high regioselectivity can be rationalized based on sterics, where nucleophilic addition at the *ortho* position is disfavored, as well as electronic effects, where the aryl anion that results from nucleophilic addition is stabilized because it is *ortho* to the inductively electron-withdrawing methoxy group. In the case of p-bromoanisole, again only one benzyne can be formed. In this case, the subsequent nucleophilic addition yields an approximately equal mixture of the *meta* and *para* products. In this case, neither the steric nor the inductive effects are as significant, so there is essentially no regioselectivity. Finally, in the case of m-bromoanisole, two possible benzynes can be formed. However, only the *meta* product is observed in this reaction (as in the case of o-bromoanisole). This result suggests that only benzyne with the triple bond adjacent to the methoxy group is formed. The result supports the notion that the inductive electron-withdrawing effects of the methoxy group favor *ortho* deprotonation.

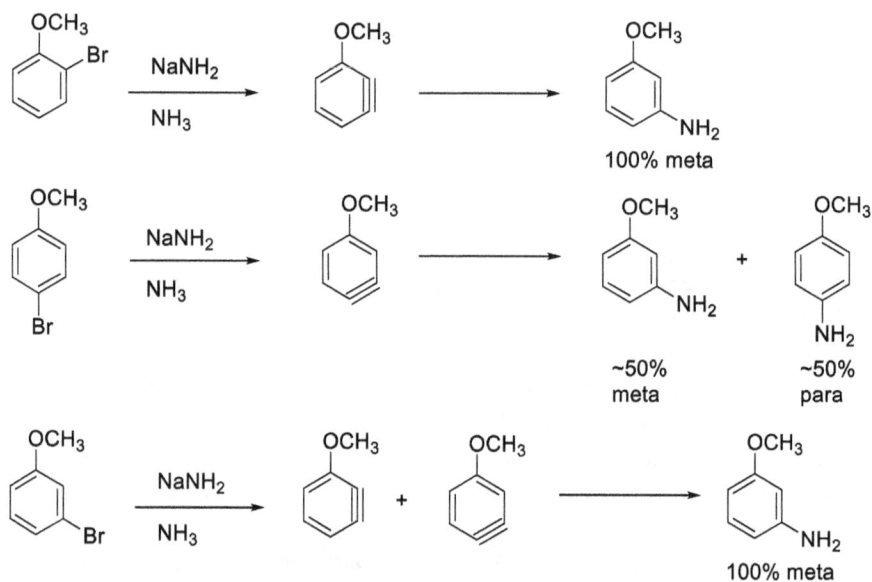

Figure 5.6: Regiochemistry of benzyne formation and reaction for bromoanisoles.

Table 5.1 provides more examples, showing how the regioselectivity varies with different substituents [9]. How does changing the substituents influence the regiochemistry of benzyne formation and the regiochemistry of the nucleophilic addition step? To consider the regiochemistry of benzyne formation, we should consider the *m*-substituted halobenzenes where more than one benzyne can be formed. In the case of a *meta*-trifluoromethyl group, only *meta*-substitution is observed, consistent with the corresponding anisole. However, when the substituent is an inductively electron-donating methyl group, a product distribution of 22% *ortho*, 56% *meta*, and 22% *para* is observed. This product distribution suggests that both benzynes are being formed and that the regioselectivity of the resulting nucleophilic attack is poor. To explore the regioselectivity of the nucleophilic attack, let us consider the *para*-substituted systems, where only one benzyne can be formed, but where there are two sites for nucleophilic attack. For p-bromoanisole we saw very little selectivity. When a more strongly electron-withdrawing group is used such as with p-bromofluorobenzene, the *para*-substituted product is preferred (80% *para*/20% *meta*). In contrast, when the electron-donating methyl group is used, the *meta* product is preferred (62% *meta*/38% *para*). These results are consistent with the stabilization of the phenyl anion that results from nucleophilic attack. Electron-withdrawing groups stabilize the anion at the closer *meta* position, leading to *para* substitution, while electron-donating substituents react at the *meta* position to place the anion at the *para* position, furthest away from the electron-donating substituent (Figure 5.7) [12].

Table 5.1: Regiochemistry of reactions of substituted halobenzenes with sodium amide.

R	X	% Ortho	% Meta	% Para
o-OCH$_3$	Br		100	
m-OCH$_3$	Br		100	
p-OCH$_3$	Br		49±1	51±1
p-F	Br		20±1	80±1
m-CF$_3$	Cl		100	
o-CH$_3$	Br	48.5±2	51.5±2	
o-CH$_3$	Cl	45±4	55±4	
m-CH$_3$	Br	22±4	56±4	22±4
m-CH$_3$	Cl	40±4	52±4	8±4
p-CH$_3$	Cl		62±4	38±4

Adapted from Roberts et al. [9].

While we can understand the factors that control regiochemistry of benzyne reactions in these systems, it is important to highlight that mixtures of products are often observed. As such, most other methods for benzyne formation are regiospecific, thereby avoiding one potential source of poor regioselectivity. Something else that is not readily apparent is that the yields of these substitution reactions are often modest, which provides the impetus for exploring alternative ways of generating benzynes.

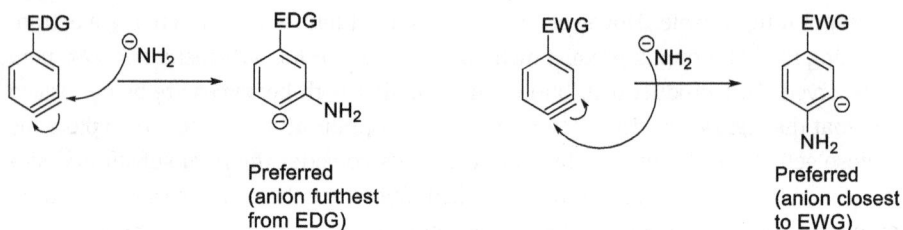

Figure 5.7: Electronic effects on nucleophilic addition on substituted benzynes.

5.2.2 Aryne formation from dihalobenzenes

An alternative approach to forming arynes is using *o*-dihaloarenes. The triple bond is formed between the carbons bearing the halogens, so the regiochemistry of benzyne formation is well defined. Another key distinction when compared to the elimination of HX discussed above is that there is no excess nucleophile present to do the addition step. This difference means that other reagents can be used, opening up the scope of reactivity of the arynes.

Early studies often made use of *o*-bromofluorobenzene. The benzyne is generated either using a lithium mercury amalgam or using magnesium to form an organometallic species that undergoes subsequent elimination of fluoride (Figure 5.8) [13].

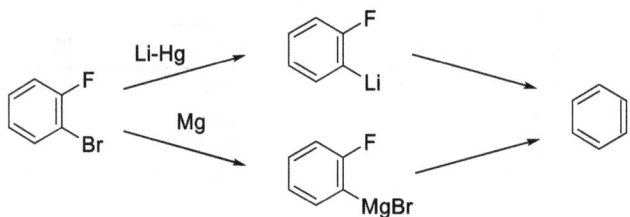

Figure 5.8: Formation of benzyne from *o*-bromofluorobenzene.

A more common alternative to bromofluorobenzenes is to use more readily available *o*-dibromo aromatic compounds. Treating these compounds with alkyllithiums such as butyllithium results in lithium–halogen exchange followed by elimination (Figure 5.9). This approach is frequently used to generate arynes.

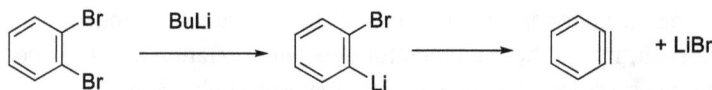

Figure 5.9: Formation of benzyne from *o*-dibromobenzene.

5.2.3 Aryne formation by elimination of small molecules

Several of the methods for benzyne formation involve in the elimination of small gaseous molecules such as N_2, CO_2, or SO_2. The most widely used of these methods involves the reaction of an o-aminobenzoic acid with sodium nitrite or isoamyl nitrite to form the diazonium salt (see Section 3.6.1) [14]. With heating, the compound loses CO_2 and N_2 to form the corresponding benzyne (Figure 5.10).

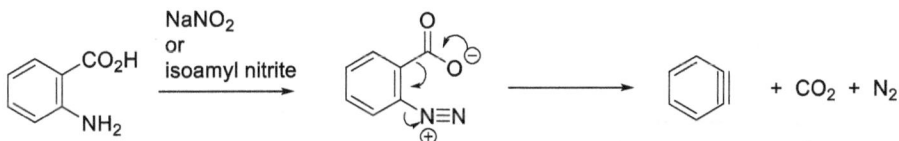

Figure 5.10: Formation of benzyne via the diazonium carboxylate [14].

In some cases, the diazonium salt is formed in situ, while in other cases it can be isolated as a solid. By virtue of the fact that the aryne can be generated thermally without the need for other reagents, the benzyne precursor can be combined with other compounds and the reaction with the aryne occurs readily once it is generated thermally.

As another example of benzyne formation by elimination of small molecules, 1-aminobenzotriazole can be oxidized to the corresponding nitrene in the presence of $Pb(OAc)_4$, which is then followed by benzyne formation accompanied by the loss of two nitrogen molecules (Figure 5.11) [15].

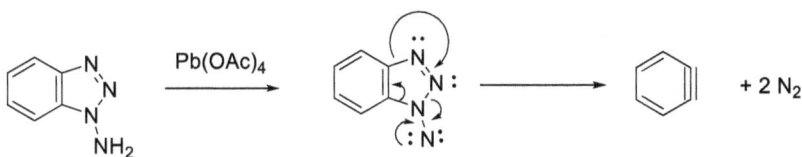

Figure 5.11: Generation of benzyne from 1-aminobenzotriazole.

The reaction has the advantage that it generates benzyne readily even at low temperatures. However, the preparation of 1-aminobenzotriazole requires several steps, so its applicability for the preparation of substituted benzynes. The synthesis of 1-aminobenzotriazole is outlined in Figure 5.12 [15]. o-Nitroaniline is reacted with sodium nitrite in acidic conditions to form the diazonium salt, which is reacted with diethyl malonate to form the corresponding hydrazone. The nitro group is then reduced by hydrogenation, and the resulting amine is converted to the diazonium salt, which undergoes an intramolecular nucleophilic cyclization with the hydrazone to give the benzotriazole. Finally, hydrolysis of the hydrazone under acidic conditions gives 1-aminobenzotriazole.

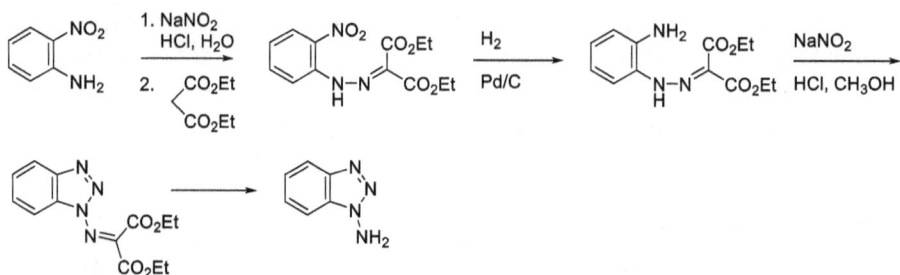

Figure 5.12: Synthesis of 1-aminobenzotriazole [15].

In a related example, benzothiadiazole-1-1-dioxide can serve as a benzyne precursor, decomposing thermally to produce benzyne, nitrogen, and sulfur dioxide (Figure 5.13) [16, 17]. As in the case of 1-aminobenzotriazole, the preparation of the benzyne precursors requires several steps, limiting their utility for the preparation of substituted arynes.

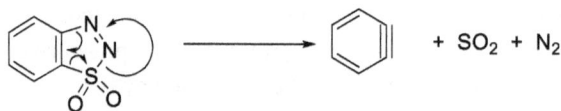

Figure 5.13: Preparation of benzyne from benzothiadiazole-1-1-dioxide.

5.2.4 Fluoride-induced aryne formation

Several of the approaches for the generation of arynes require the harsh reaction conditions such as strongly basic conditions or high temperatures, which can pose problems in terms of functional group compatibility. More recently, several methods for generating arynes under mild conditions have been developed. Most notably, Kobayashi reported the preparation of arynes from o-silyl aryl triflates in the presence of fluoride (Figure 5.14) [18, 19].

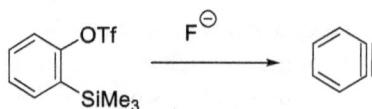

Figure 5.14: Aryne formation from o-silyl aryl triflates in the presence of fluoride.

Other similarly mild approaches include the reaction of o-bis(trimethylsilyl) benzene with (diacetoxyiodo)benzene and triflic acid to form (phenyl)[o-(trimethylsilyl)phenyl]iodonium triflate, which can be used to generate benzyne in the presence of fluoride Figure 5.15) [20].

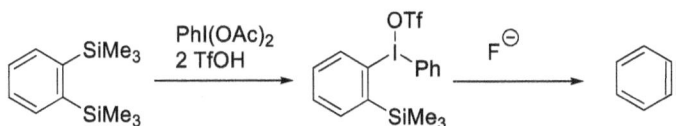

Figure 5.15: Benzyne formation via the (phenyl)[o-(trimethylsilyl)phenyl]iodonium triflate [20].

o-Silyl aryl triflates are now the most widely used aryne precursors because the aryne can be formed under mild conditions (a fluoride source) that are very tolerant to a variety of functional groups [19]. The o-silyl aryl triflates are typically prepared from the corresponding o-bromophenols. In Kobayashi's original report, o-bromophenol was treated twice with n-BuLi and trimethylsilyl chloride to silylate the phenol and then perform a lithium–halogen exchange to silylate the benzene ring. Subsequently, cleavage of the trimethylsilyl ether formed the phenoxide, which was trapped with triflic anhydride (Figure 5.16) [18].

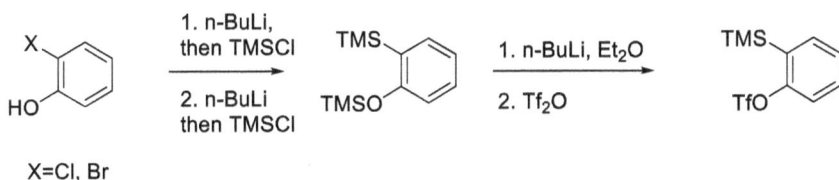

Figure 5.16: Kobayashi's synthesis of o-silyl aryl triflates [18].

A modified approach was developed by Guitián and coworkers, where they silylated o-bromophenol using hexamethyldisilazane [21]. Treatment with n-BuLi resulted in a lithium–halogen exchange, which was followed by a silyl migration (analogous to a retro-Brook reaction), generating the phenoxide which could be trapped with triflic anhydride (Figure 5.17).

In 2009, Garg and co-workers reported an alternative approach for preparing o-silyl aryl triflates from the corresponding phenol using a directed *ortho* metalation (DoM) approach (Figure 5.18) [22]. They prepared a carbamate from phenol which was used in a DoM reaction and trapped with trimethylsilyl chloride. The carbamate directing group was then removed and the resulting phenol was trapped in situ to form the triflate using PhNTf$_2$.

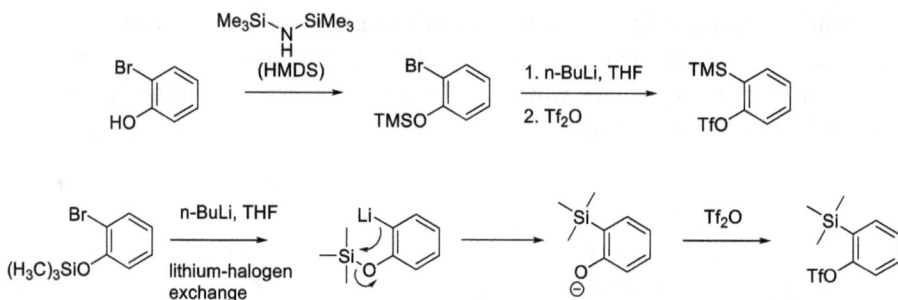

Figure 5.17: Synthesis of aryne precursors via a silyl migration [21].

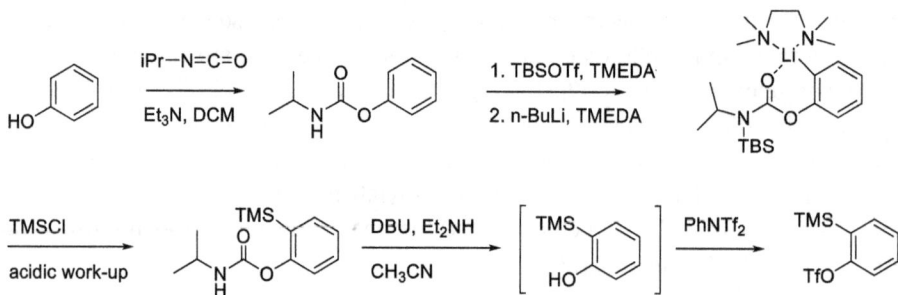

Figure 5.18: Synthesis of o-silyl aryl triflates via directed *ortho* metalation [22].

The most common fluoride sources for the generation of arynes from o-silyl aryl triflates are KF, CsF, or tetrabutylammonium fluoride (TBAF), with the latter two being the most commonly used. The choice of fluoride source depends in part on the solvent being used for the reaction. The most common solvents include CH_3CN, THF, or CH_2Cl_2. A common combination is CsF in CH_3CN, where the solubility of CsF is sufficient to allow aryne formation. For reactions carried out in THF, inorganic fluoride salts have low solubility, so TBAF is often used, or KF in the presence of 18-crown-6 ether. In these reactions, the goal is not rapid and quantitative aryne formation in solution, but rather to generate the aryne at a steady rate at low concentration [19].

While in most cases the solvent does not affect the reaction outcome, there are some cases where solvents that are thought to be "aryne compatible" actually participate in the reaction. For example, Suh and Chenoweth reported an unexpected reaction to the solvent when either CH_2Cl_2 or CH_3CN were used, leading to a solvent adduct as the major product and the desired product in very low yield (Figure 5.19) [23]. In this example, following initial nucleophilic attack by the phthalizine on the benzyne, the phenyl anion intermediate reacts with the solvent in an acid–base reaction. The solvent anion then serves as a nucleophile, attacking the heteroaromatic ring. While in many cases, reactions with arynes are readily carried out without interference by the solvent, this example does serve as a cautionary tale.

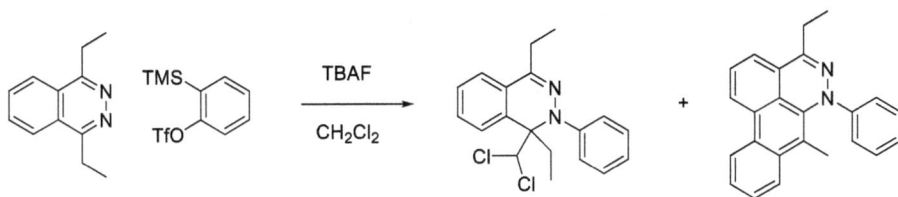

Figure 5.19: Unexpected reaction of arynes involving dichloromethane [23].

5.3 Reactions of arynes

Arynes can undergo a wide variety of reactions, ranging from cycloadditions, nucle-ophilic additions, insertion reactions, and a number of transition metal-catalyzed reactions (Figure 5.20).

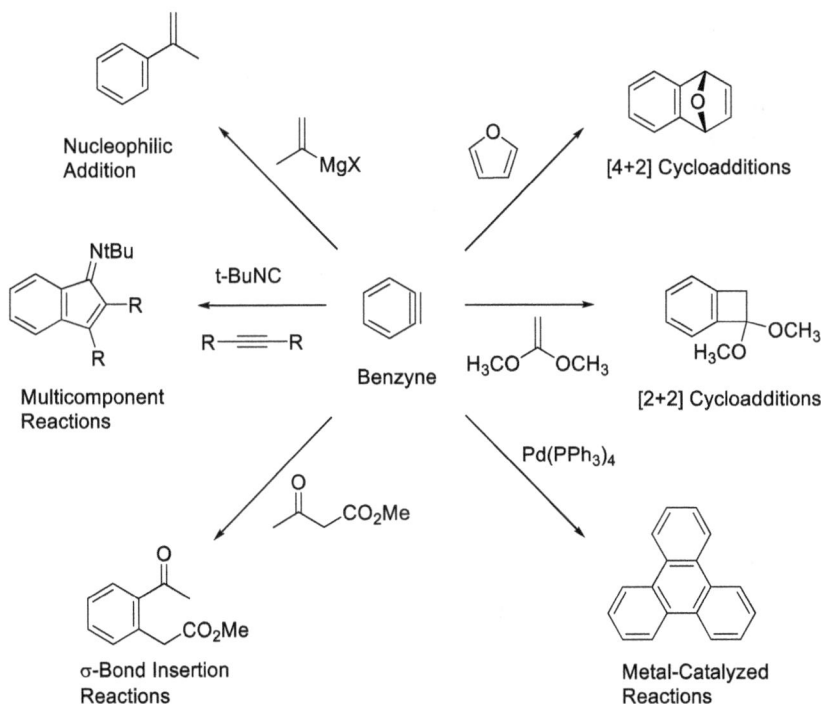

Figure 5.20: Summary of reactions of arynes.

5.3.1 Cycloaddition reactions

5.3.1.1 [4 + 2] Cycloaddition reactions

One of the most prevalent and widely studied reactions of arynes is as dienophiles in Diels–Alder reactions. For example, arynes will react with dienes such as furan or and isobenzofurans to form bicyclic adducts (Figure 5.21) [24].

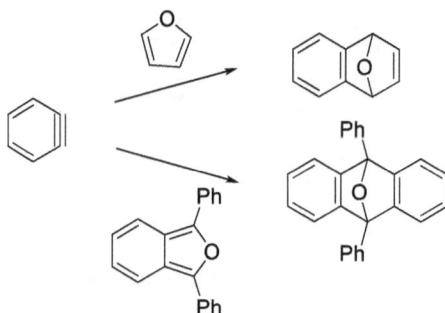

Figure 5.21: Examples of benzyne as a dienophile in Diels–Alder reactions.

The synthesis of bicyclic adducts via aryne cycloaddition reactions has been used for the synthesis of natural products. For example, the reaction of the substituted benzyne with 2-methoxyfuran gave the substituted hydroxynaphthalene, which was used for the synthesis of gilvocarcin natural products (Figure 5.22) [25, 26]. Because 2-methoxyfuran was used as the diene, the initial bicyclic adduct undergoes a ring opening and aromatization to give the hydroxynaphthalene. It is also noteworthy that the aryne cycloaddition reaction occurs regioselectively to place the methoxy group near the alkoxy group of the aryne. We will explore more details of this synthesis in Chapter 6.

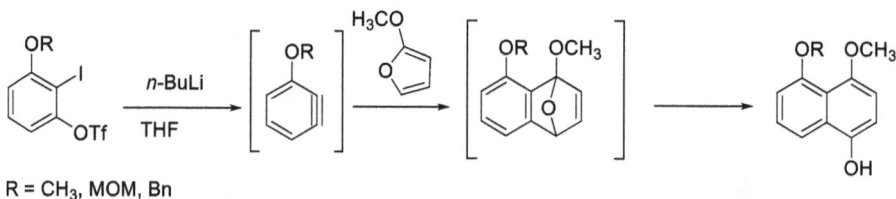

R = CH$_3$, MOM, Bn

Figure 5.22: [4+2] Cycloaddition reaction of arynes with 2-methoxyfuran to form substituted naphthalenes [25].

Diels–Alder reactions of benzynes have also been used to prepare polycyclic aromatic compounds. For example, the naphthalyne generated from 2,3-bis(trimethylsilyl) naphthalene reacts with tetraphenylcyclopentadienone to form the corresponding tetraphenylanthracene (Figure 5.23) [27]. We will explore these reactions used to construct aromatic rings and in particular for the synthesis of polycyclic aromatic hydrocarbons in Chapters 8–10.

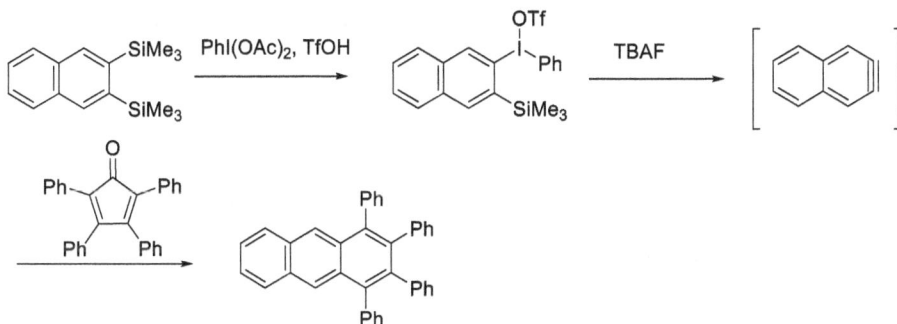

Figure 5.23: Diels–Alder reaction of naphthalyne with tetraphenylcyclopentadienone [27].

Benzyne can also react with anthracene in a [4+2] cycloaddition to form triptycene. Indeed, one of the early reported syntheses of triptycene was reported by Wittig using benzyne formed from an *ortho*-fluoro Grignard reagent (Figure 5.24) [28].

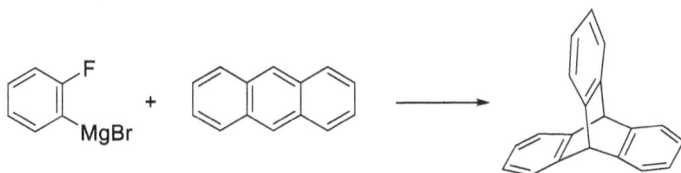

Figure 5.24: Reaction of benzyne with anthracene to form triptycene reported by Wittig [28].

As an example of the Diels–Alder reaction of arynes applied to the synthesis of natural products, Guitián and coworkers used a benzyne cycloaddition for the formation of the polycyclic framework of dynemicin A (Figure 5.25) [29]. The substituted pyrone served as the diene in a reaction with benzyne, which had been generated by decomposition of the diazonium carboxylate. The initial adduct undergoes a cheletropic reaction where CO_2 is lost and aromatization occurs to give the fused naphthalene ring.

Arynes can also react with acyclic dienes to form dihydronaphthalene derivatives. Depending on the substitution pattern of the diene, this reaction leads to the formation of chirality centers. Using a chiral auxiliary allowed the reaction to be carried out in a highly diastereoselective manner (Figure 5.26) [30].

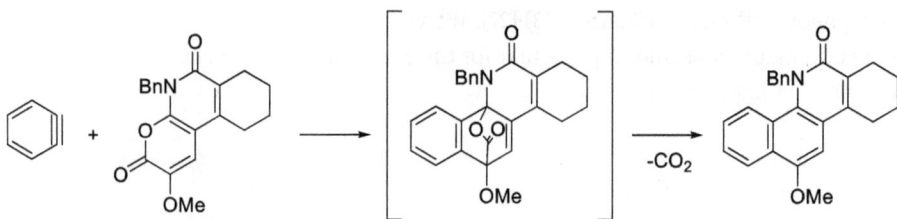

Figure 5.25: Benzyne cycloaddition in the synthesis of the framework of dynemicin A [29].

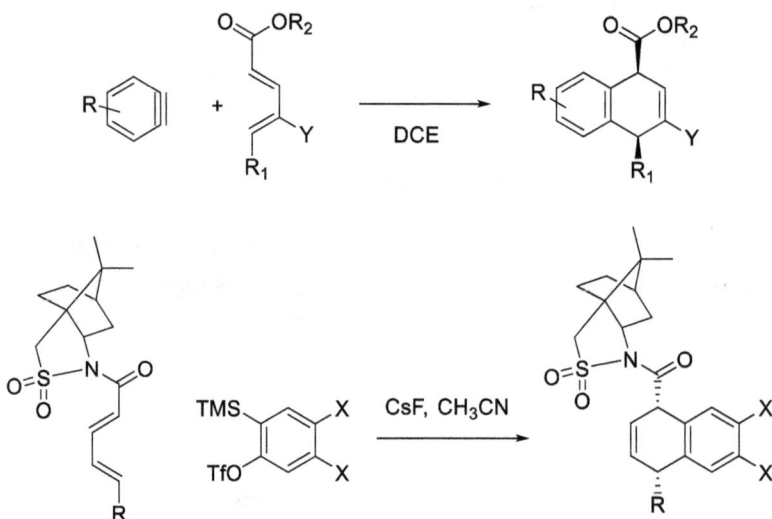

Figure 5.26: Diastereoselective benzyne cycloaddition [30].

5.3.1.2 Intramolecular Diels–Alder reactions

Arynes have also been used for intramolecular cycloaddition reactions for the construction of polycyclic natural products. For example, an intramolecular aryne Diels–Alder reaction was used to from the isoquinolone framework found in lycorine natural products (Figure 5.27) [31]. In this example, the aryne is generated by elimination of HX using an excess LDA, which undergoes a hetero-Diels–Alder reaction with the aryl amide, followed by oxidation to rearomatize and produce the tetracyclic structure, which could subsequently be converted into lycorine natural products.

Similarly, an intramolecular aryne Diels–Alder has been used to form the indole skeleton in the synthesis of ergot alkaloids such as lysergic acid (Figure 5.28) [32]. Again, the benzyne was generated by elimination of HX using LDA and then underwent an intramolecular Diels–Alder reaction – this time with a vinyl pyridine

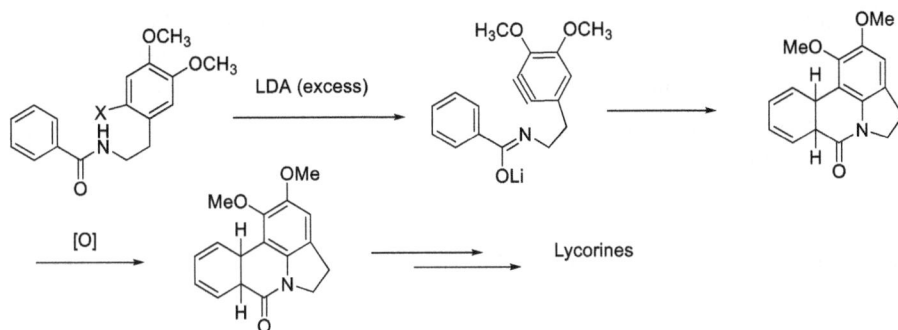

Figure 5.27: Intramolecular aryne cycloaddition reaction strategy for the preparation of lycorines [31].

group as a diene. The Diels–Alder reaction resulted in the formation of the desired carbon skeleton, which was converted into lysergic acid in several steps.

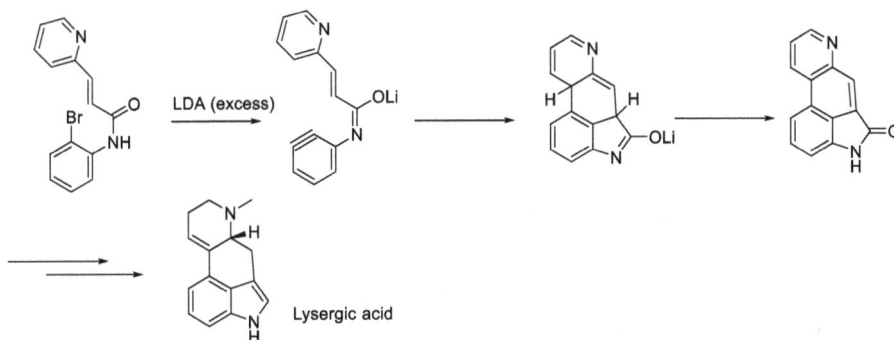

Figure 5.28: Intramolecular Diels–Alder reaction of an aryne for the preparation of lysergic acid [32].

5.3.1.3 [2+2] Cycloadditions

Arynes can undergo [2+2] cycloadditions. For example, arynes can dimerize to form the corresponding o-biphenylenes (Figure 5.29) [33].

Figure 5.29: [2+2] dimerization of benzyne to form biphenylene.

Arynes are also known to undergo [2+2] cycloadditions with electron-rich alkenes to produce benzocyclobutenes, which are useful intermediates for further transformation (e.g., by ring-opening reactions). For example, benzynes generated by treatment of bromobenzenes with sodium amide undergo [2+2] cycloaddition with

1,1-dimethoxyethylene to produce the corresponding dimethoxy benzocyclobu-tenes (Figure 5.30) [34, 35]. In the examples below, where the benzynes formed are substituted, there is the potential for the formation of more than one regioisomer. The formation of the observed products can be rationalized by considering the di-methoxyalkene as a nucleophile, where there is at least some polar character to the reaction. Indeed, one could draw the reaction as a polar mechanism with nu-cleophilic attack followed by intramolecular trapping of the aryl anion intermedi-ate. Regardless of whether the mechanism is a concerted reaction or a stepwise polar reaction, the regiochemistry is governed by the stabilization of developing partial negative charge during the reaction, similar to the discussion of regiochem-istry of benzyne reactions in Section 5.2.1.

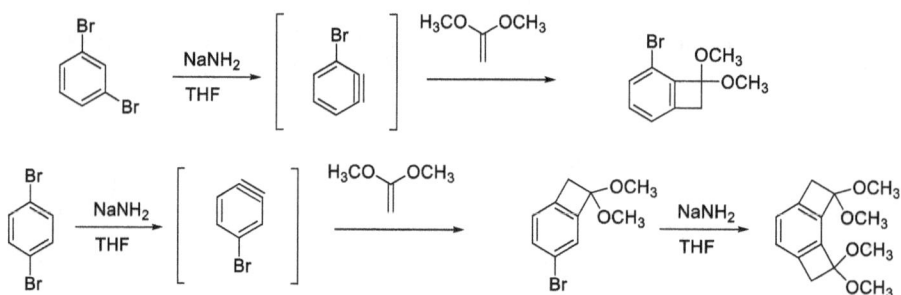

Figure 5.30: Examples of [2+2] cycloadditions of arynes with electron-rich alkenes.

Arynes have been shown to undergo [2+2] cycloaddition reactions with enamines. For example, benzyne generated from the o-fluorophenylmagnesium bromide reacts with the cyclohexyl enamine shown to give a mixture of products (Figure 5.31) [36].

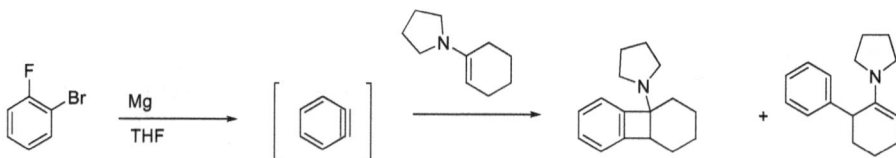

Figure 5.31: Reaction of benzyne with an enamine [36].

The first product is the result of an apparent [2+2] cycloaddition, while the second product could result from an ene reaction. Alternatively, these products could result from a polar mechanism involving nucleophilic addition of the enamine. The polar mechanism and concerted cycloaddition reaction mechanisms are outlined in Fig-ure 5.32. Gingrich and coworkers suggest that the [2+2] cycloaddition is not con-certed, while the formation of the substituted enamine is the result of a concerted ene reaction [36].

concerted pathways

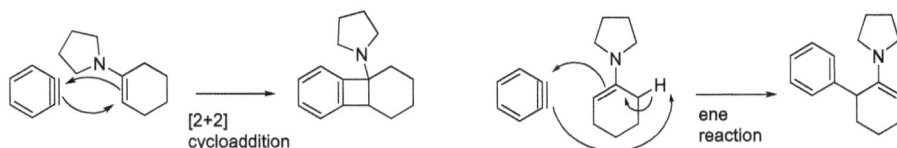

polar pathways

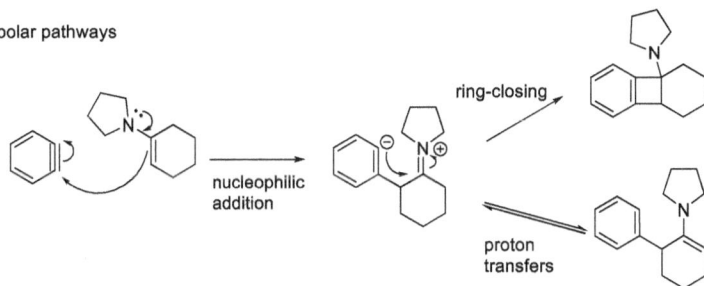

Figure 5.32: Possible mechanistic pathways the reactions of arynes with enamines.

5.3.1.4 1,3-Dipolar cycloadditions

Arynes can also react in 1,3-dipolar cycloadditions to form five-membered hetero-cycles. Two examples of 1,3-dipolar cycloadditions to give five-membered benzo-fused heterocycles are shown in Figure 5.33 [37, 38].

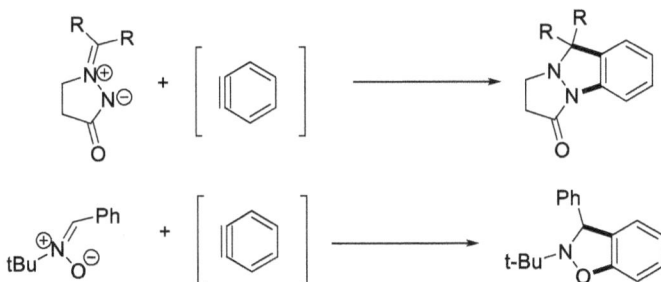

Figure 5.33: Examples of 1,3-dipolar cycloadditions of arynes.

The groups of Larock and Yamamoto showed that 1,3-dipolar acycloadditions of ben-zynes with diazo compounds could be used to prepare indazoles and *N*-arylindazoles (Figure 5.34) [39, 40]. The aryne was generated from the *o*-trimethylsilyl aryl triflate in the presence of fluoride and could react with a variety of diazo compounds.

Interestingly, the either of the products could be favored depending on the reac-tion conditions: when TBAF or KF/18-crown-6 was used as the fluoride source in THF, the indazole was favored, while the use of CsF in acetonitrile favored the *N*-aryl-indazoles. The mechanism of the reaction consists of the 1,3-dipolar cycloaddition,

Figure 5.34: Formation of indazoles by 1,3-dipolar cycloaddition of diazo compounds with an aryne.

followed by a hydrogen shift. Under conditions where there is a high concentration of benzyne, the initial indazole product can react with a second equivalent of the benzyne in a nucleophilic addition reaction, which upon proton transfer gives the *N*-arylindazole (Figure 5.35) [40].

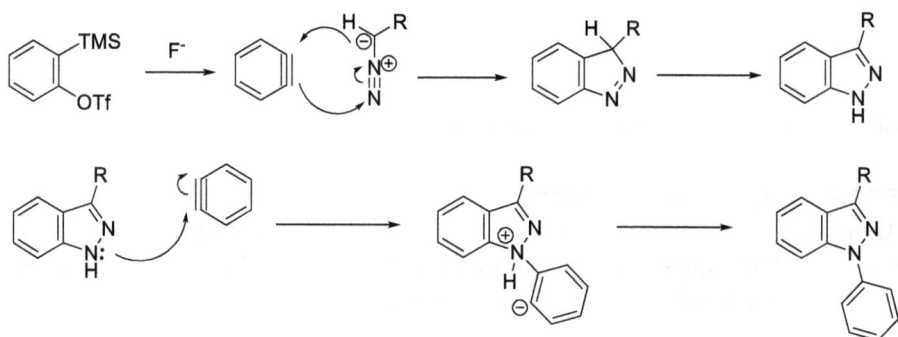

Figure 5.35: Mechanism of indazole formation.

Larock and coworkers showed that this type of reaction could be applied for the synthesis of benzotriazoles by reacting arynes with organic azides [41]. Again, the arynes were generated under mild conditions from the *o*-trimethylsilyl aryl triflates using fluoride and reacted with a variety of azides to form the substituted benzotriazoles (Figure 5.36). Zhang and Moses extended this approach by generating the azide and benzyne in situ in one pot from the corresponding aniline and anthranilic acid in the presence of *t*-butyl nitrite and $TMSN_3$ [42].

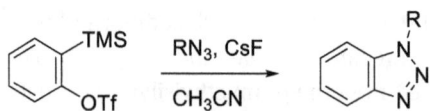

Figure 5.36: Formation of benzotriazoles by reaction of arynes with azides [41].

5.3.2 Reactions of arynes with nucleophiles

Arynes are electrophilic and as such are susceptible to attack by nucleophiles. The result of nucleophilic addition to the aryne is an aryl anion intermediate. Depending on the reaction, the aryl anion can be quenched with a proton to give simple nucleophilic addition, can react with an external electrophile to form two new bonds in a single step (referred to as a multicomponent reaction), or can undergo an intramolecular reaction with an electrophile (referred to as an insertion reaction). Both multicomponent and insertion reactions are particularly appealing because they allow the formation of two new bonds in a single step. These three scenarios are shown in Figure 5.37.

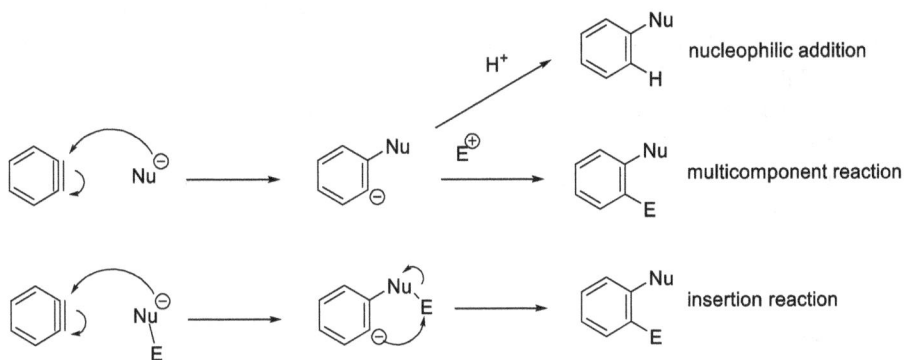

Figure 5.37: General description of reactions of arynes with nucleophiles.

5.3.2.1 Nucleophilic addition reactions

We have already seen that arynes formed by elimination of H–X react with nucleophiles such as sodium amide, resulting in an overall nucleophilic substitution. In these reactions, the nucleophilic attack on the aryne generates an aryl anion intermediate that is quenched with a proton. Because of the potential for formation of different regioisomers and the availability of other synthetic methods, intermolecular nucleophilic addition to arynes are not that widely used. However, the nucleophilic addition to arynes has been used as a strategy for intramolecular cyclizations in the synthesis of polycyclic natural products [6]. There are numerous examples of this strategy in natural products synthesis, so here we will consider only a few representative examples.

One of the earliest examples of the use of nucleophilic additions to arynes in the synthesis of natural products was Kametani's synthesis of cryptaustoline and cryptowoline [43]. A key step in the synthesis involved an intramolecular nucleophilic addition of an amine onto a benzyne that was generated by elimination of HBr using NaNH$_2$ (Figure 5.38).

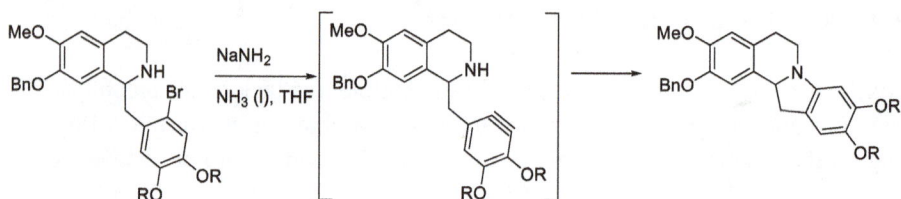

Figure 5.38: Intramolecular nucleophilic addition to a benzyne during the synthesis of cryptaustoline and cryptowoline [43].

In a more recent example, Sanz et al. made use of an aryllithium nucleophile that was generated by lithium–halogen exchange for the synthesis of trisphaeridine and N-methylcrinasiadine [44]. The approach is illustrated for the preparation of N-methylcrinasiadine in Figure 5.39. The starting material shown is treated with 3.5 equivalents of t-butyllithium, which results in benzyne formation by elimination of HF, as well as concomitant lithium–halogen exchange to generate the aryllithium nucleophile. Nucleophilic attack leads to cyclization, and quenching with methanol protonates the resulting aryl anion. Finally, oxidation in air leads to the formation of the target natural product.

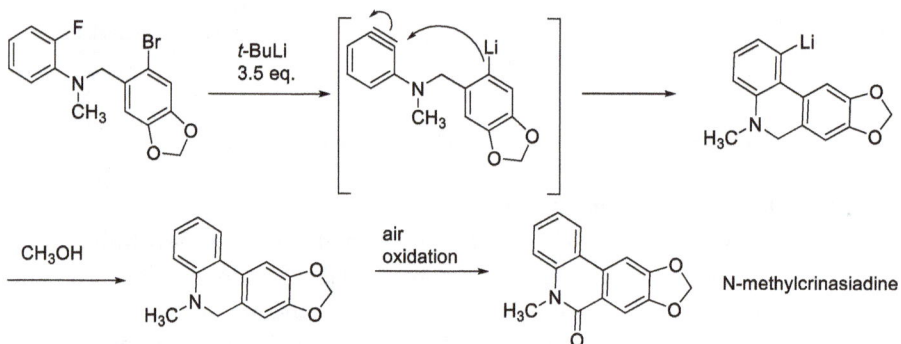

Figure 5.39: Synthetic approach to N-methylcrinasiadine [44].

As another example, Garg and coworkers used an intramolecular nucleophilic addition of an enolate onto an indolyne as part of their synthesis of welwitindolinone alkaloids (Figure 5.40) [45]. In the presence of sodium amide, the indolyne is formed, as well as the enolate. The enolate then acts as the nucleophile in an intramolecular attack on the indolyne, which is followed by protonation of the aryl anion. In this reaction, attack by the enolate oxygen is also observed as a minor product.

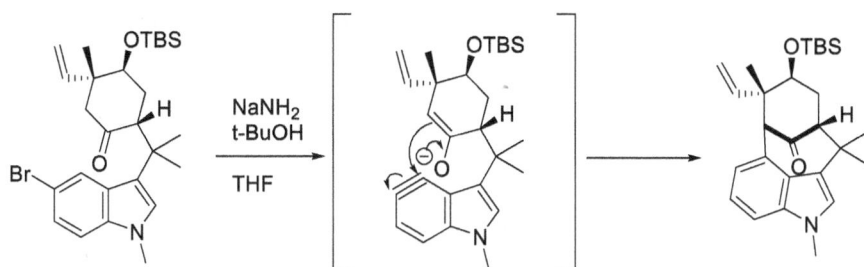

Figure 5.40: Intramolecular nucleophilic addition to an indolyne [45].

5.3.2.2 Multicomponent reactions of arynes

We have seen that nucleophilic addition to an aryne produces an aryl anion intermediate. In the examples we have examined thus far, the aryl anion is trapped with a
proton. However, if an electrophile is also present, the aryl anion can be trapped, ultimately introducing two substituents in one step (Figure 5.41). Because the reaction
involved at least three species (aryne, nucleophile, and electrophile), it is often referred to as a multicomponent reaction. The reaction presents an attractive approach
for the preparation of o-disubstituted benzenes. The scope of the nucleophile is fairly
broad, including amines, imines, N-heteroaromatics, phosphines, and isocyanides.
The electrophile is often a carbonyl compound such as an aldehyde or ketone.

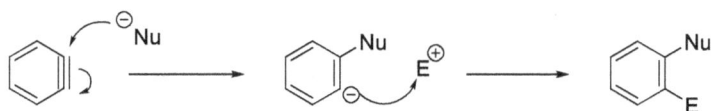

Figure 5.41: General concept of a multicomponent reaction of an aryne.

A multicomponent aryne reaction was used as part of Barrett's synthesis of ent-
clavilactone B [46]. In this reaction, an aryne was generated by elimination of HF
using butyllithium, which was reacted with a Grignard reagent, and the intermediate
reaction with an aldehyde electrophile, resulting in the formation of two aryl–carbon
bonds in one step (Figure 5.42).

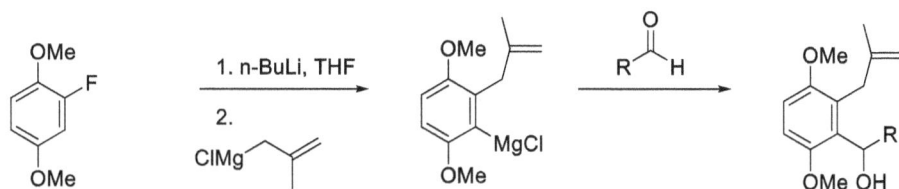

Figure 5.42: A multicomponent aryne reaction as part of the synthesis of ent-clavilactone B [46].

An interesting variation on the typical multicomponent reaction of arynes was reported by Biju and coworkers, who demonstrated a multicomponent reaction of a benzyne with a tertiary aniline nucleophile and aldehyde or ketone electrophile can be followed by an aryl migration similar to a Smiles rearrangement (Figure 5.43) [47].

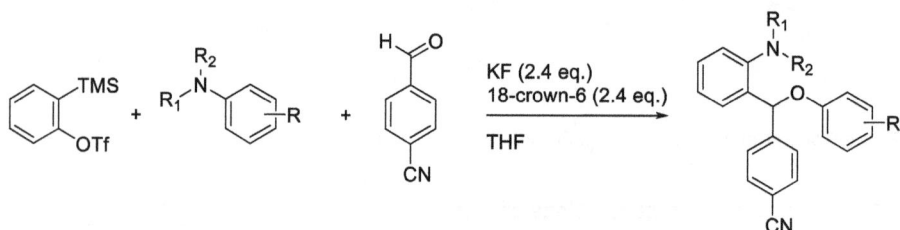

Figure 5.43: A multicomponent aryne reaction with an aniline and an aldehyde.

The proposed mechanism begins with a standard multicomponent aryne reaction where the aniline nucleophile attacks the benzyne to generate the aryl anion intermediate, which reacts with the electrophilic aldehyde to form an oxy-anion intermediate (Figure 5.44). The oxy-anion then serves as the nucleophile for the Smiles-type rearrangement. An interesting feature of this reaction is that it proceeds even in the absence of electron-withdrawing groups to stabilize the Meisenheimer complex. The reaction is limited to tertiary anilines; when a secondary amine is used, the multicomponent reaction occurs, but no rearrangement is observed. This observation is attributed to a proton transfer from the ammonium group to the phenolate, essentially deactivating the nucleophile [47].

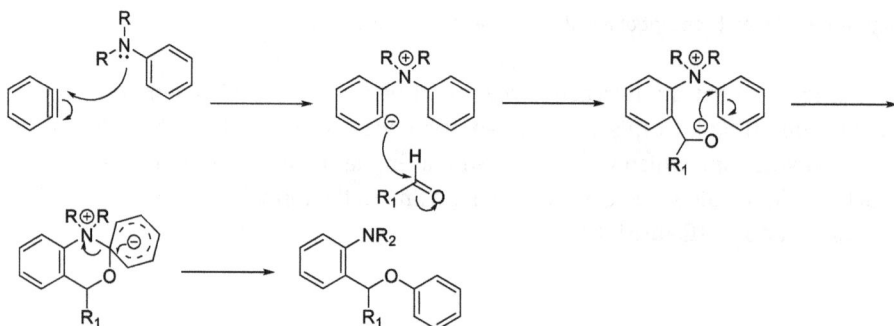

Figure 5.44: Proposed mechanism of multicomponent reaction with arynes, tertiary amines, and aldehydes with an amino-group migration [47].

5.3.2.3 Aryne insertion reactions

As described above, aryne insertion reactions essentially result from a reaction of an aryne with a nucleophile, followed by an *intramolecular* trapping of the aryl anion intermediate. Depending on the reactants, this may also be accompanied by further changes in bonding.

An early example of a formal σ-bond insertion reaction using a benzyne involved the reaction of 2-bromoanisole and diethyl malonate in the presence of sodium amide. The product of the reaction was converted into the natural product melleine in four steps (Figure 5.45) [48].

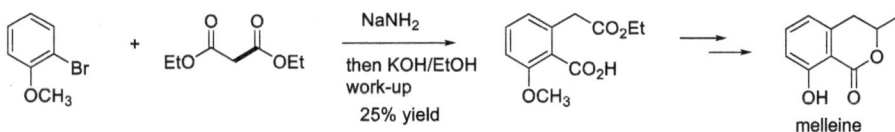

Figure 5.45: Aryne σ-bond insertion in the synthesis of melleine [48].

Formally, the reaction can be viewed as an insertion of the benzyne generated from 2-bromoanisole across the bond between the carbonyl carbon and alpha carbon of diethyl malonate. Mechanistically, it is likely a nucleophilic attack of the enolate of diethyl malonate, followed by an intramolecular nucleophilic acyl substitution (Figure 5.46) [48]. The low yield of the reaction can be attributed to the harsh reaction conditions, as well as low regioselectivity of the addition to the benzyne.

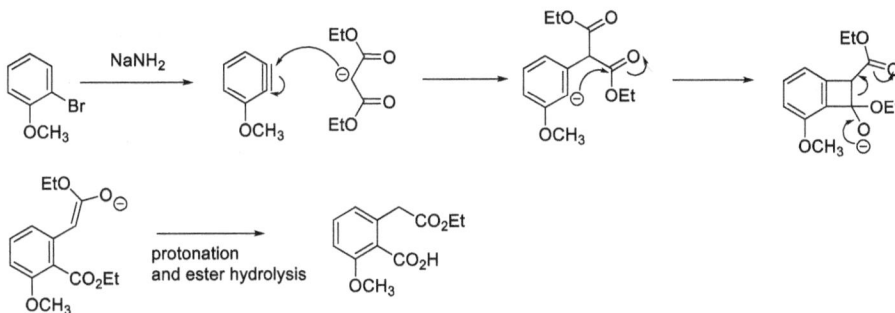

Figure 5.46: Mechanism of σ-bond insertion with diethyl malonate [48].

Another example of benzyne insertion into the σ-bond between a carbonyl carbon and the α-carbon is the reaction of benzyne with 2-methylcyclopentanone (Figure 5.47) [49].

This reaction takes place in the presence of excess base, where one equivalent is used for benzyne formation, while the second equivalent is used to generate the enolate. While formally this reaction is viewed an insertion of the benzyne onto the σ-bond between the carbonyl carbon and the α-carbon, mechanistically it consists of the enolate nucleophile attacking the benzyne to form the aryl anion intermediate, which then serves as the nucleophile for an intramolecular addition onto

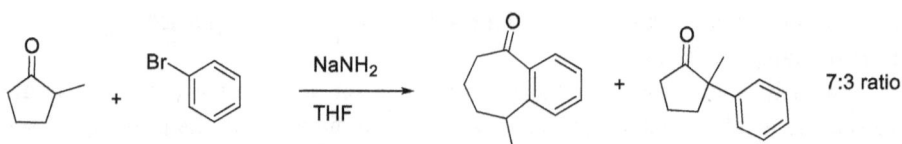

Figure 5.47: A σ-bond insertion reaction of benzyne with 2-methylcyclopentanone [49].

the carbonyl carbon (Figure 5.48). Reformation of the carbonyl π-bond and ring-opening of the cyclobutene gives the ring-expanded benzylic anion, which is protonated to yield the major product. The observed minor product is the result of protonation after the nucleophilic attack of the enolate on the aryne.

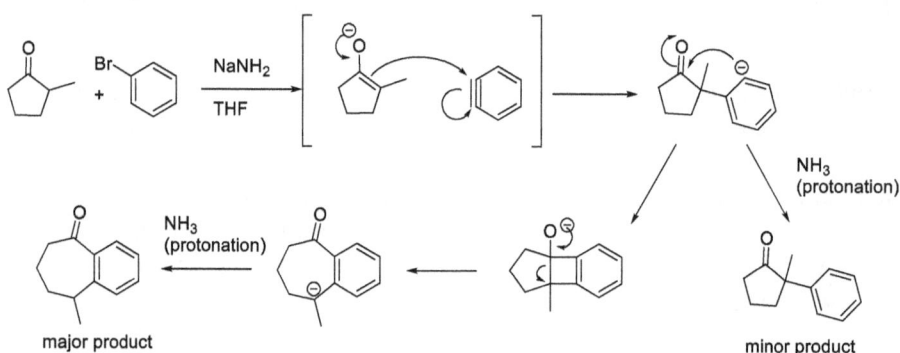

major product   minor product

Figure 5.48: Mechanism of σ-bond insertion reaction of benzyne with 2-methylcyclopentanone.

This type of formal σ-bond insertion is now improved with the use of aryl trimethylsilyl triflates that allows the benzynes to be generated under mild conditions (Figure 5.49) [50].

As another example of a σ-bond insertion reaction involving an aryne, consider the following synthesis of a substituted anthraquinone (Figure 5.50) [51].

In this reaction, LDA is used to deprotonate both reactants, leading to benzyne formation for one of the reactants and a nucleophilic anion α to the nitrile on the other (Figure 5.51). The nucleophilic anion adds to the benzyne and the resulting aryl anion attacks the ester carbonyl group. Finally, elimination of cyanide leads to the formation of the anthraquinone. The reaction is a variant of the Hauser annulation, where the aryne serves in place of a Michael acceptor [52]. The reaction mechanism highlights two important considerations. The first is that even though two regioisomeric benzynes could be formed, the benzyne formation is regioselective, giving only the benzyne adjacent to the methoxy group. This regioselectivity can be explained by the inductive electron-withdrawing effects of the methoxy group which favor deprotonation in the *ortho*-position. The second is that even though the final product does not reflect it clearly, the reaction still

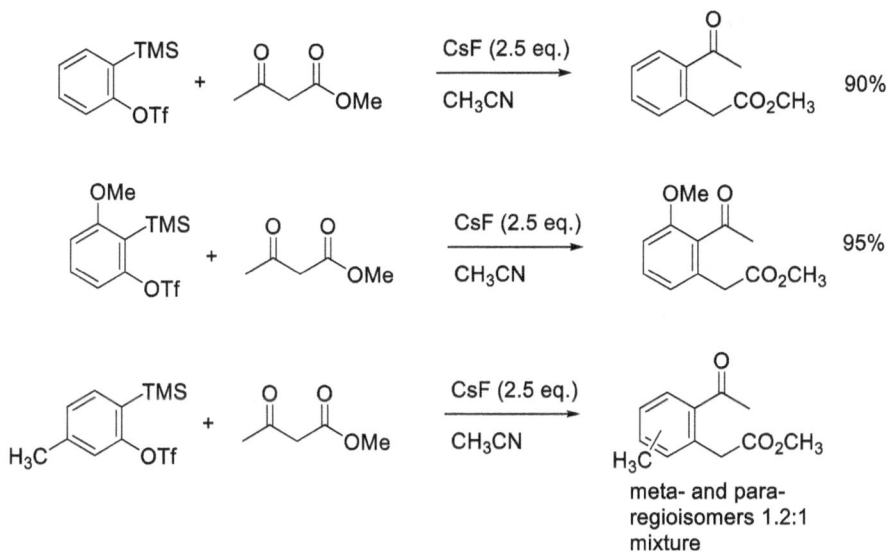

Figure 5.49: More examples of σ-bond insertions from trimethylsilylaryl triflates [50].

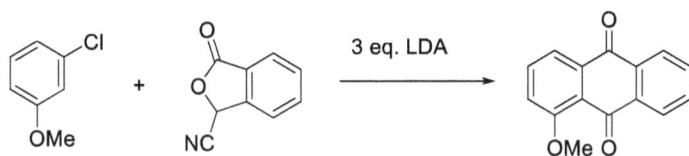

Figure 5.50: Synthesis of an anthraquinone by σ-bond insertion reaction of an aryne.

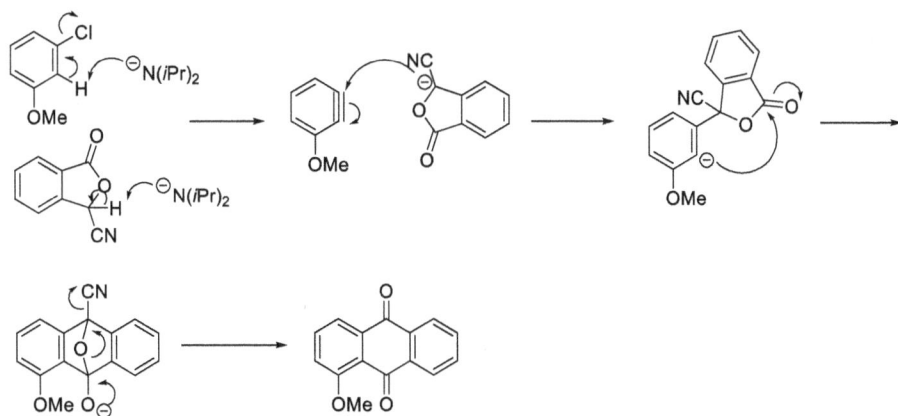

Figure 5.51: Mechanism of anthraquinone formation by aryne σ-bond insertion [51].

proceeds via nucleophilic addition to the benzyne, followed by intramolecular electrophilic trapping of the resulting aryl anion.

This methodology has been used for the preparation of some anthraquinone natural products. For example, it has been used for the preparation of rubiadin and damnacathol [53], we well as the synthesis of mumbaistatin (Figure 5.52) [54].

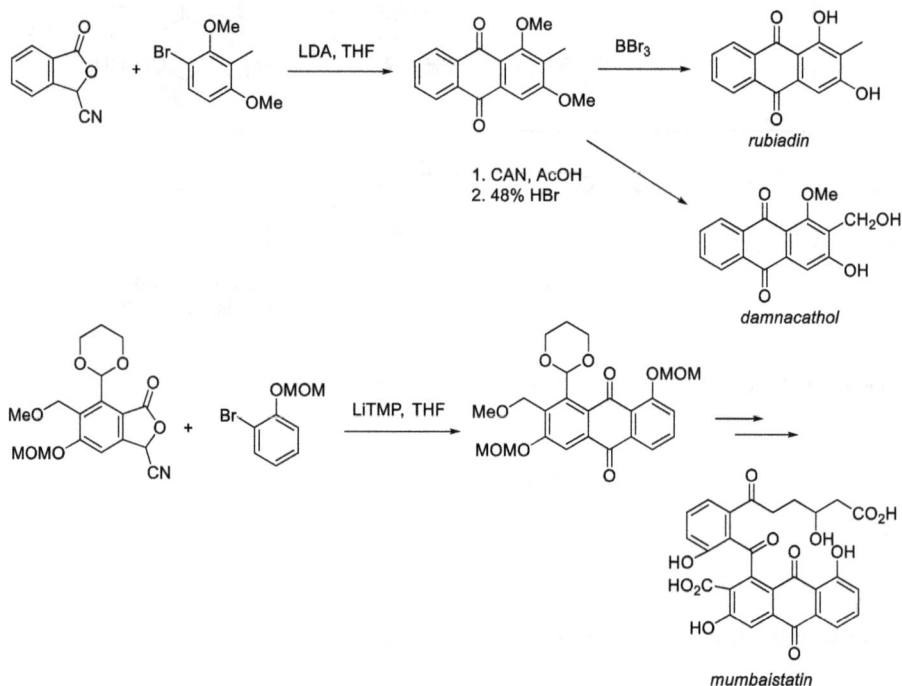

Figure 5.52: Synthetic approach to anthraquinone natural products via aryne insertion.

In an interesting variation on the reaction of nucleophiles with arynes, Greaney and coworkers reported the application of aryne chemistry in conjunction with the Truce–Smiles rearrangement (see Section 3.2.6) in order to form biaryls (Figure 5.53) [55]. This strategy presents a metal-free alternative for the formation of biaryls, in contrast with the transition metal-catalyzed coupling of organometallic species with aryl halides (which we will explore in Chapter 6). The overall reaction is shown below.

Mechanistically, nucleophilic addition of the sulfonamide nitrogen to the benzyne generates an aryl anion, which acts as the nucleophile in the Truce–Smiles reaction. In this case, collapse of the Meisenheimer intermediate is also accompanied by loss of SO_2 to yield the amino-substituted biaryl (Figure 5.54) [55].

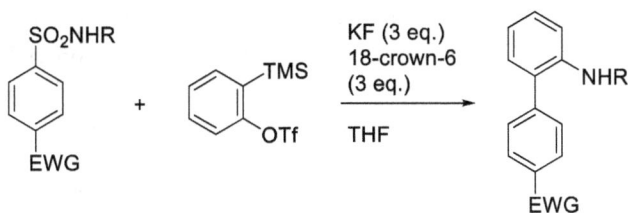

Figure 5.53: An example of the Truce–Smiles rearrangement with arynes [55].

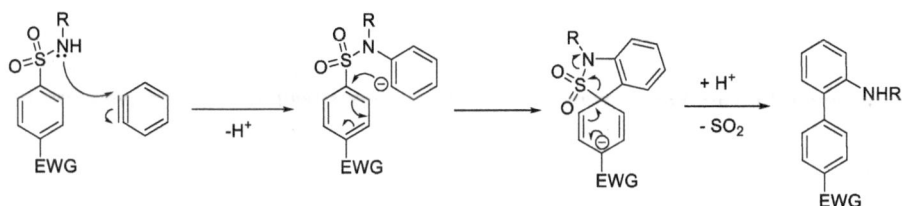

Figure 5.54: Proposed mechanism of the aryne Truce–Smiles rearrangement [55].

5.3.3 Transition metal-mediated reactions of arynes

A number of transition metal complexes with arynes have been reported (see examples and scope). These reactions suggest that arynes could potentially be involved in transition metal-mediated reactions.

In 1998, Peña et al. reported a palladium-catalyzed aryne cyclotrimerization to form fused polycyclic aromatic systems [56]. Conceptually, this reaction is analogous to a Vollhardt alkyne cyclotrimerization – a reaction we will explore in Chapter 8 [57]. Using an aryne in the presence of a Pd(0) catalyst such as Pd(PPh$_3$)$_4$, they showed that arynes underwent cyclotrimerization in good yields (Figure 5.55). In the case of benzyne, triphenylene is produced. This reaction can be conducted with benzynes generated from *ortho*-dibromobenzenes in the presence of butyllithium, but better yields were reported when the *o*-silyl aryl triflates were treated with fluoride.

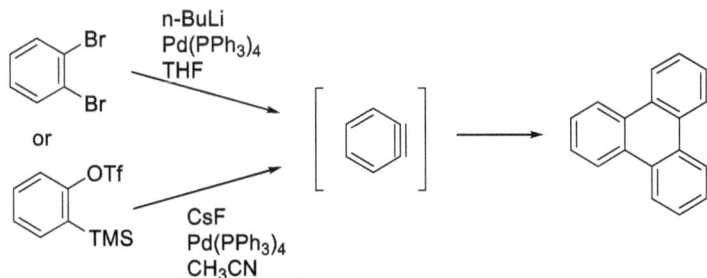

Figure 5.55: Palladium-catalyzed benzyne cyclotrimerization to form triphenylene [56].

The mechanism is thought to be similar to transition metal-catalyzed alkyne cyclotrimerization – a reaction we will see in Chapter 8. An overview of the proposed mechanism is outlined in Figure 5.56. The reaction involves coordination of two equivalents of the aryne to the palladium catalyst, which is followed by an intramolecular oxidative coupling to form the palladacycle. At this stage, there are two conceivable mechanisms. One plausible mechanism is that the palladacycle undergoes a cycloaddition with a third equivalent of the aryne, followed by loss of palladium and rearomatization. Alternatively, a third equivalent of the aryne may coordinate and insert between one of the carbon–palladium bonds to form a palladacycloheptatriene. This intermediate can then undergo an electrocyclic ring closing, followed by reductive removal of palladium to give the triphenylene and regenerate the palladium(0).

Palladium-catalyzed aryne cyclotrimerization has been used extensively for the preparation of substituted polycyclic aromatic hydrocarbons. We will see more examples of the applications of aryne cyclotrimerization for the synthesis of PAHs in Chapters 9–11.

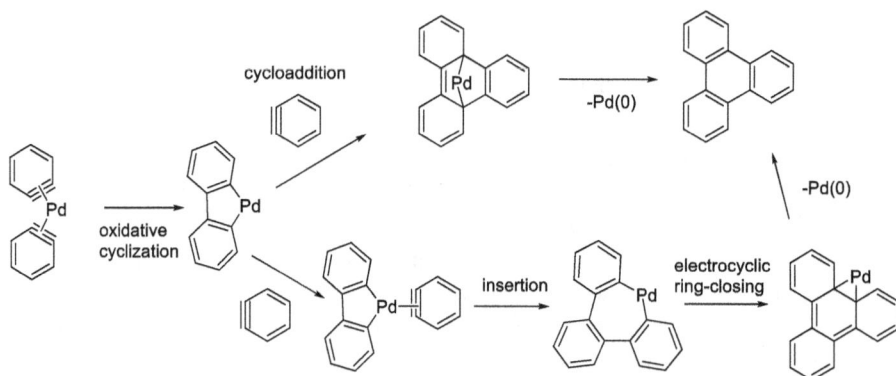

Figure 5.56: Potential mechanism for aryne cyclotrimerization. (Note: ancillary ligands on Pd have been removed for simplicity.)

The aryne cyclotrimerization was also shown to work in the presence of alkynes in a co-cyclotrimerization to produce either substituted naphthalenes or substituted phenanthrenes (Figure 5.57) [58, 59]. With five equivalents of dimethyl acetylenedicarboxylate, the reaction was found to favor the phenanthrene product when Pd(PPh$_3$)$_4$ was used (84:7 ratio), but favor the naphthalene product when Pd$_2$(dba)$_3$ was used (10:83 ratio).

This aryne–alkyne cocyclotrimerization has been applied as a key step in the total synthesis of taiwanin natural products (Figure 5.58) [60]. In the key step, a substituted aryne was generated from the trimethylsilyl aryl triflate and reacted with a diyne to form a substituted naphthalene framework, which was further elaborated to prepare taiwanins C and E.

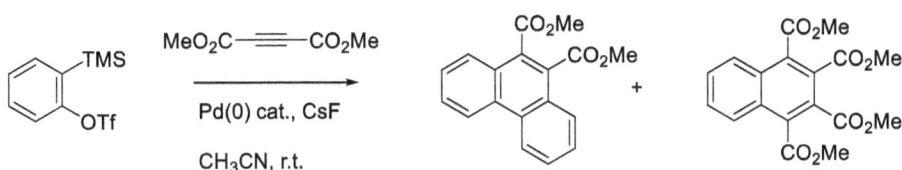

Figure 5.57: Aryne/alkyne co-cyclotrimerization to form phenanthrenes or naphthalenes [58].

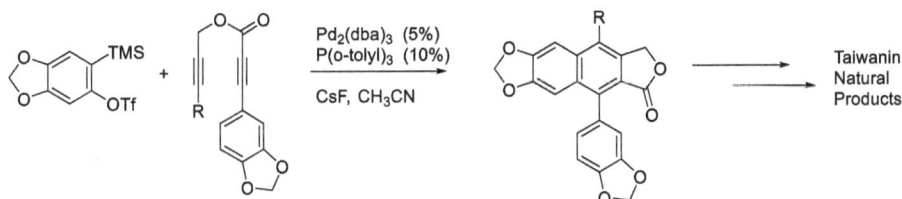

Figure 5.58: Aryne–alkyne co-cyclotrimerization as part of the synthesis of taiwanin natural products [60].

As another example of a transition metal-catalyzed reaction of arynes, they can undergo an annulation reaction with an o-halobenzaldehyde in the presence of a palladium catalyst to form substituted fluorenones (Figure 5.59) [61]. The reaction is carried out using a trimethylsilyl aryl triflate as the aryne precursor and an *ortho*-bromo- or iodo-benzaldehyde in the presence of CsF, Pd(dba)$_2$ and P(o-tolyl)$_3$ and can produce a variety of substituted fluorenones in moderate to good yields. One disadvantage of the reaction is that it uses an excess (five equivalents) of the aryne precursor to achieve good yields, which limits the utility to simple aryne precursors.

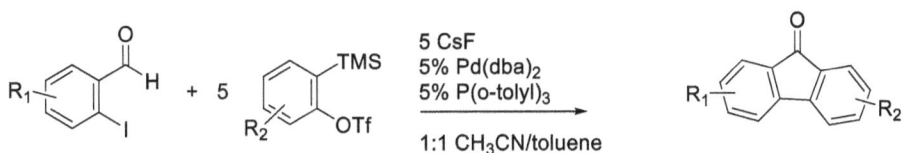

Figure 5.59: Palladium-catalyzed formation of fluorenones from arynes ando-halobenzaldehydes [61].

The mechanism of the reaction is not established, but Larock and coworkers have proposed several plausible mechanistic pathways [61]. For the sake of simplicity, one of the mechanistic pathways is shown in Figure 5.60. The mechanism involves oxidative addition of the aryl halide to Pd(0), given a Pd(II) intermediate. At this point, aryne coordination and migratory insertion leads to the formation of the aryl–aryl bond. Addition of the C–Pd bond across the carbonyl gives the palladium (II) alkoxide, which undergoes β-hydride elimination and reductive elimination to give the fluorenone product and regenerate the Pd(0) species. It is also possible

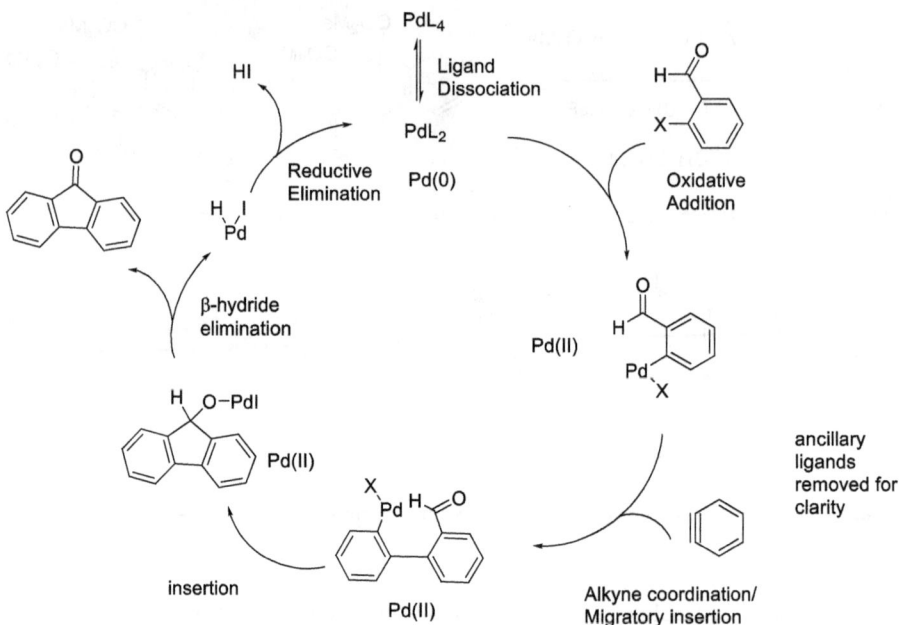

Figure 5.60: A plausible mechanistic pathway for formation of fluorenones from arynes and o-haloaldehydes. (Note: ancillary ligands on Pd are omitted for clarity.)

that the reaction first involves aryne coordination and oxidative insertion first to form a palladacycle, which then undergoes oxidative addition with the aryl halide to form a Pd(IV) intermediate. For a discussion of the other mechanistic pathways, refer to Waldo et al. [61].

In a very similar reaction, arynes react with o-haloamides in the presence of Pd (OAc)$_2$ to give substituted phenanthridine derivatives (Figure 5.61) [62]. Despite the use of a palladium(II) species as the initial catalyst, the reaction mechanism likely involves palladium(0) generated by in situ reduction of the palladium(II) species. As with the previous example of the fluorenone synthesis, the mechanism is not clearly established, and the mechanistic pathways proposed bear close resemblance to the previous example [62].

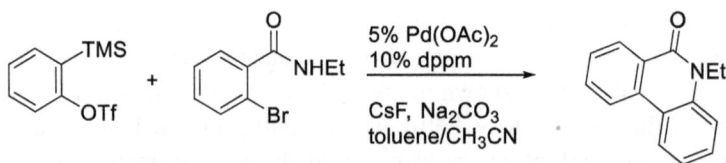

Figure 5.61: Palladium-catalyzed annulation of arynes and o-haloamides to yield phenanthridines [62].

5.4 *Meta*- and *para*-didehydrobenzenes (*m*- and *p*-arynes)

As we have discussed, benzyne is formally an *ortho*-didehydrobenzene, which is best represented as having a triple bond between adjacent carbons. It is also possible to have didehydrobenzenes where the hydrogen atoms are removed from *meta*- and *para*-substitutions. These species are sometimes referred to as *m*- and *p*-benzynes, although they are not readily represented as closed-shell species and viewed as diradical species (Figure 5.62).

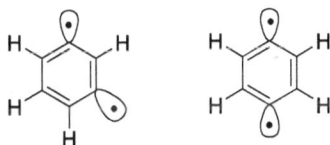

Figure 5.62: Representations of *m*- and *p*-benzynes.

The 1,4-diradical species is significant because it is implicated in the Bergman cyclo-aromatization reaction [63]. The Bergman cyclization involves a *cis*-enediyne, which upon heating can undergo a rearrangement. This reaction was first shown for the deuterated enediyne which undergoes scrambling of the deuterium labels upon heating (Figure 5.63) [64]. It was also shown that in the presence of a hydrogen atom source such as 1,4-cyclohexadiene, the 1,4-diradical could be trapped to form the corresponding benzene ring [63].

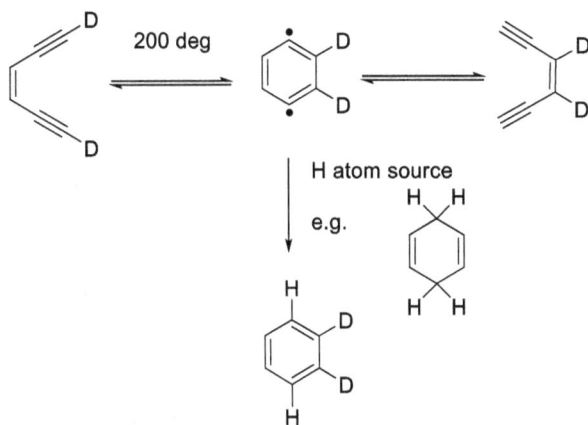

Figure 5.63: The Bergman cyclization reaction [63, 64].

The Bergman cyclization has not been used widely in synthesis despite its potential to form substituted benzenes from acyclic precursors. In one noteworthy exception, it was used to form naphthalenes and anthracenes from the corresponding diethynylbenzenes. Specifically, a Bergman cyclization of 1,2-bis(bromoethynyl)benzenes in the presence of 1,4-cyclohexadiene yielded the corresponding dibromonaphthalenes. Subsequent elaborations allowed the naphthalenes to be converted into the corresponding anthracenes (Figure 5.64) [65].

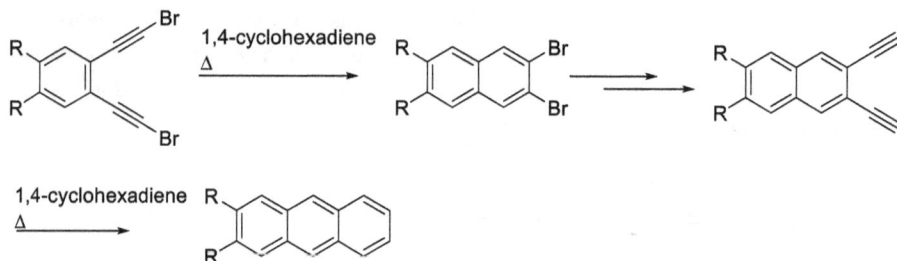

Figure 5.64: Synthesis of naphthalenes and anthracenes via Bergmann cyclization [65].

The significance of the Bergman cyclization is that it is implicated in the mechanism of cytotoxicity of a class of antibiotics that contains the enediyne motif. There are several natural products that contain an enediyne structure, including the dynemicins (Figure 5.65), calicheamicins, and esperamicins. The mechanism of cytotoxicity is thought to involve intercalation of the enediyne moiety into DNA, followed by Bergman cyclization to give the 1,4-diradical, which then reacts with DNA and leads to oxidative damage [66–68].

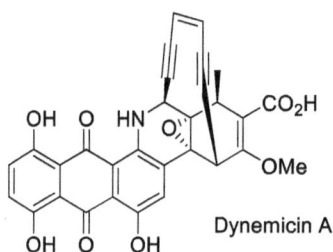

Dynemicin A

Figure 5.65: Dynemicin A – an example of an enediyne natural product.

References

[1]   Wittig G. Phenyllithium, the key to a new chemistry of organometallic compounds. Naturwissenschaftern. 1942;30:696–703.
[2]   Roberts JD, Simmons HE, Carlsmith LA, Vaughan CW. Rearrangement in the reaction of chlorobenzene-C-14 with potassium amide. J Am Chem Soc. 1953;75:3290–91.
[3]   Laing JW, Berry RS. Normal coordinates, structure, and bonding of benzyne. J Am Chem Soc. 1976;98(3):660–64.
[4]   Wenk HH, Winkler M, Sander W. One century of aryne chemistry. Angew Chem Int Ed. 2003;42 (5):502–28.

[5] Pellissier H, Santelli M. The use of arynes in organic synthesis. Tetrahedron. 2003;59:701–30.

[6] Tadross PM, Stoltz BM. A comprehensive history of arynes in natural product total synthesis. Chem Rev. 2012;112:3550–77.

[7] Gampe CM, Carreira EM. Arynes and cyclohexyne in natural product synthesis. Angew Chem Int Ed. 2012;51:3766–78.

[8] Wu D, Ge H, Liu SH, Yin J. Arynes in the synthesis of polycyclic aromatic hydrocarbons. RSC Adv. 2013;3:22727–38.

[9] Roberts JD, Vaughan CW, Carlsmith LA, Semenow DA. Orientation in animations of substituted halobenzenes. J Am Chem Soc. 1956;78:611–14.

[10] Huisgen R, Sauer J. Nucleophilic aromatic substitutions. XVII. Nucleophilic aromatic substitutions through arynes. Angew Chem. 1960;72:91–108.

[11] Bergstrom FW, Wright RE, Chandler C, Gilkey WA. The action of bases on organic halogen compounds. I. The reaction of aryl halides with potassium amide. J Org Chem. 1936;1:170–78.

[12] Bunnett JF, Kim JK. Relative reactivities of ammonia and amide ion in addition to 4-chlorobenzyne. J Am Chem Soc. 1973;95(7):2254–59.

[13] Wittig G, Pohmer L. The intermediate formation of dehydrobenzene. Chem Ber. 1956;89:1334–51.

[14] Friedman L, Logullo FM. Benzynes via aprotic diazotization of anthanilic acids: A convenient synthesis of triptycene and derivatives. J Am Chem Soc. 1963;85:1549.

[15] Campbell CD, Rees CW. Reactive intermediates. Part 1. Synthesis and oxidation of 1- and 2-aminobenzotriazole. J Chem Soc C. 1969;742–47.

[16] Wittig G, Hoffmann RW. Dehydrobenzene from 1,2,3-benzothiadiazole 1,1-dioxide. Chem Ber. 1962;95:2718–28.

[17] Wittig G, Hoffmann RW. 1,2,3-benzothiadiazole 1,1-dioxide. Org Syn. 1967;47:4–9.

[18] Himeshima Y, Sonoda T, Kobayashi H. Fluoride-induced 1,2-elimination of o-trimethylsilyl-phenyl triflate to benzyne under mild conditions. Chem Lett. 1983;12(8):1211–14.

[19] Shi J, Li L, Li Y. O-Silylaryl Triflates: A journey of Kobayashi Aryne precursors. Chem Rev. 2021;121(7):3892–4044.

[20] Kitamura T, Yamane M. (Phenyl)[o-(trimethylsilyl)phenyl]iodonium triflate. A new and efficient precursor of benzyne. J Chem Soc Chem Commun. 1995;983–84.

[21] Peña D, Cobas A, Pérez D, Guitián E. An efficient procedure for the synthesis of ortho-trialkylsilylaryl triflates: Easy access to precursors of functionalized arynes. Synthesis (Stuttg). 2002;(10):1454–8.

[22] Bronner SM, Garg NK. Efficient synthesis of 2-(trimethylsilyl)phenyl trifluoromethanesulfonate: A versatile precursor to o-benzyne. J Org Chem. 2009;74 (22):8842–43.

[23] Suh SE, Chenoweth DM. Aryne compatible solvents are not always innocent. Org Lett. 2016;18(16):4080–83.

[24] Harrison R, Heaney H, Lees P. Aryne chemistry-XI. Trapping agents for arynes produced from Grignard- and organolithium reagents. Tetrahedron. 1968;24(12):4589–94.

[25] Matsumoto T, Hosoya T, Katsuki M, Suzuki K. New efficient protocol for aryne generation. Selective synthesis of differentially protected 1,4,5-naphthalenetriols. Tetrahedron Lett. 1991;32(46):6735–6.

[26] Hosoya T, Takashiro E, Matsumoto T, Suzuki K. Total synthesis of the gilvocarcins. J Am Chem Soc. 1994;116(3):1004–15.

[27] Kitamura T, Fukatsu N, Fujiwara Y. (Phenyl)[3-(trimethylsilyl)-2-naphthyl]-iodonium triflate as a new precursor of 2,3-didehydronaphthalene. J Org Chem. 1998;63(23):8579–81.

[28] Wittig G, Ludwig R. Triptycene from anthracene and dehydrobenzene. Angew Chem. 1956;68:40.

[29] Escudero S, Pérez D, Guitián E, Castedo L. A new convergent approach to the polycyclic framework of dynemicin A. J Org Chem. 1997;62(10):3028–29.

[30] Dockendorff C, Sahli S, Olsen M, Milhau L, Lautens M. Synthesis of dihydronaphthalenes via aryne Diels-Alder reactions: Scope and diastereoselectivity. J Am Chem Soc. 2005;127 (43):15028–29.

[31] González C, Pérez D, Guitián E, Castedo L. Synthesis of lycorines by intramolecular aryne cycloadditions. J Org Chem. 1995;60(20):6318–26.

[32] Gomez B, Guitian E, Castedo L, New A. Approach to the basic skeleton of ergot alkaloids by intramolecular aryne cycloaddition. Synlett. 1992;(11):903–04.

[33] Logullo FM, Seitz AH, Friedman L. Benzenediazonium-2-carboxylate and biphenylene. Org Syn. 1968;48:12.

[34] Stevens RV, Bisacchi GS. An efficient and remarkably regioselective synthesis of benzocyclobutenones from benzynes and 1,1-dimethoxyethylene. J Org Chem. 1982;47 (12):2393–96.

[35] Maurin P, Ibrahim-Ouali M, Santelli M. Reinvestigation relative to the regioselectivity of the aryne cycloaddition. Synthesis of the tricyclo[6 2 0 02,5]-1,5,7-triene-3,10-dione. Tetrahedron Lett. 2001;42(46):8147–49.

[36] Gingrich HL, Huang Q, Morales AL, Jones M. Reactions of enamines with dehydro aromatic compounds. J Org Chem. 1992;57(14):3803–06.

[37] Taylor EC, Sobieray DM. "Bicyclobenzodiazepinones" from 3-oxo-1,2-diazetidinium hydroxide, inner salts. Tetrahedron. 1991;47(46):9599–620.

[38] Matsumoto T, Sohma T, Hatazaki S, Suzuki K. On the regiochemistry of cycloaddition of unsymmetrical aryne with nitrone remarkable effect of trialkylsilyl substituent. Synlett. 1993; (11):843–46.

[39] Jin T, Yamamoto Y. An efficient, facile, and general synthesis of 1H-indazoles by 1,3-dipolar cycloaddition of arynes with diazomethane derivatives. Angew Chem Int Ed. 2007;46 (18):3323–25.

[40] Liu Z, Shi F, Martinez PDG, Raminelli C, Larock RC. Synthesis of indazoles by the [3+2] cycloaddition of diazo compounds with arynes and subsequent acyl migration. J Org Chem. 2008;73(1):219–26.

[41] Shi F, Waldo JP, Chen Y, Larock RC. Benzyne click chemistry: Synthesis of benzotriazoles from benzynes and azides. Org Lett. 2008;10(12):2409–12.

[42] Zhang F, Moses JE. Benzyne click chemistry with in situ generated aromatic azides. Org Lett. 2009;11(7):1587–90.

[43] Kametani T, Ogasawara K. Benzyne reaction. Part 1. Total syntheses of (2)-cryptaustoline and (+-)-cryptowoline by the benzyne reaction. J Chem Soc C. 1967;2208–12.

[44] Sanz R, Fernández Y, Castroviejo MP, Pérez A, Fañanás FJ. Functionalized phenanthridine and dibenzopyranone derivatives through benzyne cyclization – Application to the total syntheses of trisphaeridine and N-methylcrinasiadine. Eur J Org Chem. 2007;(1):62–69.

[45] Huters AD, Quasdorf KW, Styduhar ED, Garg NK. Total synthesis of (-)-N-methylwelwitindolinone C isothiocyanate. J Am Chem Soc. 2011;133(40):15797–99.

[46] Larrosa I, Da Silva MI, Gómez PM, Hannen P, Ko E, Lenger SR, et al. Highly convergent three component benzyne coupling: The total synthesis of ent-clavilactone B. J Am Chem Soc. 2006;128(43):14042–43.

[47] Bhojgude SS, Baviskar DR, Gonnade RG, Biju AT. Three-component coupling involving arynes, aromatic tertiary amines, and aldehydes via aryl–aryl amino group migration. Org Lett. 2015;17(24):6270–73.

[48] Guyot M, Molho D. Nouvelle methode de synthese d'acides homophtaliques et d'homophtalimides par voie arynique. Application a la Synthese de la Melleine. Tetrahedron Lett. 1973;14(36):3433–36.

[49] Danheiser RL, Helgason AL. Total synthesis of the phenalenone diterpene salvilenone. J Am Chem Soc. 1994;116(21):9471–79.

[50] Tambar UK, Stoltz BM. The direct acyl-alkylation of arynes. J Am Chem Soc. 2005;127 (15):5340–41.

[51] Khanapure SP, Reddy RT, Biehl ER. The preparation of anthraquinones and anthracyclinones via the reaction of haloarenes and cyanophthalides under aryne-forming conditions. J Org Chem. 1987;52(26):5685–90.

[52] Mal D, Pahari P. Recent advances in the Hauser annulation. Chem Rev. 2007;107 (5):1892–918.

[53] Zhao H, Biehl E. Preparation of naturally occurring anthraquinones using the aryne reaction. J Nat Prod. 1995;58(12):1970–74.

[54] Kaiser F, Schwink L, Velder J, Schmalz HG. Studies towards the total synthesis of mumbaistatin: Synthesis of highly substituted benzophenone and anthraquinone building blocks. Tetrahedron. 2003;59(18):3201–17.

[55] Holden CM, Sohel SMA, Greaney MF. Metal free bi(hetero)aryl synthesis: A benzyne Truce-Smiles rearrangement. Angew Chem Int Ed. 2016;55(7):2450–53.

[56] Peña D, Escudero S, Pérez D, Guitián E, Castedo L. Efficient palladium-catalyzed cyclotrimerization of arynes: Synthesis of triphenylenes. Angew Chem Int Ed. 1998;37 (19):2659–61.

[57] Vollhardt KPC. Transition-metal-catalyzed acetylene cyclizations in organic synthesis. Acc Chem Res. 1977;10(1):1–8.

[58] Peña D, Pérez D, Guitián E, Castedo L. Palladium-Catalyzed cocyclization of arynes with alkynes: Selective synthesis of phenanthrenes and naphthalenes. J Am Chem Soc. 1999;121 (24):5827–28.

[59] Pena D, Perez D, Guitian E, Castedo L. Selective palladium-catalyzed cocyclotrimerization of arynes with dimethyl acetylenedicarboxylate: A versatile method for the synthesis of polycyclic aromatic hydrocarbons. J Org Chem. 2000;65(21):6944–50.

[60] Sato Y, Tamura T, Mori M. Arylnaphthalene lignans through Pd-catalyzed [2+2+2] cocyclization of arynes and diynes: Total synthesis of taiwanins C and E. Angew Chem Int Ed. 2004;43(18):2436–40.

[61] Waldo JP, Zhang X, Shi F, Larock RC. Efficient synthesis of fluoren-9-ones by the palladium-catalyzed annulation of arynes by 2-haloarenecarboxaldehydes. J Org Chem. 2008;73 (17):6679–85.

[62] Lu C, Dubrovskiy AV, Larock RC. Palladium-catalyzed annulation of arynes by o-halobenzamides: Synthesis of phenanthridinones. J Org Chem. 2012;77(19):8648–56.

[63] Bergman RG. Reactive 1, 4-Dehydroaromatics. Acc Chem Res. 1973;6(1):25–31.

[64] Jones RR, Bergman RG. p-benzyne. Generation as an intermediate in a thermal isomerization reaction and trapping evidence for the 1,4-benzenediyl structure. J Am Chem Soc. 1972;94:660–61.

[65] Bowles DM, Anthony JE. A reiterative approach to 2,3-disubstituted naphthalenes and anthracenes. Org Lett. 2000;2(1):85–87.

[66] Lee MD, Dunne TS, Siegel MM, Chang CC, Morton GO, Borders DB. Calichemicins, a novel family of antitumor antibiotics. 1. Chemistry and partial structure of calichemicin .gamma.1I. J Am Chem Soc. 1987;109(11):3464–66.

[67] Lee MD, Dunne TS, Chang CC, Ellestad GA, Siegel MM, Morton GO, et al. Calichemicins, a novel family of antitumor antibiotics. 2. Chemistry and structure of calichemicin y1. J Am Chem Soc. 1987;109(11):3466–68.

[68] Zein N, Sinha AM, Mcgahren WJ, Ellestad GA. Calicheamicin gamma-1: An antitumor antibiotic that cleaves double-stranded DNA site specifically. Science. 1988;240(4856):1198–201.

6 Transition-metal-mediated C–C bond forming reactions of aromatic compounds

6.1 Introduction

Carbon–carbon bond forming reactions to aromatic rings, in particular for the formation of biaryls, have become an indispensable tool for the preparation of substituted aromatic compounds [1]. Biaryl motifs are commonly seen in a number of active pharmaceutical ingredients [2] and are also prevalent in the development of materials such as liquid crystalline compounds. Figure 6.1 shows some examples of compounds containing a biaryl motif and includes examples not only of pharmaceuticals, but also of liquid crystalline materials of the type that are found in liquid crystal displays.

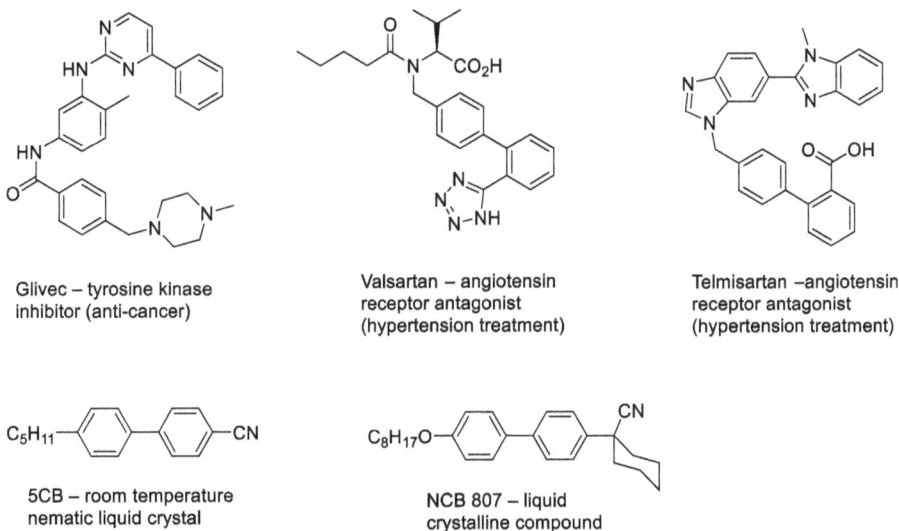

Glivec – tyrosine kinase inhibitor (anti-cancer)

Valsartan – angiotensin receptor antagonist (hypertension treatment)

Telmisartan –angiotensin receptor antagonist (hypertension treatment)

5CB – room temperature nematic liquid crystal

NCB 807 – liquid crystalline compound

Figure 6.1: Representative compounds containing biaryls.

In Chapter 3, we saw the formation of biaryls from diazonium salts (the Gomberg–Bachmann reaction and the Pschorr reaction). This chapter will focus on the preparation of biaryls using metal-mediated reactions, considering homocoupling and cross-coupling reactions. The chapter will have a particular emphasis on transition-metal catalyzed reactions for cross-coupling. We will also examine other closely related metal-mediated carbon–carbon bond forming reactions involving aromatic compounds.

https://doi.org/10.1515/9783110562682-006

6.2 Preparation of biaryls via homocoupling

6.2.1 Copper-mediated coupling: the Ullmann reaction

The homocoupling of an aryl halide (usually an aryl iodide or bromide) in the presence of copper to form a biaryl is referred to as the Ullmann reaction [3]. The initial conditions for the Ullmann reaction involved stoichiometric copper metal and were typically carried out at high temperatures in solvents such as DMF. While the mechanism of the Ullmann coupling is not clearly established, one proposed mechanism involves oxidative addition of the aryl halide to copper, disproportionation with another equivalent of copper, followed by another oxidative addition to generate a diaryl copper (III) intermediate that undergoes reductive elimination to form biaryl (Figure 6.2) [4, 5]. As shown, this mechanism requires stoichiometric amounts of copper, consuming one equivalent for every equivalent of aryl halide.

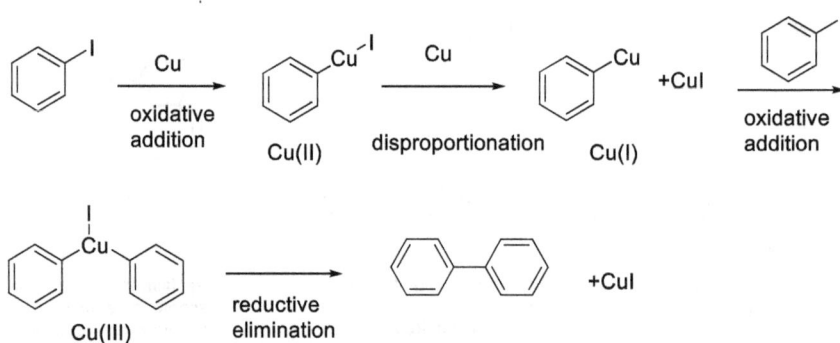

oxidative addition — Cu(II) — disproportionation — Cu(I) +CuI — oxidative addition

reductive elimination — Cu(III) — +CuI

Figure 6.2: Proposed mechanism for the copper-mediated Ullmann reaction to form biaryls.

Reactivity of the aryl halide toward Ullmann couplings depends on the nature of the halogen, with aryl iodides being the most reactive, followed by aryl bromides, and then aryl chlorides [3]. This trend in reactivity is consistent with the expected trend based on the relative reactivity toward oxidative addition. The reactivity of the substrate in Ullmann couplings is also highly influenced by substituents, although in unusual ways. For example, electron-withdrawing groups such as nitro or carboxylic ester groups increase reactivity, especially when these groups are *ortho* to the halogen. An illustration of the importance of electron-withdrawing groups in the *ortho*-position is seen in the coupling of 2,5-dibromonitrobenzene (Figure 6.3). In this reaction, coupling occurs only at the bromo positions *ortho* to the nitro group, leaving the bromo *meta* to the nitro group unreactive [6]. Similar reactions with 2,4-dibromonitrobenzene show that the reaction is favored at the *ortho*-position over the *para*-position.

However, electron-donating groups such as methyl groups and methoxy groups increase reactivity of the aryl halide, especially when positioned *meta* or *para* to

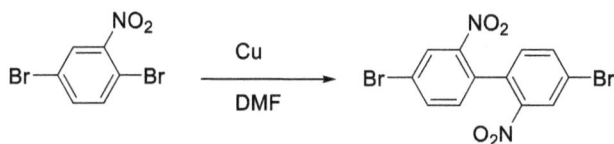

Figure 6.3: Regioselectivity in the Ullmann reaction of 2,5-dibromonitrobenzene.

the halide. In contrast, phenols, amines, and carboxylic acids prevent the Ullmann coupling. These substituents may be favoring to competing reactions such as dehalogenation.

Indeed, one of the most common and problematic side reactions observed in Ullmann reactions is dehalogenation [3]. This reaction may be attributed to protonolysis of the aryl copper intermediate. Mechanistically, it can be viewed as an electrophilic protonation, followed by loss of copper, analogous to the electrophilic removal of substituents seen in Section 2.6 (Figure 6.4).

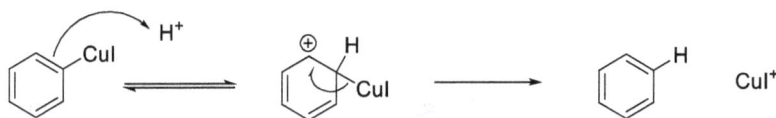

Figure 6.4: Proposed mechanism for dehalogenation side products in Ullmann couplings.

The traditional Ullmann coupling is heterogeneous, using copper metal, and often requires high temperatures. Alternative conditions have been developed that make use of Cu(I) salts instead of copper metal. These reactions take place under homogeneous conditions at lower temperatures. For example, Cohen and Cristea showed that aryl halides undergo homocoupling in the presence of copper(I) triflate in aqueous ammonia at room temperature (Figure 6.5) [5]. Similarly, Liebeskind showed that copper(I) thiophene-2-carboxylate (CuTC) can be used for efficient coupling of aryl halides at room temperature [7]. It has been proposed that the thiophene-2-carboxylate stabilizes the product of the oxidative addition via a chelate involving the carboxylate and the sulfur of the thiophene.

The Ullmann reaction is generally limited to homocoupling for the preparation of symmetric biaryls. This limitation is significant because many target molecules, including biologically active molecules, require dissymmetric biaryls. There are examples of cross-coupling using the Ullmann reaction, but these generally involve highly activated (electron-deficient) aryl halides reacting with a nonactivated aryl halide (usually an aryl iodide like iodobenzene). For example, 2,4-dibromonitrobenzene reacts with iodobenzene to produce the cross-coupled biaryl in 55% yield as well as the homocoupled product in 10% yield (Figure 6.6) [3, 8].

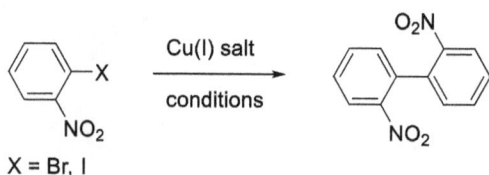

Conditions A: Cu(OTf), aq. NH₃, acetone, r.t.

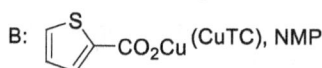

B: (thiophene)–CO₂Cu (CuTC), NMP

Figure 6.5: Homogeneous Ullmann reactions using Cu(I) salts.

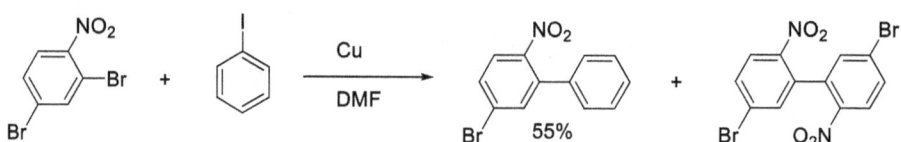

Figure 6.6: An example of an Ullmann cross-coupling [8].

While there are examples of the Ullmann coupling being used to prepare dis-symmetric biaryls, the structural requirements for effective cross-coupling limit the scope significantly, and in many cases mixtures of homocoupled and heterocoupled products are observed. For this reason, the Ullmann coupling is used primarily for homocoupling reactions and has mostly been supplanted by modern palladium-catalyzed cross-coupling reactions that we will explore in Section 6.3.

One advantage of the Ullmann coupling is that it is fairly tolerant of substituents in the *ortho*-positions, allowing the preparation of highly substituted, sterically hindered biaryls. These reactions will be discussed in more detail in Section 6.6. An example of the utility of the Ullmann coupling is highlighted in the synthesis of biphyscion, a bianthraquinone natural product featuring a sterically demanding tetra-*ortho*-substituted biaryl bond (Figure 6.7) [9]. The synthesis began with the electrophilic iodination of the aryl ester, giving the desired regioisomer as the major product. The Ullmann coupling of the aryl iodide took place with copper–bronze and heat to give the desired biaryl in good yield. The biaryl was converted to the corresponding tetrathiophenyl derivative by benzylic deprotonation using LDA and trapping with (PhS)₂. Treatment with aqueous trifluoroacetic acid resulted in the thiophenylated biphthalide. The sulfides were converted to the corresponding sulfones. The sulfone was deprotonated with lithium *t*-butoxide and reacted with the cyclohexenone to undergo a conjugate addition and condensation to yield the crude hydroanthracene derivative, which underwent oxidation to give the desired anthraquinone. Finally, a regioselective demethylation with MgI₂ gave the target compound.

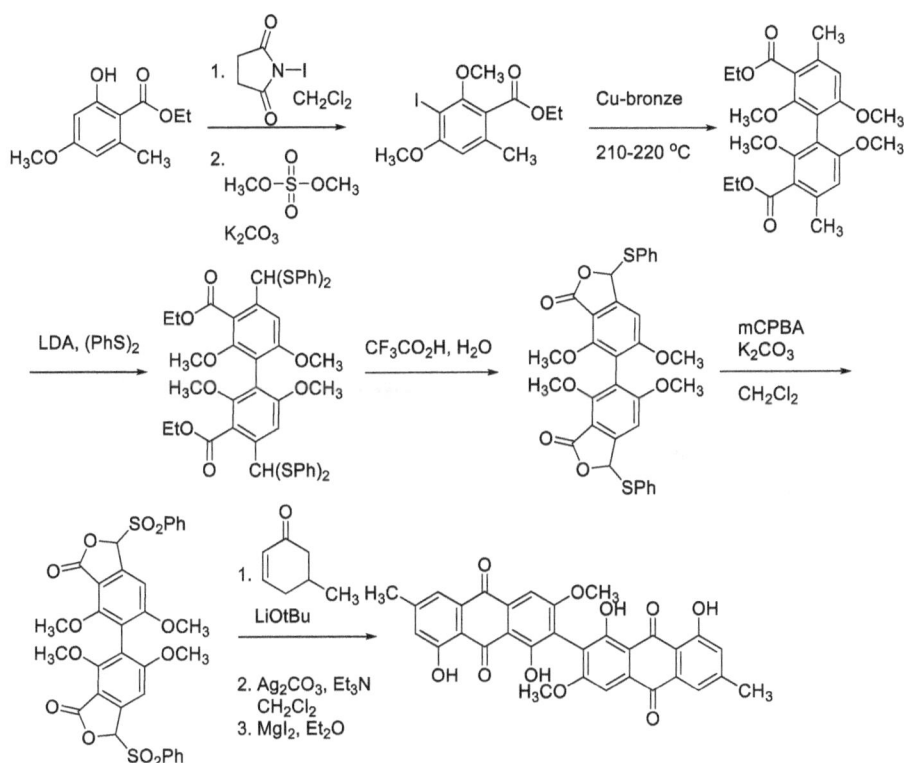

Figure 6.7: Synthesis of biphyscion [9].

6.2.2 Nickel-mediated homocoupling reactions

Nickel-mediated homocoupling of aryl halides was first reported by Semmelhack and coworkers in 1971, who showed that a variety of aryl halides reacted with Ni(COD)$_2$ in DMF (Figure 6.8) [10]. The reaction worked best with aryl bromides and iodides, and tolerated a variety of electron-donating and electron-withdrawing groups on the ring, with some limitations. Nitrobenzenes yielded no product, nor did aryl halides bearing acidic functional groups such as carboxylic acids, alcohols, and phenols. Despite these limitations, the reaction has an overall broader substrate scope than the Ullmann coupling. One drawback is that the method uses a stoichiometric amount of an air-sensitive nickel(0) complex, such as Ni(COD)$_2$ or Ni(PPh$_3$)$_4$.

The reaction is often carried out in the presence of other ligands, such as PPh$_3$ or 2,2′bipyridine (bpy), which presumably form a more active metal complex in situ. For example, with one equivalent of bpy, the Ni(COD)$_2$ reacts forms Ni(COD)(bpy), which serves as the reagent in for the aryl–aryl coupling.

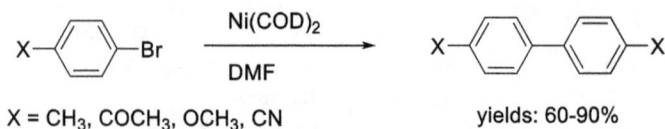

X = CH$_3$, COCH$_3$, OCH$_3$, CN yields: 60-90%

Figure 6.8: Nickel(0)-mediated homocoupling of aryl halides to form biaryls.

The mechanism of the reaction is thought to proceed as outlined in Figure 6.9 [11]. The Ni(COD)(bpy) complex undergoes oxidative addition with the aryl bromide to form a nickel(II) intermediate. Then, two equivalents of this intermediate undergo a disproportionation to yield Ni(bpy)Br$_2$ and Ni(bpy)Ar$_2$, which are both nickel(II) species. The latter diaryl complex undergoes reductive elimination to regenerate a nickel (0) species and form the biaryl. Despite the fact that a nickel(0) species is regenerated at the end of this sequence, one equivalent of Ni(bpy)Br$_2$ is also produced, meaning that a stoichiometric amount of nickel(0) complex is needed.

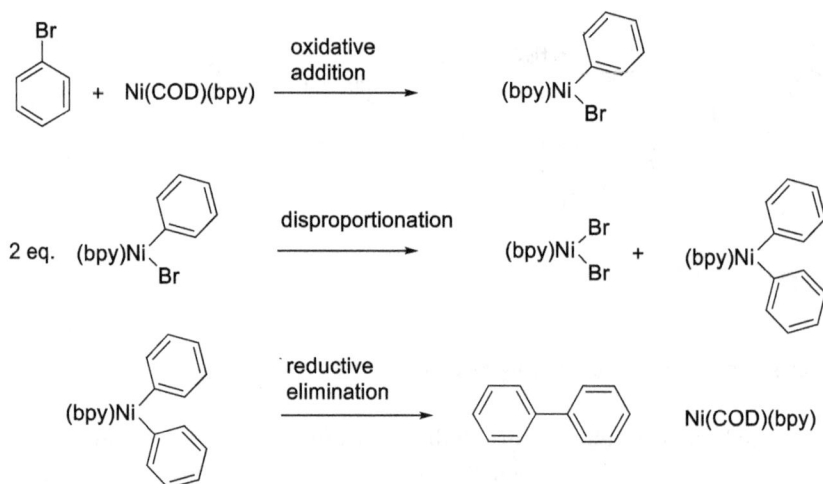

Figure 6.9: Proposed mechanism for the nickel(0)-mediated homocoupling of biaryls [11].

Subsequently, Kende and coworkers showed that the reaction could be performed using air-stable nickel(II) salts in the presence of a reducing agent, generating the active Ni(0) species in situ [12]. Specifically, they used bis(triphenylphosphine)nickel (II) dichloride with zinc metal and triphenylphosphine to generate tris(triphenylphosphine)nickel(0), which is thought to be the catalytically active species. This reaction avoids the use of air-sensitive nickel complexes, but still makes use of a stoichiometric amount of both the nickel complex and the zinc.

The nickel(0)-mediated homocoupling of aryl halides has been successfully applied to the synthesis of conjugated polymers, including polythiophenes and polyphenylenes when aryl dihalides are used (Figure 6.10) [13].

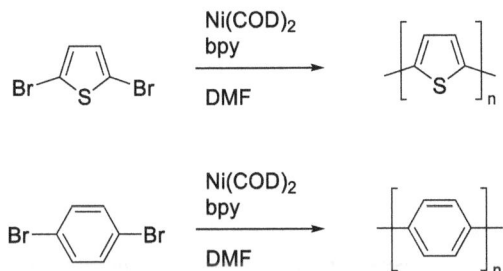

Figure 6.10: Preparation of conjugated polymers via nickel-mediated homocoupling.

Yamamoto also showed that *ortho*-dibromoarenes could undergo homocoupling to produce cyclic oligomers such as triphenylene and tetrathiophene (Figure 6.11) [14]. This nickel-mediated coupling to prepare polymers and cyclic oligomers is often referred to as the Yamamoto coupling. Later, we will see examples of how this reaction has been used to prepare polycyclic aromatic hydrocarbons.

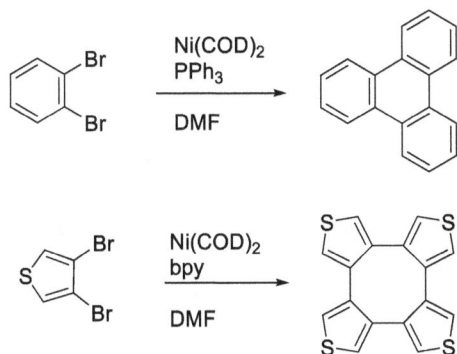

Figure 6.11: Preparation of triphenylene and tetrathiophene via nickel-mediated coupling of *ortho*-dibromoarenes [14].

The disadvantage of these nickel reactions is the need for a stoichiometric amount an air-sensitive nickel(0) complex. As we saw above, in some cases, an air-stable Ni(II) complex has been used in the presence of a stoichiometric reductant such as Zn to generate the nickel(0) species in situ. However, this reaction has usually been carried out with a stoichiometric amount of the nickel complex. In principle, this could be carried out with a catalytic amount of the nickel complex. Indeed, there are a few examples where nickel-catalyzed homocoupling of aryl halides have been carried out to prepare conjugated polymers [15, 16]. In these reactions, dibromoarenes were coupled to form polymers using catalytic $NiCl_2$ in the presence of bipyridine, triphenylphosphine, and an excess of zinc as the stoichiometric reductant. However, these methods have not been developed extensively.

6.2.3 Oxidative coupling reactions

Electron-rich aromatic compounds can undergo a dehydrogenative coupling reaction in the presence of oxidants such as $FeCl_3$, copper(II) salts, as well as other oxidants [17, 18]. In general, the mechanism is thought to proceed via a radical cation as outlined in Figure 6.12, although in some cases, an arenium cation may be implicated as an intermediate. The radical pathways involve a one-electron oxidation of the electron-rich aromatic ring, followed by C–C bond formation with another molecule. A second one-electron oxidation yields a dication, which upon deprotonation at each ring gives the biaryl. Overall, the reaction consists of the formation of an aryl–aryl bond and the formal loss of hydrogen [19].

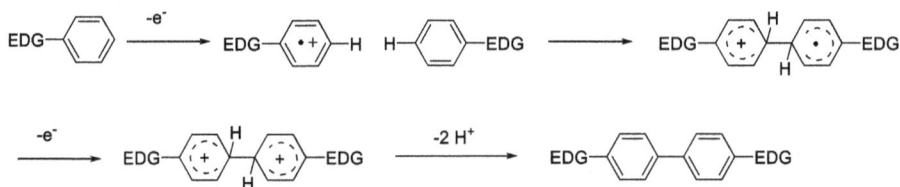

Figure 6.12: Radical mechanism for oxidative coupling [19].

For example, reaction of 1,4-dimethoxybenzenes with $FeCl_3$ yields the corresponding biphenyls in modest to good yield (Figure 6.13) [20]. Interestingly, the reaction with 1,4-dialkoxybenzene gives a lower yield of the biphenyl product than the brominated analog, likely because the dialkoxybenzene undergoes further coupling to produce oligomers or polymers, while the bromo substituents block further reaction.

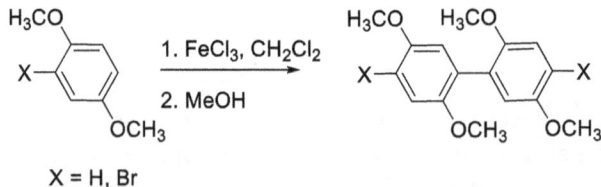

X = H, Br

Figure 6.13: Representative example of an oxidative coupling to form biphenyl [20].

In the previous example, we saw that the oxidative coupling of 1,4-dimethoxy-benzene showed low yields because it could react further to form oligomers and polymers. Indeed, the possibility of polymerization has been exploited to prepare polymers from electron-rich heteroaromatic monomers such as thiophenes [21]. For example, 3-alkyl thiophenes form the corresponding poly(3-alkylthiophenes) upon treatment with $FeCl_3$.

Oxidative couplings to form biaryls is also successful for hindered substrates such as 2-naphthol, which forms the corresponding 2,2′-binaphthol in the presence of $FeCl_3$ or $CuCl_2$ (Figure 6.14). These can also be carried out using $K_3[Fe(CN)_6]$ or in the solid state with $FeCl_3 \cdot 6H_2O$ [22]. These types of binaphthyls are of interest as chiral ligands for asymmetric synthesis, which will be explored in more detail in Section 6.6.

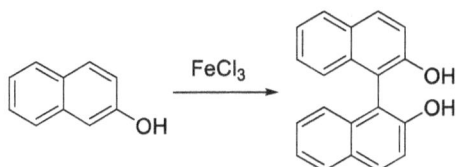

Figure 6.14: Synthesis of 2,2′-binaphthol via oxidative coupling.

Oxidative coupling has also been used in an *intramolecular* fashion to carry out cyclization reactions. These reactions have been used extensively to prepare polycyclic aromatic hydrocarbons using oxidants such as $FeCl_3$. These reactions will be explored in more detail in Chapter 9.

6.3 Preparation of biaryls via cross-coupling reactions

6.3.1 Metal-catalyzed cross-coupling

While homocoupling methods are effective for preparing symmetric biaryls, the coupling of two different aromatic compounds to make dissymmetric biaryls is much more versatile, providing access to a much greater diversity of compounds. Most of these cross-coupling reactions are catalyzed by palladium complexes, although in some cases nickel complexes are used as catalysts. They involve an aryl halide or triflate as well as and organometallic species (such as organoboron, organotin, or Grignard reagent). These cross-coupling reactions are not restricted to coupling to aromatic rings, but can also be used to couple aryl compounds with other unsaturated systems such as vinyl groups (Figure 6.15) [23–25].

Although the reaction conditions vary widely in terms of the reagents and catalysts used, they all proceed by the same general mechanism (Figure 6.16) [26]. The

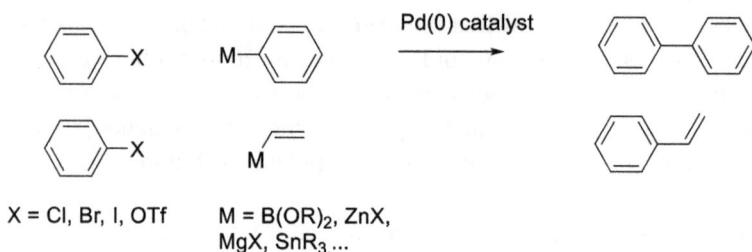

X = Cl, Br, I, OTf   M = B(OR)$_2$, ZnX, MgX, SnR$_3$,...

Figure 6.15: General scheme of a palladium-catalyzed cross-coupling.

active catalyst is a 14-electron Pd(0) species, which is generated from the corresponding 18-electron Pd(0) complex by ligand dissociation. The highly reactive 14-electron complex undergoes oxidative addition with the aryl halide to form a Pd(II) complex. This species then reacts with the organometallic reagent, undergoing a transmetallation, where the other aryl group of the organometallic reagent is transferred to the palladium. This species then undergoes a reductive elimination to form the aryl–aryl bond and regenerate the active Pd(0) catalyst.

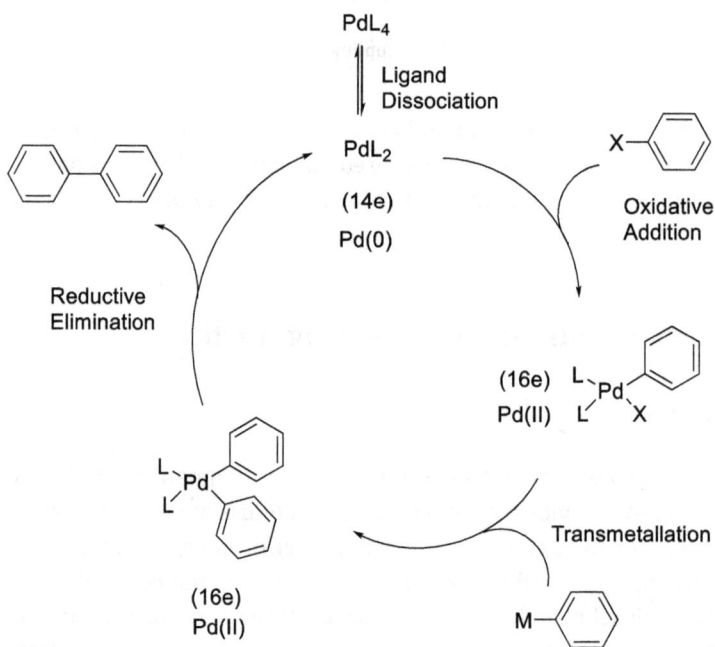

Figure 6.16: General mechanism of palladium-catalyzed cross-coupling.

Pd(0) complexes that are needed for these cross-coupling reactions are prone to oxidation, making them somewhat air sensitive. Because of this, some care is needed when handling these catalysts. Cross-coupling reactions are usually carried out under inert atmosphere and make use of degassed solvents. Some of the most common palladium complexes used for these cross-coupling reactions are Pd(PPh$_3$)$_4$ or Pd$_2$(dba)$_3$. The former complex is air and light sensitive, so longer term storage under inert atmosphere in the dark is needed.

It is also possible to generate a Pd(0) catalyst in situ by reduction of a Pd(II) complex. This approach has the advantage that the Pd(II) complexes are air stable, making them easier to handle and store. For example, the necessary palladium(0) species can be generated from Pd(OAc)$_2$ in the presence of a phosphine such as triphenylphosphine [27, 28]. As shown in Figure 6.17, ligand coordination of triphenylphosphine to Pd(OAc)$_2$ is followed by a reductive elimination of PPh$_3$ and acetate. The resulting Pd(0) can undergo ligand exchange and dissociation to give the catalytically active species. At the same time, the triphenylphosphine-acetate adduct can react with acetate anions to give acetic anhydride and triphenylphosphine oxide, or if it reacts with water, will give acetic acid and triphenylphosphine oxide.

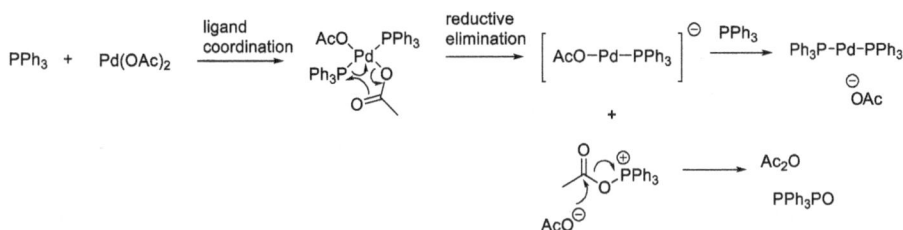

Figure 6.17: In situ reduction of Pd(OAc)$_2$ [27, 28].

Similarly, Pd(PPh$_3$)$_2$Cl$_2$ can lead to the formation of a Pd(0) species in situ in the presence of trimethylamine. The palladium complex can undergo ligand exchange with the amine, which is followed by a β-hydride elimination, followed by a reductive elimination of HCl to give the Pd(0) complex (Figure 6.18) [29].

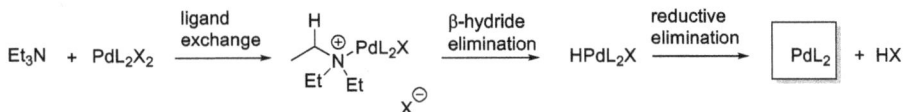

Figure 6.18: In situ reduction of Pd(PPh$_3$)$_2$Cl$_2$.

The reactivity of the palladium catalyst is highly influenced by the ligands used. As such, considerable effort has been devoted to the synthesis of new ligands and their corresponding complexes in order to produce highly active catalysts for cross-coupling [30]. The design of ligands to produce catalysts with high reactivity generally focuses on preparing electron-rich ligands that are bulky and sterically demanding. Electron-rich ligands promote oxidative addition, while sterically bulky ligands promote ligand dissociation to form coordinatively unsaturated catalytically active species. These ligands are often phosphines or *N*-heterocyclic carbenes. Figure 6.19 shows some representative ligands that are used for transition metal-catalyzed reactions. These ligands can be used to carry out coupling reactions with unreactive substrates such as aryl chlorides, or sterically demanding substrates.

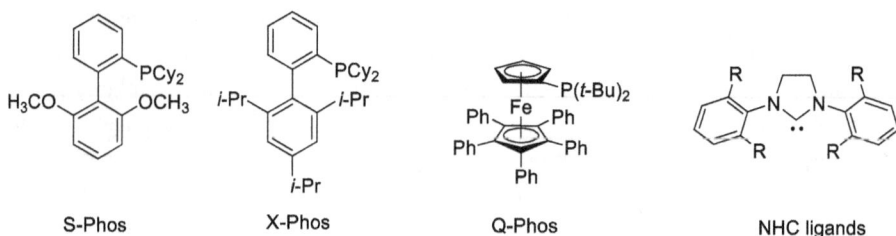

Figure 6.19: Some representative ligands used to form highly active catalysts for aryl coupling reactions.

6.3.1.1 Ar–X cross-coupling precursors

A common reactant of the cross-coupling reactions to form aryl–aryl bonds is an aryl halide or triflate. The reactivity of the aryl halides depends on their propensity to undergo oxidative addition. Overall, aryl iodides are the most reactive and aryl chlorides are least reactive, with aryl bromides and triflates displaying intermediate reactivity (Figure 6.20). Aryl fluorides, on the other hand, are unreactive. There are several other leaving groups that have been used in cross-coupling reactions, including phosphates [31] and sulfonates [32]. However, halogens and triflates remain the most widely used, in part because they are readily prepared. Chlorides are significantly less reactive than the other halides. However, effective coupling using aryl chlorides has been achieved using palladium catalysts with highly activating ligands such as hindered phosphines [33–35].

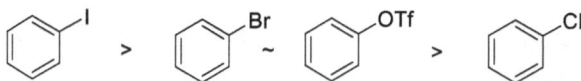

Figure 6.20: Overall reactivity trend for aryl halides (and pseudo-halides) in cross-coupling reactions.

Since the cross-coupling reactions all make use of aryl halides or triflates, it seems pertinent to review some of the approaches for preparing these cross-coupling precursors. Figure 6.21 shows some of the approaches we have seen for the preparation of aryl halides, including electrophilic halogenation (Section 2.3.4), formation via the diazonium salt (Section 3.6), bromo- or iodo-desilylation (Section 2.6.2), or halogenation of an aryllithium species (Section 4.1). Aryl triflates are readily prepared from the corresponding phenol by treatment with triflic anhydride.

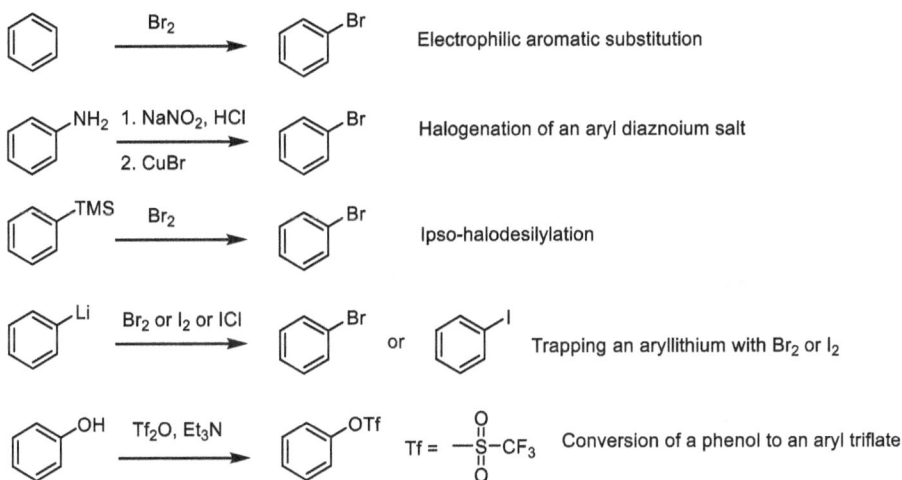

Figure 6.21: Summary of synthetic approaches to aryl halides and triflates used in cross-coupling reactions.

6.3.1.2 Suzuki–Miyaura cross-coupling (aryl boronic acids or boronate esters)

The palladium catalyzed cross-coupling of an aryl halide with an aryl boron species is referred to as the Suzuki–Miyaura coupling [36, 37]. The reaction typically involves an aryl boronic acid as the cross-coupling partner, although aryl boronates and other types of boron reagents can be used. Suzuki–Miyaura cross-coupling reactions are among the most broadly used cross-coupling methods because the boronic acids are isolable and air stable, and because the reaction is compatible with water, so rigorous exclusion of moisture is not necessary. Boronic acids and boronates themselves are not very reactive, so the Suzuki coupling requires the presence of water and a base to activate them toward transmetallation. Studies have suggested that the base in the presence of water produces hydroxide, which reacts with the palladium intermediate that results from oxidative addition to form a palladium hydroxide intermediate, which more reactive to transmetallation (Figure 6.22) [38]. The hydroxide may also play a role in promoting reductive elimination. Hydroxide can also add to the boronic acid to form an anionic $ArB(OH)_3^-$ species, which is less reactive than the corresponding boronic acid. So, the role of the base is somewhat complex, having both a positive effect on reactivity as well as an antagonistic effect.

Aryl boronic acids are the most common aryl boron species used for Suzuki couplings. They are usually prepared by reacting the corresponding organollithium species (prepared either by deprotonation or via lithium–halogen exchange – see Section 4.1) with $B(OCH_3)_3$ or $B(OiPr)_3$. Aryl boronate esters such as the pinacol

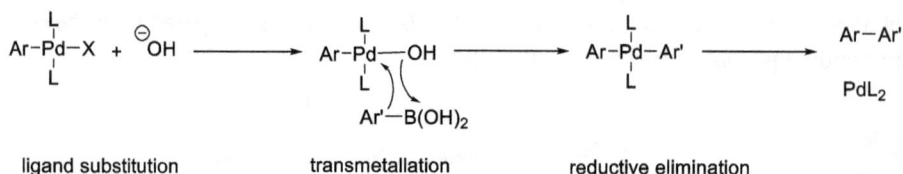

ligand substitution   transmetallation   reductive elimination

Figure 6.22: Proposed mechanism for the role of the base in Suzuki couplings.

boronate also react readily via a Suzuki coupling, although they are slightly less reactive than the corresponding boronic acid. One approach for preparing an aryl boronate ester is by reacting the boronic acid with pinacol in the presence of a dehydrating reagent (Figure 6.23). We will see other methods for preparing arylboronate esters in Chapter 7. The pinacol boronate ester has the advantage that it can be more readily purified by column chromatrography than the acid. The pinacol ester can be cleaved under aqueous acidic conditions or oxidatively with sodium periodate ($NaIO_4$) [39, 40].

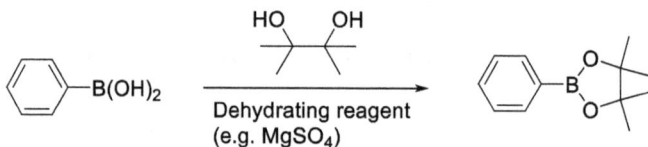

Figure 6.23: Preparation of pinacol boronate esters from the corresponding boronic acids.

More recently, other types of organoboron cross-coupling precursors have been developed. One particularly noteworthy example is potassium aryl trifluoroborate, which can be produced readily by reaction of potassium hydrogen fluoride (KHF_2) with organoboron species such as dihaloboranes, boronic acids, or boronate esters (Figure 6.24) [41–43].

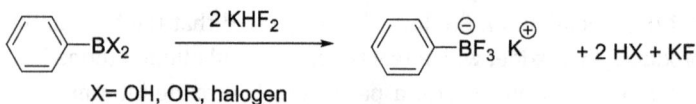

X= OH, OR, halogen

Figure 6.24: Preparation of potassium aryl trifluoroborate for Suzuki couplings.

Potassium aryl trifluoroborates can be prepared on a large scale, and are air-stable solids. They are less prone to protodeborylation and exhibit higher reactivity than the corresponding boronic acids.

For an example of the use of the Suzuki coupling to prepare active pharmaceutical ingredients, let us consider the synthesis of losartan, a nonpeptide angiotensin II receptor antagonist used to control hypertension. A key step in the synthesis is the coupling of phenyl tetrazole with a benzyl imidazole derivative to form the biaryl bond (Figure 6.25) [44]. Phenyltetrazole was first protected with a trityl group, which was followed by directed *ortho* metalation (DoM, see Chapter 4) with butyllithium. The aryllithium was trapped with triisopropyl borate, followed by aqueous work up to give the boronic acid. This boronic acid was then subjected to Suzuki coupling with the aryl bromide shown, in the presence of Pd(OAc)$_2$, PPh$_3$, and K$_2$CO$_3$ in THF/water/dimethoxymethane, giving the desired biaryl in good yield. Finally, removal of the trityl protecting group gave losartan.

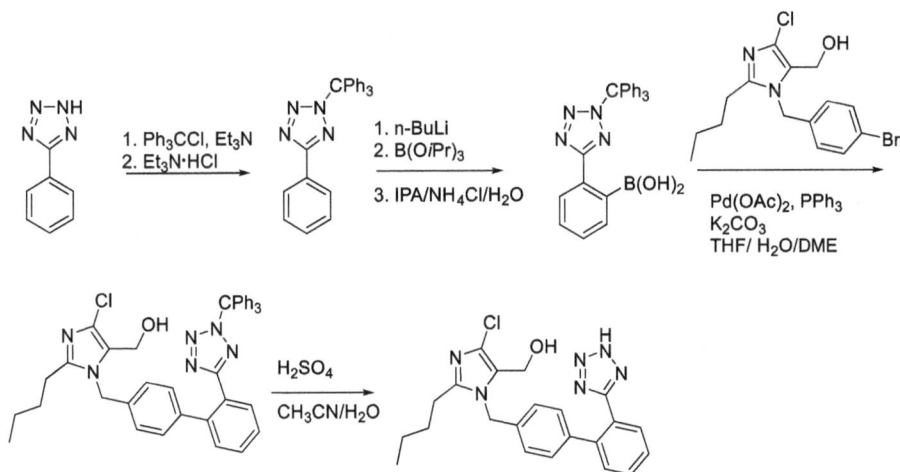

Figure 6.25: Synthesis of losartan [44].

Suzuki cross-coupling reactions have also been used for the preparation of natural products featuring biaryl linkages. For an example, we will consider the preparation of the naturally occurring fluorenone derivative dengibsinin (Figure 6.26) [45]. This synthesis makes use of directed *ortho* metalation (DoM), cross-coupling, and directed remote metalation (DreM) to prepare the fluorenone natural product. The boronate ester was prepared by directed *ortho*-metalation of the *meta*-isopropoxybenzamide derivative followed by trapping with tributylborate. A Suzuki coupling with the aryl iodide gave the corresponding biaryl. Following this, treatment with LDA led to directed remote metalation to yield the fluorenone. Selective cleavage of the isopropoxy protecting groups using BCl$_3$ gave dengibsinin.

6.3.1.3 Negishi cross-coupling (organozinc)
The metal-catalyzed cross-coupling of an aryl zinc species with an aryl halide is referred to as the Negishi coupling (Figure 6.27) [46, 47]. Palladium catalysts are usually used, although the reaction can also be catalyzed by nickel complexes [48]. The aryl zinc species is generated in situ, typically by reaction of an aryllithium

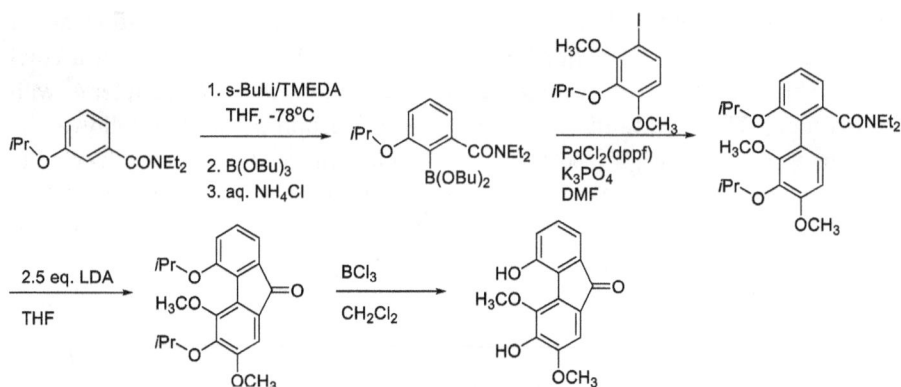

Figure 6.26: Synthesis of dengibsinin [45].

species with $ZnCl_2$. One of the drawbacks of this reaction is that the aryl zinc species is moisture sensitive. Consequently, it is not used as often as Suzuki couplings.

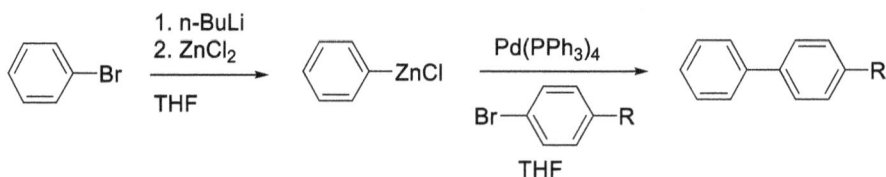

Figure 6.27: The Negishi coupling.

Despite the drawback of the Negishi coupling, it has been used in the large scale synthesis of pharmaceutically active compounds. For example, the biaryl in Figure 6.28 is a nonsteroidal ligand for glucocorticoid receptors developed by Abbott, and its synthesis made involved a Negishi coupling [49]. The aryl zinc reactant was generated in situ by deprotonation, followed by treatment with $ZnCl_2$. Subsequent addition of the aryl bromide and palladium catalyst then gave the substituted biphenyl, which was further elaborated to prepare the desired compound.

6.3.1.4 Stille cross–coupling (organotin)
The Stille coupling is a palladium-catalyzed cross-coupling involving an aryl stannane as the organometallic agent and an aryl halide (Figure 6.29) [50, 51]. Unlike the Suzuki cross-coupling, no base is needed to activate the organometallic reagent.

The aryl stannane is typically prepared from the corresponding aryllithium species by reaction with tributyltin chloride. The aryl stananne has the advantage that it is isolable and readily purified by column chromatography. The Stille coupling has the distinct disadvantage that it the organotin species are highly toxic. Consequently,

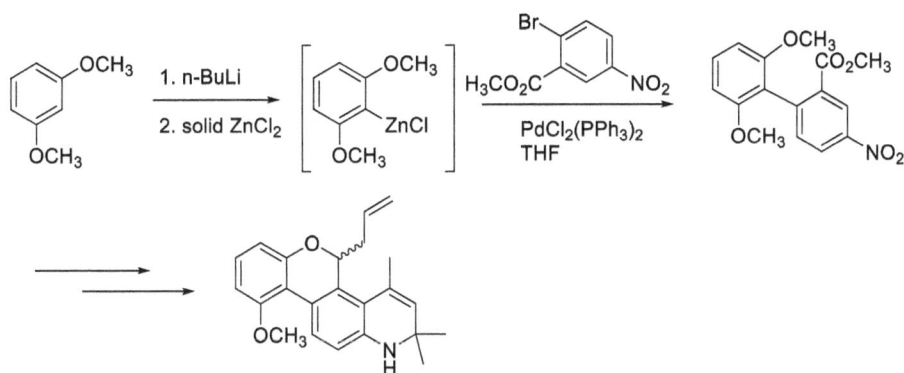

Figure 6.28: Synthesis of a glucocorticoid receptor target [49].

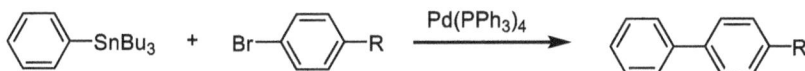

Figure 6.29: The Stille coupling of an aryltin reagent with an aryl halide.

it is not as widely used as some of the other cross-coupling methods. Nevertheless, it has found use in the synthesis of biaryls and is particularly effective for the coupling of thiophenes.

Because of the toxicity of organotin reagents, the Stille coupling has not been used much for the synthesis of pharmaceuticals. There are nonetheless some examples where Stille coupling have proven more effective than other cross-coupling methods. For example, it was used in the large scale synthesis of an imidazole–thienopyridine which was being investigated as a potential antitumor agent by researchers at Pfizer (Figure 6.30) [52]. The chloro-substituted thienopyridine was first iodinated by lithiation and trapping with iodine. The product was then reacted with tributylstannyl imidazole in the presence of $Pd(PPh_3)_4$ in a Stille coupling to form the imidazole–thiophene bond. Finally, nucleophilic aromatic substitution at the chloropyridine using an aminoindole gave the target compound.

As an example of the use of the Stille coupling for installing a vinyl group, let us consider the synthesis of defucogilvocarin V (Figure 6.31) [53]. Defucogilvocarin V is the aglycone of the corresponding gilvocarin, a natural product isolated from Streptomyces that displays cytotoxic activity. The synthesis also features directed metalation chemistry and a Suzuki coupling. The synthesis begins by metalation of the substituted naphthalene *ortho* to the carbamate DMG, followed by trapping with iodine to give the aryl iodide. Following this, a Suzuki coupling with the boronic acid gave the biaryl in good yield. Treatment of this biaryl with LDA resulted in directed remote metalation, followed by an anionic Fries rearrangement and cyclization to give the lactone. Under these reaction conditions, the MOM protecting

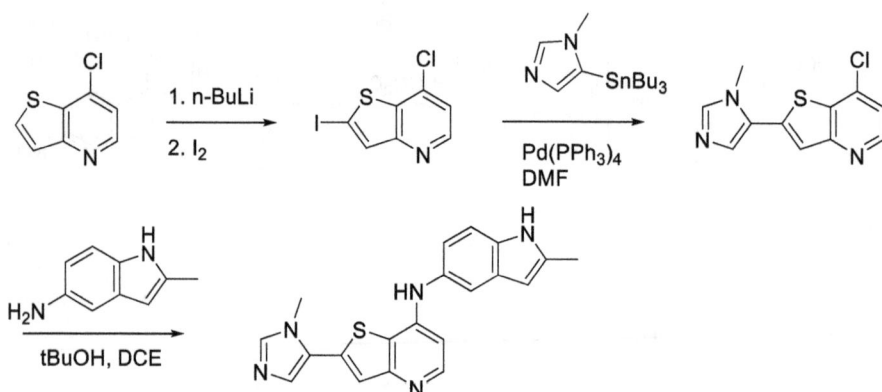

Figure 6.30: Synthesis of an antitumor agent via a Stille coupling [52].

group was cleaved, allowing the phenol to be converted into the corresponding triflate. At this point, a Stille coupling with vinyltributyltin using a palladium catalyst installed the vinyl group. Finally, selective deprotection of the isopropyl group gave the target defucoglivocarin V.

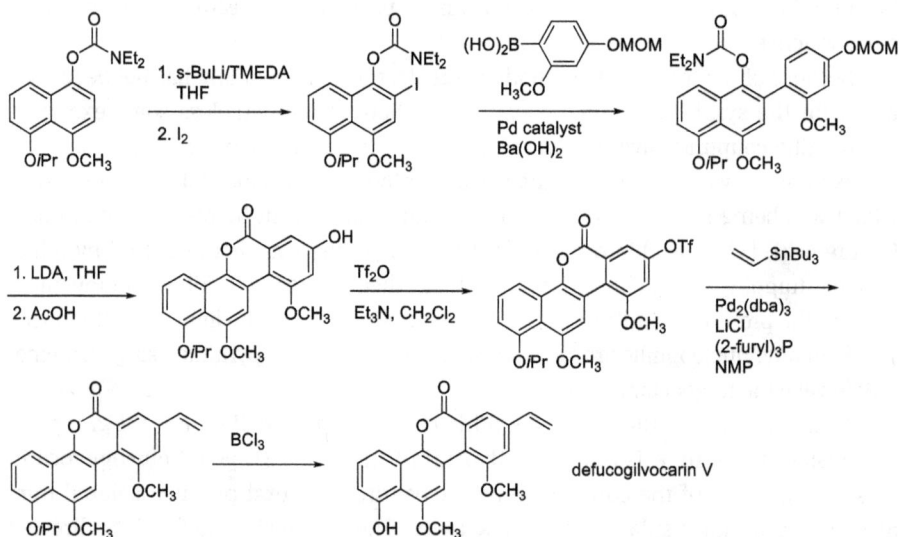

Figure 6.31: The synthesis of defucogilvocarin V [53].

6.3.1.5 The Hiyama cross-coupling (organosilanes)

An alternative to the use of toxic organostannanes is the use of organosilanes. This reaction is referred to as the Hiyama coupling [54]. Silanes are relatively unreactive to transmetallation, but when they are activated by the addition of fluoride to form a pentacoordinate silyl anion, they will undergo transmetallation readily. As such, the reaction is typically carried out with an aryl halide, an organosilane, and fluoride in the presence of a palladium catalyst. For example, Hiyama's original report showed that a variety of aryl iodides, vinyl iodides, or vinyl bromides reacted with vinyl or alkynyl silanes with catalytic allylpalladium chloride dimer and tris(dimethylamino)sulfonium difluorotrimethylsilicate as the fluoride source (Figure 6.32) [54]. Based on the role of fluoride in increasing reactivity of the organosilanes, Denmark used a silyloxide anion instead of a silane, eliminating the need for fluoride [55, 56]. This modification is often referred to as the Hiyama–Denmark reaction.

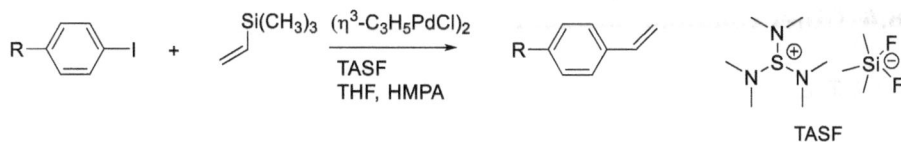

Figure 6.32: Representative example of the Hiyama coupling.

6.3.1.6 Kumada–Corriu cross-coupling (Grignard reagents)

The Kumada–Corriu coupling is the nickel- or palladium-catalyzed reaction of a Grignard reagent with an aryl halide to form a new aryl carbon bond [57, 58]. The reaction can be used to prepare biaryls, but can also be extended to other Grignard reagents to allow alkyl or alkenyl groups to be coupled to aryl rings. The ability to couple saturated hydrocarbons to aromatic rings is an advantage because several other cross-coupling methods are not readily amenable to using saturated systems due to competing β-hydride elimination reactions. A disadvantage of the Kumada–Corriu reaction is the reactivity of the Grignard reagents, which limits the functional group tolerance of the reaction. For this reason, it is not used as widely as other cross-coupling reactions such as the Suzuki–Miyaura cross-coupling.

The Kumada coupling has been used on a large scale for the preparation of active pharmaceutical ingredients. For example, it was used for the kilogram-scale preparation of the substituted phenyl pyridine and an analogous phenyl thiazole (Figure 6.33), which were used as intermediates in the synthesis of candidates for HIV protease inhibitors [59]. In both cases, the reaction was carried out with the phenyl Grignard reagent and the bromo-substituted heteroaromatic in the presence of a nickel catalyst.

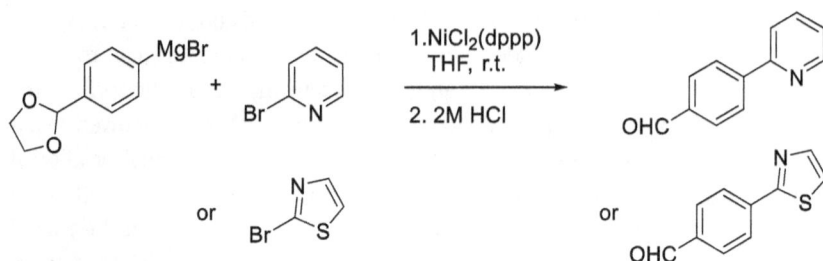

Figure 6.33: Examples of biaryl synthesis via Kumada coupling [59].

6.4 Other coupling reactions

6.4.1 The Heck reaction

Aryl halides and triflates also undergo palladium-catalyzed reactions with alkenes (usually terminal alkenes) in the presence of a base to form a new carbon–carbon bond and eliminate HX [60–63]. This coupling is referred to as the Heck reaction. The general reaction is shown in Figure 6.34.

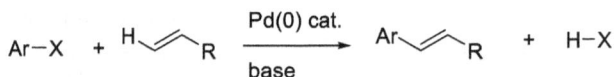

Figure 6.34: The Heck reaction.

Mechanistically, the catalytic cycle begins with ligand dissociation to form a 14-electron complex, followed by oxidation of the aryl halide or triflate to give a Pd(II) complex, as we saw in the cross-coupling reaction mechanism (Figure 6.35) [61, 62]. At this stage, the alkene coordinates to the palladium, followed by a migratory insertion reaction, where the carbon–carbon bond is formed. Subsequently, β-hydride elimination leads to the product formation. Finally, base-promoted reductive elimination of HX regenerates the active catalyst. Many procedures for the Heck reaction make use of Pd(OAc)$_2$ or other Pd(II) salts in the presence of a phosphine ligand and an amine base. In these cases, the active Pd(0) catalyst is generated in situ, as described previously.

The Heck reaction is an effective way to couple an aryl halide or triflate with an alkene. Most examples involve monosubstituted alkenes bearing electron-withdrawing groups, where the reaction takes place with high selectivity to generate the *trans*-substituted aryl alkene product (Figure 6.36). As with the other coupling reactions we have seen, aryl iodides are the most reactive substrates, followed by aryl triflates and aryl bromides. This difference in reactivity can be used to achieve selective Heck reactions, as shown in Figure 6.36 [23, 64].

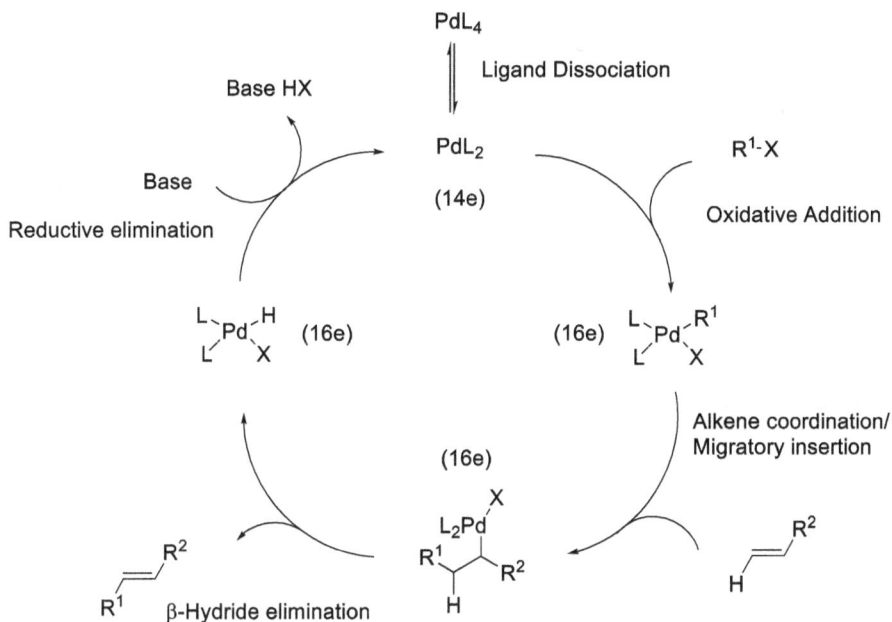

Figure 6.35: Catalytic cycle for the Heck reaction.

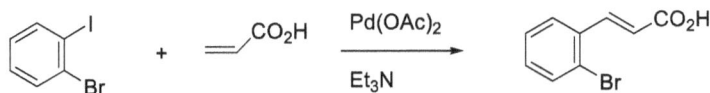

Figure 6.36: Representative Heck reaction demonstrating higher reactivity of aryl iodides [64].

When a cyclic alkene is used, isomerization of the double bond is observed (Figure 6.37). This outcome is due to the stereochemical outcome of the migratory insertion, which places the aryl group syn to the palladium, and the fact that β-hydride elimination takes place in a syn manner [63, 65]. Since the hydrogen on the carbon bearing the aryl group is not syn to the palladium, the β-hydride elimination takes place at the other position. This issue is observed primarily with cyclic alkenes because acyclic alkenes can undergo a bond rotation to place the hydrogen in a syn orientation. Isomerization by shifting the position of the double bond can also take place through reversible hydropalladation/dehydropalladation (essentially reversible β-hydride elimination) when the alkene is part of a ring or saturated chain.

Isomerization of the double bond in the Heck reaction is also often observed for allylic alcohols, because the β-hydride elimination is reversible, it is under thermodynamic control and can lead to the more stable enol product. For example, in the synthetic approach to the antiasthma agent Singulair™, a Heck reaction of an aryl halide with an allylic alcohol gave the corresponding saturated ketone, via the enol

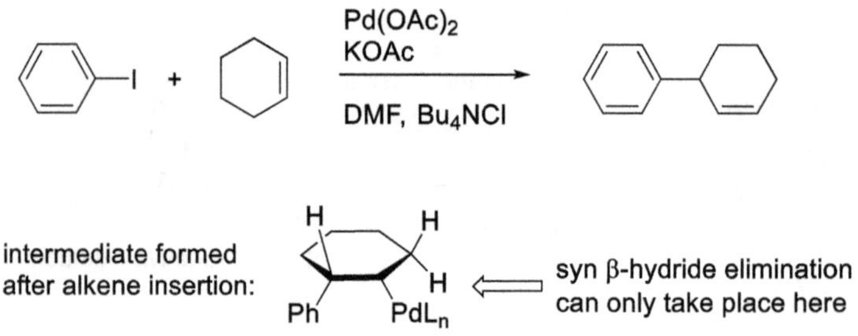

intermediate formed
after alkene insertion:

syn β-hydride elimination
can only take place here

Figure 6.37: Heck reaction of an aryl halide with a cyclic alkene.

intermediate (Figure 6.38) [66]. In this reaction, the β-hydride elimination favors the
formation of the enol.

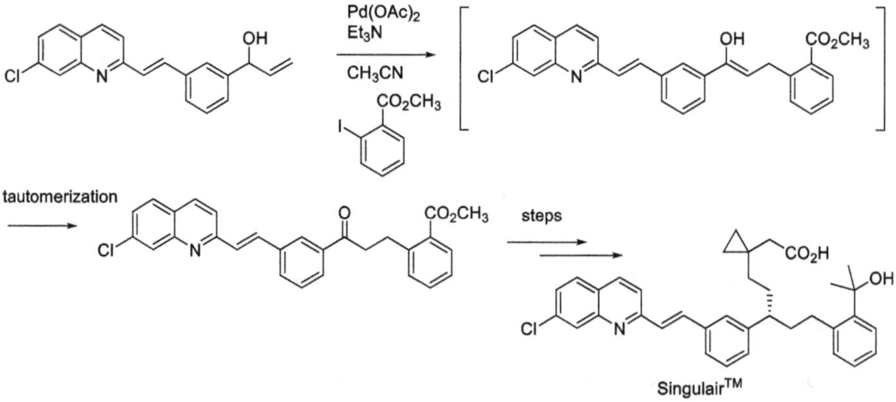

Figure 6.38: Heck reaction in the synthetic approach for the antiasthma agent Singulair™ [66].

The Heck reaction has been used on an industrial scale for the synthesis of several
different fine chemicals [67]. For example, a Heck reaction using Pd/C as a heterogeneous catalyst in the presence of sodium carbonate in NMP was used in the synthesis
of 2-ethylhexyl p-methoxy-cinnamate, a commonly used compound in sunscreen
(Figure 6.39) [67].

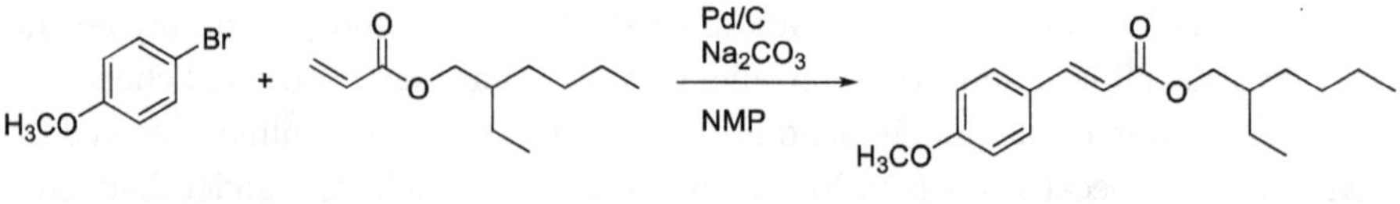

Figure 6.39: Synthesis of a sunscreen agent via the Heck reaction [67].

In another example, a synthetic process for Naproxen™ was developed that used a Heck reaction as a key step (Figure 6.40) [67]. In this synthesis, 2-bromo-6-methoxynaphthalene was reacted with ethylene at 30 bar in the presence of PdCl$_2$ and neomenthyldiphenylphosphine as a ligand in the presence of trimethylamine to form the vinyl naphthalene. From the vinyl naphthalene, a hydroxycarboxylation of the alkene using CO, PdCl$_2$, CuCl$_2$, and aqueous HCl gave the target naproxen.

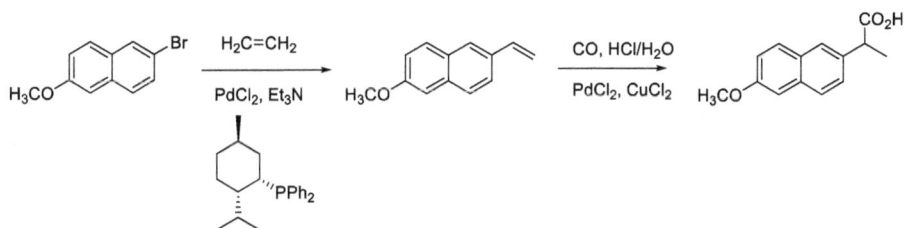

Figure 6.40: Application of the Heck reaction for the synthesis of Naproxen™ [67].

6.4.2 The Sonogashira reaction

Aryl iodides can react with copper acetylides to form aryl alkynes. The reaction is known as the Castro–Stephens reaction and involves oxidative addition of the aryl iodide to the copper acetylide, followed by reductive elimination to form the aryl–alkyne bond (Figure 6.41) [68, 69]. The copper acetylide is generated in situ from the corresponding alkynyl lithium or Grignard reagent by treatment with a copper (I) salt such as CuI. The Castro–Stephens requires a stoichiometric copper and elevated temperatures.

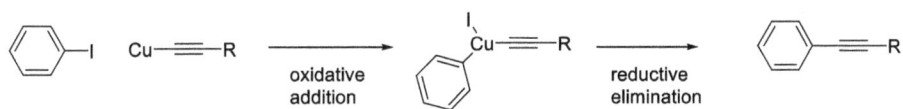

Figure 6.41: The Castro–Stephens reaction [68, 69].

A palladium-catalyzed variation of the Castro–Stephens reaction reacts the terminal acetylene with the aryl halide using catalytic CuI. This reaction is known as the Sonogashira reaction and is used more widely because it uses catalytic copper and lower temperatures (Figure 6.42) [70–72]. The copper acetylide is generated in situ by reaction of the terminal alkyne with CuI in the presence of an amine base.

The catalytic cycle for the Sonogashira coupling is depicted in Figure 6.43 [71]. In many respects, the mechanism is thought to proceed via the same general mechanism as some of the other cross-coupling reactions: The active catalyst is a coordinatively unsaturated Pd(0) species which reacts with the aryl halide in an oxidative

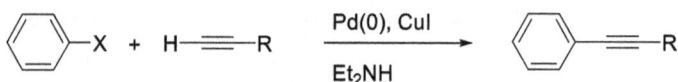

Figure 6.42: The Sonogashira reaction of an aryl halide with a terminal alkyne.

addition reaction. A transmetallation with the copper acetylide is followed by a reductive elimination to give the aryl acetylene and regenerate the Pd(0). The copper acetylide is formed when alkyne reacts with the copper(I) iodide in the presence of the amine. During the transmetallation step, the Cu(I)X is reagenerated, allowing the reaction to proceed with both catalytic amounts of the palladium catalyst and copper(I) iodide.

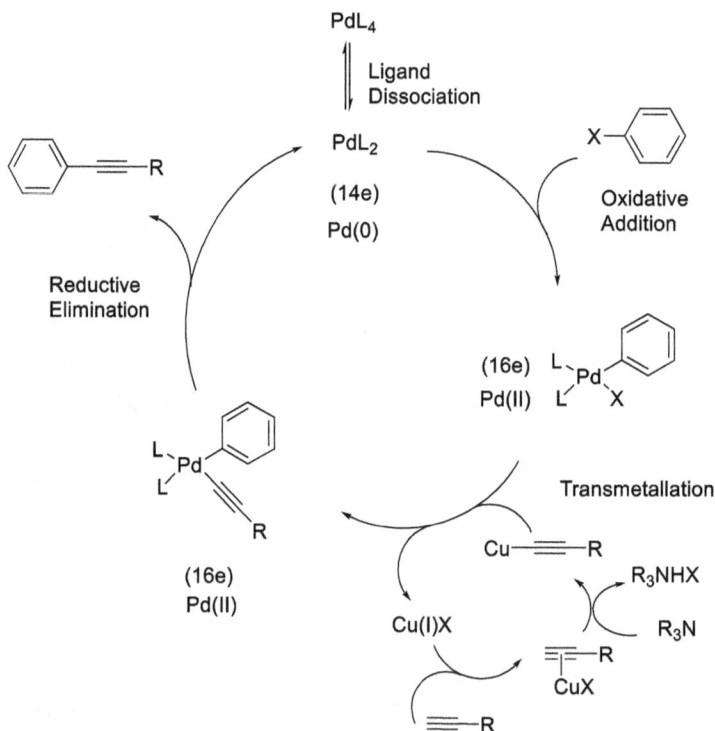

Figure 6.43: Catalytic cycle of the Sonogashira reaction.

The trend in reactivity of aryl halides in the Sonogashira coupling is similar to that of the cross-coupling reactions described previously: aryl iodides are the most reactive, followed by aryl bromides and triflates. Aryl chlorides are relatively unreactive and require highly activated catalysts. This difference in reactivity can be exploited to control the Sonogashira reactions in substrates bearing more than one halogen. At room temperature, aryl iodides will undergo Sonogashira reactions while aryl bromides react very slowly. At higher temperatures, aryl bromides react readily.

When diaryl acetylenes or terminal aryl acetylenes are the target, protected acetylenes are often used, rather than acetylene itself. Among the most commonly used reagents is trimethylsilyl acetylene as well as other related silyl acetylenes (Figure 6.44) [73]. The trimethylsilyl groups can be readily removed using fluoride, or under slightly basic (such as catalytic K_2CO_3 in methanol/dichloromethane). In some cases, in situ deprotection can be used to carry out two sequential Sonogashira coupling in one pot to yield a diaryl acetylene. Another common reagent for Sonogashira couplings is 2-methyl-3-butyn-2-ol [74, 75]. The protecting group can be removed under more strongly basic solutions of hydroxide, liberating acetone in the process (Figure 6.44). While 2-methyl-3-butyn-2-ol is less expensive than trimethylsilyl acetylene, the conditions for removing the protecting group are harsher.

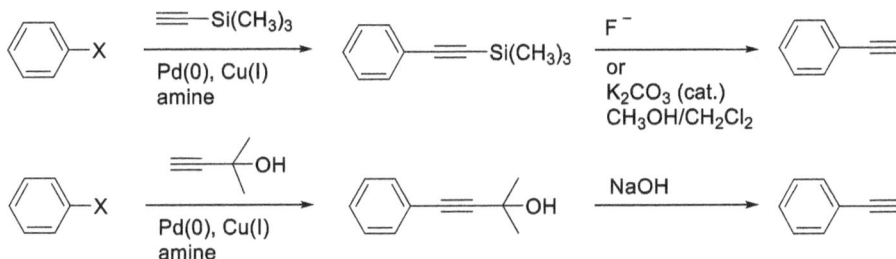

Figure 6.44: Examples of common reagents for the introduction of acetylenes via Sonogashira coupling.

The Sonogashira coupling has been used widely for the preparation of conjugated compounds that are luminescent or possess desirable properties for applications in organic electronics. For example, Yamaguchi and coworkers reported the preparation of luminescent oligo(phenyleneethynylenes) via sequential Sonogashira coupling reactions (Figure 6.45) [76]. This sequence demonstrates how the difference in reactivity between iodides and bromides can be used for sequential Sonogashira reactions and also makes use of an in situ deprotection of the TMS protecting group.

The Sonogashira reaction has also been used by the groups of Moore and Haley for the synthesis of phenylene ethynylene macrocycles and related dehydrobenzannulenes [77–80]. For an example, we will look at the synthesis of a functionalized dehydrobenzo[18]annulene (Figure 6.46) [80]. The synthesis begins by electrophilic iodination of 4-nitroanline. The aniline was then diazotized and converted to the dialkyltriazene, which we saw in Section 3.6.3 could be used as a masked aryl iodide. The iodo-substituted triazene derivative was subjected to a Sonogashira coupling with (triisopropylsilyl)acetylene. The triazene was converted to the corresponding aryl iodide using iodomethane [81], which allowed a second Sonogashira coupling, this time with (trimethylsilyl)butadiyne. In the following step, the TMS groups were selectively removed

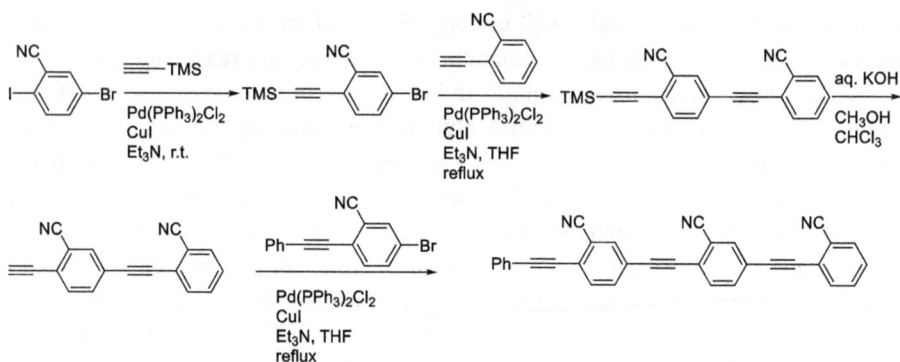

Figure 6.45: Synthesis of a conjugated oligo(phenyleneethynylene) via sequential Sonogashira reactions [76].

using KOH in situ, allowing a twofold Sonogashira coupling with 4,5-diiodoveratrole. The TIPS groups were removed using TBAF, and the terminal alkynes were subjected to copper-mediated oxidative dimerization to give the desired macrocycle.

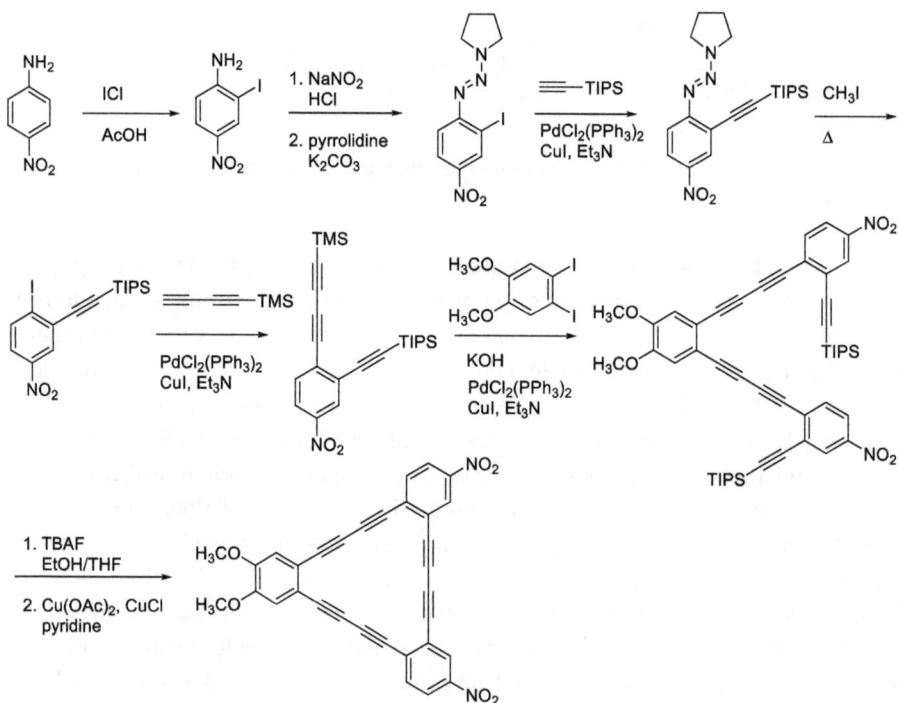

Figure 6.46: Synthesis of a functionalized didehydrobenzo[18]annulene [80].

As another illustrative example of the use of the Sonogashira coupling in multistep synthesis, we will turn to the synthesis of an anthropomorphic molecule referred to as "Nanokid," reported by Tour in 2003 (Figure 6.47) [82]. This synthesis also serves to highlight some of the reactions discussed in earlier chapters. The synthesis begins with electrophilic iodination of 1,4-dibromobenzene to give 1,2-dibromo-2,5-diiodobenzene. A Sonogashira coupling with two equivalents of 3,3′dimethylbutyne using PdCl$_2$(PPh$_3$)$_2$ and CuI with trimethylamine and THF resulted in the alkynes reacting at the iodo positions selectively to give the dialkynyl benzene. Subsequently, lithium–halogen exchange at one of the bromines and trapping with DMF gave the aldehyde, which was converted to the corresponding cyclic acetal with ethylene glycol in the presence of acid. To increase the reactivity of the remaining bromo group, it was converted into the corresponding iodide by lithium–halogen exchange and trapping with 1,2-diiodoethane. This gave "Fragment A," the upper body of the Nanokid.

Turning to the lower body, 4-nitroanaline underwent twofold bromination *ortho* to the amino group. The amino group was then removed by converting it to the diazonium salt with sodium nitrite and reaction with ethanol. The nitro group was then reduced using tin(II) chloride to give 3,5-dibromoaniline. This compound was again converted to the diazonium salt and trapped with KI to give the corresponding iodide. Another Sonogashira coupling with trimethylsilyl acetylene took place selectively at the iodo position to give the protected aryl acetylene. Another twofold Sonogashira coupling using 1-pentyne attached the "legs," resulting in the triethynyl benzene. The trimethylsilyl protecting group was then removed to give the lower body with the terminal alkyne. The terminal alkyne of the lower body was then reacted with fragment A under Sonogashira coupling to give the Nanokid as the final product.

It should be noted that Tour extended their synthetic approach to prepare various different anthropomorphic molecules (termed "nanoputians") [82], and also used a synthetic approach relying heavily on Sonogashira couplings to synthesize single-molecule "nanocars" [83].

Beyond the synthesis of anthropomorphic molecules, the Sonogashira reaction has been used in the synthesis of active pharmaceutical ingredients. As an example, researchers at Novartis made use of the Sonogashira coupling in the synthesis of an antimitotic agent which was investigated for the treatment of specific skin disorders. They made use of a stepwise one-pot reaction to prepare the diarlyacetylene precursor to their target compound (Figure 6.48) [84]. Specifically, 2-bromo-1,4-dimethoxybenzene reacted with 2-methyl-3-butyn-2-ol in the presence of PdCl$_2$, PPh$_3$, and CuI with i-Pr$_2$NH in toluene to carry out the first Sonogashira coupling. A phase transfer catalyst and aqueous base were added to deprotect the alkyne, which was then treated with the bromo-substituted quinazoline to give the desired diaryl acetylene. The target compound was accessed by hydrogenation of the acetylene. This example showcases the possibility of conducting stepwise Sonogashira

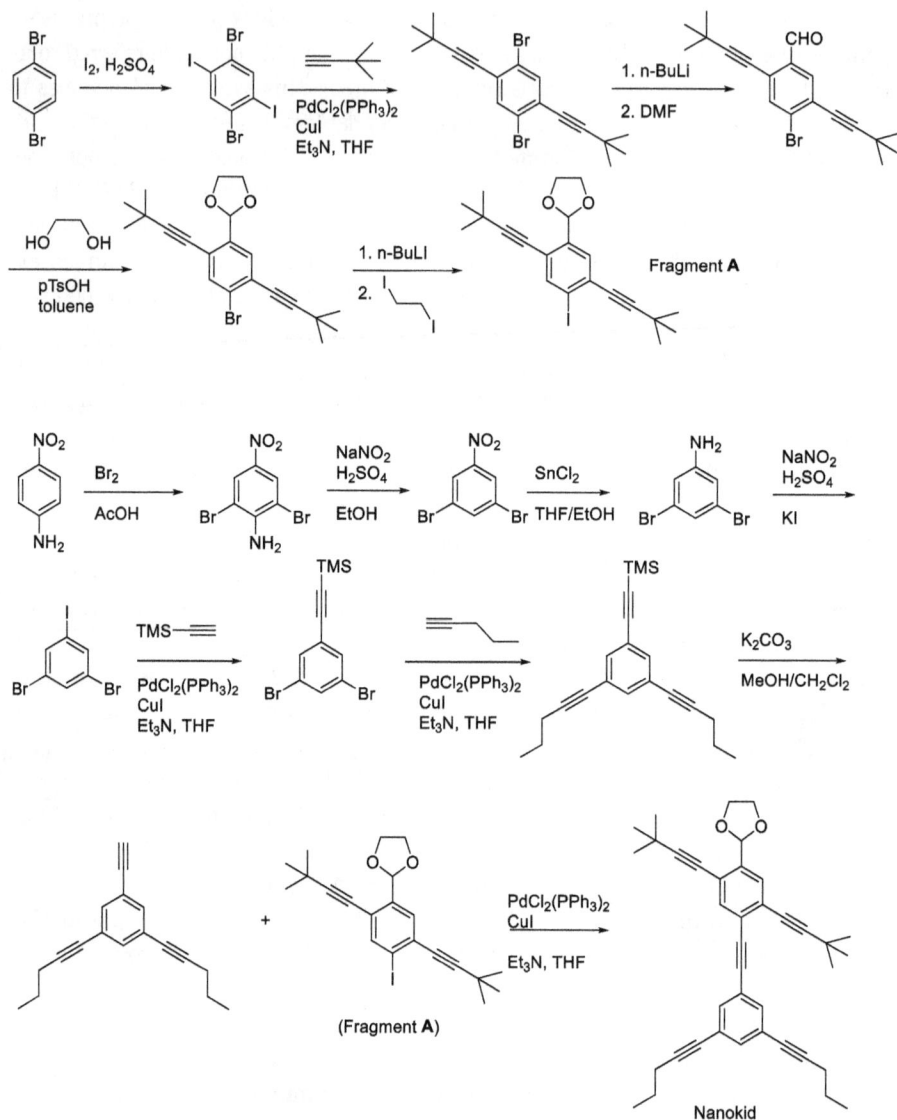

Figure 6.47: Synthesis of Nanokid [82].

couplings in one pot. It also demonstrates that the Sonogashira coupling can be used for the indirect introduction of saturated carbon chains.

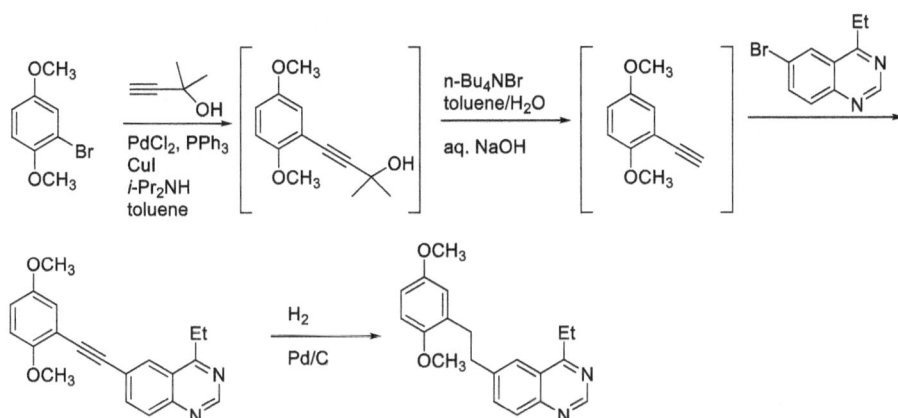

Figure 6.48: A one-pot twofold Sonogashira reaction for the preparation of an API [84].

6.4.3 Transition metal-mediated cyanation

We have seen that cyano groups can be introduced onto aromatic rings via the diazonium salt in Chapter 3. An alternative approach that could be used is nucleophilic aromatic substitution. However, substitutions at unactivated aryl halides can be challenging, these types of transformations can be achieved mediated or catalyzed by metals. For example, aromatic halides undergo substitution in the presence of copper metal or copper salts. A long-established example of this type of reaction is the preparation of aryl nitriles from the corresponding aryl halides using cuprous cyanide, known as the Rosendmund–von Braun reaction (Figure 6.49) [85].

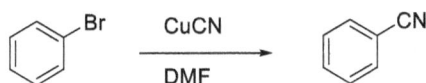

Figure 6.49: Rosendmund–von Braun reaction.

While one could be tempted to view this as an S_NAr reaction, the reactivity of the aryl halide is in the order of Ar–I > Ar–Br > Ar–Cl. Furthermore, the reaction requires Cu(I) CN – other cyanides such as potassium or sodium cyanide do not react. Instead, the reaction is thought to involve oxidative addition of the aryl halide to the copper(I) salt, which then undergoes reductive elimination. While the reaction can in principle be catalytic because copper (I) is regenerated, in this particular reaction, the copper salt also serves as the source of the cyanide nucleophile, so it is used in stoichiometric quantities. Furthermore, the reaction usually requires high temperatures. Alternative approaches use palladium catalysts with cyanide sources such as $Zn(CN)_2$ or KCN to prepare aryl nitriles from the corresponding aryl halides or triflates [85, 86]. Palladium-catalyzed cyanations have even been extended to aryl chlorides, which do not react readily in Rosendmund–von Braun reaction conditions [87].

These palladium-catalyzed cyanation reactions have been used for the preparation of active pharmaceutical ingredients containing nitriles. For example, in the synthesis of taranabant, a candidate for the treatment of obesity, a palladium-catalyzed cyanation was used to convert the aryl bromide to the nitrile (Figure 6.50) [88].

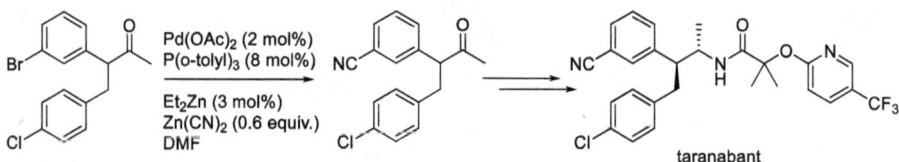

Figure 6.50: Palladium-catalyzed cyanation for the synthesis of taranabant.

As another example, a palladium-catalyzed cyanation was also used for the preparation of citalopram, which is used as a treatment for depression. During development, this reaction was carried out under microwave conditions using $Zn(CN)_2$ as the cyanide source and a catalyst system of $Pd_2(dba)_3$ and Xantphos as the ligand (Figure 6.51) [89].

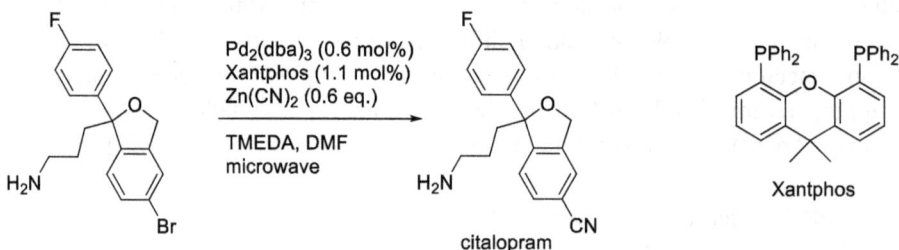

Figure 6.51: Palladium-catalyzed cyanation in the synthesis of citalopram.

6.5 Direct arylation (C–H functionalization)

6.5.1 Introduction to direct arylation

The transition metal-mediated cross-coupling reactions discussed thus far require an aryl halide (or pseudohalide such as a triflate) and an organometallic species. An attractive alternative to these reactions is a coupling directly at a C–H site of an aromatic ring. This direct arylation approach has the advantage that it avoids the need for prior functionalization of the aromatic ring. One potential challenge with this type of approach is that regiochemistry may be difficult to control where there are several

C–H sites available for reaction. Despite the potential regiochemical challenges, direct arylation has emerged as a viable tool for the formation of aryl–aryl bonds.

Direct arylation can be classified into three general types: (1) direct arylation with aryl halides, (2) oxidative direct arylation with organometallic reagents, and (3) direct arylation with arenes, which is sometimes referred to as dehydrogenative arylation (Figure 6.52) [90, 91].

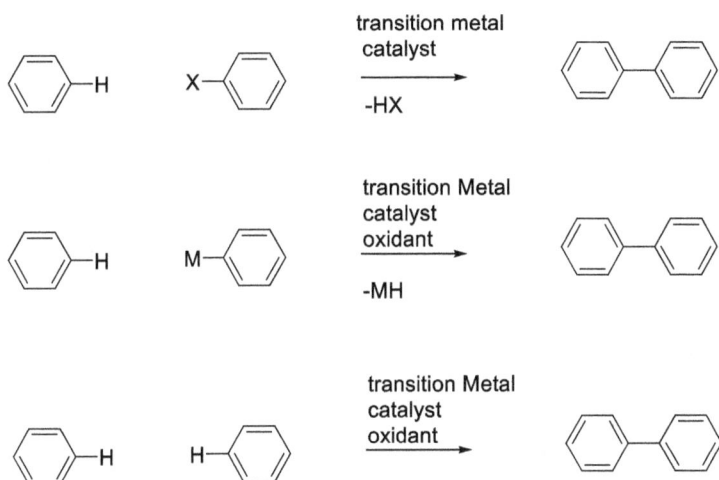

Figure 6.52: General types of direct C–H arylation.

Direct arylation reactions usually make use of Pd catalysts, although Rh and Ru have also been used. These reactions are usually carried out in polar aprotic solvents such as DMF, DMA, DMSO, or CH_3CN at elevated temperatures. The reaction usually involves a base such as a carbonate or pivalate. The role of the base is not always clear, although in some cases there is evidence that it is involved directly in the C–H insertion step.

6.5.2 Direct arylation with aryl halides

The direct C–H functionalization using aryl halides or pseudohalides have been extensively explored. The general mechanism involves oxidative addition to a Pd(0) catalyst, followed by a C–H insertion to generate a palladium biaryl species, which undergoes reductive elimination to yield the biaryl and regenerate the Pd(0) (Figure 6.53) [90, 91].

The mechanism of the C–H insertion step is central to the reaction and may vary depending on the reaction conditions and substrates, and is not clearly established for all reactions. Some of the proposed mechanisms for C–H insertion include:

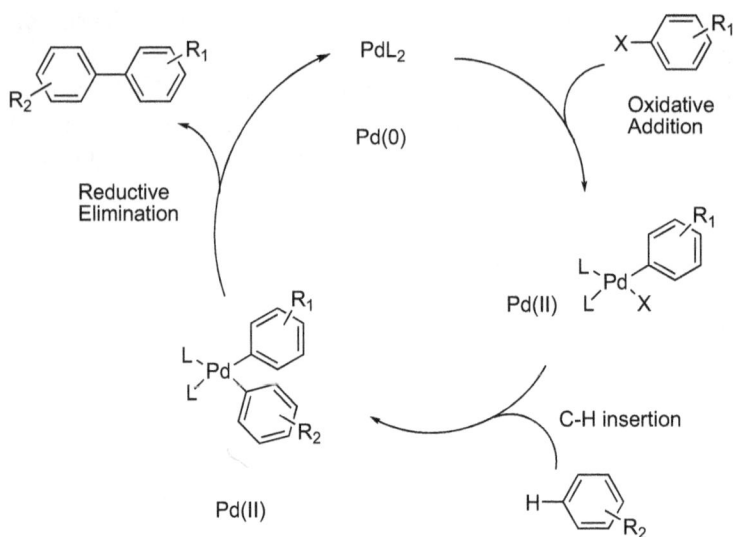

Figure 6.53: General mechanism of direct arylation involving aryl halides with arenes.

(i) electrophilic aromatic substitution of the metal, (ii) C–H bond oxidative addition (which in the case of palladium catalysts would involve a Pd(IV) intermediate), (iii) a Heck-type reaction, (iv) a σ-bond metathesis reaction, or a concerted S_E3 reaction [90].

6.5.2.1 Intramolecular direct arylation reactions with aryl halides

As described above, one of the potential challenges of direct arylation is the regiochemistry of arylation, especially when multiple aryl C–H groups are available. In the case of intramolecular direct arylation reactions, the regiochemistry is largely determined by the cyclic metallocycle intermediate in the reaction, with five- and six-membered rings being the preferred final products. Some of the earliest examples of direct arylations were intramolecular cyclization reactions using Pd(OAc)$_2$ and base, conducted at elevated temperatures to yield benzofurans, carbazoles, and fluorenone (Figure 6.54) [90, 92].

Many of the early examples of intramolecular direct arylation made use of aryl iodides or aryl bromides. Fagnou and coworkers showed that direct arylations could be carried out with aryl chlorides, bromides, and iodides using Pd(OAc)$_2$ with PCy$_3$-HBF$_4$ as a ligand in the presence of K$_2$CO$_3$ [93]. These reaction conditions were used to prepare a wide variety of cyclized products with five- or six-membered rings. For some selected examples, see Figure 6.55. Interestingly, they observed that aryl iodides reacted poorly under these conditions, which they attributed to iodide poisoning the catalyst. This problem was avoided by using silver carbonate to scavenge the iodide.

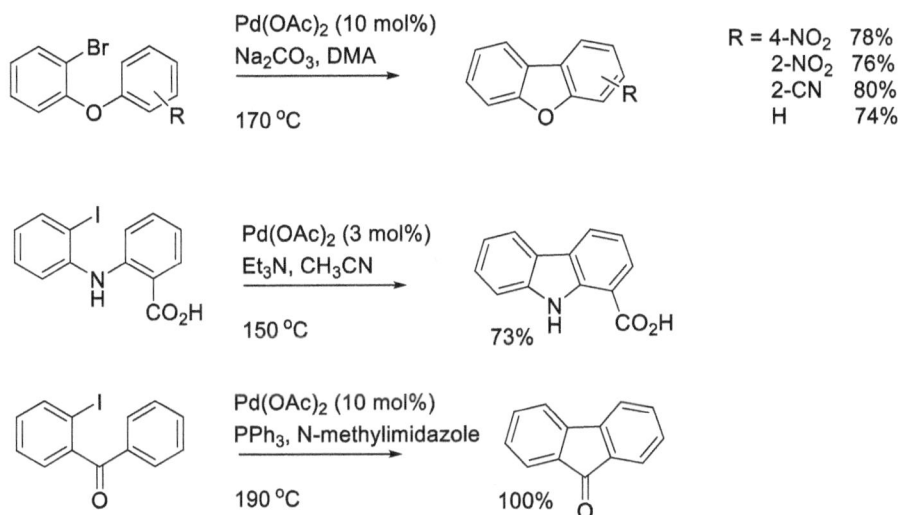

Figure 6.54 reaction schemes:

Scheme 1 (top):
Pd(OAc)$_2$ (10 mol%)
Na$_2$CO$_3$, DMA
170 °C

R = 4-NO$_2$ 78%
    2-NO$_2$ 76%
    2-CN 80%
    H   74%

Scheme 2 (middle):
Pd(OAc)$_2$ (3 mol%)
Et$_3$N, CH$_3$CN
150 °C
73%
(product with CO$_2$H)

Scheme 3 (bottom):
Pd(OAc)$_2$ (10 mol%)
PPh$_3$, N-methylimidazole
190 °C
100%

Figure 6.54: Representative examples of intramolecular direct arylation reactions.

For an example of the application of an intramolecular direct arylation in total synthesis, we will consider the synthesis of gilvocarin M. Recall that earlier in this chapter we considered the synthesis of defucogilvocarin, a gilvocarin structure is missing the sugar unit (Figure 6.31). In that synthesis, the target compound was prepared by a sequence involving directed *ortho*-metalation (DoM, Section 4.2.1), followed by Suzuki coupling (Section 6.3.1.4), and an anionic Fries rearrangement (Section 4.2.3). Suzuki and coworkers reported an alternative approach to gilvocarin M that uses an intramolecular direct arylation for the formation of the biaryl (Figure 6.56) [94]. First, the naphthol unit was prepared from the iodoaryl triflate by treatment with *n*-BuLi to form the benzyne, which underwent a Diels–Alder reaction with 2-methoxyfuran. The cycloadduct aromatized under the reaction conditions to form the substituted naphthol. The naphthol was acylated with the aryl acid chloride, which was followed by direct arylation using (PPh$_3$)$_2$PdCl$_2$ with sodium acetate in *N*,*N*-demethylacetamide to give the tetracyclic product. Finally, cleavage of the benzyl groups gave (+)-gilvocarin M.

6.5.2.2 Direct arylation using directing groups

While intramolecular direct arylations usually proceed with good regioselectivity, intermolecular direct arylations can be more challenging. One approach to control the regiochemistry of direct arylation reactions is to use directing groups. Typically, these directing groups possess a lone pair of electrons to coordinate to the transition metal catalyst. The C–H insertion then takes place near the directing group, usually proceeding through a five- or six-membered metallacycle intermediate. Examples of directing groups that have been used in these direct arylations include

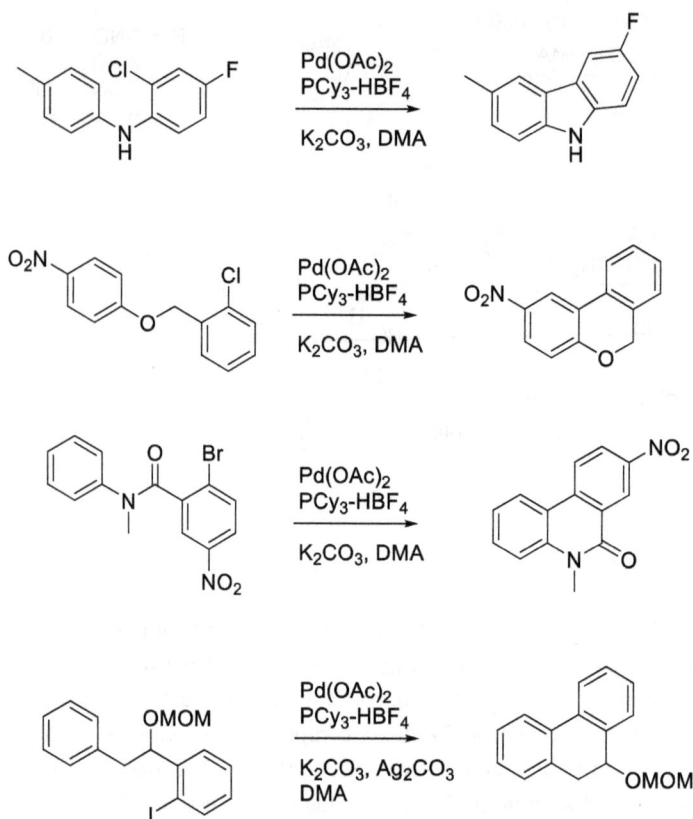

Figure 6.55: Representative intramolecular direct arylations of aryl chlorides, bromides, and iodides [93].

phenols, carbonyls, imines, pyridines, oxazolines, and related five-membered *N*-heteroaromatic compounds.

For example, Miura reported the regioselective arylation of compounds such as 2-phenylphenol and 1-naphthol (Figure 6.57) with aryl iodides [95]. These reactions both use catalytic Pd(OAc)$_2$, cesium carbonate as the base, and take place in DMF at elevated temperatures. For both reactions, the aryl group is added selectively to the adjacent ring nearest the phenol group.

Mechanistically, this reaction is thought to proceed by in situ formation of the Pd(0) complex, which undergoes oxidative addition with the aryl iodide. The Pd(II) complex then coordinates to the phenoxide, followed by a C–H insertion on the adjacent ring and then reductive elimination. Cesium carbonate was essential for the reaction, likely because its high solubility in DMF facilitated the C–H insertion step by facilitating deprotonation. As an extension of the previous example, multiple arylations can take place when an excess of the aryl halide is used. For example,

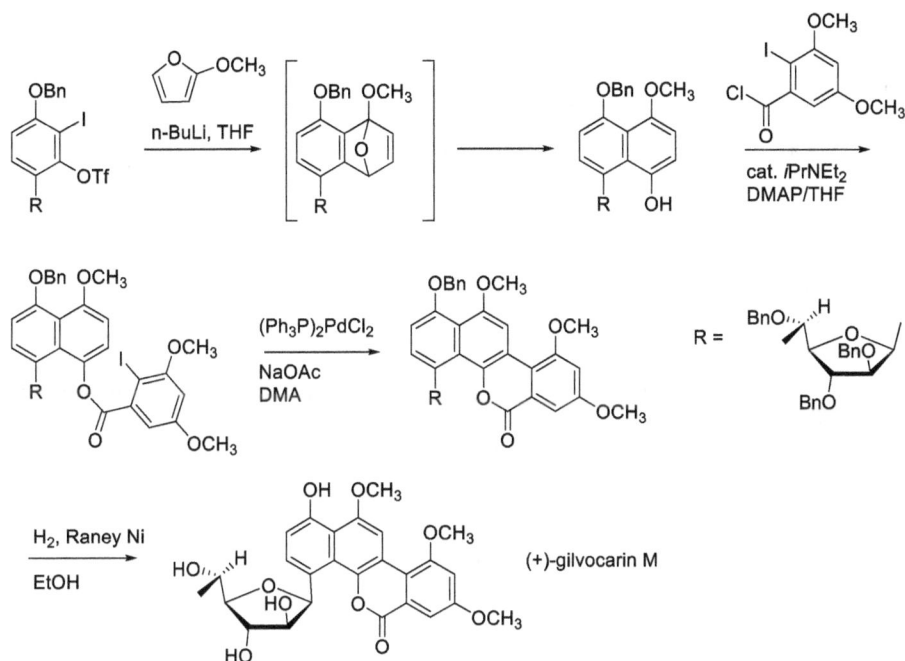

Figure 6.56: Synthesis of gilvocarin using an intermolecular direct arylation reaction [94].

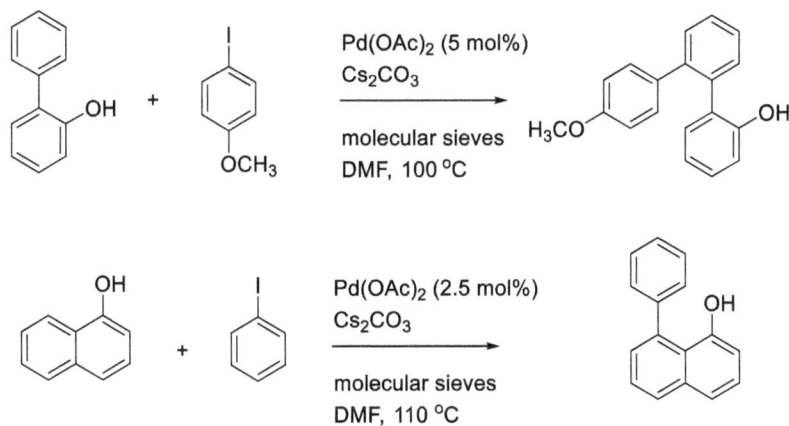

Figure 6.57: Intermolecular direct arylations using phenols as directing groups [95].

phenol can undergo direct arylation and when this reaction is carried out with a large excess of the aryl halide, can result in pentaarylation (Figure 6.58) [96]. *Ortho* arylation of phenols with aryl bromides has also been carried out under similar conditions using rhodium(I) catalysts such as RhCl(PPh$_3$)$_3$ or [RhCl(COD)]$_2$.

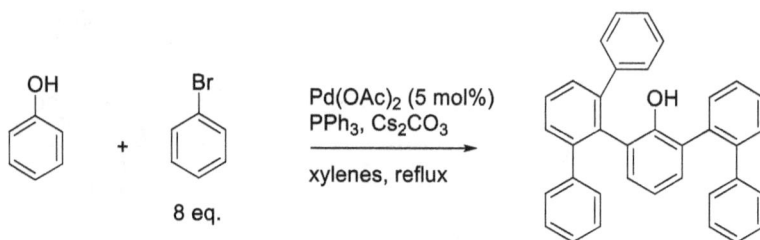

Figure 6.58: A polyarylation reaction in palladium-catalyzed direct arylation.

6.5.2.3 Direct arylation without directing groups

Among the early examples of direct arylations lacking a directing group, Fagnou and coworkers reported the direct arylation of electron-deficient fluorinated benzenes using aryl halides (Figure 6.59) [97]. The reaction works well for electron-rich and electron-deficient aryl halides. For polyfluorinated benzenes where more than one C–H site is available, in some cases diarylation is observed. However, diarylation can be minimized by using an excess of the fluorobenzene. In these systems, the reactivity depended on the acidity of the C–H group, suggesting that anionic ligands or bases are involved in the C–H bond breaking.

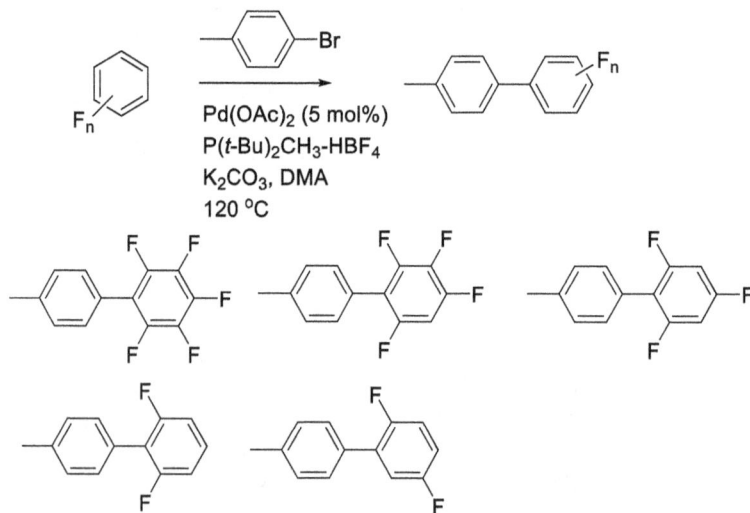

Figure 6.59: Direct arylation reactions of polyfluorinated benzenes.

Following this study, Fagnou and coworkers extended this study to the direct arylation of benzene using aryl halides in the presence of Pd(OAc)$_2$, a activating phosphine ligand such as DavePhos, potassium carbonate, and pivalic acid as a co-catalyst (Figure 6.60) [98]. These conditions furnished the corresponding biaryls in good yields. Under these conditions, it is thought that the pivalate binds to the palladium and plays an important role in the C–H insertion step by facilitating the deprotonation of the benzene. Competition experiments with benzene and electron-rich anisole or electron-deficient fluorobenzene show that the electron-deficient arene is more reactive. However, depending on the substrate, controlling regiochemistry of the direct arylation in these systems can still be a challenge.

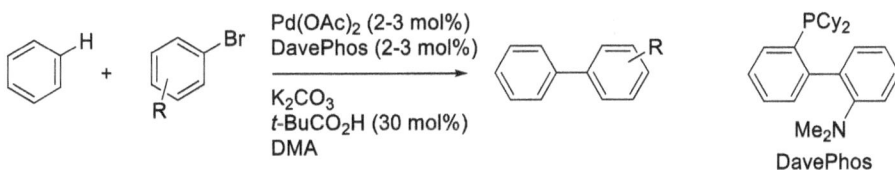

Figure 6.60: Palladium-catalyzed direct arylation of benzene using an activated phosphine ligand and pivalic acid as a co-catalyst.

6.5.2.4 Direct arylation of heteroaromatic compounds

Direct arylation of heteroaromatic compounds has been explored extensively. There are many examples of intramolecular direct arylation reactions involving heteroaromatic rings [90]. As with other intramolecular direct arylations, the regiochemistry of intramolecular direct arylations with heteroaromatic compounds is often determined by the size of the ring being formed. In contrast, for intermolecular direct arylation of heteroaromatics, the structure of the heteroaromatic system has a significant influence on the regiochemistry of arylation, although the catalyst, ligand, and other reaction conditions can also be important. Here, we will briefly consider the regiochemistry of direct arylation for some of the common heteroaromatic systems.

Pyrroles, furans, and thiophenes, which are all electron-rich aromatic consisting of five-membered rings, generally react at the 2-positions, adjacent to the heteroatom (Figure 6.61). Similarly, benzofurans and benzothiophenes undergo arylation at the 2-position. Some of these reactions are sensitive to electronics, and will work most effectively with electron-deficient aryl halides [90].

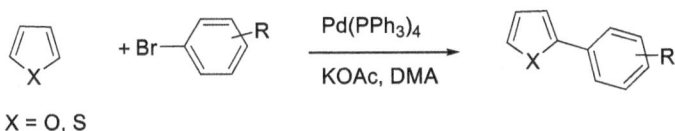

X = O, S

Figure 6.61: General example of direct arylation of furans and thiophenes.

Conjugated thiophene derivatives have been the focus of considerable attention due to their potential applications as fluorophores and in organic electronic materials. Given that direct arylation methods introduce aryl groups at the 2-positions of thiophenes, this presents an alternative approach as compared with transition metal-catalyzed cross-coupling using organometallic reagents. Indeed, Schipper and Fagnou have shown that direct arylation is synthetically useful for the preparation of several thiophene-based materials, often offering advantages over the typical cross-coupling methods [99]. For example, Figure 6.62 shows the synthesis of a conjugated compound that was investigated as a fluorescence marker for amyloid aggregate associated with Alzheimer's disease [100]. The original synthesis formed the aryl thiophene bond by preparation of the tributylstannyl thiophene, followed by a Stille coupling. Schipper and Fagnou showed that this transformation could be achieved in one step via a direct arylation, avoiding the use of toxic organotin species [99]. Following this, the target compound could be accessed following the original report: the formyl group was installed by lithiation and trapping with DMF. Condensation with malononitrile, followed by deprotection gave the desired product.

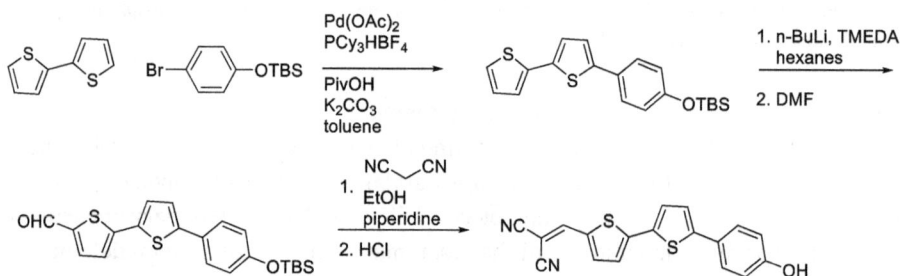

Figure 6.62: Preparation of a fluorescent biomarker using direct arylation.

The regiochemistry of direct arylation of indoles depends on the structure of the indole substrate, the reacting alkyl halide, and the reaction conditions and generally favors substitution at either the 2- or 3-position of the indole ring [101–103]. Early studies on the direct arylation of indoles showed that *NH* and *N*-alkyl indoles undergo direct arylation at the 2-position, while *N*-tosyl indoles undergo arylation at the 3-position, especially when the aryl halide is bulky (Figure 6.63) [101].

The differences in regiochemistry can be rationalized by considering the proposed mechanism of the arylation reaction (Figure 6.64) [103]. Following oxidative addition of the aryl halide, the reaction is thought to proceed via an electrophilic metalation to yield a cationic intermediate with the metal at the 3-position. At this stage, two possible pathways can occur: 1,2-migration of the palladium, followed by deprotonation and reductive elimination to give the 2-arylindole, or deprotonation instead of 1,2-migration to give the 3-metalloindole, followed by reductive elimination give the 3-arylindole. For unhindered substrates, the 1,2-migration occurs readily, favoring the 2-arylindole, while for bulky aryl groups, the 1,2-migration is slow, so the 3-arylindole is favored.

There are not many reported examples of intermolecular direct arylation of pyridines. Direct arylation at the 2-position of pyridine was achieved with aryl halides in the presence of Pd/C and zinc, although the yields were modest [104]. Fagnou

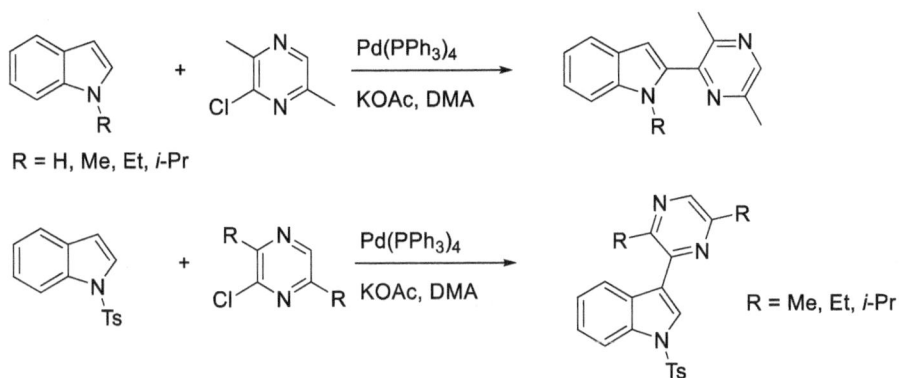

Figure 6.63: Direct arylation of indoles with chloropyrazines [101].

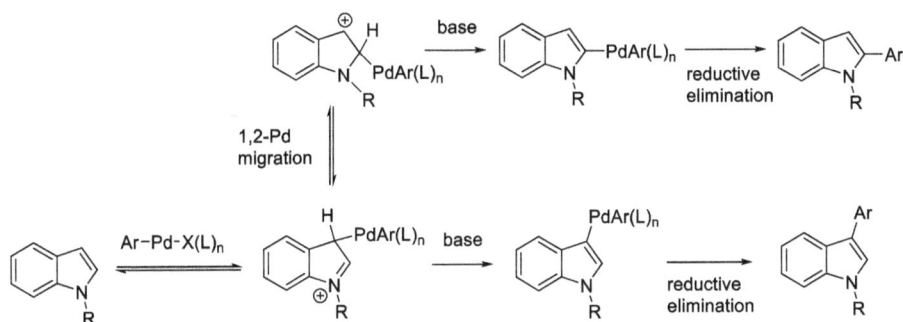

Figure 6.64: Proposed mechanism of indole arylation showing the formation of 2- and 3-arylindoles.

reported a higher-yielding direct arylation of pyridine *N*-oxides with aryl bromides and showed that reaction took place selectively at the 2-positions in good yields (Figure 6.65) [105]. The pyridine-*N*-oxide could readily be converted into the corresponding pyridine using ammonium formate with Pd/C. This approach provides a convenient route to 2-arylpyridines, which is important because 2-pyridyl organometallic reagents such as boronates and stannanes that could be used for traditional cross-coupling reactions tend to be unstable.

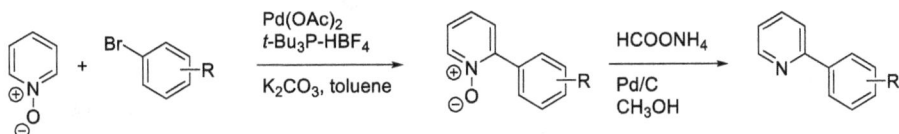

Figure 6.65: Direct arylation of pyridine *N*-oxides to yield 2-arylpyridines.

One example of the use of direct arylation of heteroarenes is in the preparation of an imidazotriazine derivative (Figure 6.66) [106]. The target compound is a GABA agonist, so researchers at Merck were interested in developing a synthesis that would permit more detailed investigation of its biological activity. The synthesis began with a Suzuki coupling of 2-fluorophenylboronic acid with 2-bromo-3-fluorobenzonitrile to give the corresponding biphenyl. Following this, electrophilic bromination was achieved using dibromodimethylhydantoin. The direct arylation reaction of the aryl bromide and the imidazotriazine was achieved using Pd(OAc)$_2$ with PPh$_3$ and KOAc in dimethylacetamide gave the target compound in good yield.

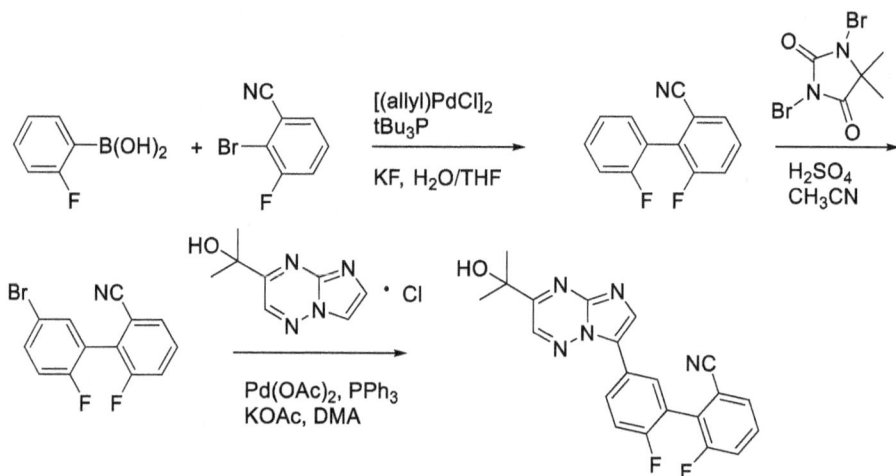

Figure 6.66: Synthesis of a biologically active imidazotriazine by direct arylation [106].

6.5.3 Direct arylation using organometallic species

Direct arylations involving organometallic species such as boronic acids have not been as extensively used as aryl halides, but nonetheless offer a complementary approach for carbon-carbon bond formation [91, 107]. Most commonly carried out using palladium or rhodium catalysts, these reactions share some common mechanistic features with the reactions involving aryl halides with two noteworthy differences: the first step of the reaction is a transmetallation reaction involving a Pd(II) catalyst, an oxidant is needed to regenerate the Pd(II) species after the reductive elimination step (Figure 6.67) [91, 107]. Thus, the catalytic cycle begins with a C–H insertion reaction on the Pd(II) catalyst, leads to loss of HX. This step is followed by a transmetallation using the organometallic reagent. Reductive elimination leads to the formation of the aryl–aryl bond and a Pd(0) species. The Pd(0) complex is then oxidized using a stoichiometric oxidant to regenerate the Pd(II) species.

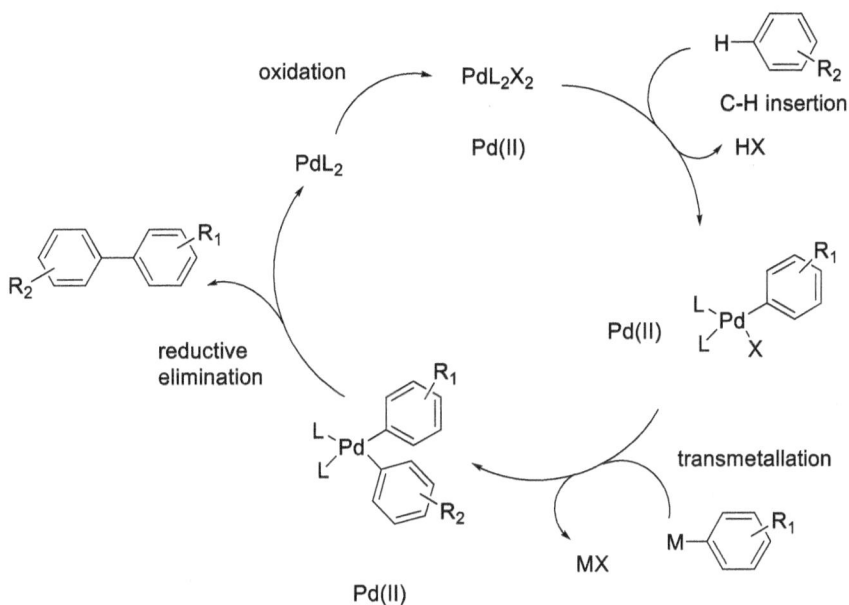

Figure 6.67: General mechanism of direct arylation involving organometallic species with arenes.

For example, boronic acids and boronate esters have been used in direct arylation reactions using palladium or rhodium catalysts in the presence of an oxidant [108–110]. The reaction is not limited to aryl boronic acids, but can also be used to introduce alkyl groups (such as methyl groups) onto aromatic rings. To control regioselectivity, pyridyl and carboxylic acid groups have been used as directing agents. Figure 6.68 shows some representative direct arylations involving boronic acids, boronates, or trifluoroborates using either palladium or rhodium catalysts, and a stoichiometric oxidant, such as silver oxide, benzoquinone, TEMPO, or oxygen [108–110].

6.5.4 Direct arylation with arenes: dehydrogenative coupling

In Section 6.2.3, we explored the oxidative coupling of aromatic compounds using metal salts. In those reactions, the metal served as a stoichiometric oxidant. It is also possible to perform transition metal-catalyzed coupling of two arenes. The reaction is usually carried out with a Pd(II) catalyst in the presence of an oxidant. Mechanistically, the reaction is thought to proceed via C–H insertion into the Pd(II) catalyst, followed by a second C–H insertion to give a diaryl palladium intermediate, which undergoes reductive elimination to yield the biaryl and a Pd(0) species (Figure 6.69) [111]. The Pd(0) species is then reoxidized to Pd(II).

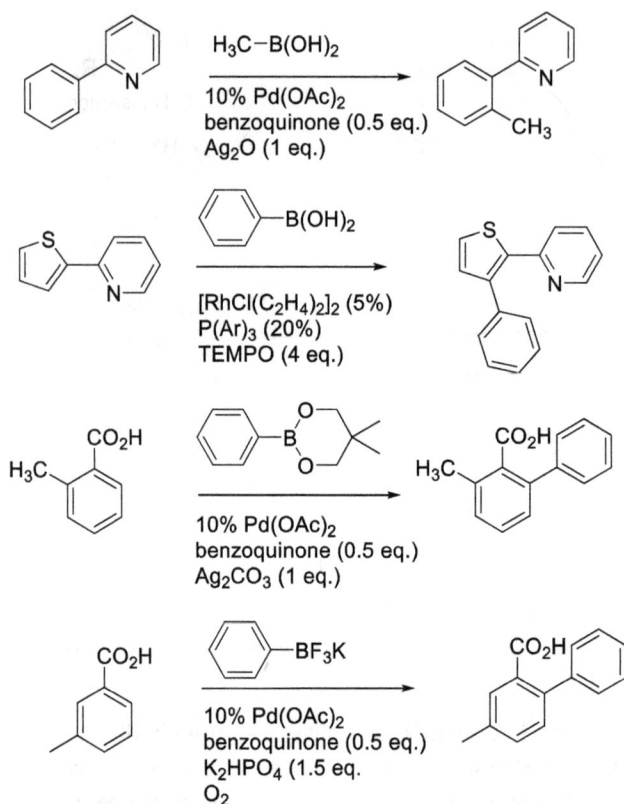

Figure 6.68: Direct arylation of boronic acids in the presence of an oxidant.

In addition to the challenge of controlling the regiochemistry of the direct arylation, achieving a direct arylation reaction between two different arenes is a challenge because a mixture of products is often observed resulting from homocoupling and heterocoupling. This challenge imposes limitations on the scope of these arylation reactions.

There are several examples of palladium-mediated intramolecular oxidative couplings. Early examples involve the formation of dibenzofurans from the corresponding diphenyl ethers using stoichiometric amounts of palladium [112]. Subsequently, the formation of carbazole derivatives with catalytic palladium(II) salts with oxidants such as Cu(OAc)$_2$ or oxygen was reported [113–115]. More recently, Fagnou and coworkers have shown that diphenylamines undergo oxidative coupling to form the corresponding carbazoles in the presence of Pd(OAc)$_2$ and potassium carbonate and pivalic acid open to air, avoiding the need for added stoichiometric oxidizing agents (Figure 6.70) [116]. Ohno and coworkers extended this approach conduct a one-pot aryl amination and oxidative coupling [117, 118].

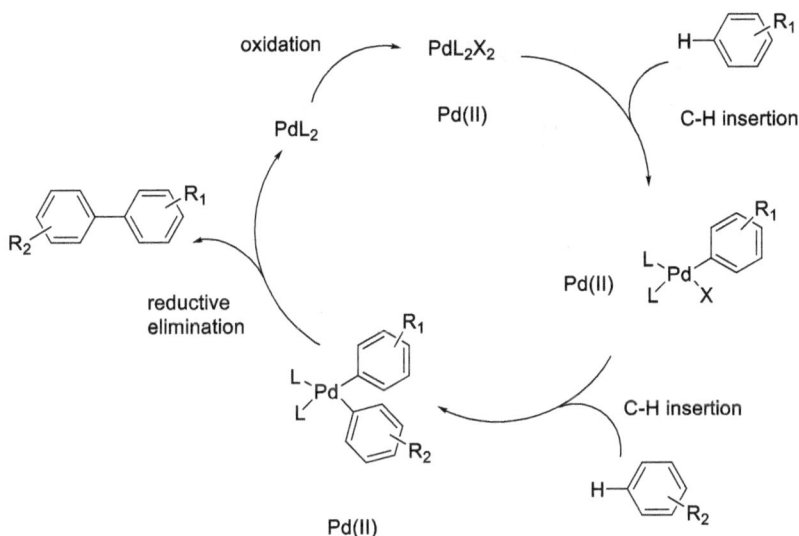

Figure 6.69: Proposed catalytic cycle for direct C–H arylation of arenes.

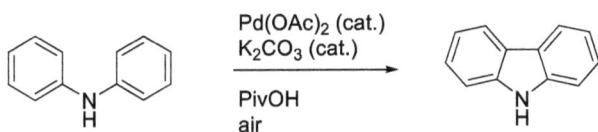

Figure 6.70: Formation of carbazole via a intramolecular dehydrogenative coupling.

One approach to achieve intermolecular cross-coupling is to control the stoichiometry of the reactants, usually using one of the coupling partners in excess [119]. Alternatively, when one of the arenes is an electron-rich heteroaromatic compound, effective heterocoupling can be achieved. For example, Stuart and Fagnou showed that N-acetyl indoles can undergoes palladium-catalyzed arylation with benzene derivatives [111], giving the 3-arylindoles as the major product, while DeBoef and coworkers reported different conditions for the reaction of benzofurans and indoles with unactivated arenes to give direct arylation at the two- or three-position of the heteroaromatic system (Figure 6.71) [120]. In the first example, Cu(OAc)$_2$ is used as the stoichiometric oxidant, and the additives such as 3-nitropyridine and cesium pivalate were found to improve the rate of reaction. In the second example, oxygen was used as the stoichiometric oxidant in the presence of catalytic heteropolymolybdovanadic acid.

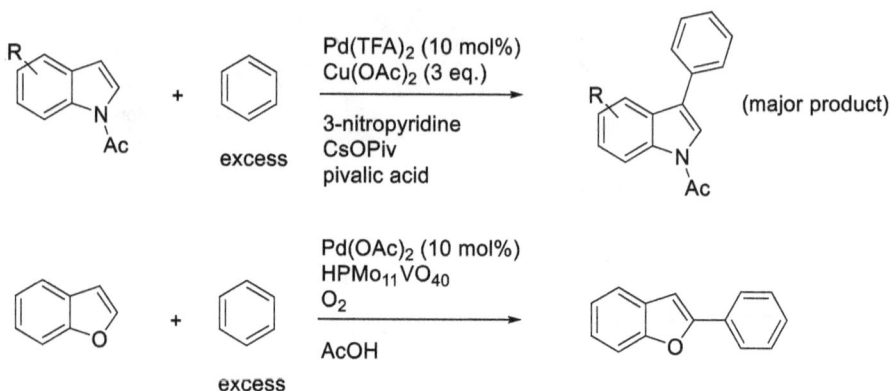

Figure 6.71: Direct C–H arylation of heteroaromatic compounds.

6.6 Axially chiral biaryls: synthetic approaches

6.6.1 Atropisomerism and axial chirality

Highly substituted biaryls bearing three or four groups *ortho* to the biaryl bond can be chiral. The chirality arises from a restricted rotation around the biaryl bond due to the presence of the *ortho*-substituents and is referred to as atropisomerism. As shown in Figure 6.72, the tetra-*ortho*-substituted biaryl is not the same as its mirror image, provided that at least some of the R groups are different (i.e., $R^1 \neq R^2$ and/or $R^3 \neq R^4$).

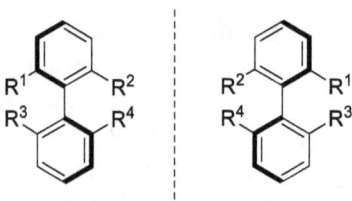

Figure 6.72: Enantiomers of atropisomeric biphenyls.

Atropisomeric biaryl motifs are found in many natural products natural products, so development of synthetic strategies for these chiral biaryls is important for the study of biologically active natural products and their synthetic analogs [121, 122]. Figure 6.73 shows korupensamine A, steganacin, and gossypol as three examples of natural products containing an atropisomeric biaryl.

Atropisomeric biaryls have also emerged as an important class of ligands for metal-catalyzed asymmetric synthesis. For example, compounds such as 2,2′-

binaphthol (BINOL), the corresponding phosphine (BINAP), and related ligands have found wide use in asymmetric catalysis (Figure 6.74).

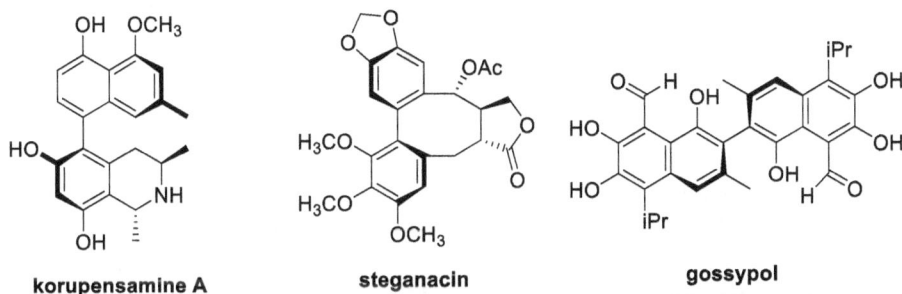

korupensamine A   **steganacin**   **gossypol**

Figure 6.73: Examples of natural products containing an atropisomeric biaryl motifs.

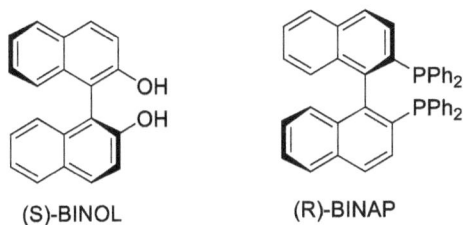

(S)-BINOL   (R)-BINAP

Figure 6.74: Atropisomeric binaphthyl derivatives.

As an example of the utility of atropisomeric biaryls as ligands for asymmetric synthesis, BINAP has been used effectively as a ligand in rhodium and ruthenium-catalyzed asymmetric hydrogenation reactions. For example, the industrial synthesis of the analgesic drug naproxen relies on an asymmetric hydrogenation using BINAP as a chiral ligand (Figure 6.75) [123, 124].

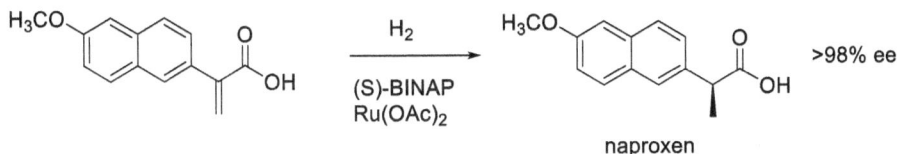

Figure 6.75: Synthesis of the analgesic naproxen using an asymmetric hydrogenation with a chiral BINAP ligand.

Axially chiral biaryls are also used as components in liquid crystalline systems to induce bulk, macroscopic properties that are essential for the applications of these systems [125–128]. Some examples of atropisomeric biaryls used to induce bulk chirality in liquid crystalline systems are shown in Figure 6.76.

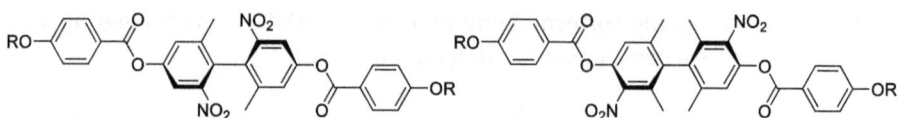

Figure 6.76: Examples of axially chiral biaryls used in liquid crystals [126, 128].

The successful synthesis of enantioenriched atropisomeric biaryls faces two challenges:

1. The biaryl bond with three or four *ortho*-substituents is very sterically congested, making the aryl–aryl bond formation potentially challenging.
2. The preparation of enantioenriched atropisomeric biaryls requires either a resolution of enantiomers, or preferably a stereoselective reaction to produce one enantiomer in high yield.

Below we will consider the general methods for preparing sterically congested biaryls in racemic form, and then explore some methods for preparing enantioenriched chiral biaryls.

6.6.2 Approaches to racemic atropisomeric biaryls

We have already seen several approaches for the synthesis of biaryl compounds. However, based on the sterically crowded nature of atropisomeric biaryls, some of these methods, including many of the standard cross-coupling approaches, are either ineffective or of more limited scope. Nonetheless, there are several approaches for the preparation of these compounds. In the absence of stereochemical control of the reaction, racemic mixtures are produced, which can be resolved classically or via methods such as chiral stationary phase HPLC to produced enantioenriched forms.

The Ullmann coupling can proceed even using highly substituted, sterically congested aryl halides (including those with two substituents *ortho* to the halogen) and can consequently be used to access atropisomeric biphenyls (e.g., Figure 6.77) [129].

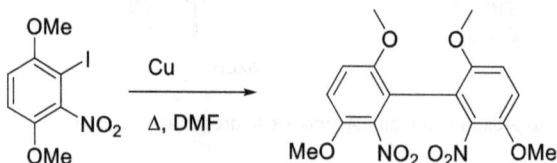

Figure 6.77: Synthesis of atropisomeric biaryls via Ullmann coupling [129].

Another common approach to access atropisomeric biaryls is an oxidative coupling using reagents such as copper salts or $FeCl_3$, a reaction that was introduced in

Section 6.2.3. As an example of oxidative coupling to prepare axially chiral biaryls, let us consider the synthesis of BINAP, a binaphthyl-derived phosphine ligand that is effective for asymmetric catalysis – especially asymmetric hydrogenations [130]. One of the syntheses of BINAP is outlined in Figure 6.78 [131]. The first step is an oxidative coupling of 2-naphthol using FeCl$_3$ to produce BINOL. BINOL was then converted to the corresponding dibromide, followed by Grignard reagent formation and treatment with Ph$_2$POCl to produce the phosphine oxide (sometimes referred to as BINAPO). At this stage, the racemic phosphine oxide is resolved using a tartaric acid derivative via fractional crystallization and treatment with base to give the optically pure phosphine oxides. Finally, reduction of the phosphine oxide with trichlorosilane gave BINAP in optically pure form.

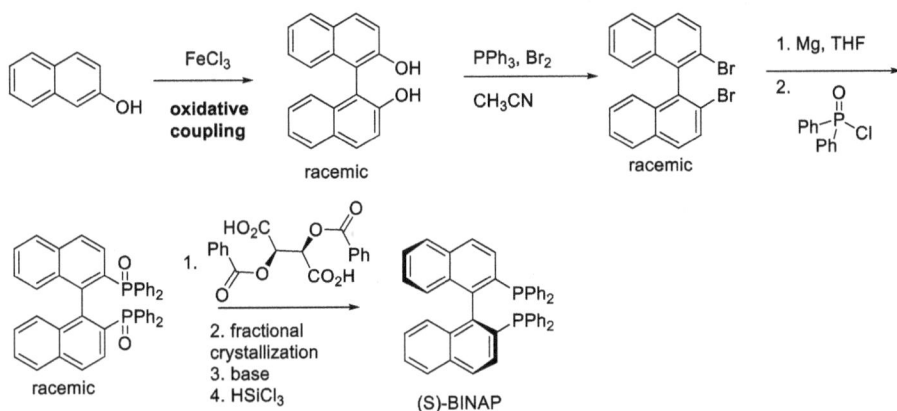

Figure 6.78: Synthesis of (*S*)-BINAP [131].

6.6.3 Stereoselective approaches to axially chiral biaryls

A desirable alternative to racemic synthesis and resolution of axially chiral biaryls is to prepare them stereoselectively [121, 122, 132]. Strategies for the stereoselective synthesis of biaryls include the use of chiral auxiliaries, chiral reagents, and chiral catalysts.

As an example of the use of chiral auxiliaries to enable the stereoselective synthesis of atropisomeric biaryls, the Ullmann coupling has been used along with chiral oxazolines as auxiliaries. As a representative example, Meyers used chiral oxazolines as chiral auxiliaries to prepare biaryls in a diastereomeric ratio of 93:3 in favor of the (*S*)-configuration biaryl (Figure 6.79) [133, 134]. The product could be further purified to achieve 100% diastereoselectivity. The stereochemical outcome avoids unfavorable steric interactions of the isopropyl groups on the oxazolines,

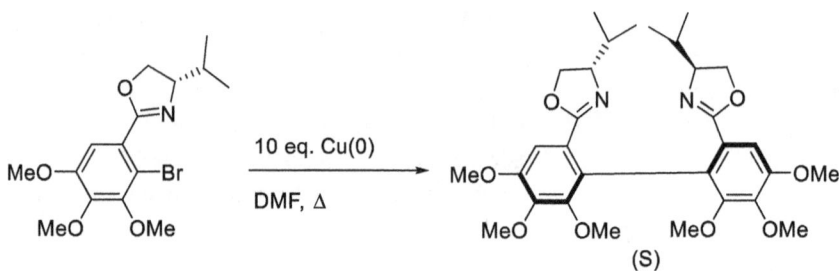

Figure 6.79: Chiral oxazoline auxiliaries for stereoselective Ullmann coupling [133].

and may be further enhanced by chelation of the oxazolines with the Cu(I) or Cu(II) salts in the course of the reaction.

As a demonstration of the application of the stereoselective Ullmann coupling, Meyers used this methodology for the stereoselective synthesis of (S)-(+)-gossypol, a compound isolated in racemic from cotton seeds [135]. Interestingly, the two enantiomers of gossypol display distinctly different biological activities. In this synthesis, the naphthoic acid derivative was converted to the corresponding chiral oxazoline using (S)-*tert*-leucinol, followed by bromination to give the bromonaphthalene (Figure 6.80). Ullmann coupling of this bromonaphthalene derivative was carried out using activated copper in DMF to give the binaphthyl in a good yield and a 17:1 diastereomeric ratio.

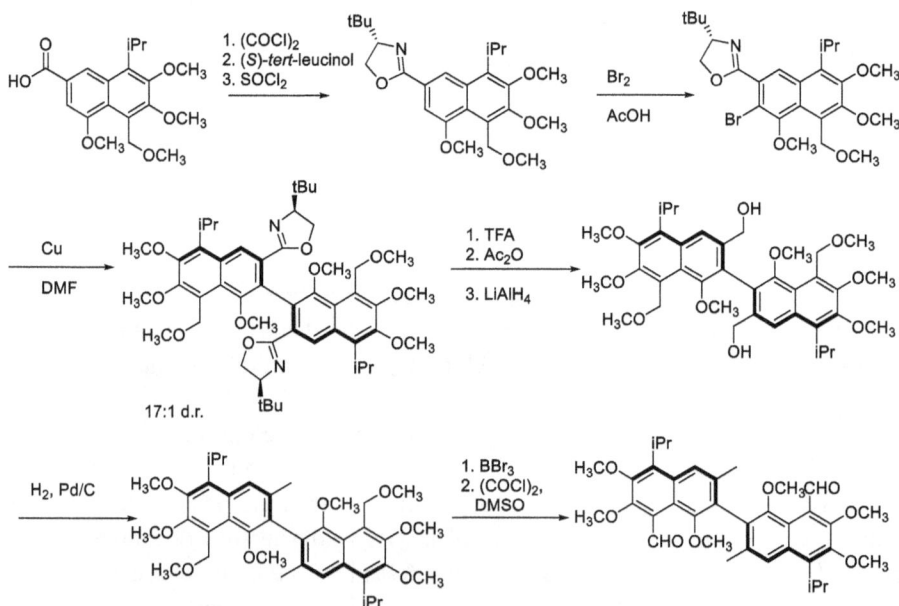

Figure 6.80: Stereoselective synthesis of (S)-(+)-gossypol [135].

After purification to give the pure diastereomer, the chiral auxiliary was removed and the diol was reduced to the corresponding dimethyl compound. The methoxy groups were cleaved using BBr$_3$, and the benzylic alcohol groups were converted to the corresponding aldehydes using Swern oxidation, to give (S)-(+)-gossypol.

Chiral auxiliaries have also been used to prepare axially chiral biaryls via Suzuki coupling. For example, Broutin and Colobert used a chiral β-hydroxy sulfoxide auxiliary appended to aryl halides to achieve diastereoselective biaryl formation (Figure 6.81) [136].

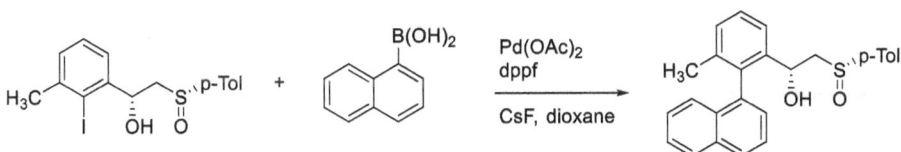

Figure 6.81: Diastereoselective Suzuki coupling using a β-hydroxy sulfoxide auxiliary [136].

Oxidative coupling reactions described in Section 6.2.3 can in some cases be carried out asymmetrically in the presence of metal salts and chiral ligands. For example, the oxidative coupling of 2-naphthol to prepare enantiomerically enriched 2,2′-BINOL has been reported using CuCl$_2$ and an excess of (S)-amphetamine (Figure 6.82) [137]. This reaction achieved a good yield and a 96% ee, with the high enantioselectivity being explained by a dynamic kinetic resolution. In a related example, a cross-coupling of 2-naphthol with the methyl ester of 3-hydroxy-2-naphthoic acid using CuCl$_2$ and sparteine as the chiral ligand proceeded in good yield, but with only a 41% ee (Figure 6.82) [138]. These reactions have the disadvantage of using stoichiometric amounts of copper and an excess of the chiral ligand.

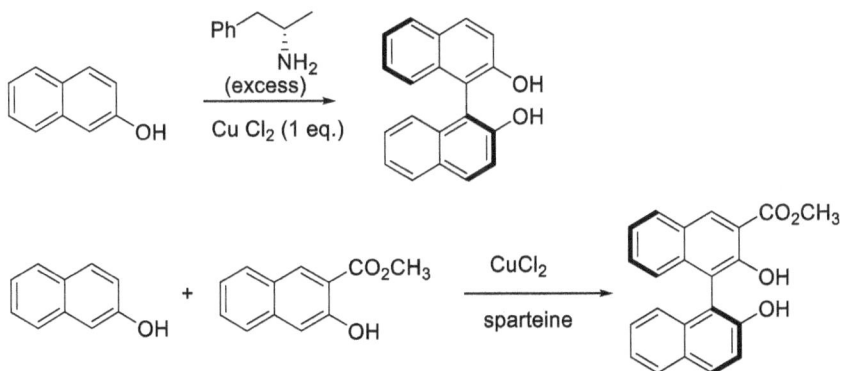

Figure 6.82: Enantioselective oxidative coupling using stoichiometric copper salts [137, 138].

Diastereoselective oxidative couplings have to form atropisomeric biaryls have also been carried out using chiral substrates. For example, a synthesis of (–)-viridi-toxin makes use of a diastereoselective oxidative coupling using 20 mol% [VO(acac)$_2$] in air to prepare the biaryl with a 3:1 diastereoselectivity (Figure 6.83) [139]. In this reaction the stereoselectivity is influenced by the existing stereocenters.

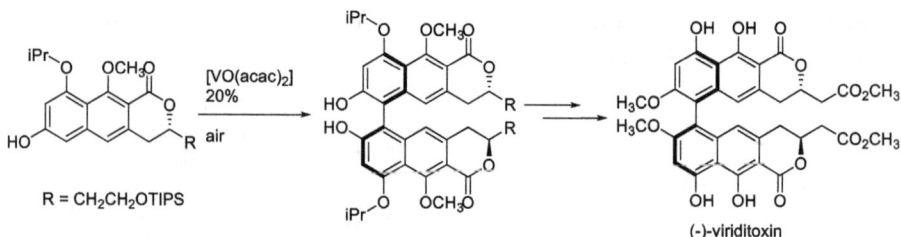

R = CH$_2$CH$_2$OTIPS

(-)-viriditoxin

Figure 6.83: Diastereoselective oxidative coupling in the synthesis of (–)-viriditoxin [139].

Given the broad utility of cross-coupling reactions, there is considerable interest in the development of asymmetric cross-couplings. Furthermore, since these reactions involve transition metal catalysts, they should be amenable to asymmetric synthesis by incorporation of chiral ligands into the catalyst. However, one significant challenge for development of asymmetric cross-couplings for the synthesis of atropisomeric biaryls is that these reactions are somewhat sensitive to sterics and often do not work well for highly substituted substrates. Nevertheless, with appropriate choice of ligand and reaction conditions, there are now numerous examples of stereoselective cross-coupling reactions (such as the Suzuki–Miyaura coupling) [132]. However, many of the examples reported focus on the synthesis of chiral tri-*ortho*-substituted biaryls. Asymmetric synthesis of tetra-*ortho*-substituted biaryls via cross-coupling methods remains a challenge.

Among the earliest examples of asymmetric cross-coupling was reported by Hayashi, Ito, and coworkers, who used chiral ferrocenylphosphine ligands in Kumada cross-coupling to prepare optically active binaphthyls [140]. This pioneering formed the basis for the development of other asymmetric cross-coupling reactions. For example, one of the first examples of an asymmetric Suzuki coupling was reported by Cammidge and Crépy, who explored the used of several different ligands for the preparation of enantioenriched binaphthyls [141]. They found that the use of palladium (II) chloride in the presence of a chiral ferrocene-based ligand gave the binaphthyl with an ee of up to 85%, although the yield was a modest 50% (Figure 6.84). In the same year, Yin and Buchwald reported an asymmetric Suzuki coupling to prepare a chiral phenyl naphthalene [142]. They used a chiral binaphthyl ligand with Pd$_2$(dba)$_3$ to couple a substituted bromonaphthalene with *ortho*-substituted phenylboronic acids and were able to achieve yields and ee of more than 90% in some cases.

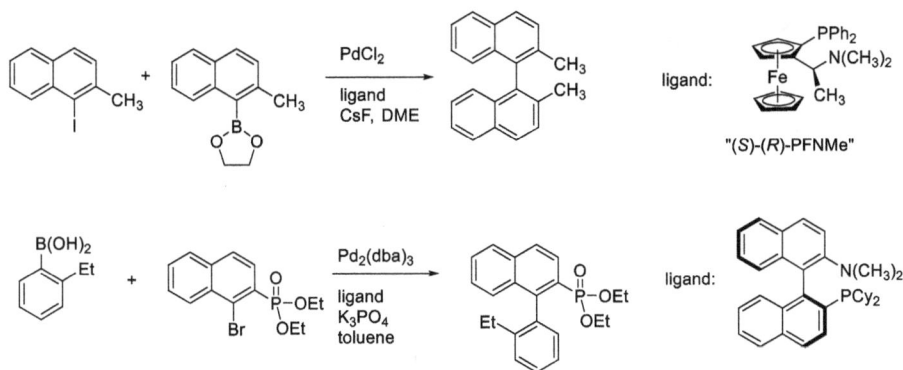

Figure 6.84: Early examples of asymmetric Suzuki couplings [141, 142].

In 2020, Tang and coworkers showed that tetra-*ortho*-substituted biaryls could be prepared via Suzuki coupling using the chiral phosphine ligand BaryPhos [143]. For example, sterically congested aryl bromide bearing two *ortho*-substituents was successfully coupled with a similarly congested aryl trifluoroborate to give the desired tetra-*ortho*-substituted biaryl in good yield and an enantiomeric excess above 90% (Figure 6.85). This method was successfully extended to a variety of biaryls bearing methoxy groups and/or formyl groups in the *ortho*-positions. Indeed, it is thought that the stereochemical control of the reaction is aided by hydrogen bonding of the hydroxyl group on BaryPhos with the oxygen of the methoxy or formyl groups.

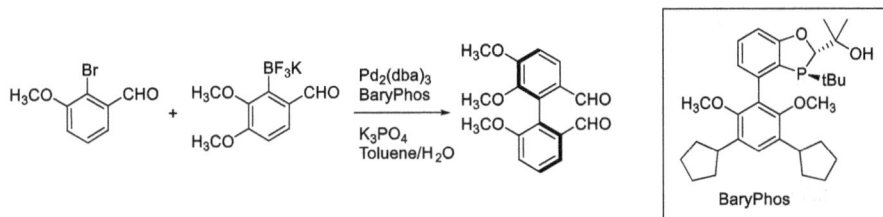

Figure 6.85: Asymmetric Suzuki coupling to prepare a tetra-*ortho*-substituted biaryl [143].

As an example of this asymmetric Suzuki coupling, Tang and coworkers used this methodology for the preparation of (−)-gossypol [143]. As an interesting modification, since the two aryl units are identical, they showed that the bromonaphthalene precursor could undergo a domino Miyaura borylation and Suzuki cross-coupling in the presence of bis(pinacolato)diboron, Pd$_2$(dba)$_3$, and BaryPhos (Figure 6.86).

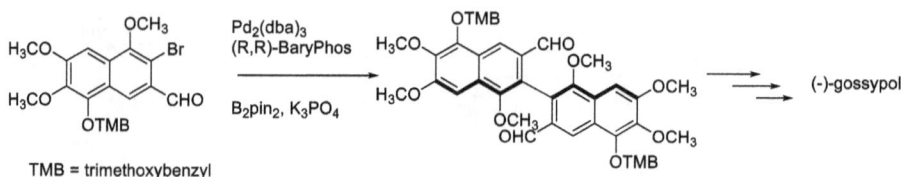

TMB = trimethoxybenzyl

Figure 6.86: Synthesis of (–)-gossypol via a domino Miyaura borylation and asymmetric Suzuki coupling [143].

References

[1] Hassan J, Sévignon M, Gozzi C, Schulz E, Lemaire M. Aryl-aryl bond formation one century after the discovery of the Ullmann reaction. Chem Rev. 2002;102(5):1359–469.

[2] Magano J, Dunetz JR. Large-scale applications of transition metal-catalyzed couplings for the synthesis of pharmaceuticals. Chem Rev. 2011;111(3):2177–250.

[3] Fanta PE. The Ullmann synthesis of biaryls, 1945-1963. Chem Rev. 1964;64(6):613–32.

[4] Sambiagio C, Marsden SP, Blacker AJ, McGowan PC. Copper catalysed Ullmann type chemistry: From mechanistic aspects to modern development. Chem Soc Rev. 2014;43:3525–50.

[5] Cohen T, Cristea I. Kinetics and mechanism of the copper(I)-induced homogeneous Ullmann coupling of o-Bromonitrobenzene. J Am Chem Soc. 1976;98(3):748–53.

[6] Corbett JF, Holt PF. Polycyclic cinnoline derivatives. Part V. Some unsymmetrical polycyclic cinnolines. J Chem Soc. 1960;3646–53.

[7] Zhang S, Zhang D, Liebeskind LS. Ambient temperature, Ullmann-like reductive coupling of aryl, heteroaryl, and alkenyl halides. J Org Chem. 1997;62(8):2312–13.

[8] Forrest J. The Ullmann biaryl synthesis. Part VI . The scope and mechanism of the reaction. J Chem Soc. 1960;594–601.

[9] Hauser FM, Gauuan PJF. Total synthesis of (±)-Biphyscion. Org Lett. 1999;1(4):671–72.

[10] Semmelhack MF, Helquist PM, Jones LD. Synthesis with zerovalent nickel. Coupling of aryl halides with Bis(1,5-cyclooctadiene)nickel(0). J Am Chem Soc. 1971;93:5908–10.

[11] Yamamoto T, Wakabayashi S, Osakada K. Mechanism of C-C coupling reactions of aromatic halides, promoted by Ni (COD), in the presence of 2, 2 ' and PPh,, to give biaryls. J Organomet Chem. 1992;428:223–37.

[12] Kende AS, Liebeskind LS, Braitsch DM. In situ generation of a solvated zerovalent nickel reagent. Biaryl formation. Tetrahedron Lett. 1975;16(39):3375–78.

[13] Yamamoto T, Morita A, Maruyama T, Zhou ZH, Kanbara T, Sanechika K. New method for the preparation of poly(2,5-thienylene), poly(p-phenylene), and related polymers. Polym J. 1990;22(2):187–90.

[14] Zhou ZH, Yamamoto T. Research on carbon-carbon coupling reactions of haloaromatic compounds mediated by zerovalent nickel complexes. Preparation of cyclic oligomers of thiophene and benzene and stable anthrylnickel(II) complexes. J Organomet Chem. 1991;414 (1):119–27.

[15] Yamamoto T, Maruyama T, Zhou ZH, Myazaki Y, Kanbara T, Sanechika K. New method using nickel(0) complex for preparation of poly(Paraphenylene), poly(2,5-thienylene) and related pi-conjugated polymers. Synth Met. 1991;41(1–2):345–48.

[16] Kobayashi M, Chen J, Chung TC, Moraes F, Heeger AJ, Wudl F. Synthesis and properties of chemically coupled poly(thiophene). Synth Met. 1984;9(1):77–86.

[17] Sarhan AAO, Bolm C. Iron(iii) chloride in oxidative C–C coupling reactions. Chem Soc Rev. 2009;38(9):2730–44.

[18] Grzybowski M, Sadowski B, Butenschön H, Gryko DT. Synthetic applications of oxidative aromatic coupling – from biphenols to nanographenes. Angew Chem Int Ed. 2020;59:2998–3027.

[19] Grzybowski M, Skonieczny K, Butenschön H, Gryko DT. Comparison of oxidative aromatic coupling and the scholl reaction. Angew Chem Int Ed. 2013;52(38):9900–30.

[20] Boden N, Bushby RJ, Lu Z, Headdock G. Synthesis of dibromotetraalkoxybiphenyls using ferric chloride. Tetrahedron Lett. 2000;41:10117–20.

[21] Leclerc M, Diaz FM, Wegner G. Structural analysis of poly(3-alkylthiophene)s. Makromol Chem. 1989;190:3105–16.

[22] Toda F, Tanaka K, Iwata S. Oxidative coupling reactions of phenols with FeCl3 in the solid state. J Org Chem. 1989;54:3007–09.

[23] Johansson Seechurn CCC, Kitching MO, Colacot TJ, Snieckus V. Palladium-catalyzed cross-coupling: A historical contextual perspective to the 2010 nobel prize. Angew Chem Int Ed. 2012;51(21):5062–85.

[24] Stanforth SP. Catalytic cross-coupling reactions in biaryl synthesis. Tetrahedron. 1998;54 (3–4):263–303.

[25] Nicolaou KC, Bulger PG, Sarlah D. Palladium-catalyzed cross-coupling reactions in total synthesis. Angew Chem Int Ed. 2005;44:4442–89.

[26] Carey FA, Sundberg RJ. Advanced Organic Chemistry Part B: Reactions and Synthesis, 5th Ed. New York: Springer; 2007.

[27] Amatore C, Jutand A, M'Barki MA. Evidence of the formation of zerovalent palladium from Pd (OAc)2 and triphenylphosphine. Organometallics. 1992;11(9):3009–13.

[28] Amatore C, Carré E, Jutand A, M'Barki MA, Meyer G. Evidence for the ligation of palladium(0) complexes by acetate ions: Consequences on the mechanism of their oxidative addition with phenyl iodide and PhPd(OAc)(PPh3)2 as intermediate in the heck reaction. Organometallics. 1995;14(12):5605–14.

[29] Strieter ER, Blackmond DG, Buchwald SL. Insights into the origin of high activity and stability of catalysts derived from bulky, electron-rich monophosphinobiaryl ligands in the Pd-catalyzed C-N bond formation. J Am Chem Soc. 2003;125(46):13978–80.

[30] Lundgren RJ, Stradiotto M. Addressing challenges in palladium-catalyzed cross-coupling reactions through ligand design. Chem -Eur J. 2012;18(32):9758–69.

[31] Chen H, Huang Z, Hu X, Tang G, Xu P, Zhao Y, et al. Nickel-catalyzed cross-coupling of aryl phosphates with arylboronic acids. J Org Chem. 2011;76(7):2338–44.

[32] Wang Z, Chen G, Shao L. N – heterocyclic carbene – palladium(II) – 1-methylimidazole complex- catalyzed Suzuki–Miyaura coupling of aryl sulfonates with arylboronic acids. J Org Chem. 2012;77:6608–14.

[33] Littke AF, Fu GC. A convenient and general method for Pd-catalyzed Suzuki cross-couplings of aryl chlorides and arylboronic acids. Angew Chem Int Ed. 1998;37(24):3387–88.

[34] Littke AF, Dai C, Fu GC. Versatile catalysts for the Suzuki cross-coupling of arylboronic acids with aryl and vinyl halides and triflates under mild conditions. J Am Chem Soc. 2000;122 (17):4020–28.

[35] Netherton MR, Fu GC. Air-stable trialkylphosphonium salts: Simple, practical, and versatile replacements for air-sensitive trialkylphosphines. Applications in stoichiometric and catalytic processes. Org Lett. 2001;3(26):4295–98.

[36] Miyaura N, Suzuki A. Palladium-catalyzed cross-coupling reactions of organoboron compounds. Chem Rev. 1995;95(7):2457–83.

[37] Kotha S, Lahiri K, Kashinath D. Recent applications of the Suzuki-Miyaura cross-coupling reaction in organic synthesis. Tetrahedron. 2002;58(48):9633–95.

[38] Amatore C, Jutand A, Le Duc G. Kinetic data for the transmetalation/reductive elimination in palladium-catalyzed Suzuki-Miyaura reactions: Unexpected triple role of hydroxide ions used as base. Chem -Eur J. 2011;17(8):2492–503.

[39] Coutts SJ, Adams J, Krolikowski D, Snow RJ. Two efficient methods for the cleavage of pinanediol boronate esters yielding the free boronic acids. Tetrahedron Lett. 1994; 35(29):5109–12.

[40] Murphy JM, Tzschucke CC, Hartwig JF. One-pot synthesis of arylboronic acids and aryl trifluoroborates by Ir-catalyzed borylation of arenes. Org Lett. 2007;9(5):757–60.

[41] Molander GA, Biolatto B. Palladium-catalyzed Suzuki-Miyaura cross-coupling reactions of potassium aryl- and heteroaryltrifluoroborates. J Org Chem. 2003;68(11):4302–14.

[42] Molander GA, Yun CS, Ribagorda M, Biolatto B. B-alkyl Suzuki-miyaura cross-coupling reactions with air-stable potassium alkyltrifluoroborates. J Org Chem. 2003;68(14):5534–39.

[43] Molander GA, Ellis N. Organotrifluoroborates: Protected boronic acids that expand the versatility of the Suzuki coupling reaction. Acc Chem Res. 2007;40(4):275–86.

[44] Larsen RD, King AO, Chen CY, Corley EG, Foster BS, Roberts FE, et al. Efficient synthesis of losartan, a nonpeptide angiotensin II receptor antagonist. J Org Chem. 1994;59(21):6391–94.

[45] Fu JM, Zhao BP, Sharp MJ, Snieckus V. Ortho and remote metalation – cross coupling strategies. Total synthesis of the naturally occurring fluorenone dengibsinin and the azafluoranthene alkaloid imeluteine. Can J Chem. 1994;72(1):227–36.

[46] Negishi E, King AO, Okukado N. Selective carbon-carbon bond formation via transition metal catalysis. 3.1 A highly selective synthesis of unsymmetrical biaryls and diarylmethanes by the nickel- or palladium- catalyzed reaction of aryl- and benzylzinc derivatives with aryl halides. J Org Chem. 1977;42(10):1821–23.

[47] Negishi E. Palladium- or nickel-catalyzed coupling. A new selective carbon-carbon formation. Acc Chem. 1982;15:340–48.

[48] Phapale VB, Cárdenas DJ. Nickel-catalysed negishi cross-coupling reactions: Scope and mechanisms. Chem Soc Rev. 2009;38(6):1598–607.

[49] Ku YY, Grieme T, Raje P, Sharma P, Morton HE, Rozema M, et al. A practical and scaleable synthesis of A-224817.0, a novel nonsteroidal ligand for the glucocorticoid receptor. J Org Chem. 2003;68(8):3238–40.

[50] Milstein D, Stille JK. Palladium-catalyzed coupling of tetraorganotin compounds with aryl and benzyl halides. Syntrhetic utility and mechanism. J Am Chem Soc. 1979;101(7):4992–98.

[51] Stille JK. The palladium-catalyzed cross-coupling reactions of organotin reagents with organic electrophiles. Angew Chem Int Ed. 1986;25(6):508–23.

[52] Ragan JA, Raggon JW, Hill PD, Jones BP, McDermott RE, Munchhof MJ, et al. Cross-coupling methods for the large-scale preparation of an imidazole – thienopyridine: synthesis of [2-(3-methyl-3H-imidazol-4-yl)-thieno[3,2-b]pyridin-7-yl] -(2-methyl-1H-indol-5-yl)-amine. Org Proc Res Dev. 2003;7(5):676–83.

[53] James CA, Snieckus V. Combined directed metalation cross coupling strategies. Total synthesis of the aglycones of gilvocarcin V, M, E. Tetrahedron Lett. 1997;38(47):8149–52.

[54] Hatanaka Y, Hiyama T. Cross-coupling of organosilanes with organic halides mediated by palladium catalyst and tris(diethylamino)sulfonium difluorotrimethylsilicate. J Org Chem. 1988;53:918–20.

[55] Denmark SE, Wehrli D, Choi JY. Convergence of mechanistic pathways in the palladium (0) - catalyzed cross-coupling of alkenylsilacyclobutanes and alkenylsilanols. Org Lett. 2000;2 (16):2491–94.

[56] Denmark SE, Regens CS. Organosilanols and their salts : Practical alternatives to boron- and tin-based methods. Acc Chem Res. 2008;41(11):1486–99.

[57] Tamao K, Sumitani K, Kumada M. Selective carbon-carbon bond formation by cross-coupling of grignard reagents with organic halides. catalysys by nickel-phosphine complexes. J Am Chem Soc. 1972;94(12):4374–76.

[58] Corriu RJP, Masse JP. Activation of grignard reagents by transition-metal complexes. A new and simple synthesis of trans-stilbenes and polyphenyls. J Chem Soc Chem Commun. 1972;144.

[59] Bold G, Fässler A, Capraro HG, Cozens R, Klimkait T, Lazdins J, et al. New aza-dipeptide analogues as potent and orally absorbed HIV-1 protease inhibitors: Candidates for clinical development. J Med Chem. 1998;41(18):3387–401.

[60] Heck RF. Palladium-catalyzed reactions of organic halides with olefins. Acc Chem Res. 1979;12(4):146–51.

[61] Beletskaya IP, Cheprakov AV. Heck reaction as a sharpening stone of palladium catalysis. Chem Rev. 2000;100(8):3009–66.

[62] Cabri W, Candiani I. Recent developments and new perspectives in the heck reaction. Acc Chem Res. 1995;28(1):2–7.

[63] Heck RF. The mechanism of arylation and carbomethoxylation of olefins with organopalladium compounds. J Am Chem Soc. 1969;91(24):6707–14.

[64] Plevyak JE, Dickerson JE, Heck RF. Selective palladium-catalyzed vinylic substitutions with bromoiodo aromatics. J Org Chem. 1979;44(23):4078–80.

[65] Abelman MM, Oh T, Overman LE. Intramolecular alkene arylations for rapid assembly of polycyclic systems containing quaternary centers. A new synthesis of spirooxindoles and other fused and bridged ring systems. J Org Chem. 1987;52(18):4130–33.

[66] Shinkai I, King AO, Larsen RD. A practical asymmetric synthesis of LTD4 antagonist. Pure Appl Chem. 1994;66(7):1551–56.

[67] De Vries JG. The heck reaction in the production of fine chemicals. Can J Chem. 2001; 79(5–6):1086–92.

[68] Castro CE, Stephens RD. Substitutions by ligands of low valent transi- tion metals. A preparation of tolanes and heterocyclics from aryl iodides and cuprous acetylides. J Org Chem. 1963;28:2163.

[69] Stephens RD, Castro CE. The substitution of aryl iodides with cuprous acetylides. A synthesis of tolanes and heterocyclics. J Org Chem. 1963;28:3313–15.

[70] Sonogashira K. Development of Pd-Cu catalyzed cross-coupling of terminal acetylenes with sp2-carbon halides. J Organomet Chem. 2002;653(1–2):46–49.

[71] Chinchilla R, Najera C. The sonogashira reaction : A booming methodology in synthetic organic. Chem Rev. 2007;107:874–922.

[72] Tykwinski RR. Evolution in the palladium-catalyzed cross-coupling of sp- and sp2-hybridized carbon atoms. Angew Chem Int Ed. 2003;42(14):1566–68.

[73] Mio MJ, Kopel LC, Braun JB, Gadzikwa TL, Hull KL, Brisbois RG, et al. One-pot synthesis of symmetrical and unsymmetrical bisarylethynes by a modification of the sonogashira coupling reaction. Org Lett. 2002;4(19):3199–202.

[74] Melissaris A, Litt MH. A simple and economical synthetic route to p-ethynylaniline and ethynyl-terminated substrates. J Org Chem. 1994;59(19):5818–21.

[75] Novák Z, Nemes P, Kotschy A. Tandem sonogashira coupling: An efficient tool for the synthesis of diarylalkynes. Org Lett. 2004;6(26):4917–20.

[76] Yamaguchi Y, Ochi T, Wakamiya T, Matsubara Y, Yoshida Z. New fluorophores with rod-shaped polycyano π -conjugated structures : Synthesis and photophysical properties. Org Lett. 2006;8(4):717–20.

[77] Zhang J, Moore JS, Xu Z, Aguirre RA. Nanoarchitectures. 1. Controlled synthesis of phenylacetylene sequences. J Am Chem Soc. 1992;114(6):2273–74.

[78] Macrocycles P, Zhang J, Pesak DJ, Ludwick JL, Moore JS, Arbor A. Geometrically-controlled and site-specifically-functionalized phenylacetylene macrocycles. J Am Chem Soc. 1994; 116(10):4227–39.

[79] Haley MM, Bell ML, English JJ, Johnson CA, Weakley TJR. Versatile synthetic route to and DSC analysis of dehydrobenzoannulenes : Crystal structure of a heretofore inaccessible [20] Annulene derivative Recei V ed No V ember 25, 1996 The synthesis and chemistry of dehydroannulenes and their benzannelated anal. J Am Chem Soc. 1997;119(12):2956–57.

[80] Pak JJ, Weakley TJR, Haley MM. Stepwise assembly of site specifically functionalized dehydrobenzo [18] annulenes. J Am Chem Soc. 1999;121(36):8182–92.

[81] Moore JS, Weinstein EJ, Wu Z. A convenient masking group for aryl iodides. Tetrahedron Lett. 1991;32(22):2465–66.

[82] Chanteau SH, Tour JM. Synthesis of anthropomorphic molecules: The nanoPutians. J Org Chem. 2003;68(23):8750–66.

[83] Shirai Y, Osgood AJ, Zhao Y, Kelly KF, Tour JM. Directional control in thermally driven single-molecule nanocars. Nano Lett. 2005;5(11):2330–34.

[84] Konigsberger K, Chen G, Wu RR, Girgis MJ, Prasad K, Repic O, et al. A practical synthesis of 6- [2- (2, 5-Dimethoxyphenyl) ethyl] –4-ethylquinazoline and the art of removing palladium from the products of Pd-catalyzed reactions. Org Proc Res Dev. 2003;7(5):733–42.

[85] Ellis GP, Romney-alexander TM. Cyanation of aromatic halides. Chem Rev. 1987;87:779–94.

[86] Sundermeier M, Zapf A, Beller M. Palladium-catalyzed cyanation of aryl halides : Recent developments and perspectives. Eur J Inorg Chem. 2003;3513–26.

[87] Jin F, Confalone PN. Palladium-catalyzed cyanation reactions of aryl chlorides. Tetrahedron Lett. 2000;41:3271–73.

[88] Wallace DJ, Campos KR, Shultz CS, Klapars A, Zewge D, Crump BR, et al. New efficient asymmetric synthesis of taranabant, a CB1R inverse agonist for the treatment of obesity abstract . Org Proc Res Dev. 2009;13(1):84–90.

[89] Pitts MR, Mccormack P, Whittall J. Optimisation and scale-up of microwave assisted cyanation. Tetrahedron. 2006;62:4705–08.

[90] Alberico D, Scott ME, Lautens M. Aryl-aryl bond formation by transition-metal-catalyzed direct arylation. Chem Rev. 2007;107(1):174–238.

[91] Ackermann L, Vicente R, Kapdi AR. Transition-metal-catalyzed direct arylation of (hetero) arenes by C-H bond cleavage. Angew Chem Int Ed. 2009;48(52):9792–826.

[92] Ames DE, Opalko A. Synthesis of dibenzofurans by palladium-catalyzed intramolecular dehydrobromination of 2-bromophenyl phenyl ethers. Synthesis (Stuttg). 1983;(3):234–35.

[93] Campeau LC, Parisien M, Jean A, Fagnou K. Catalytic direct arylation with aryl chlorides, bromides, and iodides: Intramolecular studies leading to new intermolecular reactions. J AmChem Soc. 2006;128(2):581–90.

[94] Matsumoto T, Hosoya T, Suzuki K. Total synthesis and absolute stereochemical assignment of gilvocarin M. J Am Chem Soc. 1992;114:3568–70.

[95] Satho T, Kawamura Y, Miura M, Nomura M. Pd-cat regioselective mono and diarylation reaction. Angew Chem Int Ed. 1997;36(16):1740.

[96] Yoshiki K, Tetsuya S, Masahiro M, Masakatsu N. Multiple arylation of phenols around the oxygen under palladium catalysis. Chem Lett. 1999;28(9):961–62.

[97] Lafrance M, Rowley CN, Woo TK, Fagnou K. Catalytic intermolecular direct arylation of perfluorobenzenes. J Am Chem Soc. 2006;128(27):8754–56.

[98] Lafrance M, Fagnou K. Palladium-catalyzed benzene arylation: Incorporation of catalytic pivalic acid as a proton shuttle and a key element in catalyst design. J Am Chem Soc. 2006;128(51):16496–97.

[99] Schipper DJ, Fagnou K. Direct arylation as a synthetic tool for the synthesis of thiophene-based organic electronic materials. Chem Mater. 2011;23(6):1594–600.

[100] Nesterov EE, Skoch J, Hyman BT, Klunk WE, Bacskai BJ, Swager TM. In vivo optical imaging of amyloid aggregates in brain: Design of fluorescent markers. Angew Chem Int Ed. 2005;44:5452–56.

[101] Akita Y, Itagaki Y, Takizawa S, Ohta A. Cross-coupling reactions of chloropyrazines with 1-substituted indoles. Chem Pharm Bull. 1989;36(6):1477–80.

[102] Lane BS, Sames D. Direct C – H bond arylation : Selective palladium-catalyzed C2-arylation of N -substituted indoles. Org Lett. 2004;6(17):2897–900.

[103] Lane BS, Brown MA, Sames D. Direct palladium-catalyzed C-2 and C-3 arylation of indoles: A mechanistic rationale for regioselectivity. J Am Chem Soc. 2005;127:8050–57.

[104] Mukhopadhyay S, Rothenberg G, Gitis D, Baidossi M, Ponde DE, Sasson Y. Regiospecific cross-coupling of haloaryls and pyridine to 2-phenylpyridine using water, zinc, and catalytic palladium on carbon. J Chem Soc Perkin Trans. 2000;2:1809–12.

[105] Campeau LC, Rousseaux S, Fagnou K. A solution to the 2-pyridyl organometallic cross-coupling problem: Regioselective catalytic direct arylation of pyridine N-oxides. J Am Chem Soc. 2005;127(51):18020–21.

[106] Gauthier DR, Limanto J, Devine PN, Desmond RA, Szumigala RH, Foster BS, et al. Palladium-catalyzed regioselective arylation of imidazo[1,2-b][1,2,4] triazine: Synthesis of an α2/3-selective GABA agonist. J Org Chem. 2005;70(15):5938–45.

[107] Chen X, Engle KM, Wang DH, Jin-Quan Y. Palladium(II)-catalyzed C-H aetivation/C-C cross-coupling reactions: Versatility and practicality. Angew Chem Int Ed. 2009;48(28):5094–115.

[108] Giri R, Maugel N, Li JJ, Wang DH, Breazzano SP, Saunders LB, et al. Palladium-catalyzed methylation and arylation of sp2 and sp 3 C-H bonds in simple carboxylic acids. J Am Chem Soc. 2007;129(12):3510–11.

[109] Wang D, Mei T, Yu J. Versatile Pd (II) -Catalyzed C – H activation / aryl – aryl coupling of benzoic and phenyl acetic acids. J Am Chem Soc. 2008;130:17676–77.

[110] Vogler T, Studer A. Oxidative coupling of arylboronic acids with arenes via Rh-catalyzed direct C-H arylation. Org Lett. 2008;10(1):129–31.

[111] Stuart DR, Fagnou K. The catalytic cross-coupling of unactivated arenes. Science. 2007;316:1172–76.

[112] Åkermark B, Eberson L, Jonsson E, Pettersson E. Palladium-promoted cyclization of diphenyl ether, diphenylamine, and related compounds. J Org Chem. 1975;40(9):1365–67.

[113] Hagelin H, Oslob JD, Åkermark B. Oxygen as oxidant in palladium-catalyzed inter- and intramolecular coupling reactions. Chem -Eur J. 1999;5(8):2413–16.

[114] Knölker HJ, O'Sullivan N. Indoloquinones – 3. Palladium-promoted synthesis of hydroxy-substituted 5-cyano-5H-benzo[b]carbazole-6, 11-diones. Tetrahedron. 1994;50 (37):10893–908.

[115] Knölker HJ, Reddy KR, Wagner A. Indoloquinones, pert 5. Palladium-catalyzed total synthesis of the potent lipid peroxidation inhibitor carbazoquinocin C. Tetrahedron Lett. 1998;39:8267–70.

[116] Liégault B, Lee D, Huestis MP, Stuart DR, Fagnou K. Intramolecular Pd(II)-catalyzed oxidative biaryl synthesis under air: Reaction development and scope. J Org Chem. 2008;73 (13):5022–28.

[117] Watanabe T, Ueda S, Inuki S, Oishi S, Fujii N, Ohno H. One-pot synthesis of carbazoles by palladium-catalyzed N-arylation and oxidative coupling. Chem Commun. 2007;(43):4516–18.

[118] Watanabe T, Oishi S, Fujii N, Ohno H. Palladium-catalyzed direct synthesis of carbazoles via one-pot N-arylation and oxidative biaryl coupling: Synthesis and mechanistic study. J Org Chem. 2009;74(13):4720–26.

[119] Li R, Jiang L, Lu W. Intermolecular cross-coupling of simple arenes via C-H activation by tuning concentrations of arenes and TFA. Organometallics. 2006;25(26):5973–75.

[120] Dwight TA, Rue NR, Charyk D, Josselyn R, DeBoef B. C-C bond formation via double C-H functionalization: Aerobic oxidative coupling as a method for synthesizing heterocoupled biaryls. Org Lett. 2007;9(16):3137–39.

[121] Bringmann G, Gulder T, Gulder TAM, Breuning M. Atroposelective total synthesis of axially chiral biaryl natural products. Chem Rev. 2011;111(2):563–639.

[122] Kozlowski MC, Morgan BJ, Linton EC. Total synthesis of chiral biaryl natural products by asymmetric biaryl coupling. Chem Soc Rev. 2009;38:3193–207.

[123] Ohta T, Takaya H, Kitamura M, Nagai K, Noyori R. Asymmetric hydrogenation of unsaturated carboxylic acids catalyzed by BINAP-ruthenium(II) complexes. J Org Chem. 1987;52 (14):3174–76.

[124] Harrington PJ, Lodewijk E. Twenty years of naproxen technology. Org Proc Res Dev. 1997; 1(1):72–76.

[125] Lemieux RP. Chirality transfer in ferroelectric liquid crystals. Acc Chem Res. 2001;34 (11):845–53.

[126] Yang K, Campbell B, Birch G, Williams VE, Lemieux RP. Induction of a ferroelectric SC* liquid crystal phase by an atropisomeric dopant derived from 4,4'-dihydroxy-2,2'-dimethyl-6,6'-dinitrobiphenyl. J Am Chem Soc. 1996;118(40):9557–61.

[127] Vizitiu D, Halden BJ, Lemieux RP. Enhanced polar ordering in ferroelectric liquid crystals induced by atropisomeric dopants. Chem Commun. 1997;1123–24.

[128] Vizitiu D, Lazar C, Halden BJ, Lemieux RP. Ferroelectric liquid crystals induced by atropisomeric biphenyl dopants: Dependence of the polarization power on core structure of the smectic C host. J Am Chem Soc. 1999;121(36):8229–36.

[129] Lai CW, Lam CK, Lee HK, Mak TCW, Wong HNC. Synthesis and studies of 1,4,5,8,9,12,13,16-octamethoxytetraphenylene. Org Lett. 2003;5(6):823–26.

[130] Miyashita A, Yasuda A, Takaya H, Toriumi K, Ito T, Souchi T, et al. Synthesis of 2,2/-bis (diphenylphosphino)-l,l/-binaphthyl (BINAP), an atropisomeric chiral bis(triaryl) phosphine, and its use in the rhodium(I)-catalyzed asymmetric hydrogenation of a-(Acylamino)acrylic acids. J Am Chem Soc. 1980;102:7932–34.

[131] Takaya H, Mashima K, Koyano K, Yagi M, Kumobayashi H, Taketomi T, et al. Practical synthesis of (R)- or (S)-2,2/-bis(diarylphosphino)-l,l/-binaphthyls (BINAPs). J Org Chem. 1986;51:629–35.

[132] Wencel-Delord J, Panossian A, Leroux FR, Colobert F. Recent advances and new concepts for the synthesis of axially stereoenriched biaryls. Chem Soc Rev. 2015;44(11):3418–30.

[133] Nelson TD, Meyers AI. The asymmetric Ullmann reaction. 2. The synthesis of enantiomerically pure C2-symmetric binaphthyls. J Org Chem. 1994;59:2655–58.

[134] Meyers AI, Price A. The unique behavior of a chiral binaphthyl oxazoline in the presence of Cu(I) and its role as a chiral catalyst. J Org Chem. 1998;63(3):412–13.

[135] Meyers AI, Willemsen JJ. The synthesis of (S) – (+) -gossypol via an asymmetric Ullmann coupling the first asymmetric total synthesis of (S) – (+) -gossypol is in a highly diastereoselective aryl – Aryl coupling. Chem Commun. 1997;1573–74.

[136] Broutin PE, Colobert F. Enantiopure β-hydroxysulfoxide derivatives as novel chiral auxiliaries in asymmetric biaryl Suzuki reactions. Org Lett. 2003;5(18):3281–84.

[137] Brussee J, Groenendijk JLG, te Koppele JM, Jansen ACA. On the mechanism of the formation of S(-)-(1,1'-binaphthalene)-2,2'Diol via copper(II) amine complexes. Tetrahedron. 1985;41 (16):3313–19.

[138] Smrčina M, Poláková J, Š V, Kočovský P. Synthesis of enantiomerically pure binaphthyl derivatives. Mechanism of the enantioselective, oxidative coupling of naphthols and designing a catalytic cycle. J Org Chem. 1993;58(17):4534–38.

[139] Park YS, Grove CI, Gonzμlez-lópez M, Urgaonkar S, Fettinger JC, Shaw JT. Synthesis of (-)-viriditoxin: A 6,6'-binaphthopyran-2-one that targets the bacterial cell division protein FtsZ. Angew Chem Int Ed. 2011;50:3730–33.

[140] Hayashi T, Hayashizaki K, Kiyoi T, Ito Y. Asymmetric synthesis catalyzed by chiral ferrocenylphosphine-transition-metal complexes. 6. practical asymmetric synthesis of 1,1'-Binaphthyls via asymmetric cross-coupling with a chiral [(Alkoxyalkyl)ferrocenyl] monophosphine/nickel catalyst. J Am Chem Soc. 1988;110(24):8153–56.

[141] Cammidge AN, Crepy KVL. The first asymmetric Suzuki cross-coupling reaction. Chem Commun. 2000;(18):1723–24.

[142] Yin J, Buchwald SL. A catalytic asymmetric Suzuki coupling for the synthesis of axially chiral biaryl compounds. J Am Chem Soc. 2000;122(48):12051–52.

[143] Yang H, Sun J, Gu W, Tang W. Enantioselective cross-coupling for axially chiral tetra-ortho-substituted biaryls and asymmetric synthesis of gossypol. J Am Chem Soc. 2020;142:8036–43.

7 Other transition-metal-mediated reactions of aromatic compounds

7.1 Introduction

In Chapter 6, the focus was on transition-metal-mediated carbon–carbon bond formation with aromatic rings. In this chapter, we will consider some important methods for
forming bonds between heteroatoms and aromatic rings. Aromatic rings bearing ether
or amine linkages are widely used, especially in the context of biologically active molecules. As such, several methods for creating carbon–heteroatom bonds using transition metals have been developed and will be explored in this chapter. In addition, the
formation of aryl carbon–boron bonds has become very important, especially for the
preparation of boronic acids and boronate esters that can be used for Suzuki–Miyaura
cross-coupling reactions or direct arylation reactions. We have already seen how arylboronic acids can be prepared from the corresponding aryllithium species. In this
chapter we will examine some of the transition-metal-catalyzed approaches that serve
as important alternatives to aryllithium species.

7.2 Aromatic C–N and C–O bond formation

7.2.1 Copper-catalyzed aryl C–N and C–O bond formation

Aromatic amines are important compounds as synthetic intermediates and as components of biologically active compounds. One of the common methods for accessing aromatic amines is by electrophilic nitration and reduction of the nitro group.
While this approach remains important and useful, electrophilic nitration has constraints in terms of regiochemistry and functional group compatibility. Similarly,
phenols and aryl ethers can be prepared by classical approaches, such as reaction
of aryl diazonium salts. Again, this reaction relies on the methods that can have
limitations of functional group tolerance. There is consequently an impetus for the
development of alternative methods for the formation of aryl–N and aryl–O bonds.

In Chapter 6, we saw the Ullmann coupling could be used to prepare biaryls
from aryl halides using stoichiometric amounts of copper. In the early 1900s, Ullmann also showed that copper could be used to couple aryl halides with amines or
phenols to form aryl amines or diaryl ethers (Figure 7.1) [1,2]. Shortly thereafter,
Goldberg showed that the arylation of amines could be performed with catalytic
amounts of copper [3]. Goldberg also demonstrated that copper could catalyze the
amidation of aryl halides [3].

The Ullmann condensation is typically carried out with either a catalytic or stoichiometric copper species in the presence of base at elevated temperatures. The

https://doi.org/10.1515/9783110562682-007

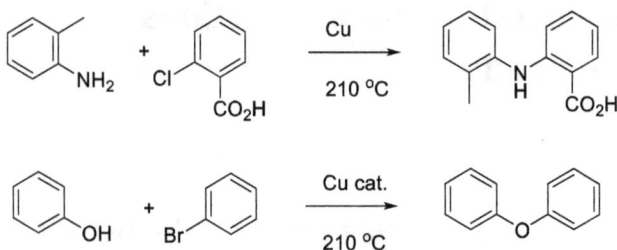

Figure 7.1: Ullman's copper-mediated synthesis of aryl amines and aryl ethers.

reaction has been carried out using copper metal, copper(I) salts, and copper(II) salts, although it is thought that copper(I) species are the active catalysts. The mechanism of the reaction is not clearly established, but may involve oxidative addition of the aryl halide to Cu(I) to give a Cu(III) intermediate, followed by amine coordination and reductive elimination. Alternatively, it may be a radical mechanism via single-electron transfer from the Cu(I) species [4].

The harsh reaction conditions of the Ullmann reaction to form ethers and aryl amines means that it is not frequently used. Instead, copper-catalyzed modifications of the reactions with ligands or palladium-catalyzed methods have mostly supplanted the traditional approach because they enable the reaction to take place under mild conditions. Nonetheless, the reaction does still find some use for the preparation of aryl–nitrogen bonds. For example, Perepichka and coworkers recently used the Ullmann reaction for the preparation of a trioxaazatriangulene, which served as a precursor for the preparation of semiconducting covalent organic frameworks (Figure 7.2) [5]. The synthesis consisted of an Ullmann reaction between 2,6-difluoroaniline and 2-iodo-1,3-dimethoxybenzene in the presence of copper, potassium carbonate, and 18-crown-6 ether to form the diarylamine. Following this, a second Ullmann coupling was carried out (at even higher temperatures) to form the substituted triarylamine. The methoxy groups were then cleaved using BBr_3 and the phenols were used for a threefold nucleophilic aromatic substitution reaction to form the trioxaazatriangulene. This compound was then nitrated to provide a synthetic handle for further functionalization in the formation of the covalent organic framework.

While the Ullmann reaction can be carried out in the absence of added ligands, the use of ligands has been shown to improve reactivity and allow the reaction to proceed at much lower temperatures. The role of the ligands in the Ullmann reaction is not clear, but may include improving solubility of the copper species or improving oxidative addition. In the case of bidentate ligands, it has been suggested that they block sites and force the aryl groups in proximity to facilitate reaction. Typical ligands used in Ullmann reactions include bidentate ligands, usually incorporating N-donors or N- and O-donors [4]. Some representative ligands used in Ullmann reactions are shown below in Figure 7.3. It should be noted that one of the advantages of the ligands

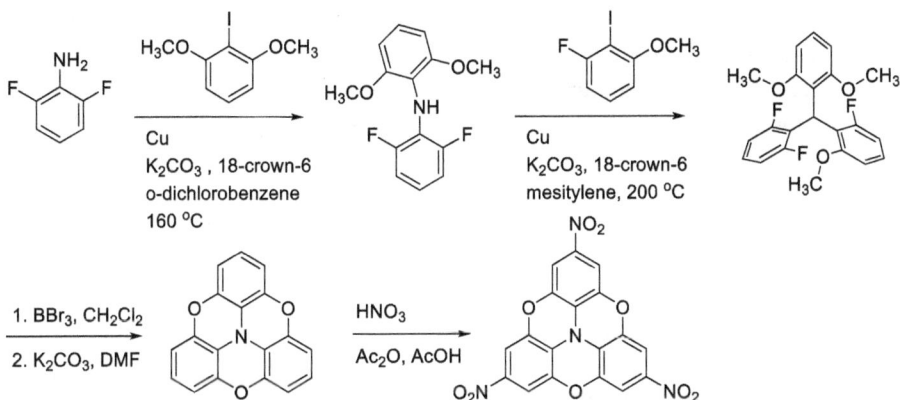

Figure 7.2: Synthesis of a trioxaazatriangulene via the Ullmann reaction [5].

used in the Ullmann reaction is that they are often much simpler than those used for some palladium-catalyzed coupling reactions.

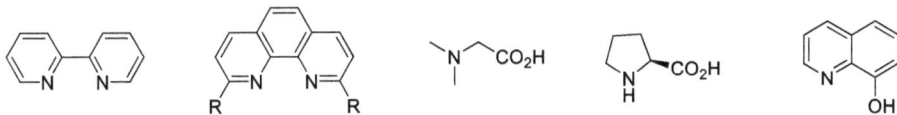

Figure 7.3: Representative ligands used for Cu-catalyzed C–N and C–O bond formation.

Copper-catalyzed amination of aryl halides has been used for the preparation of natural products and other biologically active molecules [6]. For example, the alkaloid martinellic acid, which was prepared using a copper-catalyzed aryl amination reaction, as outlined in Figure 7.4 [7].

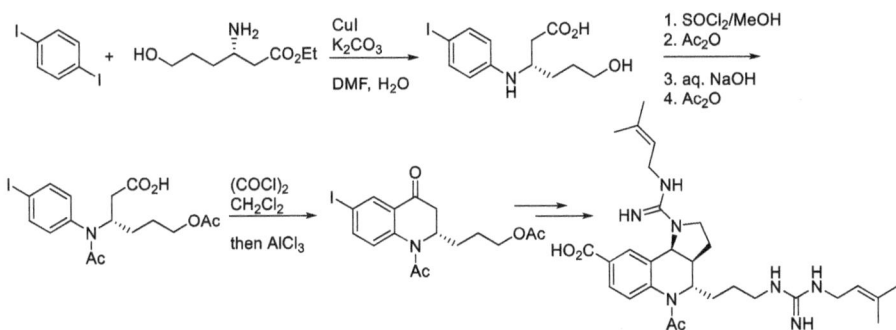

Figure 7.4: Copper-catalyzed aryl amination as part of the synthesis of martinellic acid [7].

Copper catalysts can also be used for the preparation of *N*-aryl amides and carbamates in Goldberg-type reactions. For example, in a synthetic approach to the antibiotic linezolid, an oxazolidinone was appended to an aryl halide using catalytic cuprous iodide and *trans*-1,2-diaminocyclohexane as a ligand in dioxane in the presence of potassium carbonate (Figure 7.5) [8].

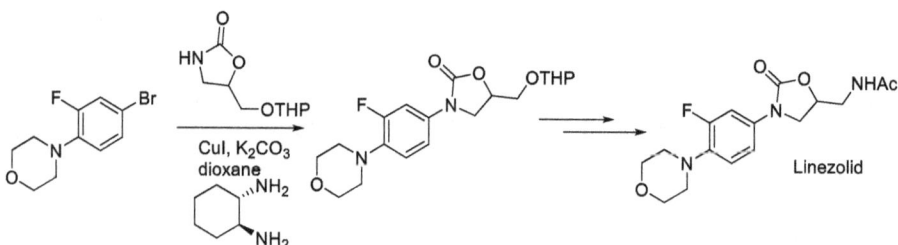

Figure 7.5: Copper-catalyzed coupling of an aryl bromide and an oxazolidinone as part of a synthesis of linezolid [8].

Triarylamines with hole-transporting properties for applications in xerography are readily accessed by copper-catalyzed amination of aryl iodides [9,10]. For example, the synthesis of the triarylamine derivatives can be achieved by reacting the diarylamine with an aryl iodide using catalytic cuprous iodide, 1,10-phenanthroline as the ligand, and an excess of potassium hydroxide (Figure 7.6) [9]. Similar reactions can be carried out using Cu(OAc)$_2$ as the catalyst, which presumably undergoes reduction to Cu(I) in situ [10].

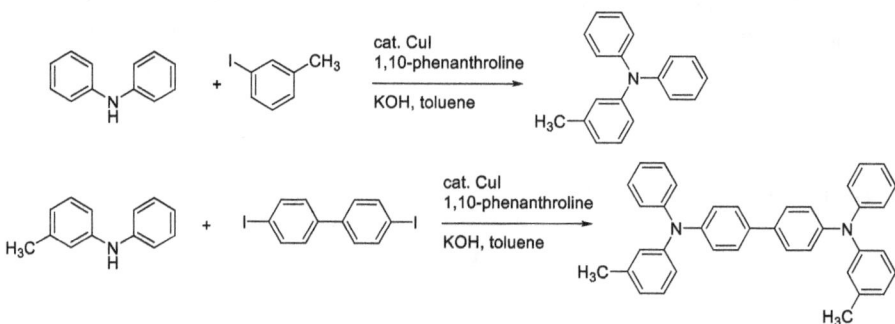

Figure 7.6: Examples of Cu-catalyzed aryl aminations for the preparation of hole-transporting materials [9].

In addition to copper-catalyzed aryl amination reactions, copper-catalyzed aryl C–O bond formation has been used for the synthesis of potential therapeutic agents. For example, a large scale synthesis of a potential anti-inflammatory agent made use of a copper-catalyzed C–O bond formation (Figure 7.7) [11]. In this reaction, 4-acetyl-2-

bromomethanesulfonanilide was brominated *ortho* to the sulfonanilide. Following this, a copper-catalyzed reaction with 2,4-difluorophenol in the presence of cuprous chloride and potassium carbonate in pyridine gave the desired diaryl ether. It should be noted that while the reaction is in principle catalytic in CuCl, a large amount of CuCl (0.75 eq.) was used.

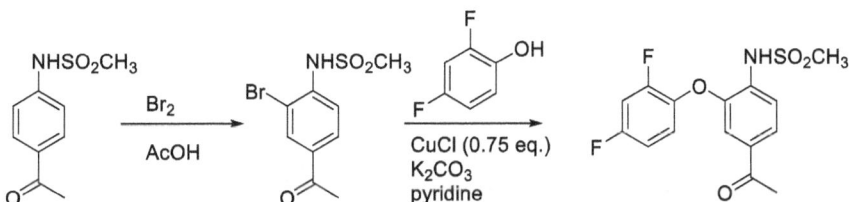

Figure 7.7: Synthesis of a potential anti-inflammatory agent via Cu-catalyzed ether formation [11].

Another example of the application of copper-catalyzed C–O formation in the synthesis of pharmaceutically active ingredients is seen in the scale-up synthesis of a leukotriene A4 hydrolase inhibitor, which was under investigation as a therapeutic for the prevention of heart attacks. In this synthesis, a copper-catalyzed ether formation between 4-chloro-1-bromobenzene and 4-methoxyphenol was carried out using cuprous iodide, cesium carbonate, and *N,N*-dimethylglycine (Figure 7.8) [12]. Unlike the previous example, which required high copper loading, this reaction took place with 10% copper iodide.

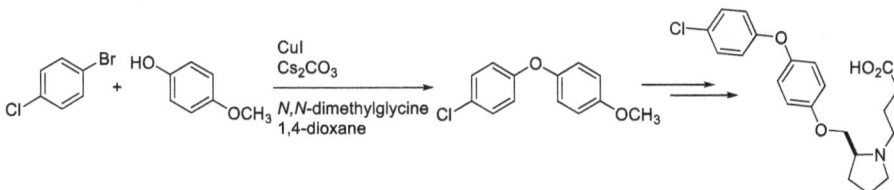

Figure 7.8: Copper-catalyzed ether formation in the synthesis of a therapeutic to prevent heart attacks [12].

7.2.2 Palladium-catalyzed aryl amination

While copper-catalyzed aryl C–N and C–O bond formation has proven useful, the reactions still often require harsh reaction conditions and limitations in the scope of the reaction. As such, there is considerable interest in the development of new methods that tolerate a wider variety of functional groups and take place under milder conditions. The first examples of a palladium-catalyzed aryl amination was reported by Migita and coworkers, who showed that aryl bromides could be converted into the corresponding *N,N*-diethylanilines using n-Bu_3SnEt_2 in the presence of palladium chloride and $P(o\text{-}tol)_3$ [13]. Since then, palladium-catalyzed aryl amination methods developed by Buchwald and Hartwig have significantly impacted synthetic chemistry by broadening the scope of aryl amination through methods that tolerate a variety of functional groups [14–17]. Indeed, palladium-catalyzed aryl amination is often referred to as the Buchwald–Hartwig reaction. In general, these reactions are carried out with an aryl halide, a primary or secondary amine, a palladium catalyst, and a strong base (Figure 7.9).

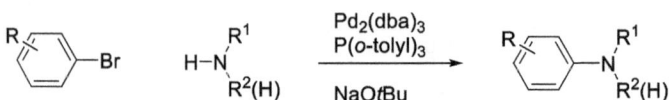

Figure 7.9: A generalized palladium-catalyzed aryl amination reaction.

The mechanism of the Buchwald–Hartwig reaction is very similar to the mechanism of the palladium-catalyzed cross-coupling reactions we explored in the previous chapter, and is outlined in Figure 7.10 [15, 18]. The first step is formation of a coordinatively unsaturated palladium(0) complex by ligand dissociation. This complex undergoes oxidative addition with the aryl halide to form a palladium(II) intermediate. At this point, the amine coordinates to the palladium and the base deprotonates the amine. Finally, reductive elimination leads to the formation of the aryl–nitrogen bond and the regeneration of the palladium(0) species.

The Buchwald–Hartwig reaction is broad in scope – aryl halides or pseudohalides can react with a variety of amines, ranging from anilines, aliphatic amines, heteroaromatic compounds, to amides. The most common base used is sodium *t*-butoxide, but other bases such as LiHMDS, Cs_2CO_3, K_2CO_3, and K_3PO_4 have also been used. The reaction is usually carried out at elevated temperatures (80–110 °C) in solvents such as dioxane or toluene. As with the cross-coupling reactions discussed in the previous chapter, the reaction requires a palladium(0) catalyst, although often palladium(II) precatalysts such as $Pd(OAc)_2$ are used and the active catalyst is generated in situ. Considerable effort has been devoted to the design of ligands that promote effective aryl amination, with the most common being dialkylbiaryl phosphines [18,19]. We saw some of these ligands in Chapter 6 in the context of palladium-catalyzed cross-coupling reactions. Figure 7.11 shows some of the different dialkyl biarylphosphines that have been developed for aryl amination reactions.

The structural features of these ligands have an impact on reactivity. For example, the phosphine typically bears large, electron-rich alkyl substituents such as *t*-butyl groups or cyclohexyl groups. The electron rich groups promote oxidative addition, while the sterically bulky groups promote reductive elimination. The adjacent aryl ring makes the ligands less susceptible to oxidation in air, while the presence of *ortho*

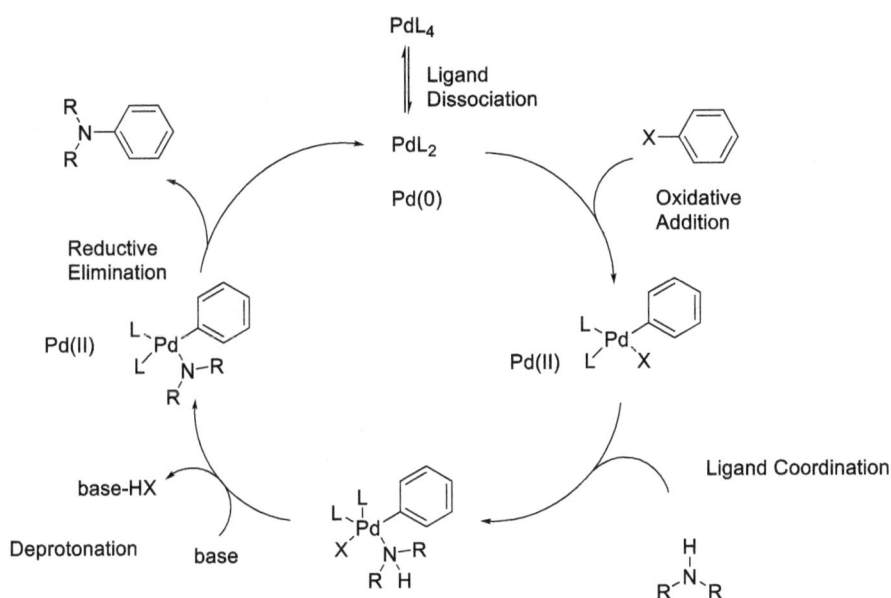

Figure 7.10: Catalytic cycle for the Buchwald–Hartwig aryl amination.

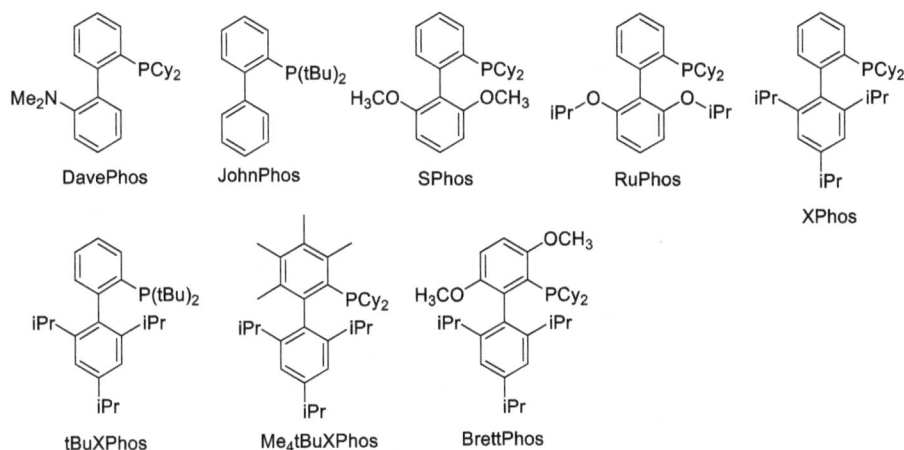

Figure 7.11: Representative dialkyl biaryl phosphine ligands for Buchwald–Hartwig aminations.

substituents on that ring enhances the dissociation of other ligands to form the coordinatively unsaturated active catalyst [18].

Despite the ligand design to prepare highly activated catalysts for aryl amination, there is no single ideal catalyst system for carrying out these reactions. Different catalytic systems are chosen depending on the amine nucleophile and the aryl halide. Surry and Buchwald provide an overview of some of the typical conditions used for different classes of reactants, highlighting the choices of different bases, ligands, and catalysts [18].

While palladium-catalyzed aryl aminations are effective with anilines, primary and secondary aliphatic amines, and even amides, they are not as effective for the preparation of anilines because ammonia can complex strongly with the palladium catalyst. Fortunately, there are surrogates for ammonia that can be used to access anilines. The most common surrogate is benzophenone imine [20,21]. It can be readily used for the aryl amination reactions and then deprotected by acidic hydrolysis, catalytic transfer hydrogenation or transamination with hydroxylamine to give the aniline (Figure 7.12) [20].

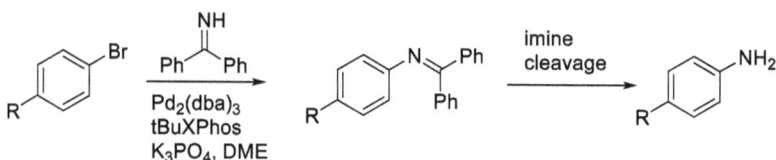

Figure 7.12: Preparation of anilines by aryl amination with benzophenone imine [20].

Given the prevalence of aryl amine bonds in natural products and in active pharmaceutical ingredients, palladium-catalyzed amination reactions have been applied extensively to the synthesis of biologically active compounds. These reactions present an alternative to other methods, such as nucleophilic aromatic substitution, which generally require electron-deficient aromatic rings. Below we will provide selected examples of applications of palladium-catalyzed aryl aminations because the utility of these reactions has been reviewed thoroughly [17,19].

As an example of the use of aryl amination in the synthesis of natural products, Campeau et al. demonstrated a concise synthesis of the carbazole-based natural product mukonine (Figure 7.13) [22]. The phenol starting material was converted to the corresponding triflate and then reacted with 2-chloroaniline in a Buchwald–Hartwig amination reaction to product the substituted diarylamine. Following this, a palladium-catalyzed intramolecular direct arylation gave mukonine. In a follow-up study, Fagnou and coworkers showed that the direct arylation using the aryl chloride could be replaced by a palladium-catalyzed dehydrogenative coupling [23].

An example of palladium-catalyzed aryl amination applied toward medicinal chemistry is shown in Figure 7.14 [24]. The final product shown served as a precursor to several compounds with activity as delta opioid receptor agonists. It was prepared by a sequence of two palladium-catalyzed arylation reactions – the first using BINAP as a ligand and the second using xantphos.

Aryl amination methods have also found use in the preparation of materials. For example, they are well-suited to the preparation of triarylamines, which we have already seen are of interest as hole-transporting materials. Harris and Buchwald showed

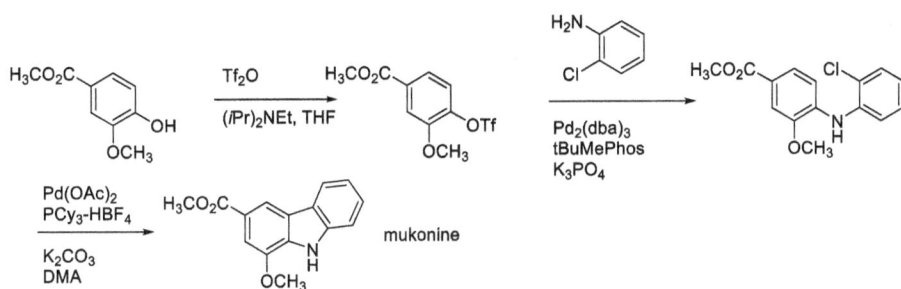

Figure 7.13: Synthesis of mukonine [22].

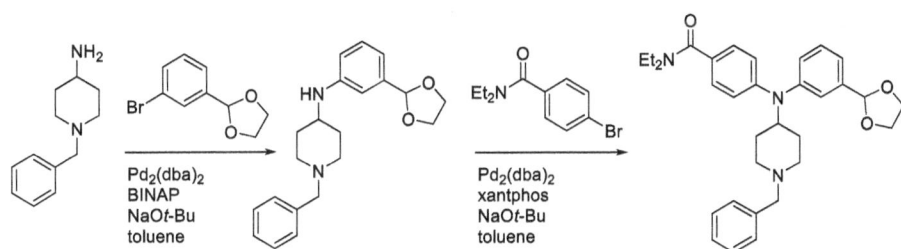

Figure 7.14: Synthesis of precursors to delta opioid receptor agonists via aryl amination [24].

that unsymmetrical triarylamnes can be prepared in a one-pot fashion from aniline precursors [25]. By reacting substituted anilines with aryl bromides and aryl chlorides in one pot, they took advantage of the differential reactivity of aryl bromides and aryl chlorides, allowing the selective synthesis of triarylamines bearing three different aryl groups. A representative example and reaction conditions are shown in Figure 7.15.

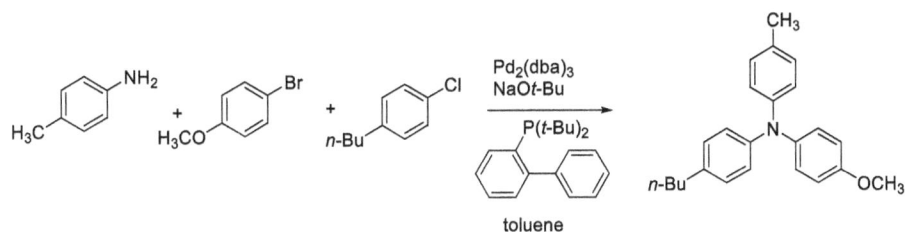

Figure 7.15: One-pot synthesis of unsymmetrical triarylamines via Pd-catalyzed aryl amination [25].

Aryl amination has also found utility in the synthesis of novel luminescent materials, including compounds designed to exhibit thermally activated delayed fluorescence (TADF). For example, a luminescent material was prepared by linking a dimethyl dihydroacridine with a dibromofluorene via a palladium-catalyzed aryl amination (Figure 7.16) [26]. The remaining bromo group on the fluorine moiety was then converted to a boronate ester via a Miyaura borylation, and then linked to a phenanthroimidazole acceptor unit via a Suzuki cross-coupling.

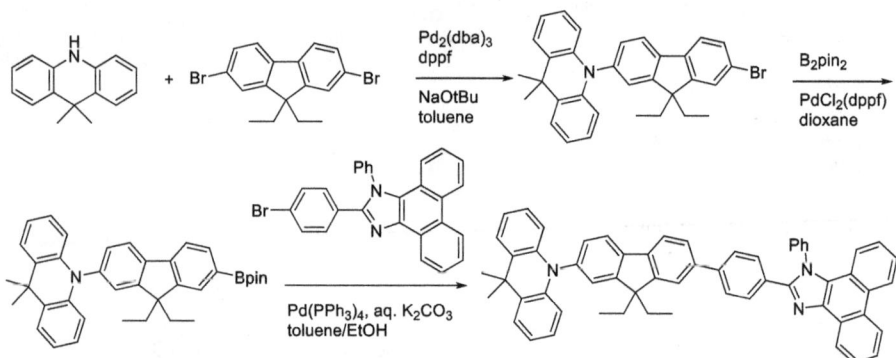

Figure 7.16: Synthesis of a luminescent material via aryl amination [26].

7.3 Transition-metal-catalyzed borylation reactions

We have seen that aryl boronic acids and boronate esters are useful precursors for Suzuki cross-coupling reactions. These reactions are widely used because boronic acids are air and moisture stable, and the reaction has excellent functional group tolerance. Arylboronic acids and boronate esters have synthetic utility that extends beyond cross-coupling reactions. They can be converted into other functional groups, as summarized in Figure 7.17 [27]. They can undergo protodeborylation, oxidation to the corresponding phenol, or conversion to the nitro- or haloarene. These transformations are not widely used because there are often simpler ways to access these groups directly. Nonetheless, these transformations can sometimes be useful.

Thus far, we have seen that arylboronic acids can be prepared by reacting organometallic reagents such as Grignard reagents or aryllithium species with trimethyl borate (Figure 7.18) [27].

While this method is effective for preparing boronic acids and versatile because aryllithium species can be prepared in a number of ways, it is not suitable for molecules that are sensitive to strongly basic conditions. Furthermore, the preparation of the aryllithium via lithium–halogen exchange is typically carried out at low temperatures, which can lead to low solubility of the reactant and therefore lower yields. Consequently, there is an incentive to find alternative approaches for the preparation of aryl boronic acids and aryl boronates. In this section, we will explore two alternative approaches for the preparation of aryl boronate esters using transition-metal-catalyzed reactions. The first is the palladium-catalyzed formation of aryl boronates from the corresponding aryl halides, while the second is the iridium-catalyzed direct borylation of an arene.

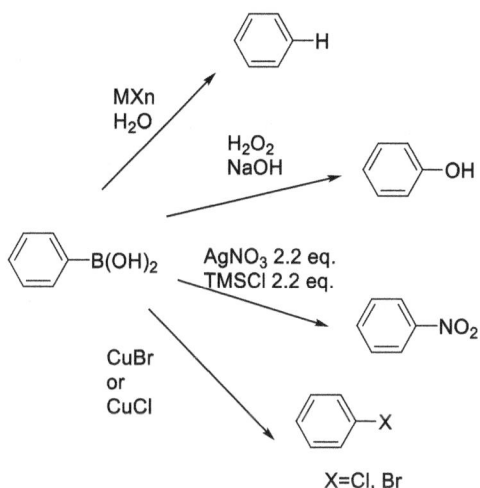

Figure 7.17: Summary of other reactions of arylboronic acids.

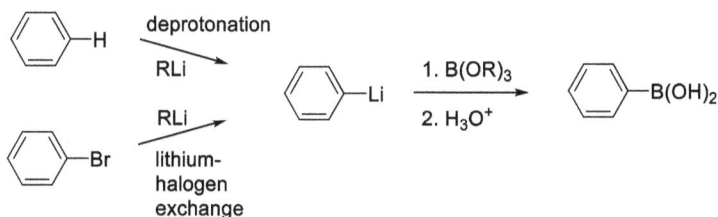

Figure 7.18: Preparation of arylboronic acids via aryllithium species .

7.3.1 Palladium-catalyzed borylation of aryl halides

One such alternative is the palladium-catalyzed borylation of aryl halides using bis(pinacolato)diboron (B_2pin_2) in the presence of potassium acetate to produce the corresponding aryl boronate ester (Figure 7.19) [28]. This reaction is often referred to as the Miyaura borylation.

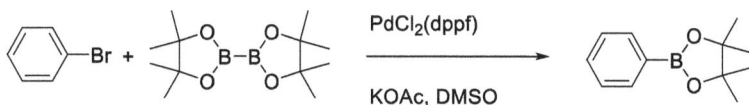

Figure 7.19: The Miyaura borylation reaction.

The reaction is typically carried out at moderately elevated temperatures in solvents such as DMSO or dioxane, which usually avoids the solubility problems that can be seen with lithium–halogen exchange. The resulting boronate ester can usually be purified by column chromatography. One disadvantage of the reaction is that boronate esters are generally less reactive than boronic acids toward Suzuki cross-couplings.

Mechanistically, the Miyaura borylation consists of an oxidative addition, transmetallation involving the diboron reagent, and reductive elimination (Figure 7.20) [28]. The active Pd(0) catalyst is generated in situ from PdCl$_2$(dppf).

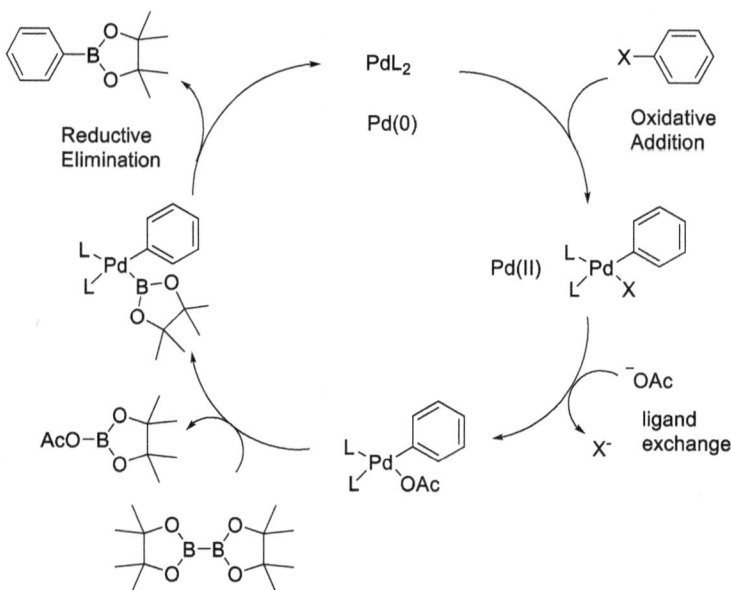

Figure 7.20: Catalytic cycle for the Miyaura borylation of aryl halides.

As seen in the mechanism outlined in Figure 7.20, using a diboron reagent such as B$_2$pin$_2$ uses only one of the two boronate groups, which is not ideal from the perspective of atom economy. Fortunately, it is possible to use the dialkoxyborane (e.g., HBpin) in the Miyaura coupling [29]. When HBpin is used in the standard Miyaura coupling conditions, dehalogenation of the aryl halide is the major product. However, when a tertiary amine such as Et$_3$N is used as a base, the desired borylation reaction is the major product and dehalogenation is suppressed.

The Miyaura borylation is used primarily for the preparation of boronate esters that serve as reactants for Suzuki cross-coupling reactions. Given the widespread use of the Suzuki cross-coupling, there are numerous examples of the use of the Miyaura coupling in multistep synthesis. Here, we will explore a few illustrative examples that target different applications.

The Miyaura borylation has been used to prepare oligophenylenes and related compounds for potential applications as in OLEDs. Specifically, a biphenyl bearing pyridyl arms was prepared for its potential as an electron-transporting material in phosphorescent OLEDs (Figure 7.21) [30]. The synthesis begins with a Suzuki cross-coupling with 1-bromo-3-iodobenzene and the pyridyl boronate ester, where the coupling takes place selectively at the more reactive iodide. The bromide is then converted into the corresponding boronate ester using a Miyaura borylation. Following this, a fourfold Suzuki coupling using the boronate ester and a tetrabromo-biphenyl gave the desired branched oligophenylene derivative.

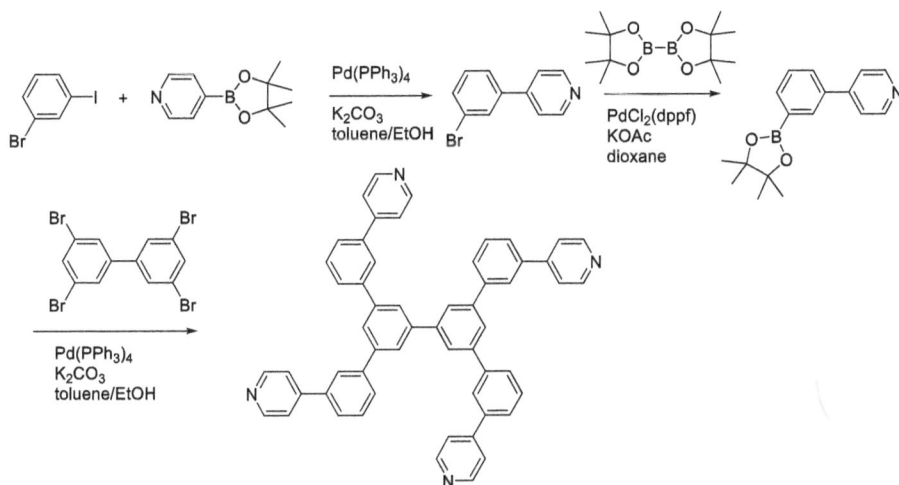

Figure 7.21: Synthesis of an electron-transporting material for phosphorescent OLEDs [30].

As another example of the application of the Miyaura borylation toward the synthesis of materials, we will consider the synthesis of a chromophore exhibiting TADF [31]. The synthesis is outlined in Figure 7.22. The synthesis began by N-protection of 3-bromocarbazole, followed by a borylation of the bromide. The resulting boronate ester underwent Suzuki coupling with 1,4-dibromobenzene to give the bis(carbazolyl) benzene. The triisopropylsilyl (TIPS) protecting groups were removed, and the carbazoles were then reacted with the fluoro-substituted phthalonitrile in a nucleophilic aromatic substitution, giving the final TADF material.

For an example of the application of the Miyaura borylation in the synthesis of pharmaceutical compounds, we consider the synthesis of abemaciclib, which was developed by Eli Lilly for the treatment of breast cancer and granted FDA approval in 2015. This synthesis highlights several of the transformations we have seen in previous chapters (Figure 7.23) [32]. In the first step, 4-bromo-2,6-difluoroaniline was treated with the Vilsmeier reagent formed from N-isopropylacetamide and POCl$_3$ to form the corresponding amidine. Unlike the Vilsmeier reactions we have considered thus far, which act as electrophiles in aromatic substitution, here the Vilsmeier reagent reacts at the aniline nitrogen. Treatment with potassium t-butoxide leads to deprotonation of the amidine nitrogen and intramolecular nucleophilic aromatic substitution to yield

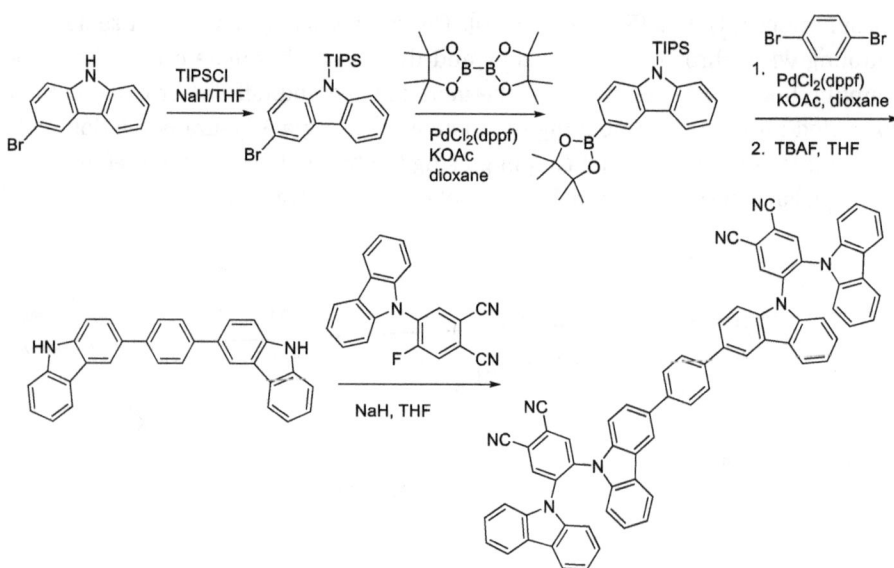

Figure 7.22: Synthesis of a TADF material [31].

the corresponding benzimidazole. Subsequent Miyaura borylation yields the aryl boronate ester, which was used for Suzuki coupling with the chloropyrimidine. Finally, a palladium-catalyzed aryl amination on the remaining chloro group yielded target compound.

7.3.2 Iridium-catalyzed direct borylation

The Miyaura borylation requires the site of borylation be functionalized as either an aryl halide or triflate. A potentially attractive alternative is a direct borylation of an arene. As with direct arylation reactions described in Chapter 6, there is a potential challenge with controlling the regiochemistry of borylation. Nevertheless, the prospect of direct borylation offers the potential advantage of a more concise synthesis. A number of different metal-catalyzed borylation reactions have been explored, including rhodium and rhenium catalyzed processes [33]. These reactions were carried out at elevated temperatures and generally had low catalytic turnover. In 2002, Ishiyama et al. developed an iridium-catalyzed direct borylation that takes place under mild conditions and is highly efficient [34,35]. This reaction, which uses bis(pinacolato)diboron as the borylating agent and an iridium catalyst such as [Ir(OMe)(cod)]$_2$ with 4,4′-di-*tert*-butyl-2,2′-dipyridyl (tbpy) (Figure 7.24). The reaction is usually carried out in hydrocarbon solvents such as cyclohexane or octane, but is also compatible with more polar solvents such as THF. This reaction has become the most widely used methods for direct borylation of arenes [33,36].

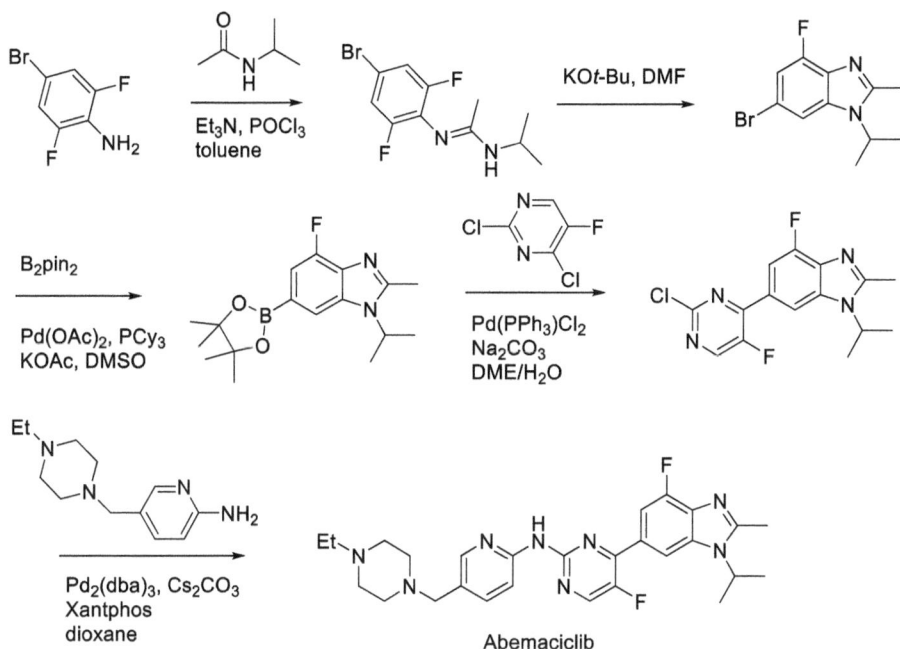

Figure 7.23: Synthesis of abemaciclib [32].

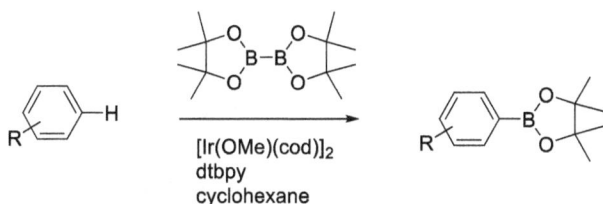

Figure 7.24: Iridium-catalyzed direct borylation.

Figure 7.25 shows a simplified version of the accepted mechanism for iridium-catalyzed direct borylation [37]. Ancillary ligands such as tbpy and COD have been omitted. The iridium complex undergoes an initial oxidative addition of B_2pin_2, followed by reductive elimination to give the iridium boronate complex. A second oxidative addition with B_2pin_2 give the iridium(III) complex with three boronate groups. Intermediate complexes such as this have been isolated, providing support for their involvement in the reaction. At this stage, the arene reacts via a C–H insertion reaction. Reductive elimination leads to the formation of the aryl boronate ester product and an iridium hydride complex. Another oxidative addition of B_2pin_2, is followed by reductive elimination of pinacolborane (HBpin) and regenerates the active iridium(III) species. It should be noted that HBpin can also undergo oxidative addition (instead of B_2pin_2), followed by reductive elimination of H_2 to regenerate the catalytically active species.

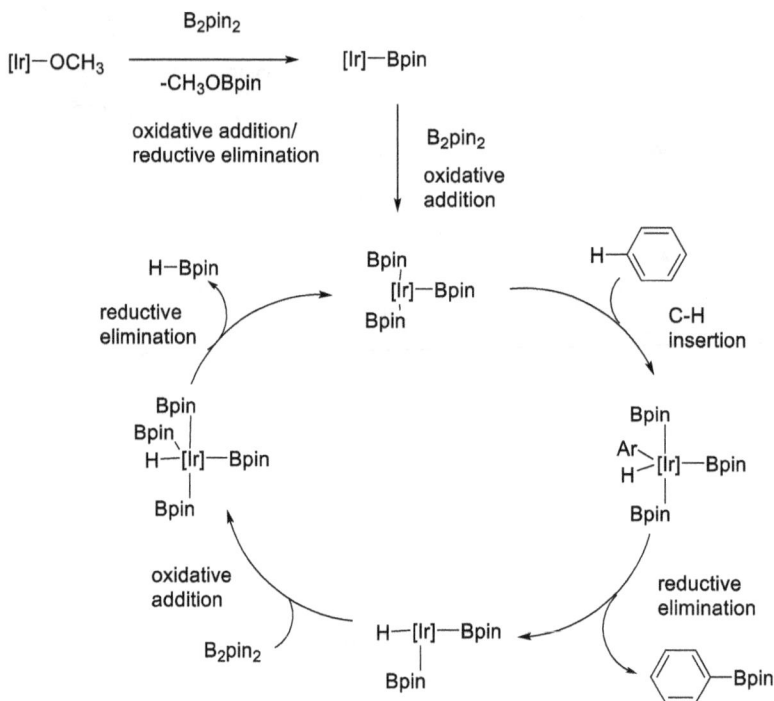

Figure 7.25: A simplified mechanism of iridium-catalyzed direct borylation.

The direct borylation avoids the need for prior functionalization such as halogenation to introduce the boronate. The regiochemistry of the borylation is determined primarily by steric considerations, where borylation occurs primarily at the least hindered site. This is explained by considering that the C–H activation occurs on the sterically crowded tris(boryl) intermediate [37, 38]. As shown in the examples in Figure 7.26, borylation *ortho* to another substituent is disfavored. In the first example, for a monosubstituted benzene, a mixture of *meta*- and *para*-products is obtained. The predominance of the *meta*, substituted product may be explained in part by probability: there are two *meta* and only one *para*-position available for reaction. It is noteworthy that changing the substituent from an electron-donating methoxy group to an electron-withdrawing trifluoromethyl group has little influence on the regiochemistry [34, 35].

The regiochemistry of Ir-catalyzed direct borylation presents the opportunity to access different substitution patterns than those obtained through electrophilic aromatic substitution or directed metalation reactions. For example, Marder and coworkers showed that pyrene, which normally undergoes electrophilic aromatic substitution at the 1, 3, 6, and 7 positions will undergo direct borylation at the least hindered 2 and 7 positions (Figure 7.27) [39]. Similarly, tetraborylation of perylene leads to substitution at the 2, 5, 8, and 11 positions (Figure 7.27). Still, when more than one sterically similar

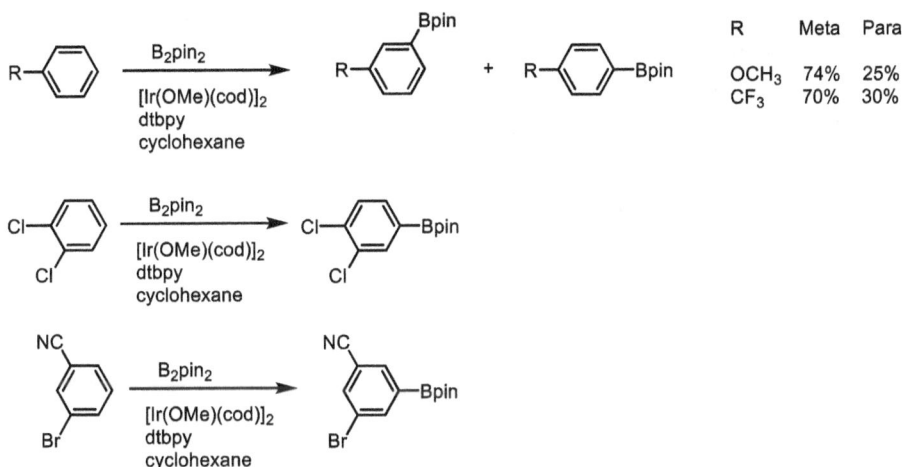

R	Meta	Para
OCH$_3$	74%	25%
CF$_3$	70%	30%

Figure 7.26: Regiochemistry of Ir-catalyzed direct borylation of substituted arenes.

site is present, a mixture of isomers is observed. For example, monoborylation of naphthalene gives exclusively 2-substituted product, but diborylation gives and approximately equal mixture of the 2,6- and 2,7-disubstituted products (Figure 7.27) [39].

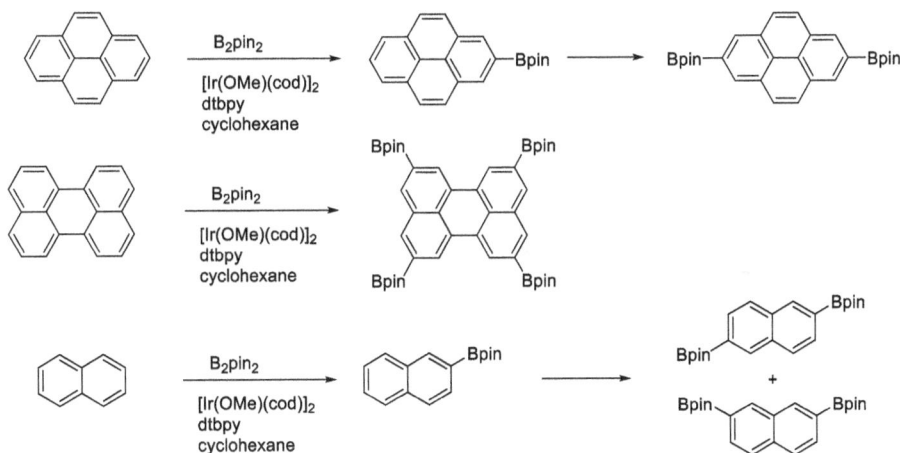

Figure 7.27: Ir-catalyzed direct borylation of some polycyclic aromatic hydrocarbons.

Direct borylation of heteroaromatic systems has also been explored. With heteroaromatic systems, the regiochemistry of borylation is influenced by the position of the heteroatom. Five-membered heteroaromatics (pyrrole, furan, and thiophene) and the corresponding benzo-fused systems (indole, benzofuran, and benzothiophene) all undergo direct borylation preferentially at the 2-position, adjacent to the heteroatom (Figure 7.28) [40]. In contrast, direct borylation of pyridine gives a 2:1 mixture of borylation products at the 3- and 4-positions. Borylation of quinoline yields borylation exclusively at the 3-position [40].

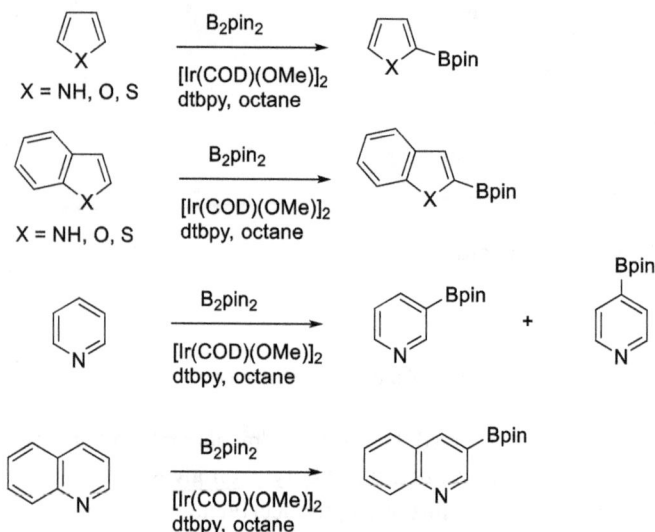

Figure 7.28: Direct borylation of some heteroaromatic compounds[40].

The regiochemistry of borylation on pyrrole and indole can be altered by introducing a bulky protecting group, such as a TIPS group. Perhaps not surprisingly, the bulky group blocks borylation at the 2-position. Instead, borylation takes place at the 3-position (Figure 7.29) [40].

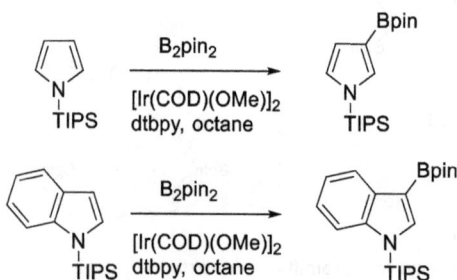

Figure 7.29: Direct borylation of N-protected pyrrole and indole [40].

If the 2-position of indole is blocked by a substituent, borylation at the 3-position becomes sterically disfavored and reaction takes place exclusively at the 7-position (Figure 7.30) [41].

Figure 7.30: Direct borylation of 2-substituted indole [41].

In general, the regiochemistry of direct borylation of heteroaromatic compounds is due to a combination of electronic effects and steric effects. In some cases, it is possible that the heteroatom directs the site of reaction by coordination [41]. However, has also been shown that the site of direct borylation in heteroaromatics correlates to the acidity of the C–H group, suggesting that position is electronically predisposed to C–H activation [42]. Since the structures of heteroaromatics are quite diverse, predicting the regiochemistry of borylation can become challenging. Hartwig and coworkers have probed the direct borylation of a variety of heteroaromatics and developed guidelines for predicting the regiochemistry of borylation in these systems [43].

To demonstrate the utility of direct borylation as a potential alternative to the Miyaura borylation, Hartwig and coworkers developed a direct borylation approach for the synthesis of a substituted pyrimidine with potential as a type II diabetes therapeutic agent (Figure 7.31) [43]. The original synthesis, which was developed by researchers at AstraZeneca, started from 5-bromo-2-chloropyrimidine [44]. The starting material reacted with a Boc-protected piperazine in a nucleophilic aromatic substitution at the 2-position. Following this, the bromide was subjected to Miyaura borylation, followed by oxidation to give the hydroxy-pyrimidine, which could be converted into the final target. The sequence of Miyaura borylation and oxidation proceeded in a modest 23% yield. In contrast, the direct borylation approach used 2-chloropyrimidine as a starting material, which is much less expensive. Direct borylation and oxidation proceeded in 91% yield (Figure 7.31). The direct borylation also proceeds at room temperature with very low catalyst loading. For this reaction, [Ir(cod)(OMe)]$_2$ was used with 3,4,7,8-tetramethyl-1,10-phenanthroline as a ligand instead of tbpy because the resulting catalyst shows much higher reactivity [45]. In this case, the direct borylation allows the use of cheaper starting materials and proceeds in higher overall yield than the Miyaura borylation.

As discussed above, one of the disadvantages of aryl boronate esters is that they are less reactive than the corresponding boronic acids. To address this potential drawback, Hartwig and coworkers developed a one-pot procedure for direct borylation, followed by conversion to the more reactive arylboronic acid or aryltrifluoroborate (Figure 7.32) [46]. The did so by using low catalyst loadings and choosing THF as a solvent, which is compatible with the subsequent conversions.

Interestingly, in the presence of a base such as potassium t-butoxide, an isomerization of the boronate groups can be observed, suggesting that the borylation is reversible under these conditions and can lead to the most favored product (Figure 7.33) [47]. This reversibility is clearly illustrated below, where borylation of the *para*-substituted

AstraZeneca approach (Miyaura borylation)

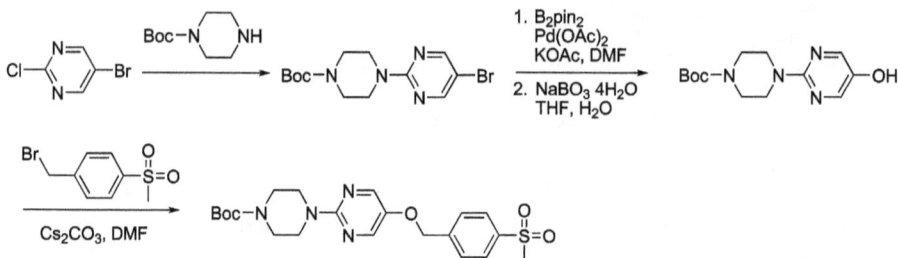

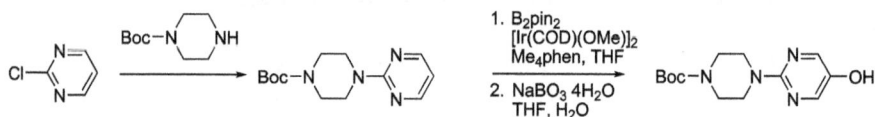

Figure 7.31: Comparison of syntheses of a potential type II diabetes therapeutic via Miyaura borylation and via direct borylation [43].

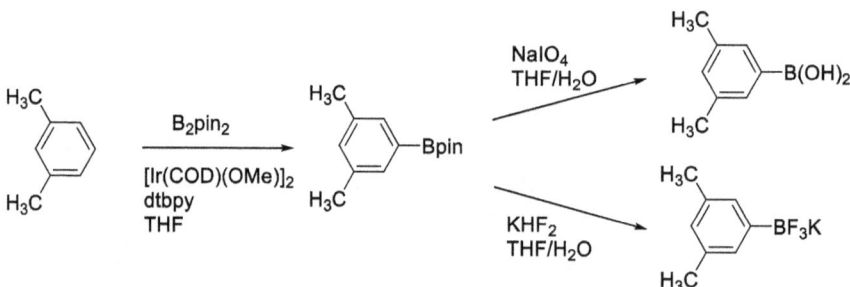

Figure 7.32: One pot direct borylation and conversion to the arylboronic acid or aryltrifluoroborate.

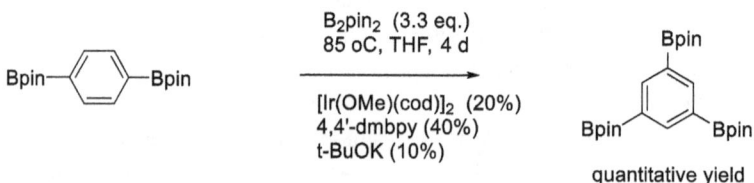

Figure 7.33: Direct borylation/isomerization in the presence of base [47].

benzene diboronate leads to isomerization and borylation to produce 1,3,5-tris(boro-nate) in good yield.

This interesting feature can be exploited to prepare polyborylated aromatic compounds with high selectivity. For example, biphenyl, pyrene, coronene, and corannulene can all be polyborylated with high regioselectivity (Figure 7.34) [47].

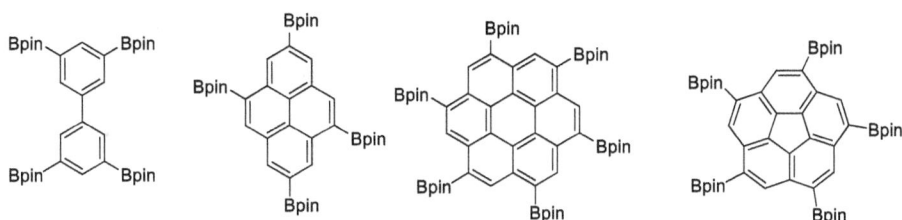

Figure 7.34: Polyborylated polycyclic aromatic hydrocarbons.

References

[1]   Ullmann F. NoUeber eine neue Bildungsweise von Diphenylaminderivaten. Ber Dtsch Chem Ges. 1903;36:2382–84.

[2]   Ullmann F, Sponagel P. Ueber die Phenylirung von Phenolen. Ber Dtsch Chem Ges. 1905;38:221–2212.

[3]   Goldberg I. Ueber phenylirungen bei Gegenwart von Kupfer als Katalysator. Ber Dtsch Chem Ges. 1906;39:1691–92.

[4]   Sambiagio C, Marsden SP, Blacker AJ, McGowan PC. Copper catalysed Ullmann type chemistry: From mechanistic aspects to modern development. Chem Soc Rev. 2014;43:3525–50.

[5]   Lakshmi V, Liu C, Rao MR, Chen Y, Fang Y, Dadvand A, et al. . A two-dimensional poly (azatriangulene) covalent organic framework with semiconducting and paramagnetic states. J Am Chem Soc. 2020;142:2155–60.

[6]   Evano G, Blanchard N, Toumi M. Copper-mediated coupling reactions and their applications in natural products and designed biomolecules synthesis. Chem Rev. 2008;108(8):3054–131.

[7]   Ma D, Xia C, Jiang J, Zhang J, Tang W. Aromatic nucleophilic substitution or CuI-catalyzed coupling route to martinellic acid. J Org Chem. 2003;68(2):442–51.

[8]   Mallesham B, Rajesh BM, Rajamohan Reddy P, Srinivas D, Trehan S. Highly efficient CuI-catalyzed coupling of aryl bromides with oxazolidinones using Buchwald's protocol: A short route to linezolid and toloxatone. Org Lett. 2003;5(7):963–65.

[9]   Goodbrand HB, Hu NX. Ligand-accelerated catalysis of the Ullmann condensation: Application to hole conducting triarylamines. J Org Chem. 1999;64(2):670–74.

[10]   Bender TP, Graham JF, Duff JM. Effect of substitution on the electrochemical and xerographic properties of triarylamines: Correlation to the hammett parameter of the substituent and calculated HOMO energy level. Chem Mater. 2001;13(11):4105–11.

[11]   Zanka A, Kubota A, Hirabayashi S, Nakamura H. Process development of a novel anti-inflammatory agent. The regiospecific bromination of 4'-acetylmethanesulfonanilide. Org Proc Res Dev. 1998;2(2):71–77.

[12]   Enache LA, Kennedy I, Sullins DW, Chen W, Ristic D, Stahl GL, et al. . Development of a scalable synthetic process for DG-051B, a first-in-class inhibitior of LTA4H. Org Proc Res Dev. 2009;13(6):1177–84.

[13]   Kosugi M, Kameyama M, Migita T. Palladium-catalized aromatic amination of aryl bromides with N,N-Diethylamino-Tributyltin. Chem Lett. 1983;12:927–28.

[14]   Hartwig JF. Recent advances in palladium- and nickel- catalyzed chemistry provide new ways to construct C-N and C-O bonds. Angew Chem Int Ed. 1998;37:2046–67.

[15] Wolfe JP, Wagaw S, Marcoux JF, Buchwald SL. Rational development of practical catalysts for aromatic carbon-nitrogen bond formation. Acc Chem Res. 1998;31(12):805–18.

[16] Hartwig JF. Carbon – heteroatom bond-forming reductive eliminations of amines, ethers, and sulfides. Acc Chem Res. 1998;31(12):852–60.

[17] Ruiz-Castillo P, Buchwald SL. Applications of palladium-catalyzed C-N cross-coupling reactions. Chem Rev. 2016;116(19):12564–649.

[18] Surry DS, Buchwald SL. Dialkylbiaryl phosphines in Pd-catalyzed amination: A user's guide. Chem Sci. 2011;2(1):27–50.

[19] Surry DS, Buchwald SL. Biaryl phosphane ligands in palladium-catalyzed amination. Angew Chem Int Ed. 2008;47(34):6338–61.

[20] Bhagwanth S, Adjabeng GM, Hornberger KR. Mild conditions for Pd-catalyzed conversion of aryl bromides to primary anilines using benzophenone imine. Tetrahedron Lett. 2009;50 (14):1582–85.

[21] Wolfe JP, Ahman J, Sadighi JP, Singer RA, Buchwald SL. An ammonia equivalent for the palladium catalyzed amination of aryl halides and triflates. Tetrahedron Lett. 1997;36 (36):6367–70.

[22] Campeau LC, Parisien M, Jean A, Fagnou K. Catalytic direct arylation with aryl chlorides, bromides, and iodides: Intramolecular studies leading to new intermolecular reactions. J AmChem Soc. 2006;128(2):581–90.

[23] Liégault B, Lee D, Huestis MP, Stuart DR, Fagnou K. Intramolecular Pd(II)-catalyzed oxidative biaryl synthesis under air: Reaction development and scope. J Org Chem. 2008;73 (13):5022–28.

[24] Griffin AM, Brown W, Walpole C, Coupal M, Adam L, Gosselin M, et al. Delta agonist hydroxy bioisosteres: The discovery of 3-((1-benzylpiperidin-4-yl){4-[(diethylamino)carbonyl]phenyl} amino)benzamide with improved delta agonist activity and in vitro metabolic stability. Bioorg Med Chem Lett. 2009;19(21):5999–6003.

[25] Harris MC, Buchwald SL. One-pot synthesis of unsymmetrical triarylamines from aniline precursors. J Org Chem. 2000;65(17):5327–33.

[26] Reddy SS, Sree VG, Gunasekar K, Cho W, Gal YS, Song M, et al. Highly efficient bipolar deep-blue fluorescent emitters for solution-processed non-doped organic light-emitting diodes based on 9,9-Dimethyl-9,10-dihydroacridine/phenanthroimadazole derivatives. Adv Opt Mater. 2016;4(8):1236–46.

[27] Hall DG editor. Boronic Acids: Preparation and Applications in Organic Synthesis and Medicine. Weinheim: Wiley-VCH Verlag; 2005.

[28] Ishiyama T, Murata M, Miyaura N. Palladium(0)-catalyzed cross-coupling reaction of alkoxydiboron with haloarenes: A direct procedure for arylboronic esters. J Org Chem. 1995;60(23):7508–10.

[29] Murata M, Watanabe S, Masuda Y. Novel palladium (0) -catalyzed coupling reaction of dialkoxyborane with aryl halides : Convenient synthetic route to arylboronates lation between arylmagnesium or -lithium reagents and or bromides 1, giving the corresponding products 3 Bearing in mind. J Org Chem. 1997;62(entry 6):6458–59.

[30] Su SJ, Tanaka D, Li YJ, Sasabe H, Takeda T, Kido J. Novel four-pyridylbenzene-armed biphenyls as electron-transport materials for phosphorescent OLEDs. Org Lett. 2008;10 (5):941–44.

[31] Chen D, Kusakabe Y, Ren Y, Sun D, Rajamalli P, Wada Y, et al. Multichromophore molecular design for thermally activated delayed-fluorescence emitters with near-unity photoluminescence quantum yields. J Org Chem. 2021;86(17):11531–44.

[32] Flick AC, Leverett CA, Ding HX, McInturff E, Fink SJ, Helal CJ, et al. Synthetic approaches to the new drugs approved during 2017. J Med Chem. 2019;62(16):7340–82.

[33] Mkhalid IAI, Barnard JH, Marder TB, Murphy JM, Hartwig JF. C-H activation for the construction of C-B bonds. Chem Rev. 2010;110:890–931.

[34] Ishiyama T, Takagi J, Ishida K, Miyaura N, Anastasi NR, Hartwig JF. Mild iridium-catalyzed borylation of arenes. High turnover numbers, room temperature reactions, and isolation of a potential intermediate. J Am Chem Soc. 2002;124(3):390–91.

[35] Ishiyama T, Takagi J, Hartwig JF, Miyaura N. A stoichiometric aromatic C?H Borylation catalyzed by Iridium(I)/2,2′-Bipyridine complexes at room temperature. Angew Chem Int Ed. 2002;41(16):3056–58.

[36] Hartwig JF. Borylation and silylation of C-H bonds: A platform for diverse C-H bond functionalizations. Acc Chem Res. 2012;45(6):864–73.

[37] Tamura H, Yamazaki H, Sato H, Sakaki S. Iridium-catalyzed borylation of benzene with diboron. Theoretical elucidation of catalytic cycle including unusual iridium(V) intermediate. J Am Chem Soc. 2003;125(51):16114–26.

[38] Nguyen P, Blom HP, Westcott SA, Taylor NJ, Marder TB. Synthesis and structures of the first transition-metal tris (boryl) complexes: (eta-6-Arene)Ir(BO2C6H4)3. J Am Chem Soc. 1993;115:9329–30.

[39] Coventry DN, Batsanov AS, Goeta AE, Howard JAK, Marder TB, Perutz RN. Selective Ir-catalysed borylation of polycyclic aromatic hydrocarbons: Structures of naphthalene-2,6-bis (boronate), pyrene-2,7-bis(boronate) and perylene-2,5,8,11-tetra(boronate) esters. Chem Commun. 2005;2172–74.

[40] Takagi J, Sato K, Hartwig JF, Ishiyama T, Miyaura N. Iridium-catalyzed C-H coupling reaction of heteroaromatic compounds with bis(pinacolato)diboron: Regioselective synthesis of heteroarylboronates. Tetrahedron Lett. 2002;43(32):5649–51.

[41] Paul S, Chotana GA, Holmes D, Reichle RC, Maleczka RE, Smith MR. Ir-catalyzed functionalization of 2-substituted indoles at the 7-position: Nitrogen-directed aromatic borylation. J Am Chem Soc. 2006;128(49):15552–53.

[42] Vanchura BA, Preshlock SM, Roosen PC, Kallepalli VA, Staples RJ, Maleczka RE, et al. Electronic effects in iridium C-H borylations: Insights from unencumbered substrates and variation of boryl ligand substituents. Chem Commun. 2010;46(41):7724–26.

[43] Larsen MA, Hartwig JF. Iridium-catalyzed C-H borylation of heteroarenes: Scope, regioselectivity, application to late-stage functionalization, and mechanism. J Am Chem Soc. 2014;136(11):4287–99.

[44] Scott JS, Birch AM, Brocklehurst KJ, Broo A, Brown HS, Butlin RJ, et al. Use of small-molecule crystal structures to address solubility in a novel series of g protein coupled receptor 119 agonists: Optimization of a lead and in vivo evaluation. J Med Chem. 2012;55(11):5361–79.

[45] Preshlock SM, Gha B, Maligres PE, Krska SW, Maleczka RE, Smith MR. High-throughput optimization of Ir-catalyzed C–H Borylation: A tutorial for practical applications. J Am Chem Soc. 2013;135:7572–82.

[46] Murphy JM, Tzschucke CC, Hartwig JF. One-pot synthesis of arylboronic acids and aryl trifluoroborates by Ir-catalyzed borylation of arenes. Org Lett. 2007;9(5):757–60.

[47] Eliseeva MN, Scott LT. Pushing the Ir-catalyzed C-H polyborylation of aromatic compounds to maximum capacity by exploiting reversibility. J Am Chem Soc. 2012;134(37):15169–72.

8 Constructing aromatic rings

8.1 Introduction

Benzene and several of its derivatives are common compounds that are readily available from petroleum feedstocks. As such, in many cases, the synthesis of aromatic compounds containing benzene rings starts from simpler precursors containing a benzene ring. There are some situations, however, where it is useful to construct a benzene ring from non-aromatic precursors. This approach can allow access to benzene substitution patterns that are not otherwise easy to access. Another important situation for the construction of an aromatic ring from non-aromatic precursors is for heteroaromatic compounds, which are important in a variety of applications, yet are not readily available. In this chapter, we will see some of the strategies for making benzene rings from non-aromatic precursors. Some of these reactions will be used in the Chapter 9–11 in the context of preparing polycyclic aromatic compounds. This chapter will also highlight some of the approaches for preparing heteroaromatic compounds. Given the number of different heteroaromatic rings and variety of synthetic methods to access them, the goal here is not a comprehensive review, but rather to highlight some of the general strategies for preparing heteroaromatics and show some of the examples of how these are applied.

8.2 Preparing benzene rings from non-aromatic precursors

8.2.1 Cycloaddition reactions

Diels–Alder reactions are well known for the construction of six-membered rings. The product of a Diels–Alder reaction is a cyclohexene (or a cyclohexadiene if the dienophile is an alkyne), and these can be converted to the corresponding benzene by oxidation (Figure 8.1).

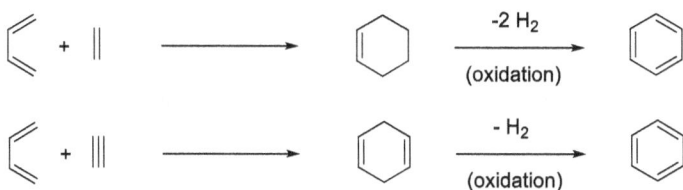

Figure 8.1: Diels–Alder reactions as an approach to benzene rings.

Many examples of Diels–Alder reactions to prepare aromatic rings involve a cyclic diene, where the initial cycloadduct undergoes elimination of a small molecule to generate the

https://doi.org/10.1515/9783110562682-008

aromatic ring. These reactions are referred to as cheletropic eliminations, and are essentially retro-cycloaddition reactions. The general concept and mechanism is shown in Figure 8.2.

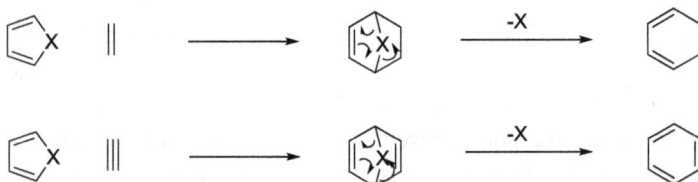

Figure 8.2: General mechanism of a cheletropic reaction.

In these reactions, loss of small molecules such as CO, CO_2, and SO_2 are common. For example, cyclopentadienones can be prepared by aldol condensation and then react with acetylenes to produce the corresponding substituted benzene [1]. Figure 8.3 shows how this approach can be used to prepare hexaphenylbenzene [2]. In this example, benzil reacts with 1,3-diphenylacetone in a twofold aldol condensation to product tetraphenylcyclopentadienone. The tetraphenylcyclopentadienone undergoes a Diels–Alder reaction with diphenylacetylene, which is followed by expulsion of carbon monoxide to form the central benzene ring. This approach can be powerful for the preparation of highly substituted benzenes that often cannot be accessed readily by cross-coupling methods because of steric constraints.

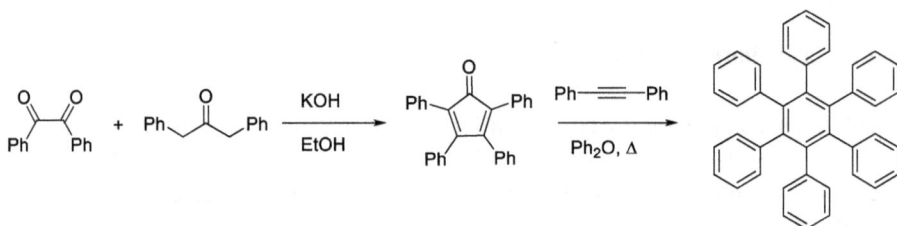

Figure 8.3: Synthesis of hexaphenylbenzene via a Diels–Alder/cheletropic reaction [2].

A potential limitation of the use of cyclopentadienones is that some have limited stability, presumably because of their partial antiaromatic character. In these cases, the aldol condensation does not go to completion, instead generating an aldol addition intermediate [3]. Fortunately, the cyclopentadienone can often be generated in situ in the presence of acid, allowing the Diels–Alder reaction to take place (Figure 8.4) [4].

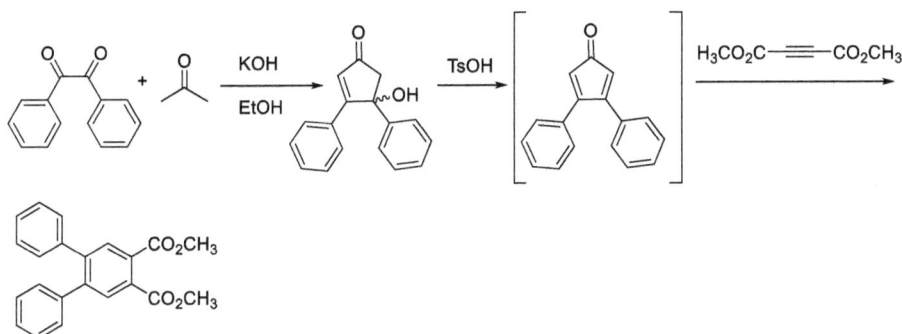

Figure 8.4: In-situ formation of a cyclopentadienone, followed by Diels–Alder reaction [4].

Thiophene-1,1-dioxides can also undergo Diels–Alder reactions and cheletropic elimination of SO_2 to produce a new benzene ring [5]. Thiophene-1,1-dioxides can be prepared by oxidation of the corresponding thiophenes, using oxidizing agents such as mCPBA, peracetic acid, or trifluoroperacetic acid (Figure 8.5). Upon treatment with an alkene, the thiophene-1,1-dioxide undergoes a Diels–Alder reaction, followed by cheletropic elimination of SO_2 to give a cyclohexadiene, which can often be converted to the benzene ring by an elimination or an oxidation reaction [5].

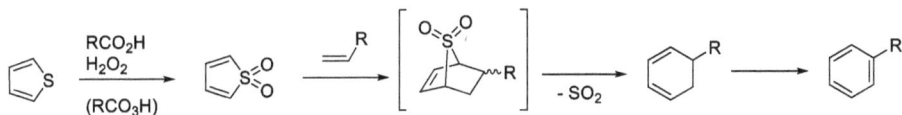

Figure 8.5: Preparation of thiophene-1,1-dioxides and subsequent Diels–Alder reaction.

For example, 3,4-di-*tert*-butylthiophene was oxidized to the corresponding dioxide using mCPBA (Figure 8.6) [6]. A Diels–Alder reaction with phenyl vinyl sulfone leads to cheletropic extrusion of SO_2, followed by elimination of phenylsulfenic acid to give *o*-di-*tert*-butylbenzene. In this case, aromatization is driven by the elimination of phenylsulfenic acid. The thiophene-1,1-dioxide could also react with an alkyne to give the aromatic ring directly after SO_2 extrusion, without the need for extra elimination step. It should be noted that this method allowed the preparation of an aromatic ring bearing two bulky *tert*-butyl groups *ortho* to one another – a substitution pattern that is not easily achieved by other methods because of steric congestion.

This reaction can also be used to prepare substituted naphthoquinones and anthraquinones with a regiochemistry that would not be seen by direct bromination (Figure 8.7) [7]. 3,4-Dibromothiophene was converted into the corresponding thiophene-1,1-dioxide using peroxytrifluoroacetic acid, which was generated from hydrogen peroxide with trifluoroacetic anhydride. This compound underwent a Diels–Alder reaction with benzoquinone, followed by expulsion of SO_2 to give the hydroquinone, which

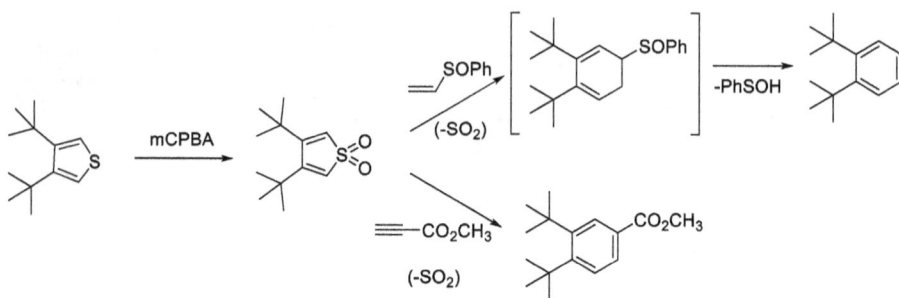

Figure 8.6: Preparation of sterically crowded benzenes via Diels–Alder reactions of thiophene-1,1-dioxides [6].

underwent oxidation under the reaction conditions to produce the desired naphthoquinone. An excess of benzoquinone was used to prevent formation of the anthraquinone, as well as a homo-Diels Alder reaction between thiophene dioxide units. This excess benzoquinone also likely served as the oxidizing agent for the quinone formation.

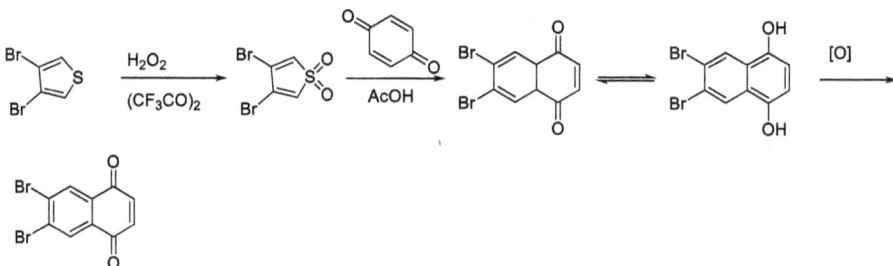

Figure 8.7: Preparation of substituted naphthoquinones from thiophene-1,1-dioxides [7].

8.2.2 Transition metal-catalyzed alkyne cyclotrimerization

Another approach to preparing benzene rings from acyclic precursors is a formal [2 + 2 + 2] cycloaddition of alkynes, shown schematically in Figure 8.8.

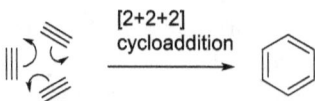

Figure 8.8: Hypothetical [2 + 2 + 2] cycloaddition of alkynes to form a benzene ring.

While this reaction should in principle be exothermic, it would be entropically unfavorable and the probability of orienting three alkynes with the appropriate geometry to undergo cycloaddition is very low. As such, thermal [2 + 2 + 2] cycloaddition reactions are not generally practical. In 1948, Reppe and coworkers reported the nickel (II)-catalyzed cyclotrimerization of acetylene to form benzene, showing that this transformation could be achieved with the help of transition metals [8]. Since then, several approaches for the preparation of aromatic rings by alkyne cyclotrimerization have been developed, using transition metal catalysts involving cobalt, nickel, and rhodium, among others [9, 10]. Among the most noteworthy is the use of cobalt catalysts for alkyne cyclotrimerization developed by Vollhardt and coworkers [9]. They first reported that $(\eta^5$-cyclopentadienyl)cobalt dicarbonyl, CpCo(CO)$_2$, could catalyzed the reaction of 1,5-hexadiyne with bis(trimethylsilyl)acetylene (BTMSA) to give 4,5-bis(trimethylsilylbenzocyclobutene) (Figure 8.9) [11].

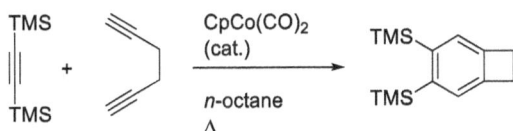

Figure 8.9: Preparation of a benzocyclobutene via cobalt-catalyzed cyclotrimerization [11].

This transformation is significant not only because it demonstrates the formal [2 + 2 + 2] cycloaddition of alkynes to form a benzene ring, but also because in this case the product is a benzocyclobutene, which can undergo electrocyclic ring-opening to form an o-quinodimethane, which is a highly reactive diene that can be used for subsequent transformations. For example, heating the benzocyclobutene in the presence of a dienophile such as maleic anhydride gives the Diels–Alder adduct in high yield (Figure 8.10) [11].

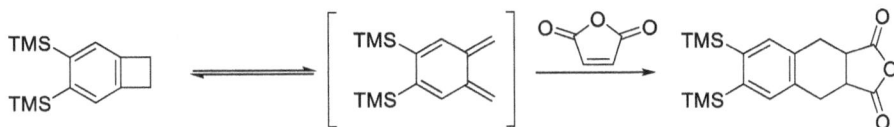

Figure 8.10: Reactions of benzocyclobutenes [11].

The mechanism of this reaction is not clearly established, but there are two plausible mechanistic scenarios, outlined in Figure 8.11. The mechanism is thought to begin by sequential dissociation of the CO ligands and coordination of two alkynes. For convenience, coordination of the alkyne moieties of the 1,5-hexadiyne is shown, but coordination of BTMSA is perhaps more likely because it is generally used in excess. With two alkyne units coordinated, the next step is an oxidative coupling, giving the metallocyclopentadiene. At this point, there are two possible pathways. The first involves a Diels–Alder reaction between the third alkyne and the metallocyclopentadiene, followed by dissociation from the metal complex. The second involves alkyne coordination, followed by a migratory insertion to give the metallocycloheptatriene, which undergoes an electrocyclic ring-closing to form the six-membered ring. Again, dissociation from the metal gives the final product.

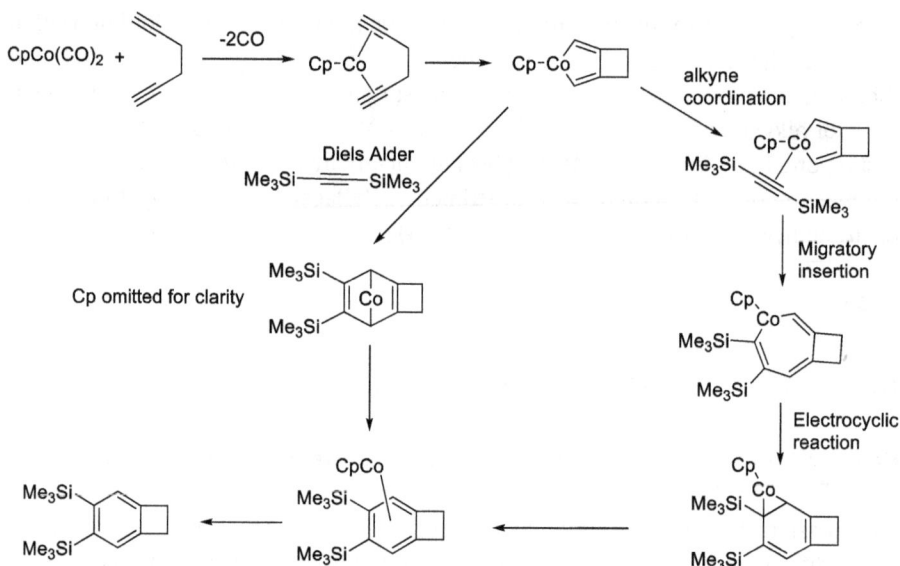

Figure 8.11: Proposed mechanism of alkyne cyclotrimerization.

Vollhardt applied this methodology to the preparation of a variety of natural products, including the synthesis of steroid hormones [12–14]. For example, Vollhardt used the alkyne cyclotrimerization for the total synthesis of estrone (Figure 8.12) [12, 14]. In this synthesis, the cyclopentanone with the pendant hexadiyne unit reacted with BTMSA in the presence of CpCo(CO)$_2$ to yield the benzocyclobutene derivative. Upon heating in decalin, the benzocyclobutene underwent an electrocyclic ring-opening to give the o-quinodimethane, which underwent an intramolecular Diels–Alder reaction with the vinyl group, resulting in in an elegant construction of the steroid skeleton. A regioselective protiodesilylation (recall Section 2.6.1), followed by an oxidation with Pb(O$_2$CCF$_3$)$_4$ gave the desired estrone in racemic form.

Vollhardt and coworkers also used the cobalt-catalyzed alkyne cyclotrimerization to access [n]phenylenes. [n]Phenylenes are compounds where two or more benzene rings are linked by two bonds *ortho* to one another, as shown in Figure 8.13.

Because the alkyne cyclotrimerization strategy can be used for the preparation of benzene rings, including strained systems such as benzocyclobutene, it is well-suited to accessing phenylenes. For example, biphenylene can be prepared by cyclotrimerization involving 1,2-bis(ethynyl)benzene and BTMSA to prepare the bis(trimethylsilyl)biphenylene, which could be protiodesilylated using trifluoroacetic acid to give biphenylene (Figure 8.14) [15]. The strategy was also extended to the preparation

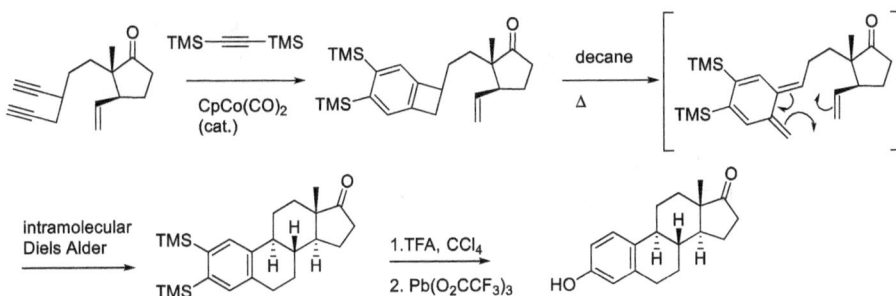

Figure 8.12: Synthesis of estrone via Vollhardt alkyne cyclotrimerization [14].

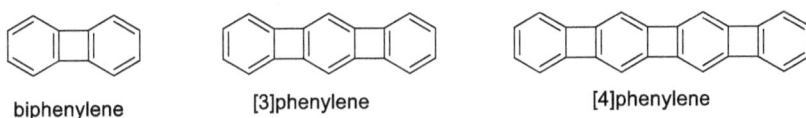

biphenylene   [3]phenylene   [4]phenylene

Figure 8.13: Representative [n]phenylene structures.

of [3]phenylene, as outlined in Figure 8.14. Starting with 1,2,4,5-tetraiodobenzene, a fourfold Sonogashira coupling gave the tetrakis(ethynyltrimethylsilyl)benzene, which could be desilylated under basic conditions to give 1,2,4,5-tetraethynelben-zene – an unstable compound that was prone to detonation upon heating. A twofold alkyne cyclotrimerization with BTMSA gave the [3]phenylene bearing TMS groups, which could be desilylated under basic conditions to give [3]phenylene [15].

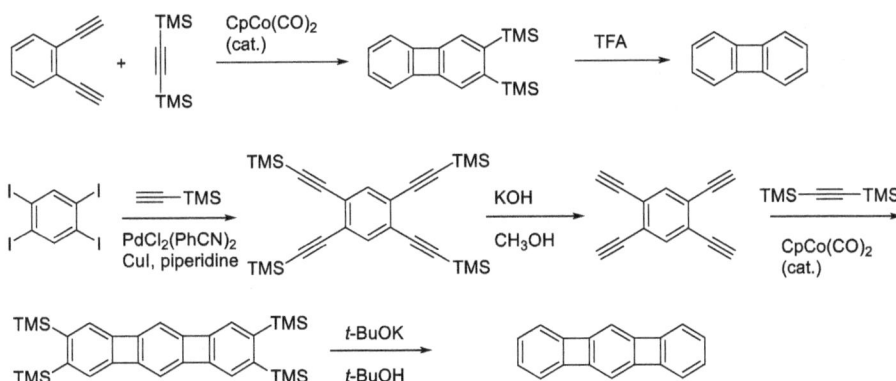

Figure 8.14: Synthesis of biphenylene and [3]phenylene via alkyne cyclotrimerization.

This same synthetic approach, using a combination of Sonogashira cross-couplings and alkyne cyclotrimerizations, could be extended to the synthesis of angular phenyl-enes, such as in the synthesis of the angular [4]phenylene and angular [5]phenylene outlined in Figure 8.15 [16]. The synthesis began with a sequence of two Sonogashira coupling reactions on *o*-bromoiodobenzene with two differentiated silyl acetylenes. Selective removal of the TMS group was followed by cobalt-catalyzed cyclotrimerization with trimethylsilyl acetylene to give the substituted biphenylene. The silyl groups were converted to the corresponding iodo groups by an *ipso* iododesilylation using ICl. A sequence to two more Sonogashira couplings gave the trialkyne, which

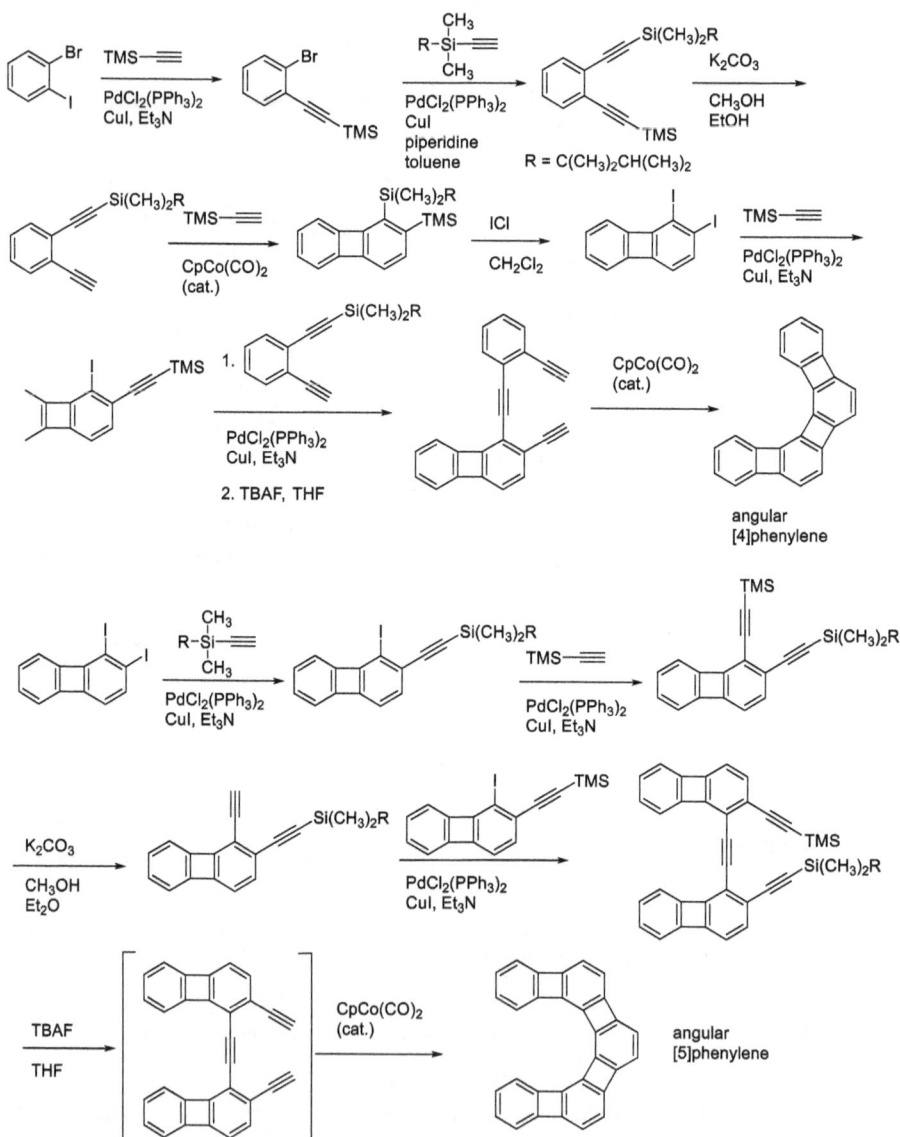

Figure 8.15: Synthesis of angular phenylenes [16].

underwent cyclotrimerization to give the angular [4]phenylene. A similar approach was followed for the preparation of the corresponding angular [5]phenylene.

This methodology could be further extended to the preparation of triangular [4] phenylenes and even [7]phenylenes, as outlined in Figure 8.16 [17, 18]. The synthesis of both of these compounds began with a sixfold Sonogashira coupling of trimethylsilyl acetylene with hexabromobenzene. A cyclotrimerization with BTMSA followed by protiodesilylation gave the triangular [4]phenylene. For the triangular [7]phenylene, cyclotrimerization with the protected triyne gave the ethynyl-substituted [4]

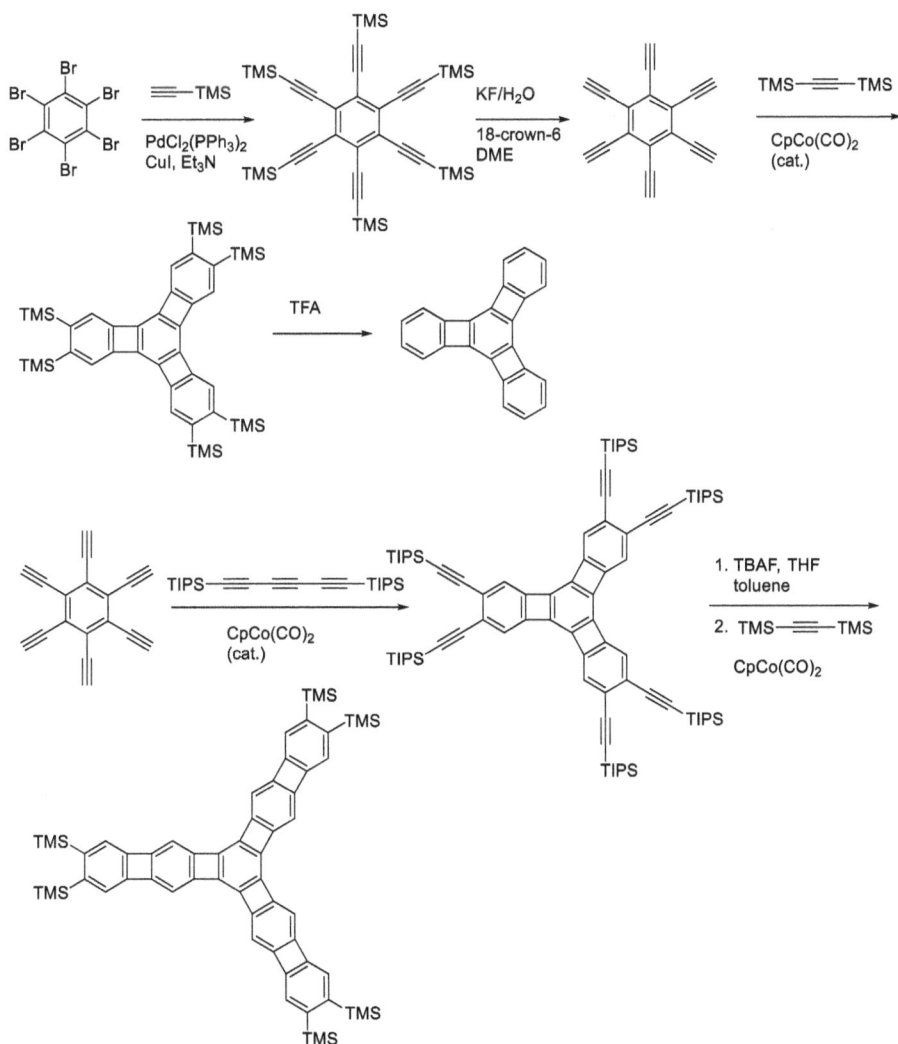

Figure 8.16: Synthesis of triangular [4]- and [7]-phenylenes [17, 18].

phenylene, which could be deprotected and subjected to another cyclotrimerization reaction with bis(trimethylsilyl)acetylene to give the TMS-substituted [7]phenylene.

Cobalt-catalyzed alkyne cyclotrimerizations are not limited to the synthesis of benzocyclobutenes or [n]phenylenes, and can also be used for the preparation of other highly substituted benzenes. For example, the reaction has been used to prepare substituted hexaphenylbenzenes by cyclotrimerization of diphenyl acetylenes using cobalt catalysts such as CpCo(CO)$_2$ or Co$_2$(CO)$_8$ (Figure 8.17) [19, 20].

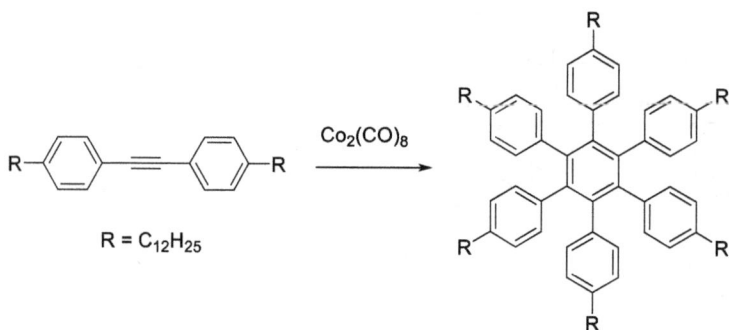

Figure 8.17: Synthesis of substituted hexaphenylbenzenes by cobalt-catalyzed alkyne cyclotrimerization [19, 20].

We saw in the previous section that hexaphenylbenzenes could be prepared by a step-wise aldol condensation to produce the cyclopentadienone, followed by a Diels–Alder reaction with diphenyl acetylene. The alkyne cyclotrimerization offers a more direct method of preparing hexaphenylbenzenes in one step from diphenyl acetylenes. On the other hand, the condensation/Diels–Alder approach allows different substituents to be installed on the phenyl rings.

8.2.3 Olefin metathesis

Ring-closing olefin metathesis (RCM) has become a widely used approach for the preparation of double bonds. The generalized reaction is shown below and uses metal alkylidene catalysts such as those developed by Schrock and Grubbs (Figure 8.18). Mechanistically, the reaction consists of a sequence of [2 + 2] cycloadditions and cycloreversions and is usually under thermodynamic control.

RCM is generally carried out under mild conditions and very tolerant of different functional groups. Olefin metathesis can also be useful for the construction of aromatic rings [21]. As an example of one strategy, acyclic octatrienes can undergo olefin metathesis to produce cyclohexadienones, which tautomerize to the corresponding phenols (Figure 8.19) [22]. This approach has also been used for the preparation of substituted anilines. This method has the advantage that it can be used

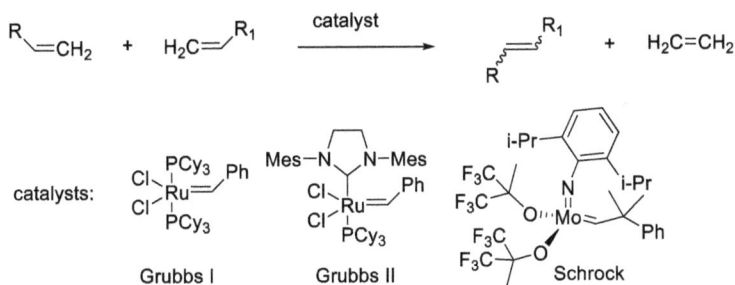

Figure 8.18: Some common catalysts for olefin metathesis.

to achieve substitution patterns that are not readily achieved using more conventional methods that focus on functionalizing existing benzene rings.

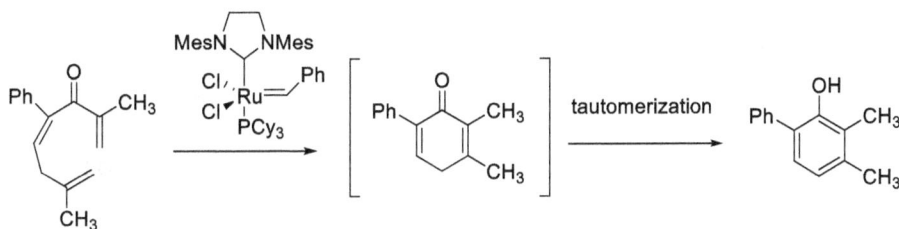

Figure 8.19: An example of olefin metathesis to prepare an aromatic ring [22].

As an example of the use of olefin metathesis to prepare an aromatic ring as part of a total synthesis, we will consider the synthesis of hasubanonine reported by Castle and coworkers (Figure 8.20) [23]. This synthesis also highlights some of the chemistry we have seen in earlier chapters. The synthesis begins with nitration of 4-benzyloxy-2,3-dimethoxybenzaldehyde. Since there are two sites on the aromatic ring that are activated, the regiochemistry of nitration is likely influenced by the steric bulk of the benzyloxy substituent. The nitro group was then reduced using $FeSO_4$ in ammonium hydroxide. The resulting aniline was converted to the diazonium salt using t-butyl nitrite and tetrafluoroboric acid and reacted with KI to give the corresponding aryl iodide. At the same time, regioselective iodination of 3,4-dimethoxybenzaldehyde was achieved by directed *ortho*-metalation using n-BuLi and N,N,N-trimethylethylenediamine, followed by phenyllithium, and ultimately trapped with 1,2-diiodoethane to yield the desired iodide. This modified DoM protocol was used to achieve good regioselectivity for the *ortho*-lithiation step [24]. The resulting aryl iodide was converted to the corresponding pinacol boronate ester using Miyaura borylation conditions. Suzuki coupling of the aryl boronate with the aryl iodide gave the desired biphenyl. The formyl groups were converted to the vinyl groups via a Wittig reaction, which set the stage for olefin metathesis using

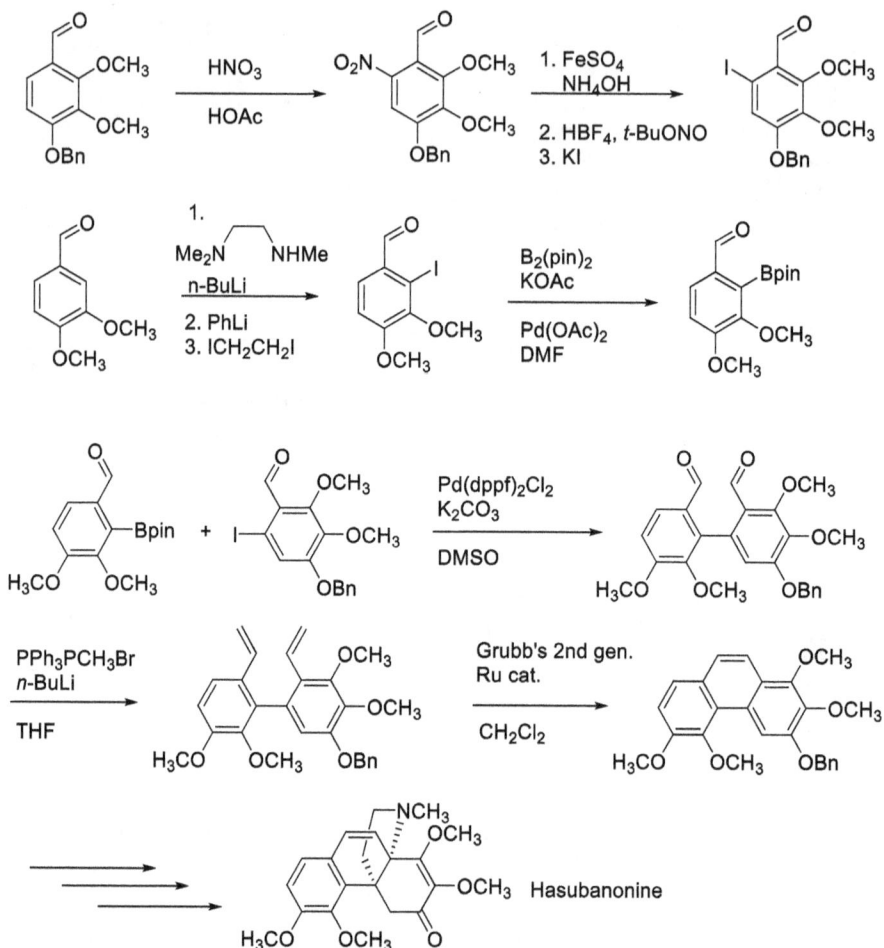

Figure 8.20: Synthesis of hasubanonine [23].

Grubb's second generation catalyst, which gave the desired phenanthrene derivative. A sequence of subsequent steps gave the natural product hasubanonine. It
should be noted that during these steps, the aromatic ring created through olefin
metathesis was ultimately lost. However, the synthesis provides an instructive summary of several of the reactions we have explored and demonstrates the use of olefin metathesis for the preparation of aromatic compounds.

8.3 Preparing aromatic heterocycles

The synthesis of aromatic heterocycles is very important because of the prevalence
of these systems in biologically active compounds ranging from natural products to

synthetic pharmaceuticals. The synthesis of heteroaromatic compounds is also diverse – much like the number of different heteroaromatic systems. A thorough exploration of the construction of heteroaromatic rings could be the subject of its own textbook [25]. Here, the goal is to highlight some of the general approaches to heteroaromatic rings and provide some illustrative examples.

8.3.1 Five-membered ring heterocycles

The simplest five-membered heteroaromatic compounds are those with one heteroatom, such as pyrrole, furan, and thiophene. One of the general synthetic approaches to all of these compounds is from the 1,4-dicarbonyl compounds (Figure 8.21).

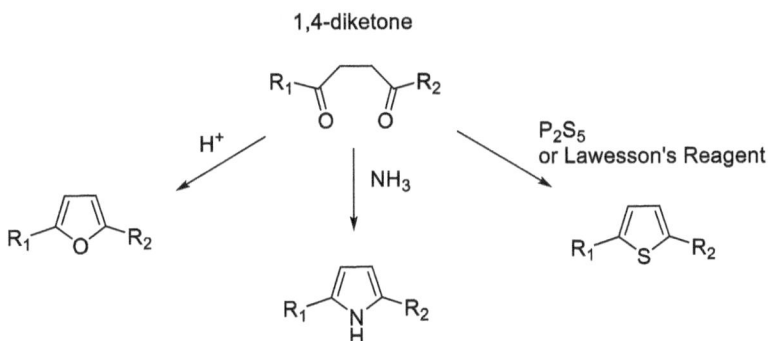

Figure 8.21: Synthetic approach to five-membered heteroaromatic compounds from 1,4-dicarbonyl compounds.

For example, the synthesis of furans from 1,4-dicarbonyls under acidic conditions is known as the Paal–Knorr furan synthesis. The mechanism of the reaction is outlined in Figure 8.22 [26].

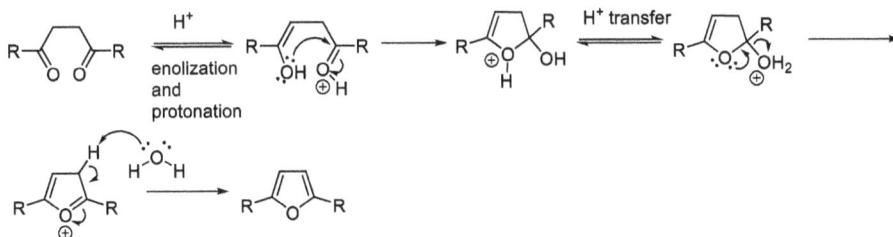

Figure 8.22: Mechanism of the Paal–Knorr furan synthesis.

The corresponding Paal–Knorr synthesis of pyrroles uses a diketone and ammonia (Figure 8.23) [25]. The mechanism is likely similar, although it involves nucleophilic attack on the carbonyl by ammonia prior to cyclization using the nitrogen as the intramolecular nucleophile. A limitation of the scope of both of these methods is that there are a limited number of readily available 1,4-dicarbonyl starting materials.

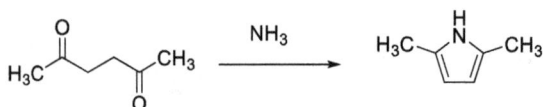

Figure 8.23: Paal–Knorr pyrrole synthesis.

A conceptually related approach is the Knorr synthesis, in which an α-aminoketone reacts with ethyl acetoacetate (Figure 8.24) [25]. The resulting imine undergoes an intramolecular condensation reaction to give the substituted pyrrole.

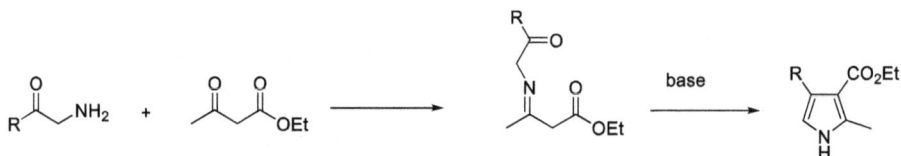

Figure 8.24: The Knorr synthesis of pyrroles.

Sometimes, the amino ketone is generated in situ from the β-keto ester (such as ethyl acetoacetate) by reaction with sodium nitrite to give the oxime, followed by reduction with zinc (Figure 8.25) [25].

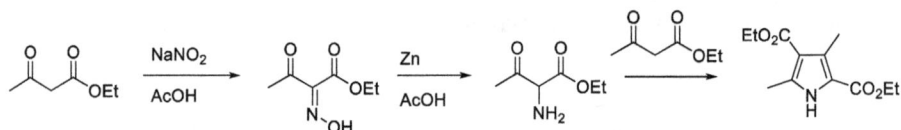

Figure 8.25: Variation of the Knorr synthesis of pyrroles.

As described above, thiophenes can be prepared in a similar way using the 1,4-dicarbonyl compounds in the presence of P_2S_5 or Lawesson's reagent. An alternative approach to thiophenes is the Hinsberg synthesis, which involves the condensation of the sulfide shown with a 1,2-dicarbonyl compounds (Figure 8.26) [27]. The product is a thiophene with ester groups attached to the 2 and 5 positions and R groups (from the 1,2-dicarbonyl) attached at the 3 and 4 positions. This compound can often be decarboxylated to give the 3,4-disubstituted thiophene, presenting an alternative to the Paal–Knorr synthesis, where substituents are attached at the 2 and 5 positions.

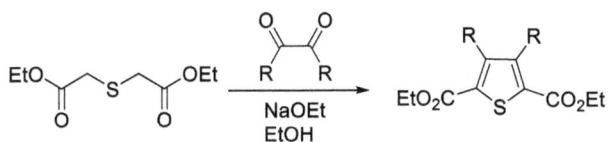

Figure 8.26: The Hinsberg thiophene synthesis.

Five-membered heteroaromatic compounds with two heteroatoms include imidazole, thiazole, oxazole, and pyrazole. Two common five-membered heteroaromatics containing more than two heteroatoms are triazole and tetrazole (Figure 8.27).

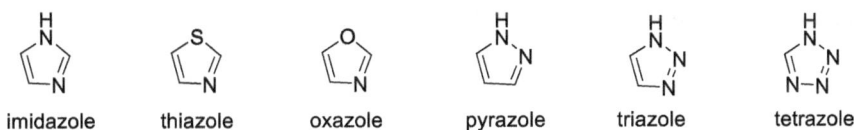

imidazole   thiazole   oxazole   pyrazole   triazole   tetrazole

Figure 8.27: Five-membered heteroaromatic compounds with two heteroatoms.

One general approach to 5-membered heteroaromatic rings with two-nonadjacent heteroatoms involves reaction of an α-bromo-carbonyl compound with an amidine or thioamide. This approach is illustrated for the synthesis of imidazoles and thiazoles (Figure 8.28).

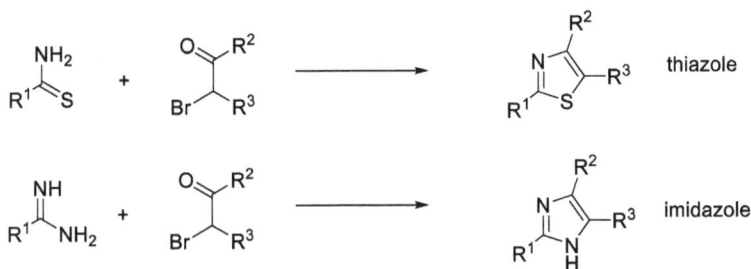

Figure 8.28: Synthesis of thiazoles and imidazoles.

In contrast, oxazoles are not prepared in the same way because amides are not reactive enough. Rather, an intramolecular cyclization of the dicarbonyl under acidic conditions, similar to the synthesis of furans (Figure 8.29).

Pyrazoles are usually prepared by the condensation of a 1,3-dicarbonyl compounds with hydrazine (Figure 8.30).

For an example of the applications of the synthesis of pyrazole, we can consider the synthesis of sildenafil, which is better known under its trade name Viagra™. The synthesis begins with pyrazole formation by condensation of a β-keto ester with hydrazine (Figure 8.31) [28]. Following this step, the pyrazole nitrogen was

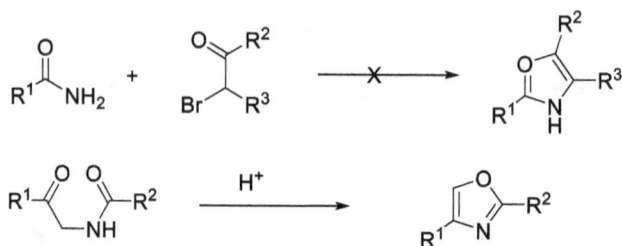

Figure 8.29: Synthesis of oxazoles.

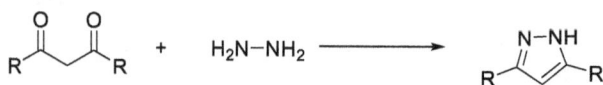

Figure 8.30: General synthesis of pyrazoles.

methylated using dimethyl sulfate, and the ester was hydrolyzed. Pyrazole underwent electrophilic nitration at the remaining available position, and the carboxylic acid was converted to the amide in a two-step sequence. Reduction of the nitro group using hydrogen and palladium on carbon gave the necessary aminopyrazole, which was transformed in a sequence of steps to sildenafil.

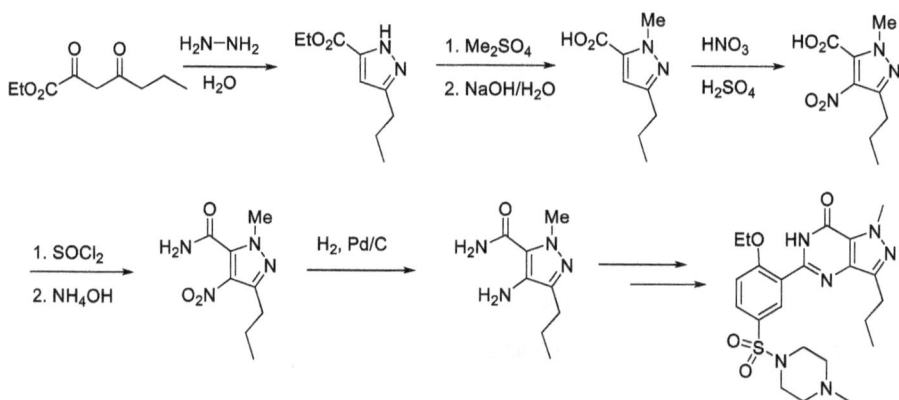

Figure 8.31: Synthesis of sildenafil [28].

8.3.2 Six-membered ring heterocycles

Six-membered *N*-heteroaromatic compounds such as pyridines, pyrazines, pyrimidines, and pyridazines (Figure 8.32) are structural motifs that are frequently seen in biologically active molecules. There are numerous approaches to these compounds, but many involve reactions of amines with carbonyl compounds as a key step.

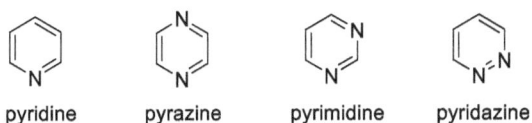

pyridine   pyrazine   pyrimidine   pyridazine

Figure 8.32: Some six-membered N-heteroaromatic compounds.

One of the classical approaches to prepare pyridines is known as the Hantzsch pyridine synthesis, which consists of the reaction of a β-keto ester with an aldehyde and ammonia (Figure 8.33) [29, 30]. The reaction consists of condensation reactions between the β-keto ester and the aldehyde, followed by imine formation and cyclization to give a dihydropyridine as the initial product. The dihydropyridine can be oxidized to the corresponding pyridine using oxidizing agents such as DDQ.

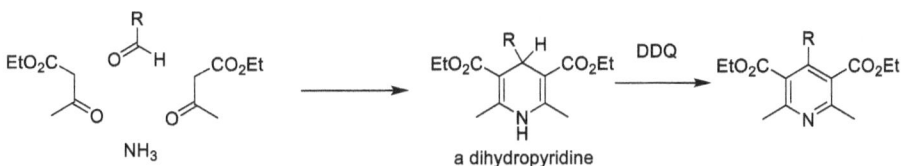

a dihydropyridine

Figure 8.33: The Hantzsch pyridine synthesis.

The Hantzsch pyridine synthesis described above is well suited to symmetrically substituted pyridines. In order to prepare dissymmetric pyridines, a modification of the Hantzsch pyridine synthesis consists of a reaction of an α,β-unsaturated carbonyl with an enamine (Figure 8.34) [29, 30]. Mechanistically, this reaction consists of a conjugate addition of the enamine with the α,β-unsaturated carbonyl, followed by a condensation between the amine and the remaining carbonyl to give a dihydropyridine.

Figure 8.34: Modified Hantzsch pyridine synthesis for unsymmetrical pyridines.

This method has been used for the preparation of felodipine, a dihydropyridine derivative that is a calcium ion transport inhibitor used to reduce blood pressure (Figure 8.35).

Six-membered heterocyclic systems with two nitrogen atoms include pyrazines, pyrimidines, and pyridazines. Pyrazines are generally prepared by condensation of 1,2-diones with 1,2-diamines, followed by oxidation (Figure 8.36). The synthesis of

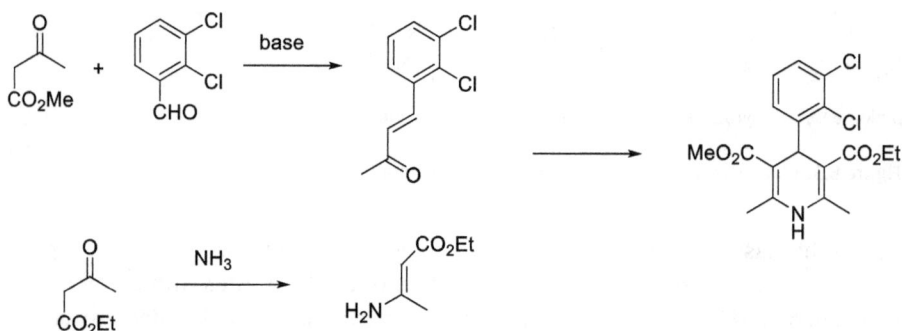

Figure 8.35: Synthetic approach to felodipine.

benzopyrazines (often refers to as quinoxalines) are prepared using a similar approach, but without the need for an oxidation step (Figure 8.36).

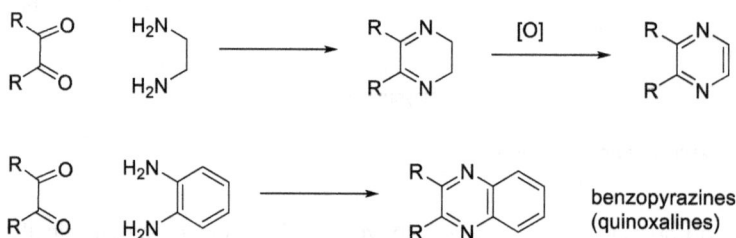

Figure 8.36: General approach to the synthesis of pyrazines and quinoxalines.

For example, a condensation of a 1,2-dicarbonyl and a 1,2-diamine can be used in the synthesis of folic acid (Figure 8.37).

Similarly, pyridazines are usually prepared by condensation of 1,4-dicarbonyls with hydrazine, followed by oxidation (Figure 8.38).

For an example, consider the synthesis of cyanamid, an herbicide used on cotton (Figure 8.39) [31]. In this case, hydrazine was condensed with a keto-ester to give the dihydropyridazinone, which was oxidized to the pyridazinone using bromine. This compound could then be converted into the chloropyridazine using $POCl_3$. A nucleophilic aromatic substitution with sodium methoxide gave the final product.

Pyrimidines can be prepared by condensation of 1,3-dicarbonyl compounds with urea, thiourea, or amidines such as guanidine. Using urea, thiourea, or guanidine results in the corresponding heteroatom-substituted pyrimidine, while when amidines are used, the corresponding hydrocarbon-substituted pyrimidine is produced (Figure 8.40).

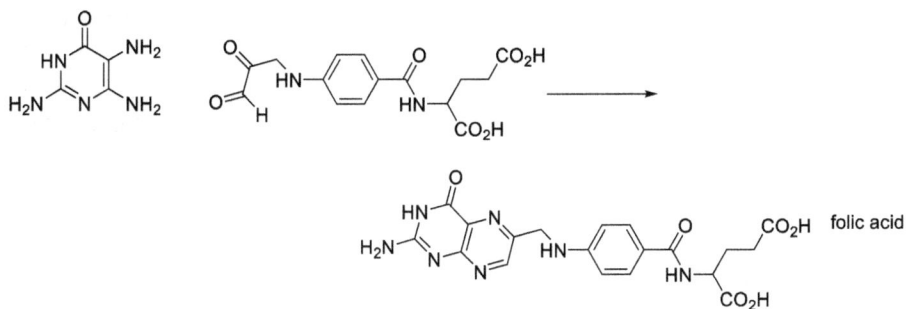

Figure 8.37: Synthesis of folic acid.

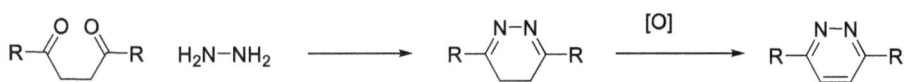

Figure 8.38: General approach to pyridazines.

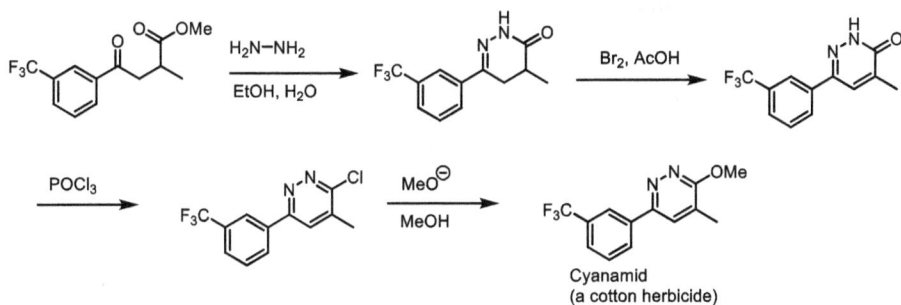

Cyanamid
(a cotton herbicide)

Figure 8.39: Synthesis of cyanamid.

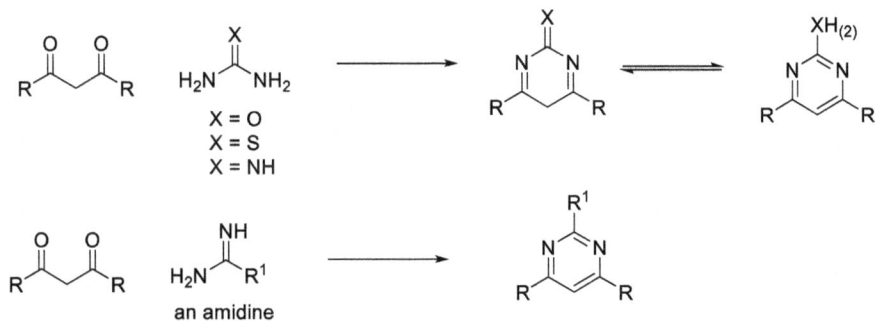

$X = O$
$X = S$
$X = NH$

an amidine

Figure 8.40: General synthetic approaches to pyrimidines.

For an example, consider a synthesis of the antibacterial agent trimethoprim, which has a diaminopyrimidine unit (Figure 8.41) [32]. Condensation of the 1,3-dicarbonyl (represented here as the enol tautomer) with guanidine results in the aminopyrimidine. The hydroxyl group was then converted into the chloride using $POCl_3$, followed by nucleophilic aromatic substitution with ammonia.

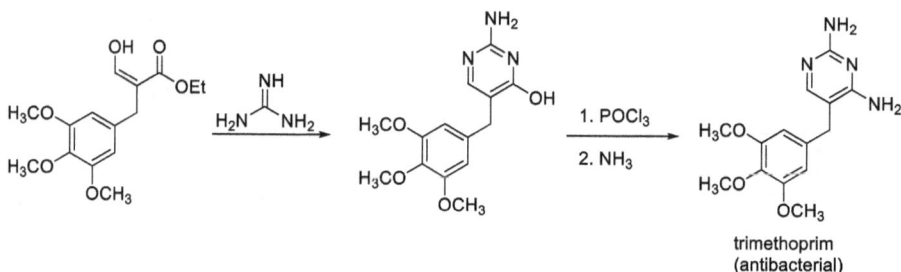

Figure 8.41: Synthesis of trimethoprim.

8.3.3 Fused heteroaromatics

There are also numerous different benzo-fused heteroaromatics that are found widely in natural products as well as synthetic compounds, usually biologically active compounds targeted for therapeutics. Here, we will explore some of the synthetic approaches for indoles, quinolines, and isoquinolines (Figure 8.42).

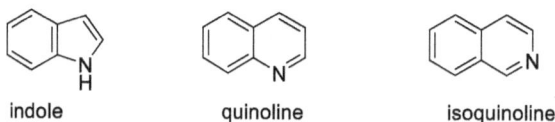

indole   quinoline   isoquinoline

Figure 8.42: Structures of simple benzo-fused heteroaromatic compounds.

8.3.3.1 Synthesis of indoles

Indoles are frequently seen in natural products – most notably the amino acid tryptophan, which serves as the starting point for the biosynthesis of other important biomolecules like serotonin. Because of their importance in natural products and biologically active molecules, there are numerous approaches for the synthesis of indoles [33, 34].

One of the best known classical approaches for the preparation of indoles is the Fischer indole synthesis, where a ketone reacts with phenylhydrazine under acidic conditions (Figure 8.43) [35].

The accepted mechanism for the Fischer indole synthesis is outlined in Figure 8.44 [35]. It consists of phenylhydrazone formation between the phenylhydrazine and ketone, followed by tautomerization. The enamine tautomer undergoes a [3,3]

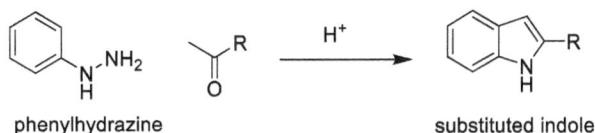

phenylhydrazine   substituted indole

Figure 8.43: Fischer indole synthesis.

sigmatropic rearrangement, resulting in carbon–carbon bond formation and a disruption of aromaticity. Rearomatization is accompanied by a cyclization where the aryl nitrogen attacks the imine intermediate. Finally, elimination of ammonia gives the indole. The reaction is typically carried out under acidic conditions, ranging from strong acids such as polyphosphoric acid, weaker acids such acetic acid, to Lewis acids such as $ZnCl_2$ or $BF_3 \cdot Et_2O$.

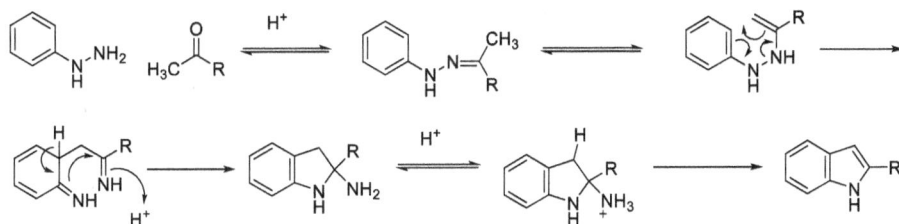

Figure 8.44: Mechanism of the Fischer indole synthesis.

Although the Fischer indole synthesis is over 100 years old, it has been used widely and continues to be a useful tool for the for the preparation of indoles. A well-known classical example of the application of the Fischer indole synthesis is in Woodward's 1954 synthesis of strychnine (Figure 8.45) [36, 37].

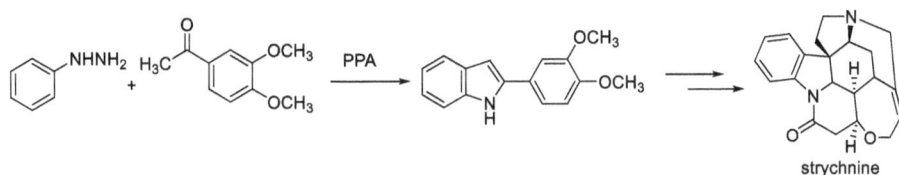

strychnine

Figure 8.45: Fischer indole synthesis as part of Woodward's strychnine synthesis.

Indoles are also used as active pharmaceutical ingredients – most notably in a class of drugs known as triptans, which are used to treat migraines [38]. The indole moiety in these compounds is often prepared by a Fischer indole synthesis. For an example, we will consider a synthesis of almotriptan, which uses a modified Fischer indole synthesis known as the Grandberg variation (Figure 8.46) [39]. In this synthesis, the aniline bearing the sulfonamide was converted to the corresponding

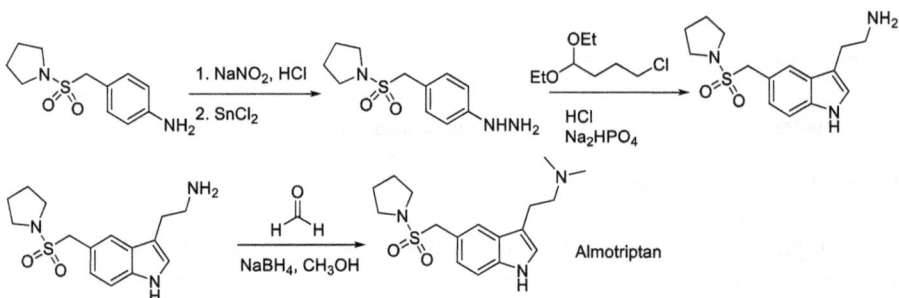

Figure 8.46: Synthesis of almotriptan [39].

phenylhydrazine by diazotiation using sodium nitrite and subsequent reduction using stannous chloride. The phenylhydrazine then reacted with chlorobutaralde-hyde diethyl acetal in HCl with Na₂HPO₄ in the modified Fischer indole synthesis to give the tryptamine derivative. In this reaction, the protected aldehyde reacts with the phenylhydrazine to form the corresponding hydrazone. Following the indole formation, the ammonia released during the reaction displaces the chloride to give the tryptamine in a single step. Finally, the amine was methylated by a twofold reductive amination using formaldehyde and sodium borohydride.

A variation of the Fischer indole synthesis is known as the Japp–Klingemann reaction, where the phenylhydrazone is accessed directly from the aryl diazonium salt without the need to prepare the phenylhydrazine. Typically, the aryldiazonium salt is treated with a β-ketoester, which undergoes deacylation and phenylhydrazone formation (Figure 8.47) [33]. The phenylhydrazone can then undergo Fischer cyclization to give the corresponding indole-2-carboxylate ester.

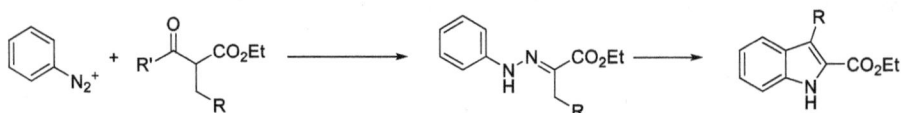

Figure 8.47: The Japp–Klingemann indole synthesis.

The Japp–Klingemann reaction has been applied to the synthesis of sumatriptan, as outlined in Figure 8.48 [40]. The aniline starting material was converted to the diazonium salt and treated with the β-ketoester shown to give the arylhydrazone. The arylhydrazone underwent Fischer cyclization in acetic acid to give the indole. Finally, ester hydrolysis was followed by decarboxylation using copper powder in quinolone to give sumatriptan.

Modern methods for the preparation of indoles have also been developed. For example, Larock and coworkers developed a palladium-catalyzed annulation reaction of iodoanilines with alkynes (Figure 8.49) [41, 42]. The reaction involves an iodoanline,

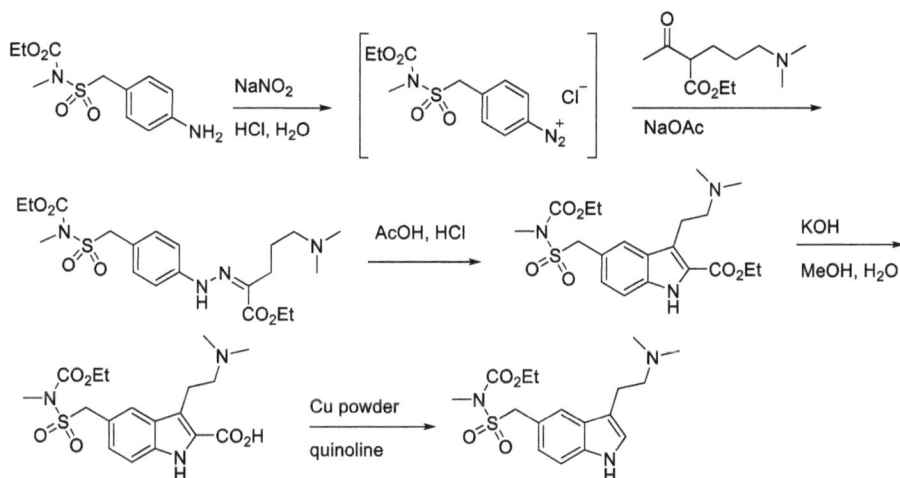

Figure 8.48: Synthesis of sumatriptan using a Japp–Klingemann reaction [40].

an alkyne, a palladium catalyst such as Pd(OAc)$_2$, weak base, and a chloride source such as LiCl. As shown in Figure 8.49, when an unsymmetrical alkyne is used, the regiochemistry favors the larger R group of the alkyne attached to the indole nitrogen.

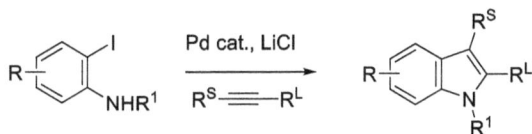

Figure 8.49: Synthesis of indoles via the Larock heteroannulation.

The proposed mechanism for this annulation reaction is outlined in Figure 8.50 [42]. The reaction is thought to proceed by reduction of Pd(OAc)$_2$ to a Pd(0) species, followed by chloride coordination. Following this, the aryl iodide underwent oxidative addition to Pd(0) and the alkyne coordinated to the Pd(II) complex. The alkyne underwent a migratory insertion into the aryl-Pd bond and the amine coordinated to the palladium, displacing the iodide to form a six-membered palladacycle. Finally, reductive elimination led to formation of the indole and regeneration of the active catalyst. The regiochemistry of the reaction is determined by the alkyne insertion step, where the larger group is positioned further from the aryl group, presumably for steric reasons.

The Larock heteroannulation reaction has been applied to the synthesis of rizatriptan, an anti-migraine triptan marketed as the benzoic acid salt under the brand name Maxalt®. The synthesis is outlined in Figure 8.51 [43]. The synthesis begins by *ortho*-iodination of the substituted aniline with ICl. Following this, the palladium-catalyzed heteroannulation reaction with the bis-TES-propargyl alcohol gave

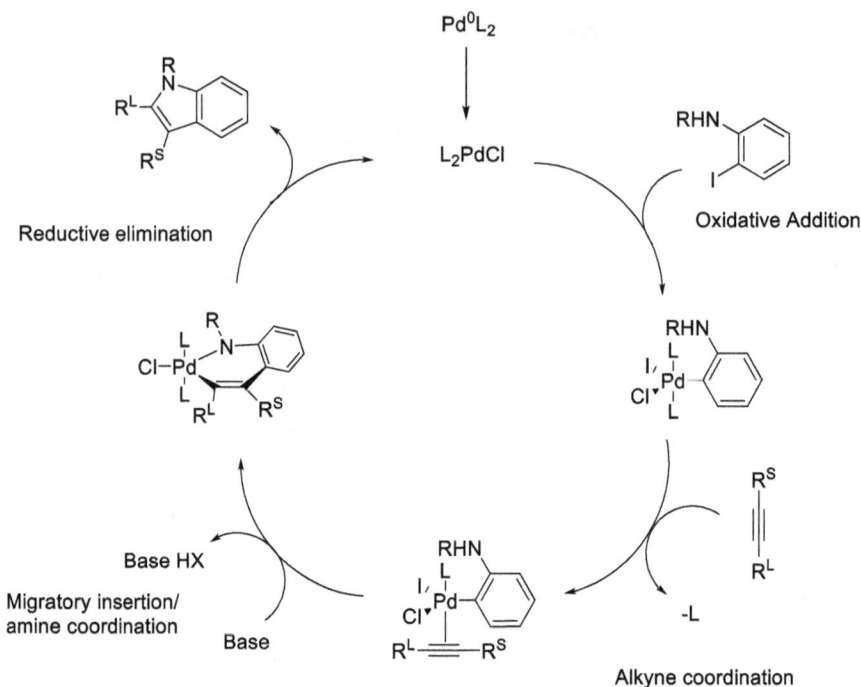

Figure 8.50: Proposed mechanism of the Larock heteroannulation [42].

the indole. The TES groups were cleaved under acidic conditions, and the alcohol was converted into the dimethylamino group via mesylation and reaction with dimethylamine to yield the desired rizatriptan.

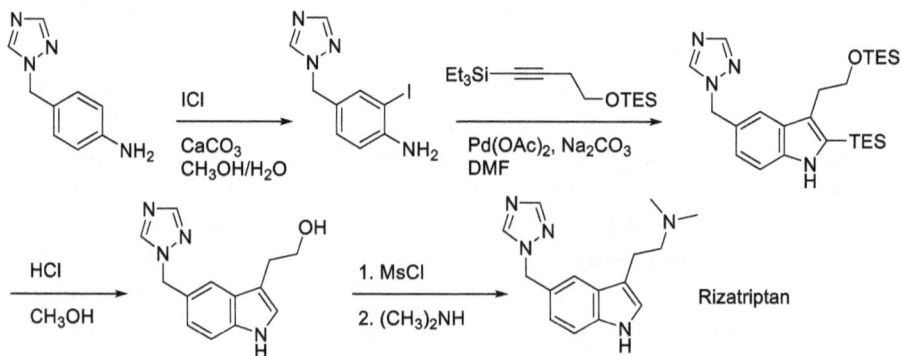

Figure 8.51: Synthesis of rizatriptan via Larock heteroannulation [43].

Another palladium-catalyzed method for preparing indoles involves an intramolecular Heck reaction. For an example of an intramolecular Heck reaction, iodoanilines can react with carbonyl compounds to form enamines, which undergo intramolecular Heck reactions to form indoles (Figure 8.52) [44].

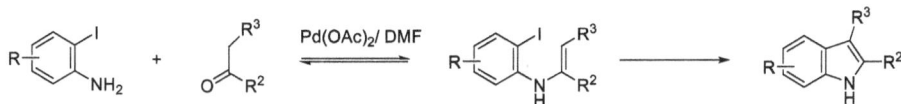

Figure 8.52: Intramolecular Heck reaction for the synthesis of indoles [44].

8.3.3.2 Synthesis of quinolines

Quinoline consists of a pyridine ring fused to a benzene ring. There are several different reactions used to prepare quinolines, but many can be generalized as consisting of imine formation and an electrophilic substitution (Figure 8.53). These can either be done sequentially or in a single synthetic step.

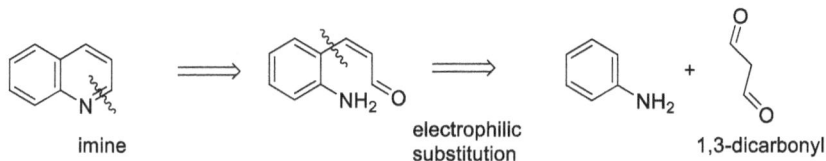

Figure 8.53: Generalized approach to quinolines.

The Combes quinolone synthesis is a good example of a one-step synthesis of quinolines from an aniline and a 1,3-dicarbonyl (Figure 8.54). This reaction is carried out in the presence and acid, and can be viewed as imine formation followed by electrophilic aromatic substitution and dehydration. Electron-donating substituents *meta* to the aniline (Z groups as shown in Figure 8.54) enhance reactivity because these will promote the electrophilic substitution step. An important consideration is that unsymmetrical 1,3-dicarbonyl compounds (where R ≠ R') will lead to a mixture of regioisomers.

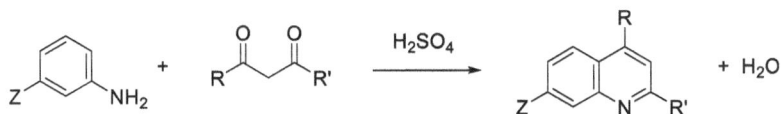

Figure 8.54: Combes quinoline synthesis.

A related reaction is the Knorr quinolone synthesis, in which an aniline and a β-keto ester react at elevated temperature to give either the 2-quinolone or 4-quinolone (Figure 8.55). Again, a mixture of regioisomers can result. Quinolones are an important

target in their own right, but can also be converted into the corresponding chloroquino-
lines using POCl$_3$.

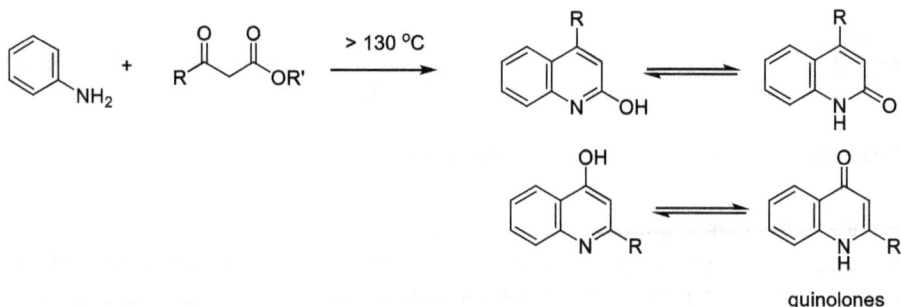

Figure 8.55: Knorr quinolone synthesis.

Another classic approach to quinolines involves the reaction of an aniline with an
α,β-unsaturated carbonyl in the presence of acid and an oxidizing agent to form a
quinoline (Figure 8.56). This reaction is known as the Skraup reaction [45].

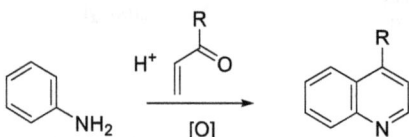

Figure 8.56: The Skraup reaction.

Often the reaction is carried out with glycerol as a reactant, which undergoes dehydration and oxidation to form acrolein in situ. The Skraup reaction typically involves
combining the aniline, glycerol or α,β-unsaturated carbonyl, acid, and oxidizing
agent together. The reaction is very exothermic and sometimes violent. Mechanistically, the reaction consists of a conjugate addition of the aniline to the α,β-unsatu-
rated carbonyl, an acid-catalyzed electrophilic cyclization, and an oxidation of the
resulting dihydroquinoline (Figure 8.57).

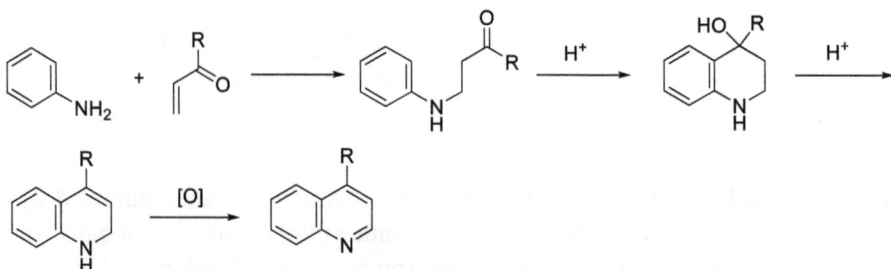

Figure 8.57: Outline of the mechanism of the Skraup reaction.

The reactions discussed thus far lead to 2- and 4-substituted quinolines and have potential challenges in terms of regioselectivity. An alternative method that addresses both of these issues is the Friedländer quinoline synthesis, which involves the reaction of an *o*-amino aldehyde with an enolizable ketone or aldehyde (Figure 8.58) [46].

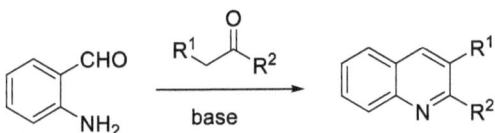

Figure 8.58: The Friedländer quinoline synthesis.

There are two plausible mechanisms for the reaction (Figure 8.59). The first mechanism involves aldol condensation, followed by intramolecular imine formation. Alternatively, it is possible that imine formation occurs first, followed by an intramolecular condensation.

Mechanism 1: Aldol condensation, then cyclization

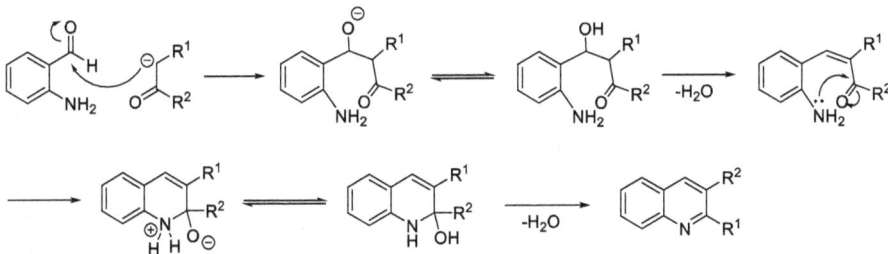

Mechanism 2: Imine formation, then intramolecular aldol

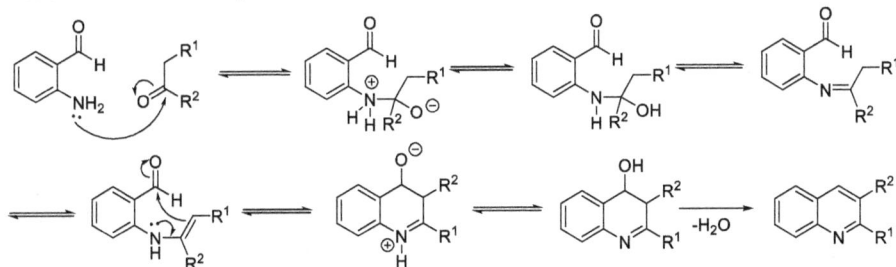

Figure 8.59: Plausible mechanisms for the Friedländer quinoline synthesis.

A variation of the Friedländer quinoline synthesis uses a nitrobenzaldehyde and carries out the aldol condensation first. The second step involves reduction of the nitro group and cyclization (Figure 8.60).

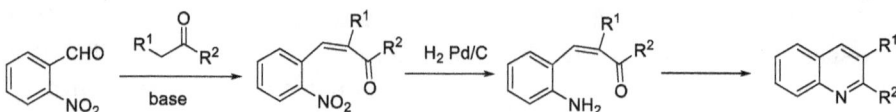

Figure 8.60: A variation of the Friedländer quinoline synthesis.

The Friedländer quinoline synthesis has been applied to the synthesis of quinoline-containing natural products and synthetic analogs. For example, camptothecin, which along with its analogs has been the subject of considerable research efforts due to its anticancer potential, was prepared by Danishefsky and coworkers using the Friedländer quinoline synthesis as a key step (Figure 8.61) [47].

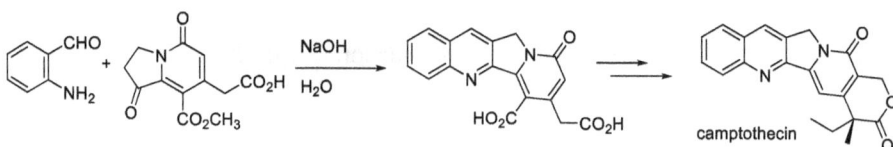

Figure 8.61: Friedländer quinoline synthesis as part of the synthesis of camptothecin [47].

8.3.3.3 Synthesis of isoquinolines

Isoquinoline derivatives are also widely found in natural products with biological activity, and as such there are numerous approaches to the syntheses of isoquinolines, as well as dihydro- and tetrahydro-isoquinolines.

One approach to isoquinolines is the Pictet–Spengler reaction, which involves reacting a 2-phenylethylamine and an aldehyde to form a tetrahydroisoquinoline (Figure 8.62) [48, 49]. The reaction involves imine formation, followed by an acid-catalyzed electrophilic cyclization. The tetrahydroisoquinoline can be converted into the corresponding isoquinoline via oxidation.

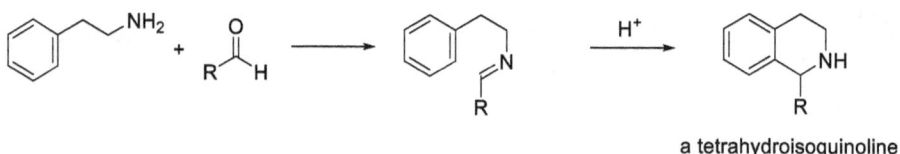

Figure 8.62: Pictet–Spengler reaction.

An alternative method is the Bischler–Napieralski isoquinoline synthesis, which uses 2-phenylethylamine and an acid chloride instead of an aldehyde (Figure 8.63) [50]. The amine and the acid chloride react to form an amide, which is then treated with $POCl_3$, which leads to an intramolecular Vilsmeier–Haack reaction. The dihydroisoquinoline product can then be oxidized to the corresponding isoquinoline.

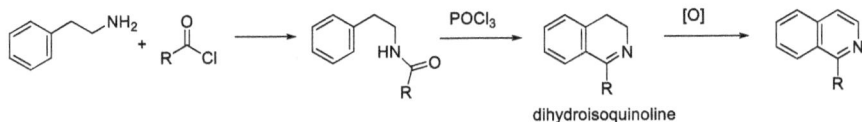

Figure 8.63: Bischler–Napieralski isoquinoline synthesis.

The Bischler–Napieralski isoquinoline synthesis has been applied to the preparation of papaverine, an opium alkaloid natural product with a variety of therapeutic
uses, along with some of its derivatives [51]. An overview of the synthetic approach is
shown in Figure 8.64 [25]. The substituted phenylethylamine reacts with the acid
chloride to give the amide. This amide is then treated with POCl₃, creating the iminium electrophile that undergoes an intramolecular electrophilic aromatic substitution to give the dihydroisoquinoline, which can be dehydrogenated using palladium
on carbon to give papaverine.

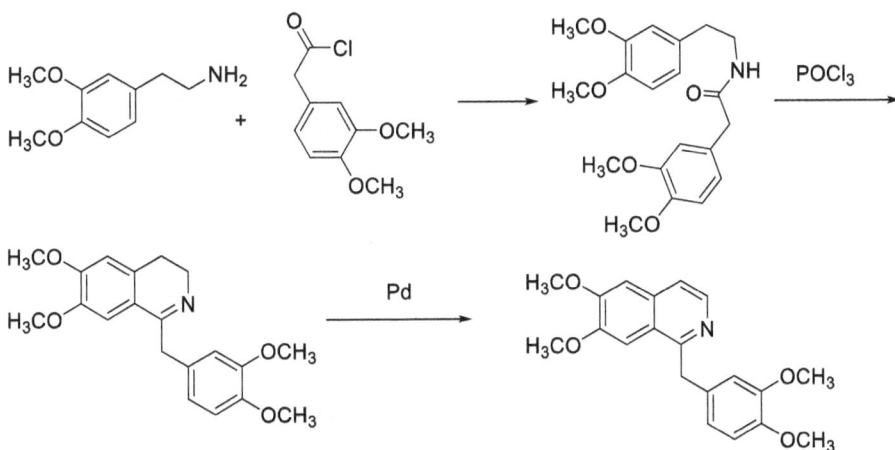

Figure 8.64: Synthesis of papaverine via a Bischler–Napieralski reaction.

Another method for preparing isoquinolines is known as the Pomeranz–Fritsch reaction [52]. It consists of the reaction of a benzaldehyde with an aminoacetalde-
hyde diethyl acetal, followed by treatment with acid (Figure 8.65). The reaction
involves imine formation, followed by an acid-catalyzed acetal hydrolysis and
electrophilic cyclization.

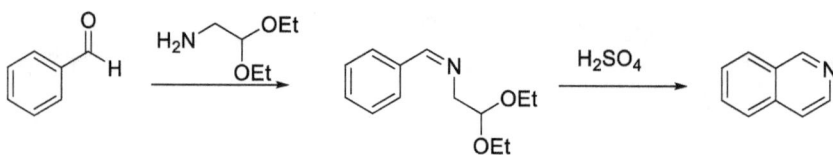

Figure 8.65: The Pomeranz–Fritsch isoquinoline synthesis.

References

[1]   Ogliaruso MA, Romanelli MG, Becker EI. Chemistry of cyclopentadienones. Chem Rev. 1965;65:261–364.
[2]   Feiser LF. Hexaphenylbenzene. Org Syn. 1966;46:44–48.
[3]   Clark TJ. A reinvestigation of the condensation of aliphatic ketones with benzil. J Org Chem. 1973;38:1749–51.
[4]   Becker H, King SB, Taniguchi M, Vanhessche KPM, Sharpless KB. New ligands and improved enantioselectivities for the asymmetric dihydroxylation of olefins. J Org Chem. 1995;60:3940–41.
[5]   Nakayama J, Sugihara Y. Chemistry of thiophene 1,1-dioxides. Top Curr Chem. 1999;205:131–95.
[6]   Nakayama J, Yamaoka S, Nakanishi T, Hoshino M. 3,4-Di-tert-butylthiophene 1,1-dioxide, a convenient precursor of o-di-tert-butylbenzene and its derivatives. J Am Chem Soc. 1988;110:6598–99.
[7]   Bailey D, Williams VE. An efficient synthesis of substituted anthraquinones and naphthoquinones. Tetrahedron Lett. 2004;45(12):2511–13.
[8]   Reppe W, Schlichting O, Klager K, Toepel T. Cyclisierende Polymerisation von Acetylen. 1. Uber Cyclooctatetraen. Justus Liebigs Ann Chem. 1948;560:1–92.
[9]   Vollhardt KPC. Transition-metal-catalyzed acetylene cyclizations in organic synthesis. Acc Chem Res. 1977;10(1):1–8.
[10] Saito S, Yamamoto Y. Recent advances in the transition-metal-catalyzed regioselective approaches to polysubstituted benzene derivatives. Chem Rev. 2000;100(8):2901–15.
[11] Aalbersberg WG, Barkovich AJ, Funk RL, Hillard III RL, Vollhardt KPC. Transition metal catalyzed acetylene cyclizations. 4,5-Bis(trimethylsilyl)benzocyclobutene, a highly strained, versatile synthetic intermediate. J Am Chem Soc. 1975;97(19):5600–02.
[12] Funk RL, Vollhardt KPC. A cobalt-catalyzed steroid synthesis. J Am Chem Soc. 1977;99 (16):5483–84.
[13] Funk RL, Vollhardt KPC. The cobalt way to dl-estrone, a highly regiospecific functionalization of 2, 3-bis(trimethylsilyl)estratrien-17-one. J Am Chem Soc. 1979;101(1):215–17.
[14] Funk RL, Vollhardt KPC. Transition-metal-catalyzed alkyne cyclizations. A cobalt-mediated total synthesis of dl-estrone. J Am Chem Soc. 1980;102(16):5253–61.
[15] Berris BC, Hovakeemian GH, Lai Y-H, Mestdagh H, Vollhardt KPC, New A. Approach to the construction of biphenylenes by the cobalt-catalyzed cocyclization of o-diethynylbenzenes with alkynes. Application to an iterative approach to [3]phenylene, the first member of a novel class of benzocyclobutadienoid hydrocarbons. J Am Chem Soc. 1985;107 (20):5670–787.
[16] Schmidt-Radde RH, Vollhardt KPC. Total synthesis of angular [4]phenylene and [5]phenylene. J Am Chem Soc. 1992;114(24):9713–15.

[17]   Diercks R, Vollhardt KPC. Tris(benzocyclobutadieno)benzene, the triangular [4]phenylene with a completely bond-fixed cyclohexatriene ring; Cobalt-catalyzed synthesis from hexaethynylbenzene and thermal ring opening to 1,2:5,6:9,10-T ribenzo-3,4,7,8,11,12-hexadehydro[12]-annulene. J Am Chem Soc. 1986;108(11):3150–52.

[18]   Boese R, Matzger AJ, Mohler DL, Vollhardt KPC. C3-Symmetric hexakis(trimethylsilyl)[7] phenylene["tris(biphenylenocyclobutadieno)cyclohexatriene"], a polycyclic benzenoid hydrocarbon with slightly curved topology. Angew Chem Int Ed. 1995;34:1478–81.

[19]   Stabel A, Herwig P, Mullen K, Rabe JP. Diode-like current-voltage curves fo a single molecule-tunneling sepctroscopy with submolecular resolution of an alkylated, pericondensed hexabenzocoronene. Angew Chem Int Ed. 1995;34(15):1609–11.

[20]   Herwig P, Kayser CW, Mullen K, Spiess HW. Columnar mesophases of alkylated hexa-perihexabenzocoronenes with remarkably large phase widths. AdvMater. 1996;8(6):510–13.

[21]   Van Otterlo WAL, De Koning CB. Metathesis in the synthesis of aromatic compounds. Chem Rev. 2009;109(8):3743–82.

[22]   Yoshida K, Imamoto T. A new synthetic approach to phenol derivatives: Use of ring-closing olefin metathesis. J Am Chem Soc. 2005;127:10470–71.

[23]   Jones SB, He L, Castle SL. Total synthesis of (±)-hasubanonine. Org Lett. 2006;8(17):3757–60.

[24]   Comins DL, Brown JD. Ortho substitution of m-anisaldehyde via a-amino alkoxide directed lithiation. J Org Chem. 1989;54(15):3730–32.

[25]   Gilchrist TL. Heterocyclic Chemistry. 3rd ed., London: Addison Wesley Longman Ltd; 1997.

[26]   Amarnath V, Amarnath K. Intermediates in the Paal-Knorr synthesis of furans. J Org Chem. 1995;60:301–07.

[27]   Wynberg H, Kooreman HJ. The mechanism of the Hinsberg thiophene ring synthesis. J Am Chem Soc. 1965;87:1739–42.

[28]   Terrett NK, Bell AS, Brown D, Ellis P. Sildenafil (Viagra(TM)), a potent and selective inhibitor of type 5 CGMP phosphodiesterase with utility for the treatment of male erectile dysfunction. Bioor Med Chem Lett. 1996;6(15):1819–24.

[29]   Eisner U, Kuthan J. The chemistry of dihydropyridines. Chem Rev. 1972;72:1–42.

[30]   Stout DM, Meyers AI. Recent advances in the chemistry of dihydropyridines. Chem Rev. 1982;82:223–43.

[31]   Speltz LM, Walworth BL Preparation of herbicidal pyridazines and method for controlling undesirable plant species. US; US 4623376; 1986.

[32]   Clayden J, Greeves N, Warren S. Organic Chemistry. 2nd ed, Oxford: Oxford University Press; 2012.

[33]   Humphrey GR, Kuethe JT. Practical methodologies for the synthesis of indoles. Chem Rev. 2006;106(7):2875–911.

[34]   Gribble GW. Recent developments in indole ring synthesis – Methodology and applications. J Chem Soc Perkin Trans. 1;2000:1045–75.

[35]   Robinson B. Recent studies on the Fischer indole synthesis. Chem Rev. 1969;69:227–50.

[36]   Woodward RB, Cava MP, Ollis WD, Hunger A, Daeniker HU, Schenker K. The total synthesis of strychnine. J Am Chem Soc. 1954;76:4749–51.

[37]   Nicolaou KC, Sorensen EJ. Classics in Total Synthesis. New York: VCH; 1996, Chapter 2.

[38]   Li -J-J, Johnson DS, Sliskovic DR, Roth BD. Contemporary Drug Synthesis. Hoboken: John Wiley & Sons; 2004, Chapter 12.

[39]   Bosch J, Roca T, Armengol, Montserrat, Fernandez-Forner D. Synthesis of 5-(sulfamoylmethyl) indoles. Tetrahedron. 2001;57:1041–48.

[40]   Pete B, Bitter I, Harsányi K, Tõke L. Synthesis of 5-substituted indole derivatives, Part II. Synthesis of Sumatriptan through the Japp-Klingemann reaction. Heterocycles. 2000;53:665–73.

[41] Larock RC, Yum EK. Synthesis of indoles via palladium-catalyzed heteroannulation of internal alkynes new methodology for the synthesis of functionalized indolizidine and quinolizidine ring systems. J Am Chem Soc. 1991;113(9):6689–90.

[42] Larock RC, Yum EK, Refvik MD. Synthesis of 2,3-disubstituted indoles via palladium-catalyzed annulation of internal alkynes. J Org Chem. 1998;63(22):7652–62.

[43] Chen C yi, Lieberman DR, Larsen RD, Reamer RA, Verhoeven TR, Reider PJ, et al.. Synthesis of the 5-HT1D receptor agonist MK-0462 via a Pd-catalyzed coupling reaction. Tetrahedron Lett. 1994;35:6981–84.

[44] Chen C, Lieberman DR, Larsen RD, Verhoeven TR, Reider PJ. Syntheses of indoles via a palladium-catalyzed annulation between iodoanilines and ketones. J Org Chem. 1997;62:2676–77.

[45] Skraup ZH. Eine Synthese des Chinolins. Ber Dtsch Chem Ges. 1880;13:316–17.

[46] Marco-Contelles J, Perez-Mayoral E, Samadi A, Carreiras MdoC, Soriano E. Recent advances in the Friedlander reaction. Chem Rev. 2009;109:2652–71.

[47] Volkmann R, Danishefsky S, Eggler J, Solomon DM, Total A. Synthesis of dl-camptothecin. J Am Chem Soc. 1971;93:5576–77.

[48] Cox ED, Cook JM. The Pictet-Spengler condensation: A new direction for an old reaction. Chem Rev. 1995;95:1797–842.

[49] Stöckigt J, Antonchick AP, Wu F, Waldmann H. The pictet-spengler reaction in nature and in organic chemistry. Angew Chem Int Ed. 2011;50:8538–64.

[50] Fodor G, Nagubandi S. Correlation of the von Braun, Ritter, Bischler-Napieralski, Beckmann and Schmidt reactions via nitrilium salt intermediates. Tetrahedron. 1980;36:1279–300.

[51] Tachikawa R. Synthesis of some isoquinoline derivatives. Tetrahedron. 1959;7:118–22.

[52] Boger DL, Brotherton CE, Kelley MD. A simplified isoquinoline synthesis. Tetrahedron. 1981;37:3977–80.

9 Fused aromatic rings – polycyclic aromatic hydrocarbons

9.1 Introduction to polycyclic aromatic hydrocarbons

Polycyclic aromatic hydrocarbons (PAHs) are a diverse class of organic molecules. They are found in fossil fuels such as crude oil, coal, and oil shale, and are also produced during incomplete combustion. PAHs also present environmental concerns and are known to be carcinogenic.

The historical motivations for the synthesis of PAHs include the production of quinone dyes, fundamental studies on the nature of aromaticity, and to further understand the carcinogenic properties of these compounds. More recently, a renewed interest in the synthesis of PAHs stems from their potential utility as organic semiconductors or light-emitting materials. By virtue of molecular properties such as a low HOMO–LUMO gap and the ability to interact via π-stacking interactions, these compounds can often transport charge, thereby serving as a potential alternative to inorganic semiconductors. As such, there is a considerable effort to develop new synthetic methods for the preparation of PAHs. In this chapter, we will explore some of the general features of PAHs, including reactivity and stability. We will also explore some of the general approaches for the synthesis of PAHs, which will form the basis for the synthetic applications discussed in Chapters 10 and 11. For a more detailed discussion of the synthesis and reactivity of PAHs, Clar and Harvey each have books focused entirely on the synthesis of PAHs [1, 2].

9.1.1 Classification and nomenclature of PAHs

PAHs are often classified based on how the rings are fused as cata-condensed systems, or as peri-condensed systems. Cata-condensed systems have ring fusions that only share two carbons (Figure 9.1). Peri-condensed systems have carbon atoms that serve as the fusion point for three rings (Figure 9.2).

In addition to their nomenclature, other descriptions of PAH structural motifs are used frequently. These descriptions of structural features are shown in Figure 9.3. The descriptions zigzag periphery, K-region, bay regions, arm-chair periphery, and fjord regions are not pertinent in terms of formal nomenclature, but are useful descriptors that are often used and have implications for reactivity.

To facilitate the discussion of the chemistry and reactivity of PAHs, it is useful to introduce some of the basic nomenclature of this diverse class of compounds. For a more detailed discussion of nomenclature of PAHs, refer to Harvey's *Polycyclic Aromatic Hydrocarbons* [1]. Figure 9.4 shows the names of some of several of the

https://doi.org/10.1515/9783110562682-009

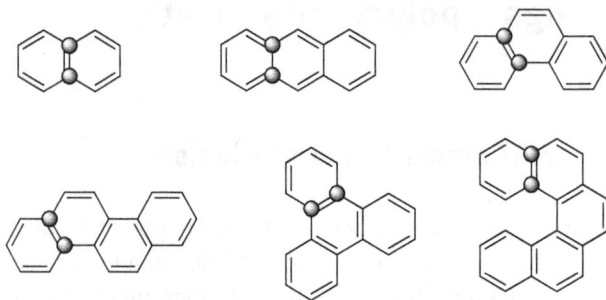

Figure 9.1: Selected examples of cata-fused PAHs. Note that the carbon atoms indicated are at the junction of two rings.

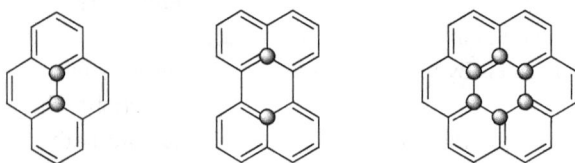

Figure 9.2: Selected examples of peri-fused PAHs. Note that the carbon atoms highlighted are at the fusion of three rings.

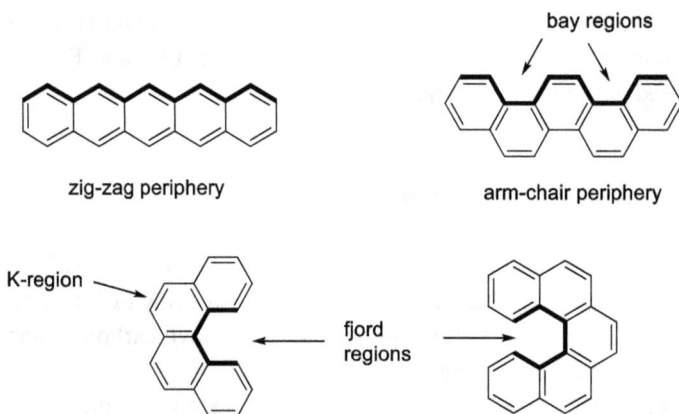

Figure 9.3: Descriptions of structural features of PAHs.

basic PAHs, whose names are used as parent structures for many of the more complex structures.

   To name a PAH, the structure is drawn to place the maximum number of rings oriented horizontally, and if more than one orientation is possible, choose the one that orients the maximum number of rings in the upper right quadrant. Carbon atoms are numbered in a clockwise direction starting with the carbon atom in the

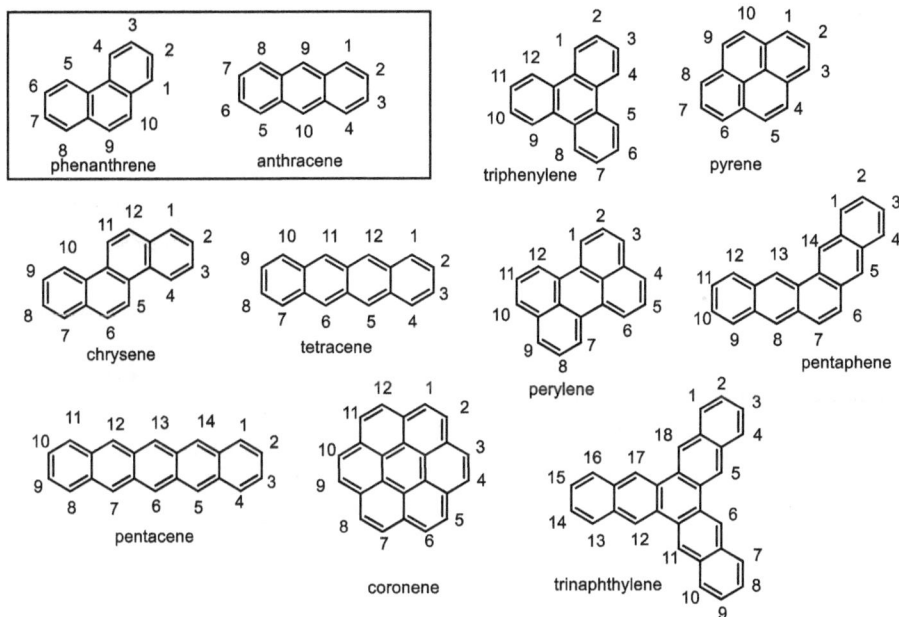

Figure 9.4: Structures of some of the parent PAH structures.

uppermost ring farthest to the right. For the purpose of numbering, carbon atoms that form a ring fusion point are omitted. An examination of the numbering shown in Figure 9.4 shows how this rule is applied, but it should be noted that the two structures in the box (phenanthrene and anthracene) are exceptions to this rule, with the central ring given the highest numbers.

The parent name is chosen to have as many rings as possible (i.e., it should be as far from the beginning of the list of structures shown in Figure 9.4). Any rings that are added to the parent structure through ring fusion should be as simple as possible and are named as prefixes (e.g., benzo, naphtho, anthra). To denote the position of fused ring components, the bonds of the parent PAH are labeled with letters alphabetically starting from the left side of the ring in the upper right quadrant and moving clockwise around the ring system. This is similar to the numbering system, but the numbers denote a carbon atom where a simple substituent can be attached, while the letters denote a bond that can be a point of ring fusion. The position of ring fusion is indicated with the corresponding letter(s) in square brackets. In Figure 9.5 are some representative examples of named compounds, where the parent PAH for the name is indicated with darker lines, and where the lettering for the substituents are indicated. Also note that these structures are drawn with the maximum number of rings in the horizontal direction, and the maximum number of rings in the upper right quadrant, where applicable. The nomenclature system is not necessarily applied consistently – especially when it comes to the choice of the parent PAH. As such, different names are sometimes applied to the same compound. However, the basic approach to naming does usually allow the structures to be deduced from the names.

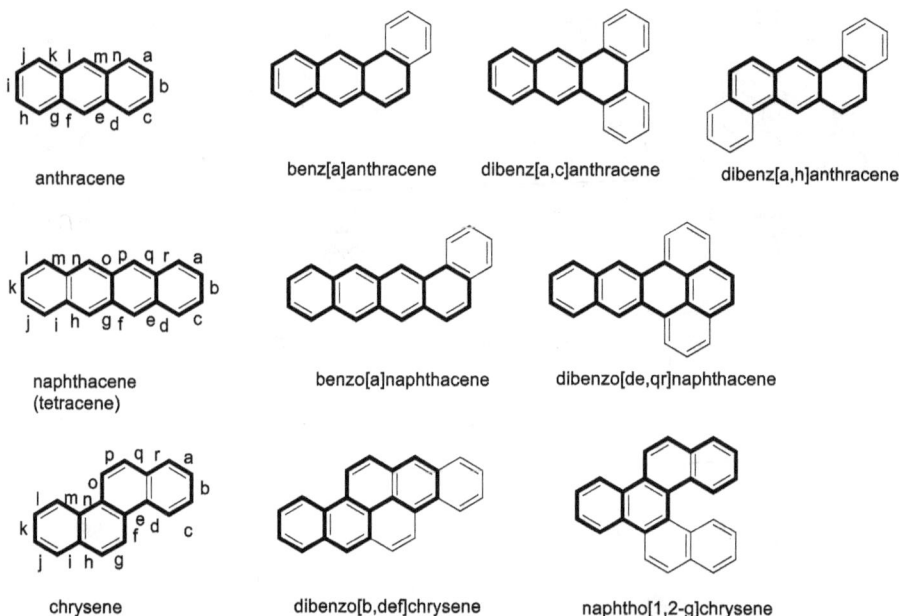

anthracene

benz[a]anthracene

dibenz[a,c]anthracene

dibenz[a,h]anthracene

naphthacene
(tetracene)

benzo[a]naphthacene

dibenzo[de,qr]naphthacene

chrysene

dibenzo[b,def]chrysene

naphtho[1,2-g]chrysene

Figure 9.5: Illustrative examples of the nomenclature of PAHs.

9.2 Stability and reactivity of PAHs

9.2.1 The aromatic sextet

In Chapter 1, we discussed the well-known Hückel rule for aromaticity, which requires $4n+2$ π-electrons in a cyclic conjugated system for aromaticity. In the case of benzene rings, this implies six π-electrons, which Armit and Robinson referred to as a "π-aromatic sextet" [3]. Hückel's rule applies only to monocyclic systems. So, how do we effectively describe aromatic stabilization in polycyclic systems? In 1972, Clar used the concept of the aromatic sextet for polycyclic aromatic systems by considering the polycyclic system as distinct aromatic sextets based on their Kekulé resonance structures [4]. Clar's rule states that the resonance structure with the largest number of aromatic sextets is the most important for describing the properties of PAHs. To clearly indicate the rings that have an aromatic sextet (often referred to as a Clar sextet), a circle is used. It is important to note that for a given Clar structure, if a ring that is fused to a sextet, that ring cannot possess a sextet because any shared π-bonds are already accounted for in the aromatic sextet. Figure 9.6 shows some of the Kekulé resonance structures of phenanthrene and anthracene and the corresponding Clar structures. If we consider phenanthrene, it can either be represented as having one aromatic sextet on the central ring, or two aromatic sextets on the terminal rings. Because the latter structure has the most sextets, it is considered the most important for describing the properties of phenanthrene. In anthracene, three Clar structures can be drawn; however, in the case of anthracene, the maximum number of Clar sextets is one. Because any one of the three rings can be drawn with an aromatic sextet, sometimes a single Clar structure is drawn with an arrow that indicates that the sextet can "migrate" between rings.

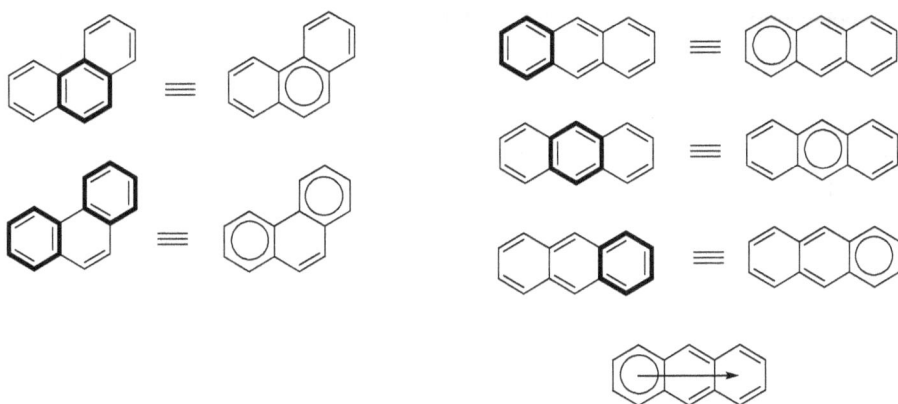

Figure 9.6: Clar structures of phenanthrene and anthracene.

If we consider the Clar structure with the maximum number of aromatic sextets, it implies differences in reactivity. For example, based on the Clar structure of phenanthrene, the 9,10-positions are more olefin-like in character. Indeed, when phenanthrene is reacted with bromine, it undergoes electrophilic addition at the central ring like an alkene rather than electrophilic aromatic substitution. Similarly, the central ring is more prone to oxidation. For anthracene, one aromatic sextet is shared among three rings, suggesting that the overall aromatic stabilization of a given ring is lower. If we consider the longer acenes, we see that even though the number of rings is extended as compared to anthracene, each structure still only has one aromatic sextet (Figure 9.7). This suggests that the longer acenes should have less benzenoid character and consequently be less stable and more reactive, which is consistent with the observed behavior of these compounds. Further, there are studies which suggest that longer acenes are can be represented as an open-shell diradical species, rather than a closed shell structure. Clar's approach provides qualitative support for this because an open shell diradical can be represented as having two Clar sextets, as shown for heptacene in Figure 9.8.

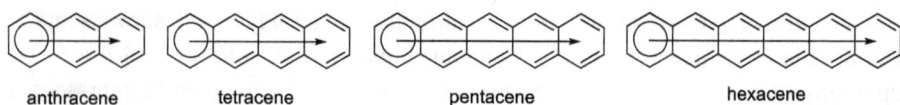

anthracene   tetracene   pentacene   hexacene

Figure 9.7: Clar structures of acenes.

Figure 9.8: Clar structure for heptacene in a closed-shell and in an open-shell diradical state.

Some more representative Clar structures of common PAHs are shown in Figure 9.9. Note that pyrene has two sextets and olefin-like K-regions. Picene is an isomer of pentacene, yet because the rings are fused in an angular fashion, picene has three sextets and is quite stable compared to pentacene. Triphenylene and tetrabenz[a,c,h,j]anthracenes are examples of structures that considered to be fully benzenoid PAHs, because all of the π-electrons are used in aromatic sextets. Fully benzenoid PAHs are predicted to be very stable.

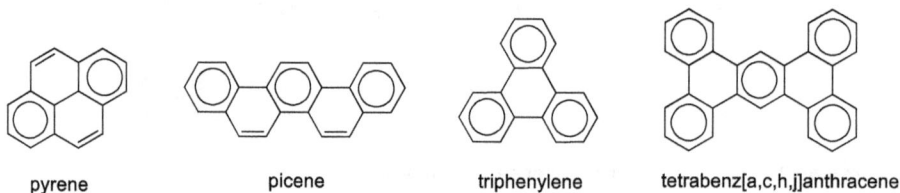

pyrene   picene   triphenylene   tetrabenz[a,c,h,j]anthracene

Figure 9.9: Clar structures of some common PAHs.

The number of Clar sextets in a structure can also be used to rationalize trends in UV-visible absorption spectra. If we consider an isomeric series of PAHs with different maximum numbers of Clar sextets, there is a qualitative correlation between the number of sextets and the absorption maxima. For example, consider pentacene and its isomers shown in Figure 9.10. The longest absorption wavelength of pentacene is approximately 575 nm. In proceeding to benzo[a]tetracene, which has two aromatic sextets, we see a substantial blue shift in the absorption maximum to ca. 450 nm. Proceeding from benzo[a]tetracene to dibenz[a,c]anthracene, which has three aromatic sextets, we see a similar large blue shift to ca. 350 nm. A comparison of dibenz[a,c]anthracene with dibenz[a,h]anthracene and benzo[g]chrysene, all of which have three aromatic sextets, shows that they have relatively similar absorption maxima (ca. 350 nm for dibenz[a,h]anthracene and ca. 335 nm for benzo[g]chrysene) [4]. Clar's aromatic sextet rule is a simple model developed nearly 50 years ago, but remains a useful tool for understanding and predicting the reactivity and properties of PAHs [5].

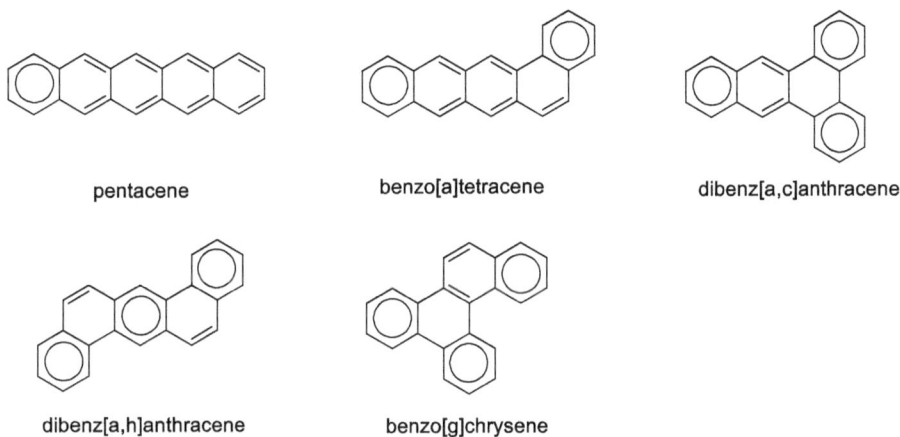

pentacene   benzo[a]tetracene   dibenz[a,c]anthracene

dibenz[a,h]anthracene   benzo[g]chrysene

Figure 9.10: Clar structures of pentacene and its isomers.

9.2.2 General reactivity of PAHs

In Chapter 2, we outlined some of the considerations for electrophilic aromatic substitution with polycyclic aromatic systems. Often they show increased reactivity toward electrophilic substitution as compared to benzene. The regiochemistry of substitution will depend on the PAH under consideration, but can usually be rationalized based on the stability or the carbocation intermediate. As with simple benzene derivatives, substituents can have activating or deactivating effects and can direct the site of electrophilic substitution.

Because PAHs often have relatively lower aromatic stabilization energies, with some rings having less aromatic character, there are several reactions that can be readily carried out on PAHs where the delocalized π-system at one of the rings is broken. We will briefly explore oxidation, reduction, and cycloaddition reactions of PAHs. We will also see that the Clar's concept of the aromatic sextet is a useful tool for understanding reactivity.

Some PAHs are susceptible to oxidation using a variety of oxidizing agents. For example, PAHs can often be oxidized to the corresponding quinones using oxidizing agents such as chromic acid or related Cr(VI) reagents. In PAHs with meso regions, the oxidation usually takes place at those sites [1]. The result is a 1,4-quinone, as shown by the oxidation of anthracene (Figure 9.11). Oxidation can also take place at the K-region to give the corresponding 1,2-quinones, as shown for the oxidation of pyrene, which, depending on conditions can be oxidized to the dione or the tetraone

(Figure 9.11). For example, K-region oxidation of pyrene to give pyrene-4,5-dione can be accomplished using RuO$_3$, which is generated in situ from RuCl$_3$ and NaIO$_4$.[6] Dihydroxylation using OsO$_4$ can also take place at the K-region to give the corresponding diols. The regiochemistry of oxidation can be explained using Clar's sextet rule. For example, in the oxidation of the meso position of anthracene, the oxidation takes place in such a way that the number of Clar sextets is increased from one to two. In the case of K-region oxidation, the number of Clar sextets remains unchanged.

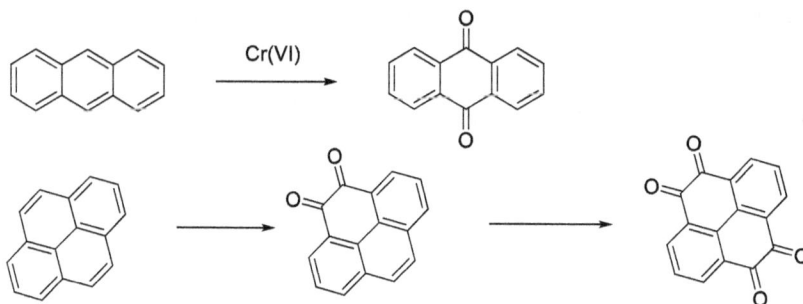

Figure 9.11: Examples of oxidation reactions.

PAHs can also be reduced by hydrogenation using H$_2$ and catalysts such as Pd/C. Often, hydrogenation will take place at the K-region, preserving the number of Clar sextets, as shown in the examples in Figure 9.12 [7, 8].

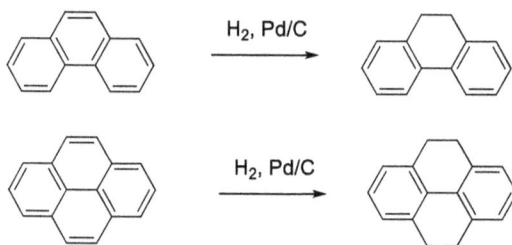

Figure 9.12: Hydrogenation of the K-region using H$_2$ and Pd/C.

Interestingly, the regiochemistry of hydrogenation depends on the catalyst used. For example, in the reduction of benz[a]anthracene, using palladium on carbon results in hydrogenation at the K-region, but if platinum is used as a catalyst, hydrogenation occurs at one of the terminal rings (Figure 9.13) [7]. The reason for the unexpected hydrogenation of the terminal ring when platinum is used is not understood.

The hydrogenation of PAHs can in some cases be exploited to control the regiochemistry of other reactions such as electrophilic aromatic substitution. For example, electrophilic aromatic substitution of pyrene typically occurs at the 1-, 3-, 6-,

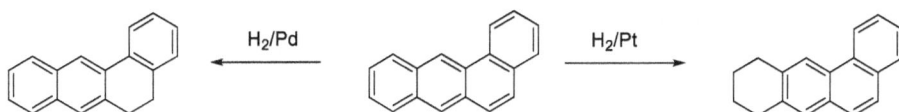

Figure 9.13: Regioselectivity of hydrogenation depends on the catalyst [7].

and 7-positions, as shown in Figure 9.14. However, hydrogenation at the K-regions to give tetrahydropyrene allows electrophilic aromatic substitution to take place at the 2- and 7-positions [9]. The tetrahydropyrene can then be oxidized using reagents such as DDQ to reform the PAH (Figure 9.14).

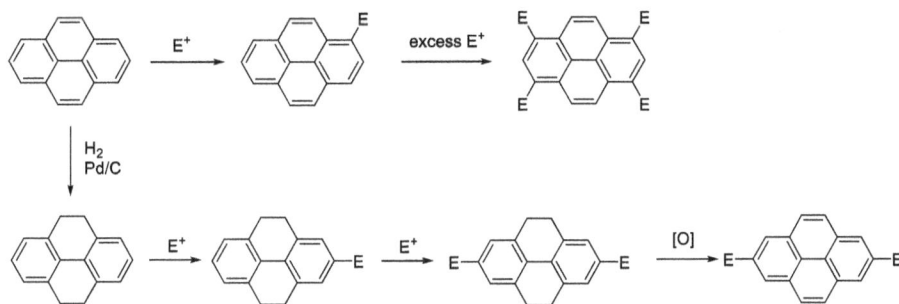

Figure 9.14: Controlling the regiochemistry of electrophilic aromatic substitution in pyrenes via hydrogenation (E = electrophile).

In addition to oxidation and reduction reactions, PAHs can also participate in cyclo-addition reactions. For example, the bay region of some PAHs can react as a diene, as shown for perylene in Figure 9.15 [10]. The initial Diels–Alder adduct can be oxidized to the corresponding aromatic system in air of in the presence of an oxidizing agent. We will see examples of how this reaction is used to prepare PAHs from smaller precursors.

Figure 9.15: Perylene as a diene in a Diels–Alder reaction [10].

Some PAHs, especially acenes such as anthracene can also participate in cycloaddition reactions [11]. For example, anthracene reacts with dienophiles such as maleic anhydride, or with benzynes to form triptycenes, which we saw in Chapter 5 (Figure 9.16). Anthracene can also undergo [4 +4] photodimerization [12] as well as photooxidation in the presence of oxygen to form an endoperoxide, which is ultimately oxidized to the corresponding quinone along with other side products (Figure 9.16) [13]. It should be noted that in each of these reactions, the reactant (anthracene) possesses only one Clar sextet, while the cycloaddition product possesses two sextets – one on each of the terminal rings. Longer acenes such as tetracene, and pentacene are even more susceptible to these types of reactions.

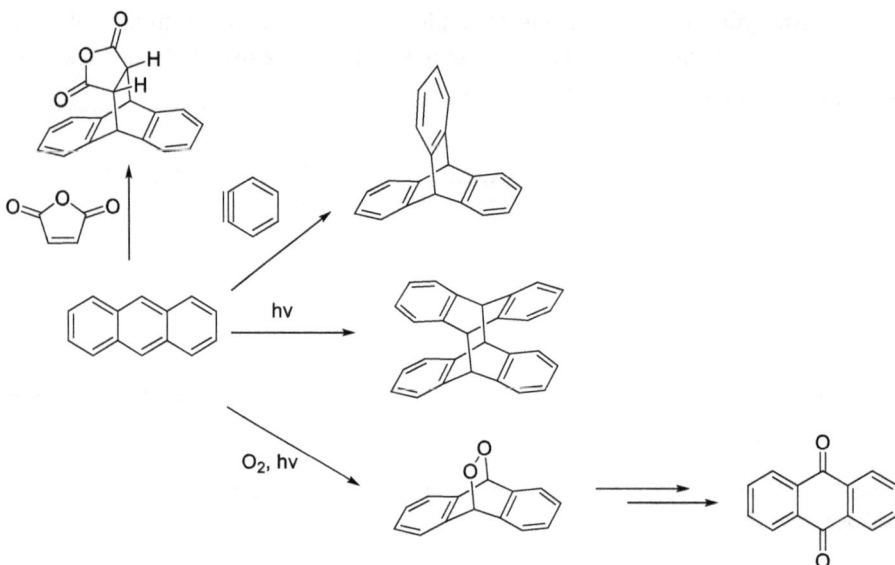

Figure 9.16: Cycloaddition reactions of anthracene.

9.3 Synthetic approaches to PAHs

In this section, we will explore some of the synthetic methods for preparing polycyclic aromatic systems from simpler precursors. This will include some of the classical approaches that are no longer as widely used, as well as some of the enduring methods and modern transition metal-catalyzed reactions. Some examples of the synthetic applications of these methods will be provided, but most of the examples will be shown in the next two chapters.

9.3.1 Haworth synthesis

One of the classic approaches to the preparation of PAHs is by annulation using the Haworth synthesis. The Haworth synthesis is a multistep approach involving Friedel–Crafts acylation using succinic anhydride, reduction, acid-catalyzed intramolecular acylation, reduction, and then dehydrogenation. Figure 9.17 shows this sequence for the preparation of phenanthrene from naphthalene [14].

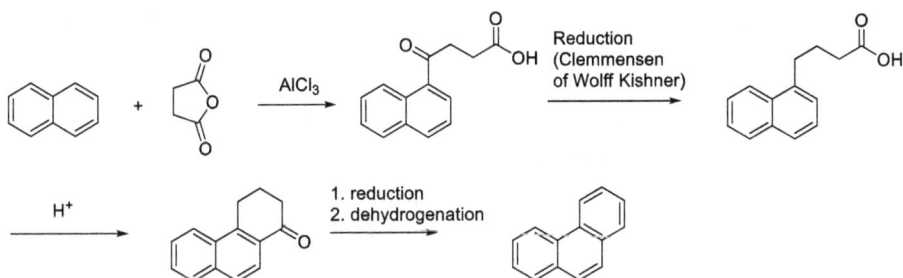

Figure 9.17: Synthesis of phenanthrene from naphthalene via a Haworth synthesis [14].

The product formed during the Haworth synthesis is determined by the regiochemistry of the initial Friedel–Crafts acylation with anhydride as well as the regiochemistry of the acid-catalyzed electrophilic cyclization. The first step is governed by the typical factors that influence electrophilic aromatic substitution touched upon in Chapter 2. The electrophilic cyclization step is also determined by the regiochemical tendencies for electrophilic substitution, but can also be influenced by the choice of reagents and steric factors. For an example of the effect of reagents, let us consider the cyclization step for 4-(2-phenanthryl)butyric acid (Figure 9.18). Electrophilic cyclization under acidic conditions using HF leads to cyclization at the 3-position of phenanthrene [15]. In contrast, converting the carboxylic acid to the corresponding acid chloride using thionyl chloride and SnCl$_4$ gave cyclization at the 1-position (Figure 9.18) [16]. The explanation for this difference in regiochemistry is that the acid-mediated electrophilic cyclization of the carboxylic acid is under thermodynamic control, while the electrophilic cyclization of the acid chloride is under kinetic control. This observation also suggests that there may be circumstances where poor regiochemical control leads to a mixture of products.

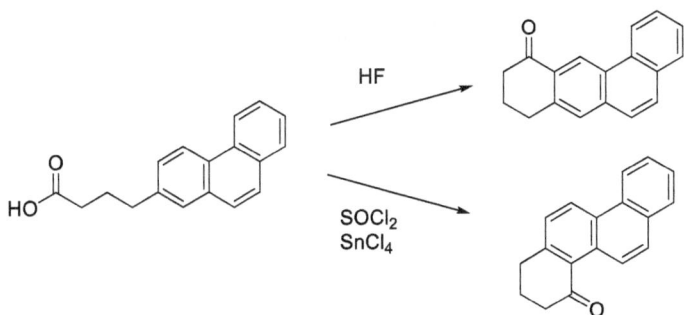

Figure 9.18: Regiochemistry of electrophilic cyclizations in the Haworth synthesis [15,16].

The classical Haworth synthesis can be extended beyond the use of succinic anhydride to other anhydrides such as phthalic anhydride or naphthalene 2,3-dicarboxylic anhydride (Figure 9.19). These reactions lead to extended PAHs, but are otherwise analogous to the traditional Haworth synthesis. In the first example, a Friedel–Crafts acylation of triphenylene with phthalic anhydride gave acylation at the 2-position. Electrophilic cyclization of the keto acid gave the quinone, which could be reduced to give dibenzo[a,c]naphthacene [17]. In the second example, the acylation of naphthalene with or naphthalene 2,3-dicarboxylic anhydride gave a mixture of regioisomers. Each of these could be cyclized to the corresponding quinones, which could be reduced to the corresponding pentacene and benzo[a]naphthacene [18].

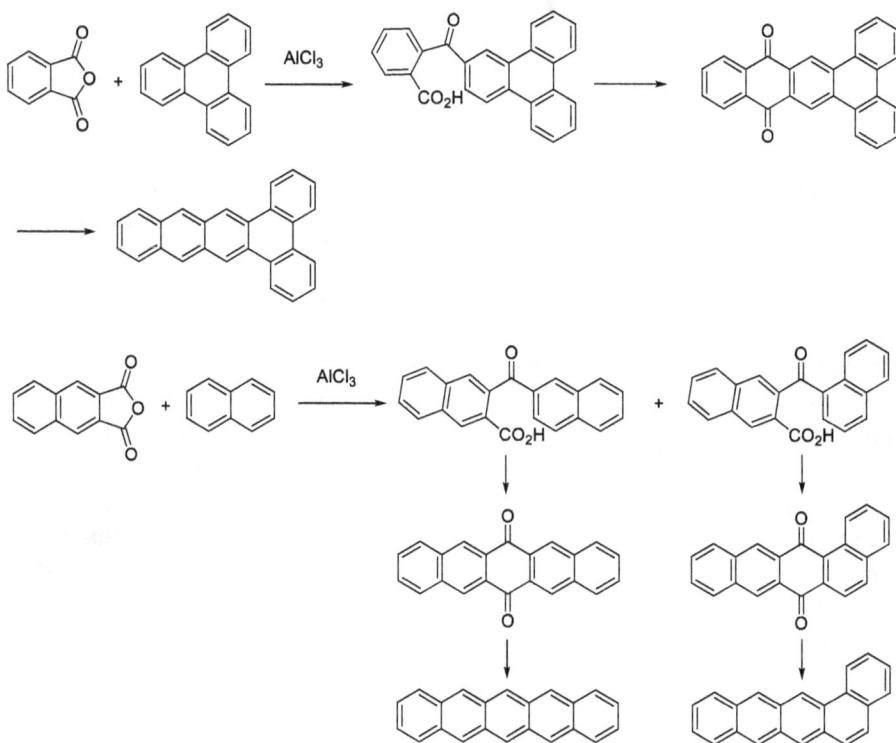

Figure 9.19: Modification of the Haworth synthesis using phthalic anhydride and naphthalene carboxylic anhydride.

As an example of a modified Haworth synthesis, we will consider the synthesis of 2-hydroxybenzo[a]pyrene, which was prepared in order to study it biological properties as a carcinogen. The synthesis of 2-hydroxybenzo[a]pyrene is outlined in Figure 9.20 [8]. The synthesis started from pyrene, which was hydrogenated at the K regions to give the corresponding tetrahydropyrene.This was done to direct the regiochemistry for the introduction of the hydroxyl group and the succinic anhydride to be used in formation of the new ring. Friedel–Crafts acylation introduced the acetyl group in the 2-position, which was converted in to a methoxy group by a sequence of Baeyer–Villiger oxidation, methanolysis of the acetate, and methylation using dimethyl sulfate. The methoxy-substituted product was then used in a Haworth approach: acylation using succinic anhydride in the presence of AlCl₃ was followed by a Clemmensen reduction. This was followed by acid-catalyzed electrophilic cyclization to give the cyclohexanone. In this case, the carbonyl was removed using a Wolff–Kishner reduction. The three rings were aromatized using palladium on carbon, and the methoxy group was cleaved using HBr to give the final product.

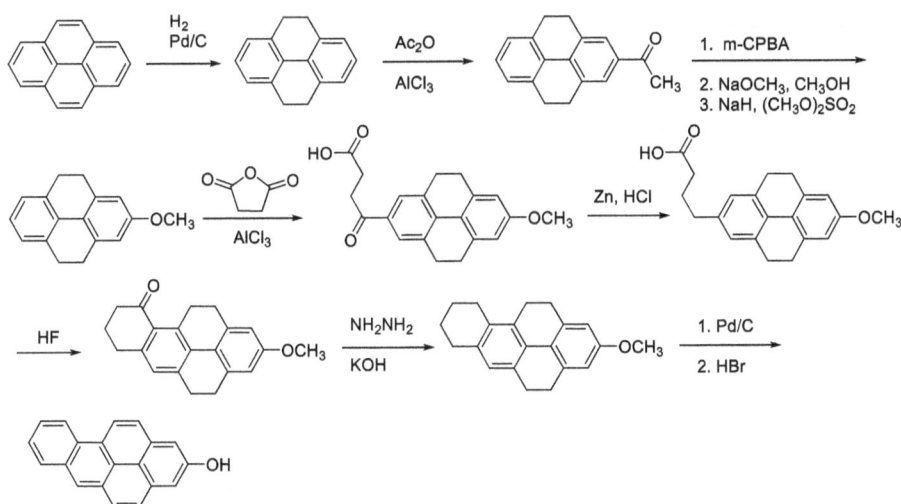

Figure 9.20: Synthesis of 2-hydroxybenzo[a]pyrene [8].

In some cases, there are challenges with the regiochemical control of the electrophilic aromatic substitution step. An alternative approach is to use a Grignard reagent (or organolithium) and react it with the anhydride. For example, using the Grignard reagent derived from 9-bromophenanthrene with phthalic anhydride avoids the formation of other regioisomers (Figure 9.21) [19]. The keto-acid can then be converted to dibenz[a,c]anthracene [20].

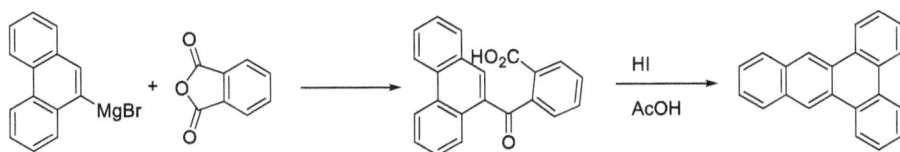

Figure 9.21: Synthesis of PAHs via a Grignard reaction with phthalic anhydride.

9.3.2 The Pschorr synthesis

The Pschorr reaction involves an intramolecular cyclization using an aryldiazonium salt with an adjacent aromatic ring (see Section 3.6.4). The reaction takes place in the presence of a copper catalyst, or with sodium iodide. The Pschorr reaction can be used to prepare PAHs such as phenanthrene from amino-substituted stilbenes [21,22]. The amino-stilbene starting material is typically prepared by a Perkin condensation of phenylacetic acid with o-nitrobenzaldehyde, followed by reduction of the nitro group. Following the Pschorr cyclization reaction, phenanthrene carboxylic acid can be decarboxylated [22]. This sequence is outlined in Figure 9.22. More recently, ferrocene was used as a homogeneous catalyst for the Pschorr cyclization [23].

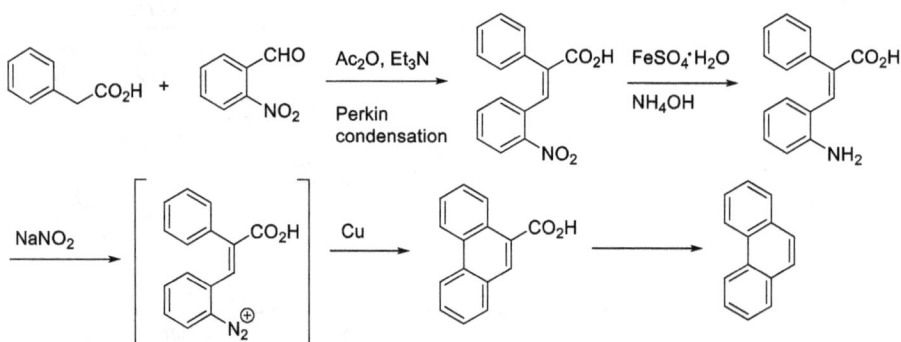

Figure 9.22: The Pschorr synthesis of phenanthrene [22].

9.3.3 The Elbs reaction

The Elbs reaction is a classic way to prepare PAHs by pyrolysis of an alkyl diaryl ketone. In 1884, Elbs demonstrated that anthracene could be prepared from o-methylbenzophenone (Figure 9.23) [24].

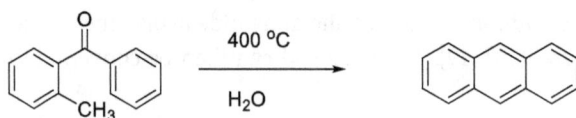

Figure 9.23: The Elbs reaction [24].

Mechanistically, this reaction can be viewed as an electrocyclic ring-closing of the enol tautomer of the diaryl ketone, followed by dehydration (Figure 9.24).

As a general synthetic strategy for the synthesis of PAHs, the Elbs reaction has not been used as widely as Haworth-type syntheses. Nonetheless, can be used to

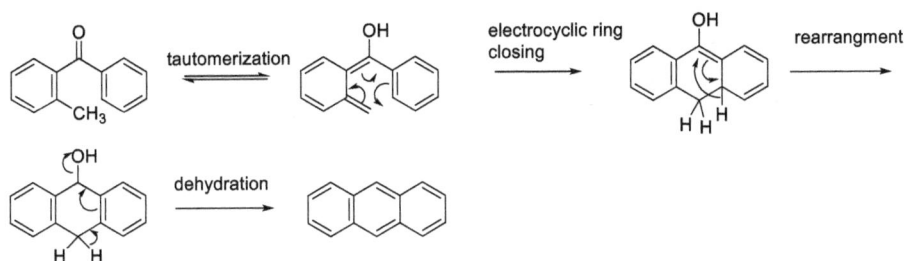

Figure 9.24: General mechanism of the Elbs reaction.

access a variety of structures, including acenes. For an example of the Elbs synthesis, we will consider the preparation of benzo[b]naphtha[1,2-k]chrysene (Figure 9.25) [25]. The synthesis begins with a Friedel–Crafts acylation of 2,6-dimethylnaphthalene using naphthoyl chloride and AlCl₃. A second Friedel–Crafts acylation using benzoyl chloride gave the diketone. Heating this compound led to a twofold Elbs reaction to give the desired benzo[b]naphtha[1,2-k]chrysene.

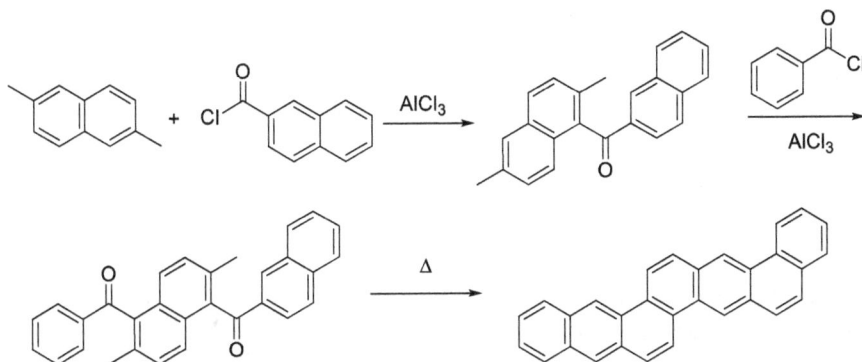

Figure 9.25: Synthesis of benzo[b]naphtha[1,2-k]chrysene involving the Elbs reaction [25].

9.3.4 Diels–Alder cycloadditions

The Diels–Alder reaction is a classic approach for forming six-membered rings which can subsequently be aromatized. Here, we will explore some of the ways in which Diels–Alder reactions have been used to form PAHs.

Quinones are ubiquitous dienophiles and these can be used for the preparation of PAHs by reduction of the adduct. For example, 1-vinylnaphthalene can undergo a Diels–Alder reaction with benzoquinone. The initial adduct oxidizes to the corresponding quinone, which can be reduced to the fully aromatic PAH using a reducing agent such as LiAlH₄ (Figure 9.26) [26].

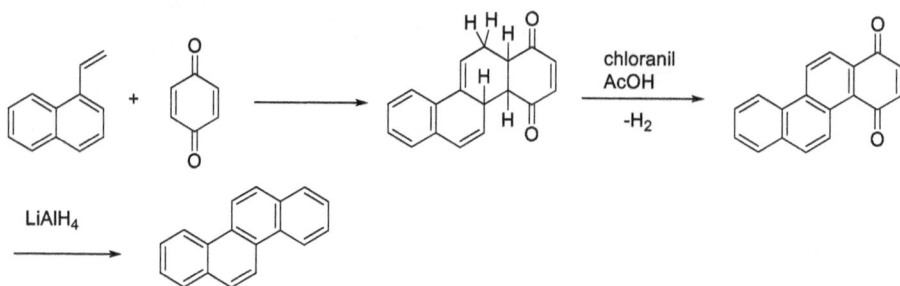

Figure 9.26: Synthesis of chrysene via a Diels–Alder reaction [26].

Maleic anhydride can also be used as a dienophile for the elaborations of PAH structures. Indeed, some PAHs themselves can react as dienes, as we saw in Section 9.2.2. For example, perylene can undergo a Diels–Alder reaction with maleic anhydride at the bay region (Figure 9.27) [10]. The resulting adduct can rearomatize and the anhydride can be removed by hydrolysis and decarboxylation.

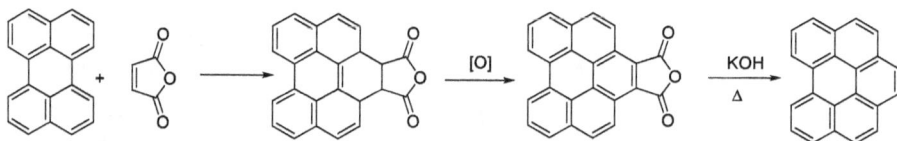

Figure 9.27: Diels–Alder reaction of perylene with maleic anhydride [10].

Arynes can also be used as dienophiles for the direct preparation of PAHs without the subsequent need to reduce the quinone. For example, in the reaction below, 1-bromo-2-fluoronaphthalene in the presence of Mg forms a Grignard reagent, which then undergoes elimination to form a naphthalyne (Figure 9.28) [27]. This naphthalyne in turn undergoes a Diels–Alder reaction with vinylnaphthalene. Again, the initial dihydro-adduct undergoes spontaneous oxidation to rearomatize and form the picene, albeit in low yield.

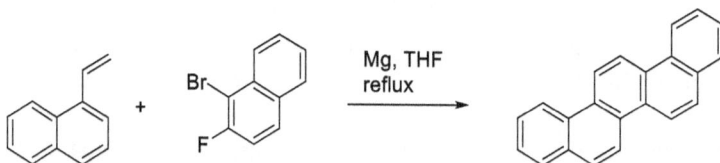

Figure 9.28: Diels–Alder reaction of an aryne with vinylnaphthalene to form picene [27].

In Chapter 5, we saw that can undergo Diels–Alder cycloaddition reactions of arynes with dienes including heteroaromatic compounds such as furan. Extrusion of the bridging heteroatom from the resulting adduct can be used to prepare the corresponding fused aromatic system as shown in Figure 9.29.

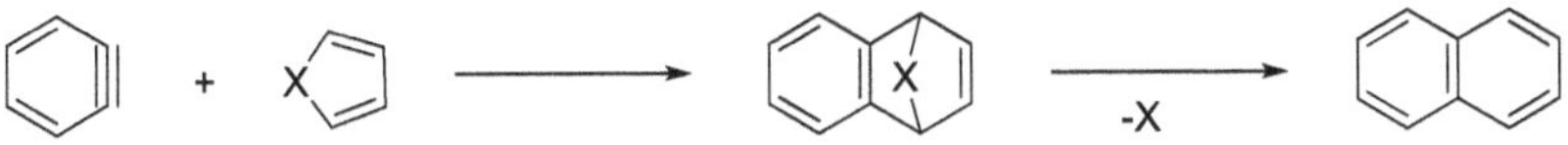

Figure 9.29: Generalized synthesis of PAHs from arynes.

This approach can be used to access a number of PAH structures, depending on the structure of both the aryne and the heteroaromatic compound. Furan is a widely used heterocycle for this type of reaction. The resulting Diels–Alder adduct is typically converted to the corresponding aromatic structure in a two-step sequence involving reduction, followed by dehydration. This approach was used in the synthesis of chrysene, shown in Figure 9.30 [28]. The substituted naphthalene can form the bis (aryne) upon treatment with phenyllithium via lithium–halogen exchange and elimination. The arynes then react with furan in a Diels–Alder reaction. The resulting adduct is reduced, and then undergoes acid-catalyzed dehydration to give chrysene.

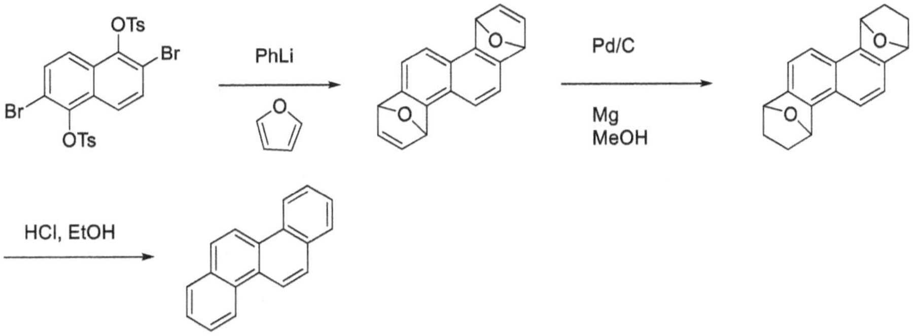

Figure 9.30: Synthesis of chrysene via aryne reaction with furan [28].

Pyrroles can undergo similar reactions with arynes, with the nitrogen bridge being removed in a subsequent step by oxidation using a peroxy acid such as mCPBA (Figure 9.31) [28].

Isobenzofurans can also serve as dienes in the synthesis of PAHs. Isobenzofurans can be generated in situ from the hydroxyl acetal under acidic conditions [29], or from the cyclic acetal in the presence of a strong base such as LDA (Figure 9.32) [30]. Both of these reactions likely proceed via the cyclic acetal and involve a 1,4-elimination to generate isobenzofuran. In the presence of dienophiles, Diels–Alder reactions take place. For example, in the presence of naphthoquinone, the isobenzofuran reacts to form tetracenequinone (Figure 9.32) [29]. Similarly, in the presence of benzoquinone, either 1,4-anthraquinone or 6,13-pentacenequinone are formed, depending on the

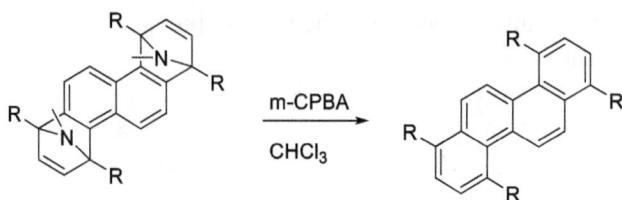

Figure 9.31: Oxidative removal of the nitrogen bridge resulting from addition of a pyrrole to an aryne [28].

stoichiometry of the reaction. In this reaction, the initial Diels–Alder adduct undergoes dehydration in the reaction conditions to form the new aromatic ring.

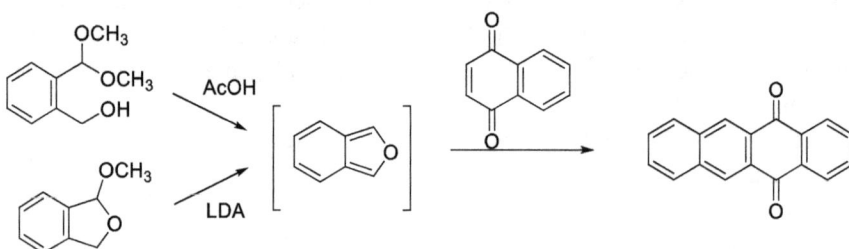

Figure 9.32: Diels–Alder reaction of isobenzofuran with naphthoquinone [29].

Isobenzofurans can also undergo Diels–Alder reactions with arynes to form new PAHs upon aromatization. For example, Rickborn and coworkers used a bis(trimethylsilyl)isobenzofuran in a reaction with benzyne, which was generated by dehydrohalogenation of bromobenzene in the presence of a strong base (Figure 9.33) [31]. The TMS groups of the Diels–Alder adduct could be removed using TBAF or under basic conditions, and the product could be reduced to the PAH system either by a two-step sequence of reduction and dehydration, or in the presence of $Fe_2(CO)_9$.

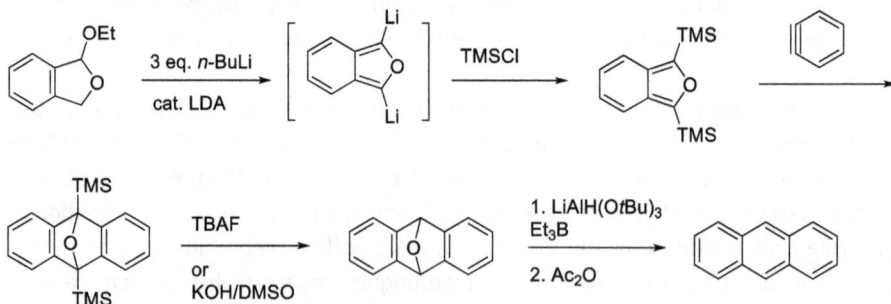

Figure 9.33: Reaction of an isobenzofuran with a benzyne to form anthracene [31].

Another cycloaddition approach for the formation of fused aromatic rings involves the reaction of o-xylylenes (also known as o-quinodimethanes) with dienophiles [32,33]. Referred to as the Cava reaction, the o-xylylene is typically generated in situ from the corresponding tetrabromide derivative via an iodide-induced 1,4-elimination of bromine. The o-xylylene then undergoes a cycloaddition reaction, followed by elimination of HBr to generate the aromatic ring (Figure 9.34).

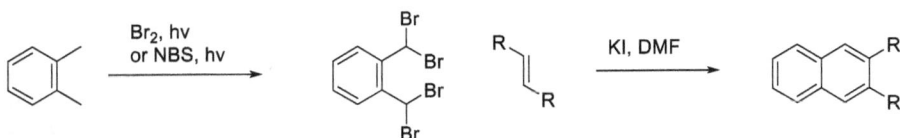

Figure 9.34: The Cava reaction.

An alternative to this approach is to use the dibromide, which yields the tetrahydro-naphthalene derivative. Depending on the substrate, this product may oxidize to the corresponding aromatic system on standing in air, or may require a subsequent oxidation step (Figure 9.35).

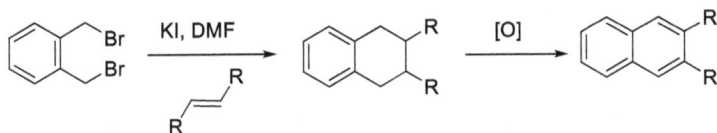

Figure 9.35: Modified Cava reaction to generate a tetrahydronaphthalene.

Several examples of the use of the Cava reaction for the preparation of PAH systems are shown in Figure 9.36. The first example shows that reaction of 1,2-dibromo-4,5-bis(dibromomethyl)benzene with simple dienophiles in the presence of KI in DMF can be used to prepare substituted naphthalenes [34, 35]. It is noteworthy that the substitution pattern is distinctly different from that typically achieved through electrophilic aromatic substitution, and the bromo substituents provide synthetic handles for further modifications, such as cross-coupling reactions. The second entry shows how reaction with quinones such as 1,4-anthraquinone to produce more extended polycyclic systems [36]. Example 3 shows how bis(benzylic bromides) can be used in the Cava reaction. In this case, the product of the initial reaction with benzoquinone undergoes a tautomerization and a subsequent air oxidation to rearomatize [36].

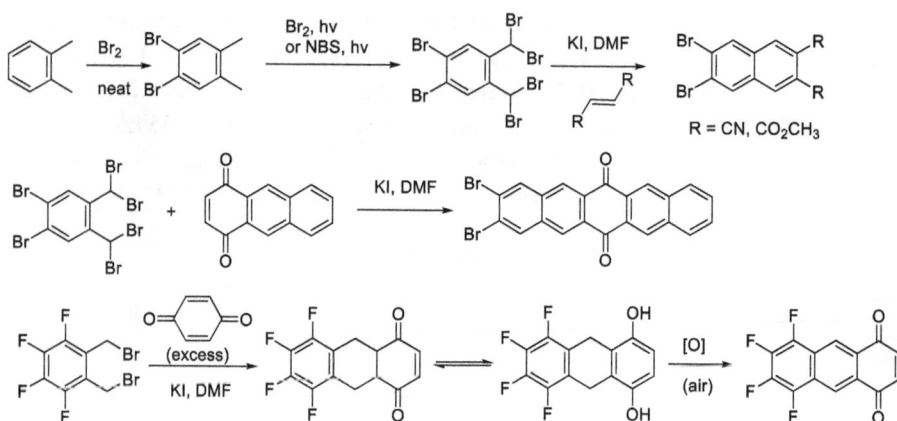

Figure 9.36: Representative examples of the Cava reaction [34–36].

9.3.5 The Wittig reaction

The Wittig reaction has been used for the preparation of extended PAHs by reaction of o-xylylene bis(triphenylphosphonium) bromide with 1,2-diones such as phenanthrene-quinone (Figure 9.37) [37]. This methodology has been used in a limited number of cases, but the reaction is typically carried out in biphasic conditions with aqueous lithium hydroxide and dichloromethane. The reaction is typically left for several days, or carried out under ultrasound conditions to accelerate the reaction [38]

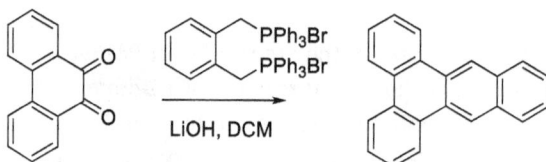

Figure 9.37: Synthesis of dibenz[a,c]anthracene via a Wittig reaction [37].

In the example above, the Wittig reaction is used directly in the formation of an aromatic ring. More commonly, the Wittig is used to prepare an aryl alkene that is converted to the corresponding aromatic ring via a subsequent cyclization. For example, consider the synthesis of 9-hydroxybenzo[a]pyrene, which was investigated as a potential metabolite of benzo[a]pyrene with carcinogenic properties (Figure 9.38) [39]. The synthesis began with a Suzuki–Miyaura cross-coupling between the naphthalene boronate ester and the bromobenzene bearing formyl groups. The biaryl product was converted into the bis(vinyl ether) using a twofold Wittig reaction. Under acidic conditions, the vinyl ether groups were converted into the corresponding aldehydes, which underwent acid-catalyzed cyclization and dehydration to dive the desired benzo[a]pyrene structure. The final step removed the methoxy protecting group to liberate the hydroxyl.

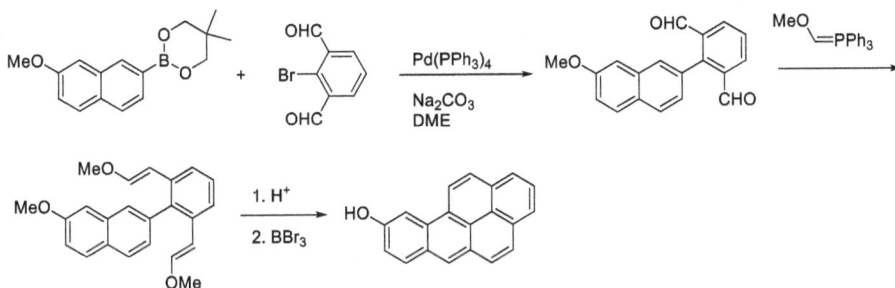

Figure 9.38: Synthesis of 9-hydroxybenzo[a]pyrene [39].

9.3.6 Photocyclization (the Mallory–Katz reaction)

A common approach for the formation of PAHs via intramolecular aryl C–C bond formation is via oxidative photocyclization of stilbene derivatives. For example, *cis*-stilbene undergoes oxidative photocyclization to form phenanthrene (Figure 9.39). The reaction proceeds via a 6π-electron photoinduced electrocyclic ring closing reaction to form a dihydrophenanthrene, which undergoes subsequent oxidation to form phenanthrene.

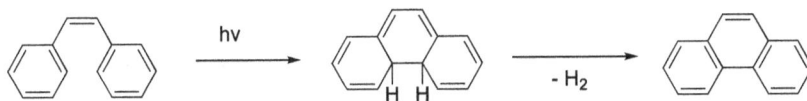

Figure 9.39: Synthesis of phenanthrene via photocyclization of *cis*-stilbene.

This reaction was discovered by Mallory and coworkers, who also extended it to the preparation of other PAHs. The reaction was typically carried out in the presence of a substoichiometric amount of iodine in the presence of air. This reaction relied on the presence of air as a formal oxidant. As shown in Figure 9.40, the reaction could be used to photocyclize stilbene derivatives as well as terphenyls. Subsequently, Katz and coworkers showed that the use of a stoichiometric amount of iodine as an oxidant, along with propylene oxide to scavenge the HI produced, improved the yields significantly [40, 41]. As an example, the preparation of benzo[c]chrysene from the corresponding alkene proceeded in quantitative yield (Figure 9.40) [42].

A limitation of the reaction is that it typically requires photochemical reactors, and often needs to be carried out at low concentrations to avoid photodimerization

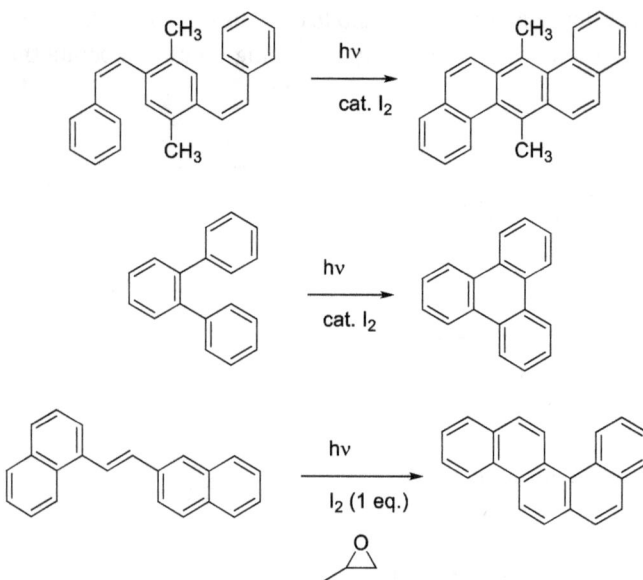

Figure 9.40: Some examples of PAH synthesis via photocyclization.

and other side reactions. Nevertheless, this reaction has found wide utility for the synthesis of PAHs, including hindered systems such as helicenes, which we will explore in Chapter 11.

9.3.7 Cyclodehydrogenation reactions

Cyclodehydrogenation refers to the direct coupling of two aromatic rings with the formal loss of hydrogen. This type of reaction has been used extensively for the formation of PAHs. One type of cyclodehydrogenation reaction is known as the Scholl reaction, which typically involves either inter- or intramolecular coupling in the presence of the Lewis acid such as $AlCl_3$ at elevated temperatures. A second type of cyclodehydrogenation is essentially an oxidative dehydrogenative coupling involving an oxidant, and is usually carried out at room temperature. We saw this latter reaction in briefly in Section 6.2.3 in the context of biaryl formation. While these two reactions were once considered as distinct, the two have become conflated in the literature, and are both commonly referred to as the Scholl reaction [43]. One potential source of confusion between the two reactions is that $FeCl_3$ is a reagent of choice, which can serve as a Lewis acid and as an oxidant. For a more detailed discussion of these two reactions, Gryko, Butenschön, and coworkers have a review focused on the comparison of oxidative aromatic coupling and the Scholl reaction [43]. For the purposes of this discussion, the Scholl reaction with a Lewis acid will be treated as mechanistically distinct from oxidative aromatic coupling. However, it is important to acknowledge that some aspects of the mechanism remain unclear and may vary depending on the substrate and the reaction conditions.

The direct coupling of two aromatic rings was first reported by Scholl and Mansfeld, who showed that the polycyclic quinone could undergo cyclization to the corresponding fused quinone in the presence of anhydrous $AlCl_3$ (neat) at 140 °C (Figure 9.41) [44].

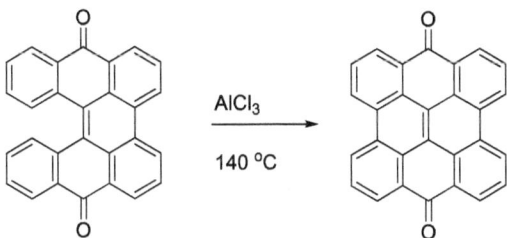

Figure 9.41: The first report of Lewis acid-mediated aromatic ring fusion by Scholl and Mansfeld [44].

The original Scholl reaction involved heating in neat $AlCl_3$, but it was also found that a 1:1 mixture of $AlCl_3$ and NaCl, which forms a liquid above 100 °C could be used to carry out the reaction. Other Lewis acids or even Brønsted acids have been used, but $AlCl_3$/NaCl is the most common conditions.

The mechanism of this reaction has been the subject of much discussion, but is generally thought to proceed via an arenium cation mechanism. Figure 9.42 outlines the proposed mechanism using o-terphenyl as a representative example. The first step of the reaction is the formation of an arenium cation by protonation or by formation of a σ-complex with the Lewis acid. In Figure 9.42, it is depicted as protonation for the sake of simplicity. The arenium cation served as an electrophile that is attacked by the other aromatic ring, generating the new carbon–carbon bond in an electrophilic aromatic substitution. The final step requires the formal loss of hydrogen to regenerate the aromatic ring and is the least understood. It is possible that this oxidation is occurring with the help of oxygen in air. It has also been proposed that carbonyl groups, which are present in many of the examples of the Scholl reaction, serve as an oxidizing agent, leading to the formation of a secondary alcohol, which is reoxidized to the ketone in the presence of air. Indeed, the Scholl reaction often works well for carbonyl-containing compounds despite the fact that the carbonyl should disfavor the formation of the arenium cation [43]. A radical mechanism for the Scholl reaction has also been proposed, however, there is experimental and theoretical support for the arenium cation mechanism, at least in cases where the reactant is a Lewis acid and not an oxidant.

The Scholl reaction has been used to access a variety of polycyclic aromatic systems, including synthetic dyes. Selected examples of the Scholl reaction are shown in Figure 9.43.

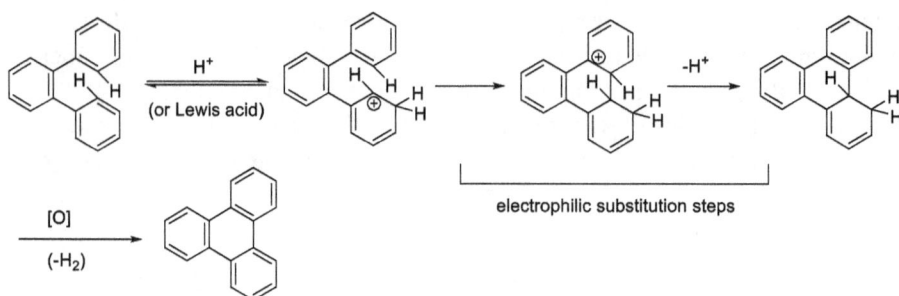

Figure 9.42: Proposed arenium cation mechanism for the Scholl reaction.

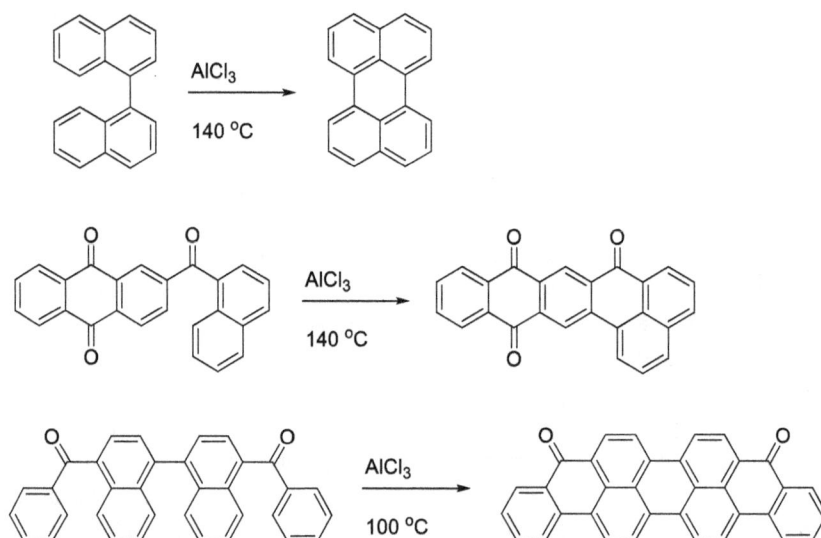

Figure 9.43: Representative examples of the Scholl reaction [45–47].

The traditional Scholl reaction has the drawback that it requires relatively harsh conditions and high temperatures, which means it has limitations in terms of practical applicability. On the other hand, oxidative cyclodehydrogenation in the presence of an oxidant is often carried out under relatively mild conditions at room temperature. The most common reaction conditions for oxidative cyclodehydrogenation are $FeCl_3$ in CH_2Cl_2. In some cases, nitromethane is also added in order to solubilize the $FeCl_3$. Other reagents that have been used for this reaction include $MoCl_5$, $Cu(OTf)_2/AlCl_3$, VOF_3 with $BF_3 \cdot Et_2O$, or DDQ with methanesulfonic acid.

The mechanism of oxidative is thought to proceed via a radical cation mechanism, although the mechanistic distinction between this mechanism and the arenium cation mechanism is not always clear. The proposed mechanism is outlined in Figure 9.44. The first step is a one-electron oxidation of one of the aromatic rings to generate a radical cation intermediate. The radical cation then undergoes a radical coupling to form the new carbon–carbon bond, resulting in a delocalized cation on one ring and a delocalized radical on the other ring. This species undergoes a second one-electron oxidation to give a dicationic species, which loses two protons in the final step to rearomatize.

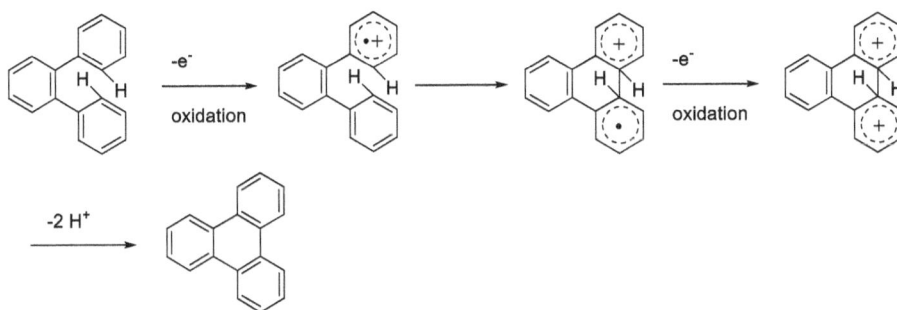

Figure 9.44: Proposed mechanism for oxidative cyclodehydrogenation.

Oxidative cyclodehydrogenation is most effective for electron-rich substrates – often alkyl- or especially alkoxy-substituted arenes. This reactivity is expected because these compounds are more prone to oxidation, and because electron-donating substituents can stabilize the radical and cationic intermediates. Figure 9.45 shows some selected examples of oxidative cyclodehydrogenation reactions.

There are some situations where oxidative cyclodehydrogenation does not give the desired product due to unwanted side reactions such as chlorination or rearrangements. Chlorination has been observed by mass spectrometry during the synthesis of large graphitic sheets via cyclodehydrogenation [51]. Structural rearrangements involving phenyl migrations have also been reported for oxidative cyclodehydrogenation of electron-rich oligophenylenes. For example, King reported that the oxidative cyclodehydrogenation of an octamethoxy quaterphenyl using (bis(trifluoroacetoxy) iodo)benzene (PIFA) and $BF_3 \cdot Et_2O$ did not yield the expected product, but rather one where a methoxy group appears to have shifted position (Figure 9.46).[52]

The product cannot be explained by simple bond rotation to a different conformer. Rather, King and coworkers proposed a phenyl shift to explain the rearrangement. Invoking an arenium cation, the proposed mechanism is shown in Figure 9.47. After protonation of one of the two inner rings, a phenyl migration takes place (here shown via a cyclopropyl intermediate) to give the rearranged carbon skeleton. At this point, a bond rotation is followed by the cyclization and oxidation steps to give the observed rearranged product.

Müllen and coworkers also reported unusual rearrangements during an attempted cyclodehydrogenation of a dimethoxy-substituted hexaphenylbenzene [53]. Rather than the expected hexa-*peri*-hexabenzocoronene, two products were

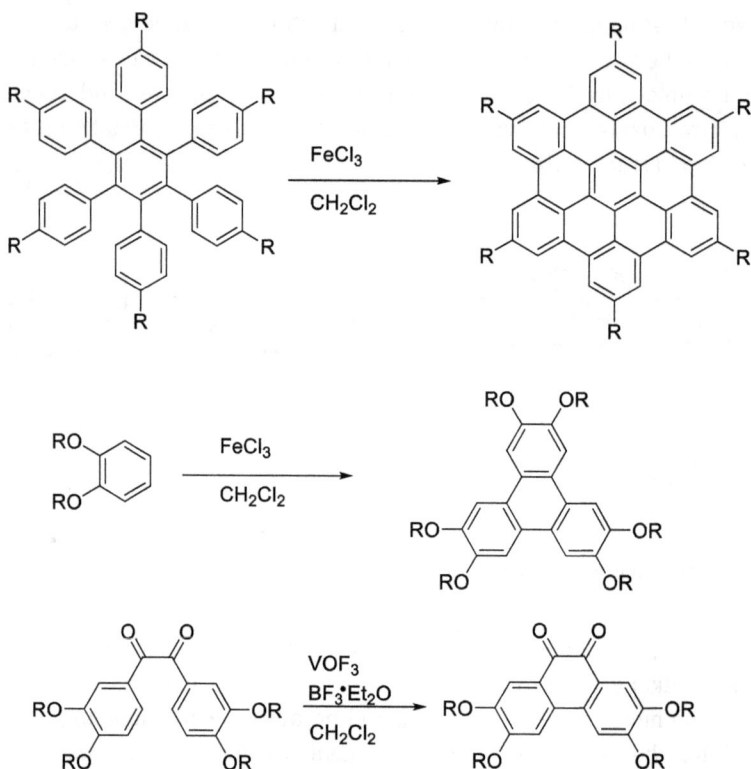

Figure 9.45: Representative oxidative cyclodehydrogenation reactions [48–50].

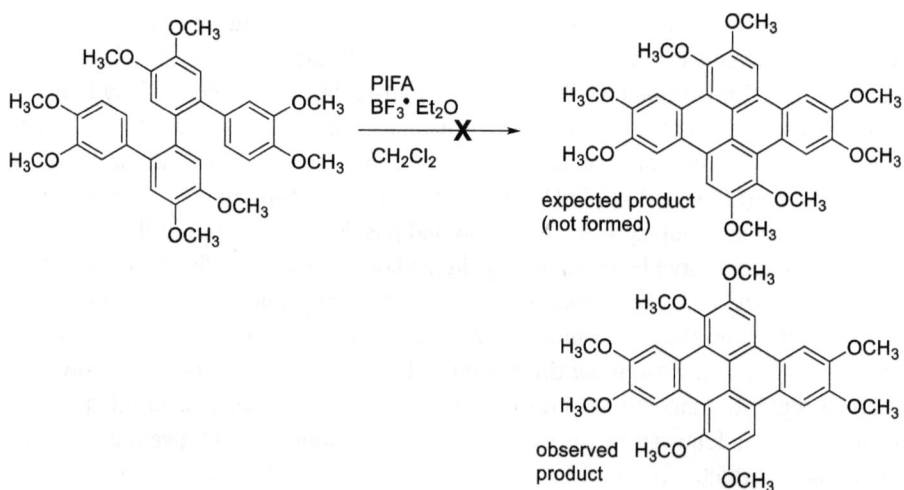

Figure 9.46: Rearrangement during the oxidatic cyclodehydrogenation of an octamethoxy quaterphenyl [52].

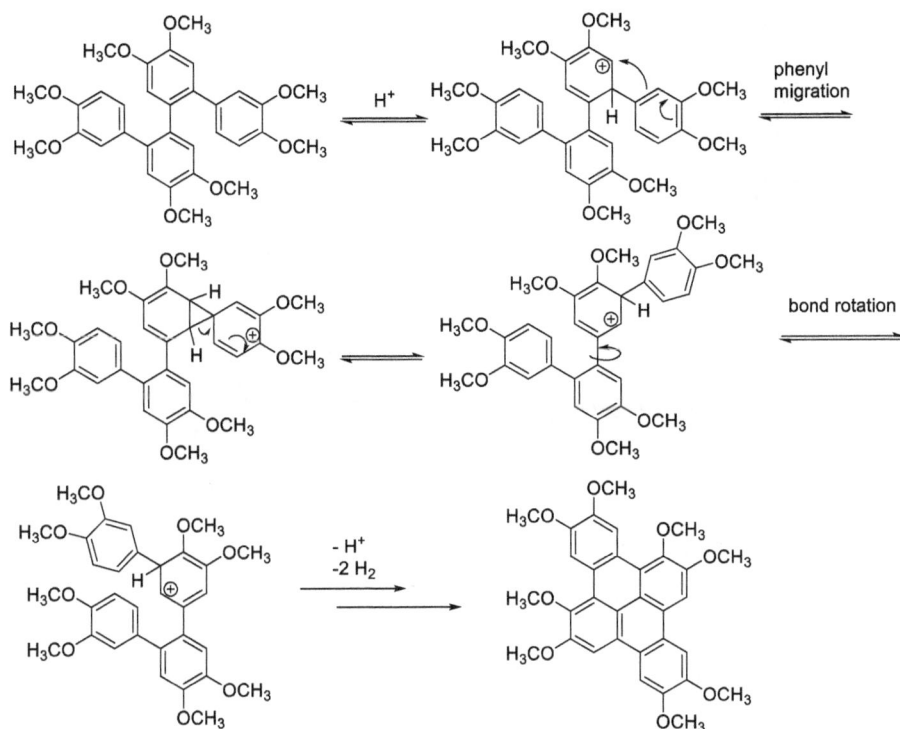

Figure 9.47: Proposed mechanism for rearrangement during oxidative cyclodehydrogenation.

isolated: an isomeric hexa-*peri*-hexabenzocoronene where one of the methoxy groups had migrated, and an unusual quinone product (Figure 9.48).

The apparent migration of the methoxy group was explained by a phenyl migration similar to that described by King and coworkers. The quinone formation, however, follows a different path. The proposed mechanism is outlined in Figure 9.49. In their proposed mechanism, following protonation (or σ-complex formation with the Fe(III)), the intramolecular carbon–carbon bond formation takes place with the carbon attached to the central ring (*para* to the methoxy group) to form the spirocyclic system. Oxidation causes rearomatization of one ring and hydrolysis leads to loss of methanol and formation of the quinone. This sequence is then repeated on the opposite side of the molecule to give the final quinone product.

Despite the fact that oxidative cyclodehydrogenation is mostly limited to electron rich substrates and does sometimes suffer from side reactions such as the rearrangements discussed above, it has become a very powerful tool for the preparation of a wide variety of PAHs, including some impressively large nanographene structures. Some of these examples will be explored in Chapter 10.

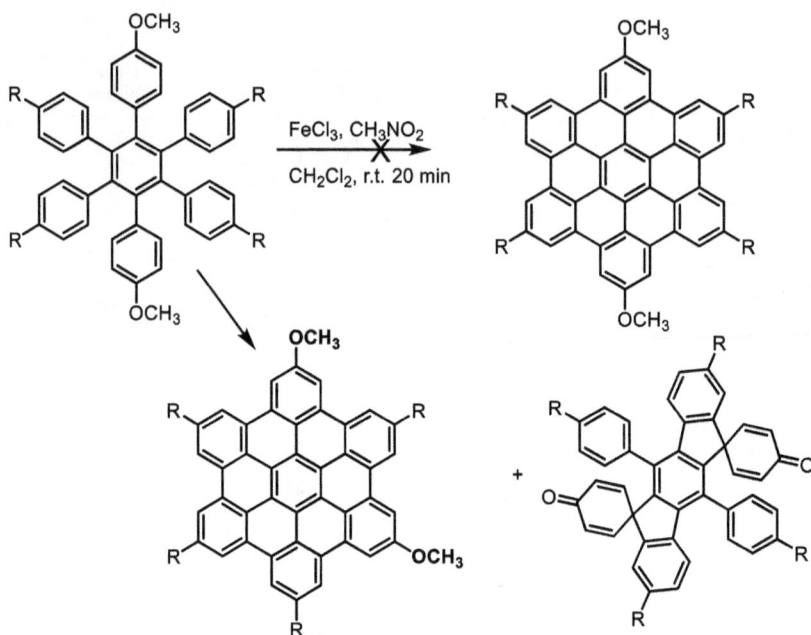

Figure 9.48: Rearrangements during the oxidative cyclodehydrogenation of a methoxy-substituted hexaphenylbenzene [53].

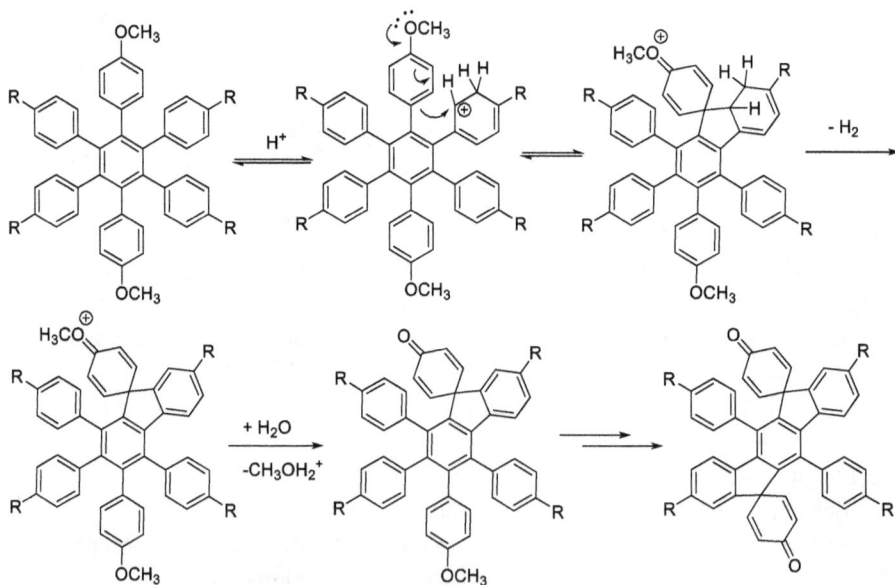

Figure 9.49: Proposed mechanism of quinone formation.

9.3.8 Cyclodehydrohalogenation reactions

An alternative to the Mallory reaction and cyclodehydrogenation reactions is the photochemical cyclodehydrochlorination reaction (sometimes referred to as a CDHC reaction). The reaction is thought to proceed through a photochemical electrocyclic ring closing, followed by elimination of HCl (Figure 9.50). It resembles the Mallory reaction, but does not rely on an oxidation of the photocyclized intermediate. Another possible mechanistic explanation for the reaction is a homolytic carbon–halogen bond cleavage, followed by a radical coupling. However, the reaction works best with aryl chlorides; if a radical reaction were occurring, one would anticipate aryl bromides and iodides to be more reactive.

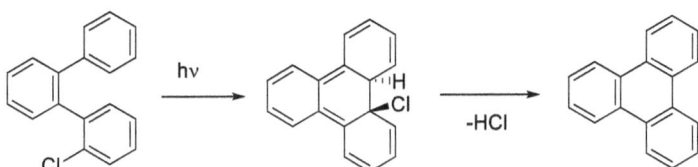

Figure 9.50: Proposed mechanism of the photochemical cyclodehydrohalogenation (CDHC) reaction.

The reaction was first reported in the early 1970s for the synthesis of dibenzonaphthacene, which involves two cyclization steps (Figure 9.51) [54, 55].

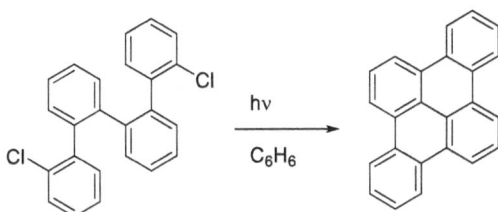

Figure 9.51: Synthesis of dibenzonaphthacene via a CDHC reaction.

More recently, the reaction has been proposed as a viable alternative to cyclodehydrogenation reactions. The CDHC reaction has the disadvantage that requires a chloro-functionalized precursor. However, it has the advantage of better control of regioselectivity. It also circumvents some of the challenges of cyclodehydrogenation reactions, such as rearrangements. For example, as shown in Figure 9.52, attempted formation of alkoxy-substituted dibenzonaphthacenes via oxidative cyclodehydrogenation gave only the phenyl-substituted triphenylene, which was the result of a rearrangement. In contrast, photochemical CDHC reaction on the corresponding chloro-substituted precursors gives the desired dibenzonaphtha-cenes in moderate yields, without any apparent rearrangements [56].

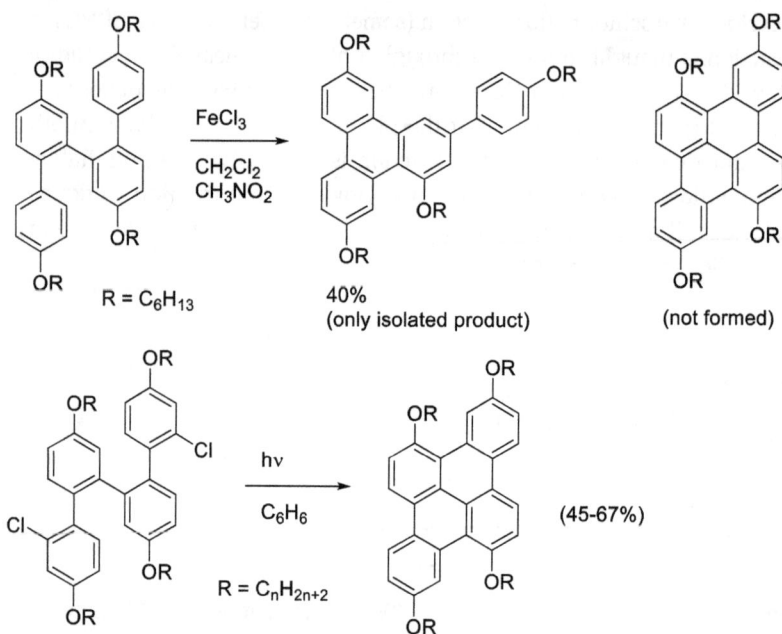

Figure 9.52: Attempted synthesis of dibenzonaphthacenes by cyclodehydrogenation reaction versus CDHC reaction [56].

The scope of this reaction has been expanded and used for a more diverse set of compounds, enabling the preparation of a number of different PAHs. These reactions are often carried out in solvents such as benzene or cyclohexane, although more recently, the reaction has been successfully performed in acetone in the presence of aqueous sodium carbonate. In these latter conditions, the base may serve to promote the elimination step, or at very least quench the HCl produced during the reaction. The following examples illustrate the scope of the reaction, which tolerates heteroatoms as well as sterically congested systems (Figure 9.53) [57].

9.3.9 Aryne cyclotrimerization and related reactions

In Chapter 5, we saw that arynes generated under mild conditions from the corresponding trimethylsilyl aryl triflate can undergo cyclotrimerization in the presence of a Pd(0) catalyst to produce triphenylenes (Section 5.3.4) [58]. This methodology can be used to access a variety of extended PAHs, including nonplanar systems. Figure 9.54 shows

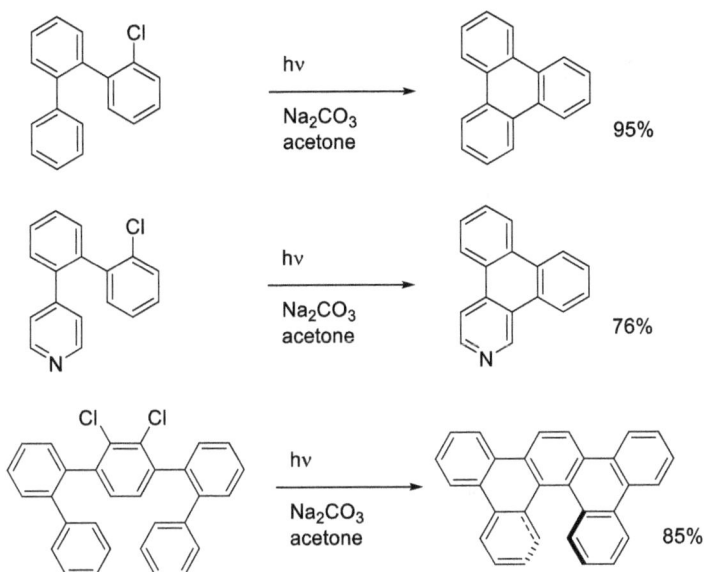

Figure 9.53: Examples of PAH formation using the CDHC reaction [57].

some representative examples of PAHs prepared by aryne cyclotrimerization. More examples will be discussed in Chapters 10 and 11.

9.3.10 Ring-closing metathesis

In Chapter 8, we saw that olefin metathesis could be used to prepare aromatic rings. This same reactivity makes it potentially useful for the preparation of polycyclic systems. While this methodology could in principle be used to prepare a wide variety of PAHs, especially angular PAHs with a K-region, the use of olefin metathesis is limited by poor solubility of some of the reactants resulting PAH products. As an example of the use of ring-closing metathesis for the preparation PAHs, Grubbs first- and second-generation catalysts have been used to prepare phenanthrenes in high yields under mild conditions (Figure 9.55) [59].

King and coworkers extended this approach to the preparation of more complex PAHs by forming multiple aromatic rings in a single step [60]. They showed that dibenz[a,j]anthracene and dibenz[a,h]anthracene could be prepared in high yields using Grubbs catalyst (Figure 9.56). Furthermore, to provide a general approach to larger PAHs with poor solubility, they showed that Schrock's catalyst with CS_2 as a solvent was effective for carrying out the same reaction [60]. The synthesis of dibenz[a,j]anthracene began by electrophilic bromination of m-xylene, followed by a twofold benzylic bromination. The bis(dibromomethyl)-dibromo benzene was converted

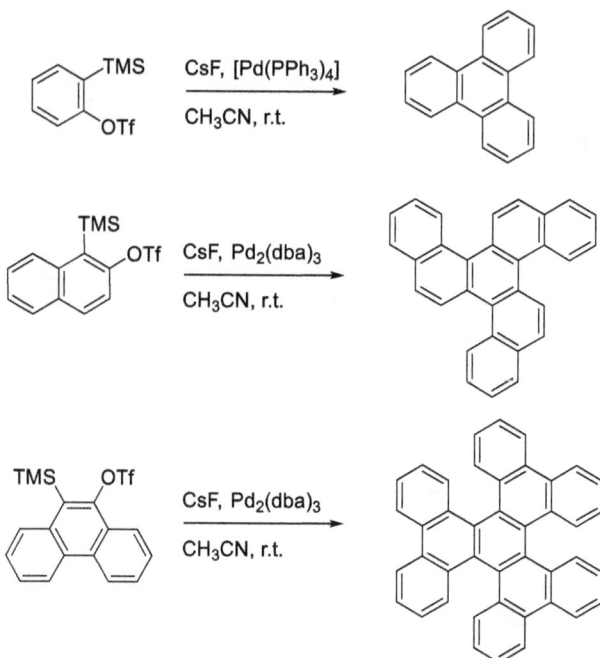

Figure 9.54: Examples of synthesis of PAHs via aryne cyclotrimerization.

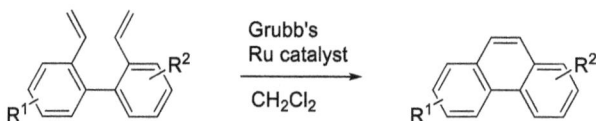

Figure 9.55: Synthesis of phenanthrene via ring-closing metathesis (RCM).

into the dialdehyde by hydrolysis with AgNO$_3$. The dialdehyde reacted with methylenetriphenyl phosphorene in a double Wittig reaction to give the divinyl benzene, which reacted in a Suzuki–Miyaura cross-coupling with *o*-styrylboronic acid to give the desired tetravinylterphenyl precursor to the olefin metathesis. The final olefin metathesis reaction could be carried out using either Grubbs or Schrock catalysts, giving the desired PAH in good yield. A similar approach was followed for the synthesis of dibenz[a,h]anthracene starting from *p*-xylene.

This approach has also been used for the preparation of other "ladder-type" oligomers using a Suzuki–Miyaura cross-coupling, followed by olefin metathesis using Grubbs second-generation catalyst (Figure 9.57) [61].

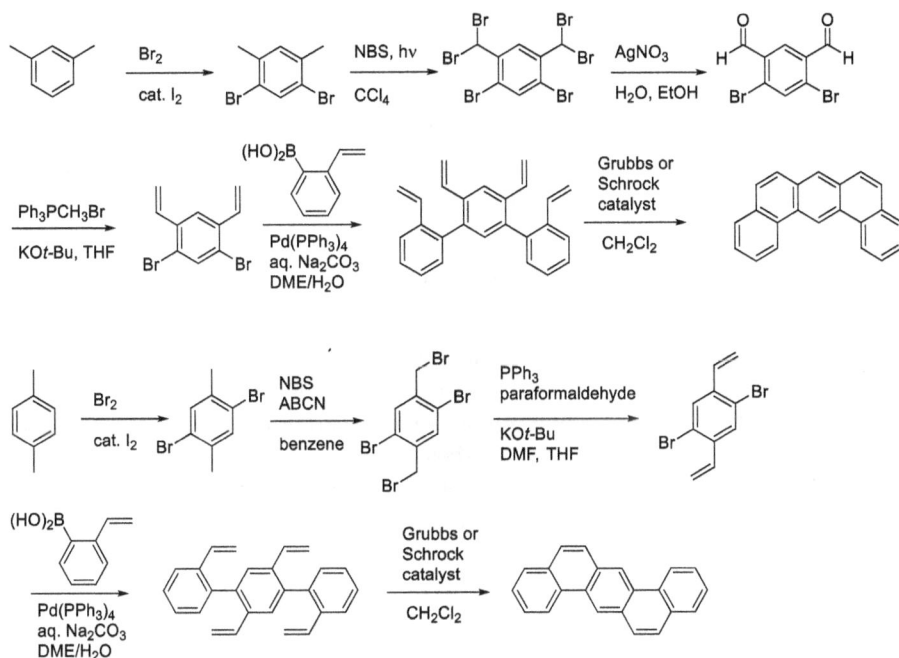

Figure 9.56: Synthesis of PAHs by olefin metathesis [60].

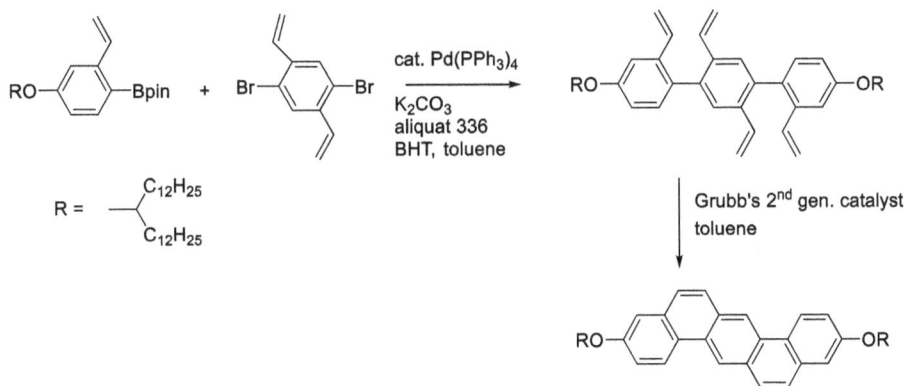

Figure 9.57: Synthesis of substituted PAHs via RCM [61].

9.3.11 Alkyne benzannulation

Cyclization of aryl alkynes can be used to prepare a wide variety of polycyclic aromatic systems [62]. These reactions are usually electrophilic cyclizations as outlined in Figure 9.58, where electrophilic addition to the alkyne is followed by cyclization and rearomatization.

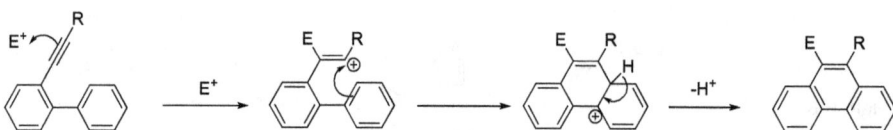

Figure 9.58: Electrophilic alkyne benzannulation.

This type of cyclization was reported by Goldfinger and Swager, who reported the twofold cyclization of bis(arylethynyl)terphenyls. The reaction was carried out using a Brønsted acid as the electrophile (trifluoroacetic acid) or using I(py)$_2$BF$_4$ as an iodine source (Figure 9.59) [63, 64]. Under these conditions, the electron-donating alkoxy groups were essential for the reaction to occur, consistent with electrophile addition as the first step, generating a carbocation that is resonance stabilized by the alkoxy group.

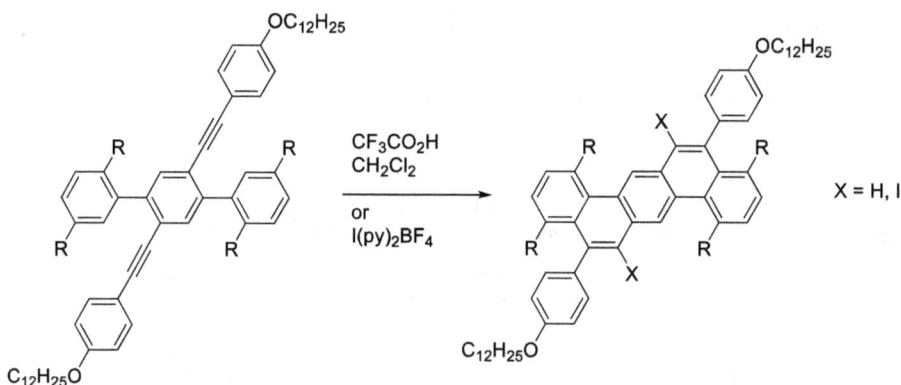

Figure 9.59: Synthesis of PAHs by electrophilic alkyne benzannulation [64].

Subsequently, the scope of the reaction has been expanded to include a broader set of electrophiles and also include substrates that are not activated by electron-donating alkoxy groups. Most notably, Larock showed that electrophiles such as ICl, I$_2$/NaHCO$_3$, NBS, and p-O$_2$NC$_6$H$_4$SCl could be used to carry out cyclization on alkynyl biaryls with a number of functional groups and heterocyclic rings [65, 66]. For example, 2-(1-alkynyl)biaryls, which were prepared via Sonogashira coupling, underwent electrophilic cyclization with ICl to produce the corresponding iodo-substituted phenanthrenes (Figure 9.60). The iodo group could then be used for further functionalization.

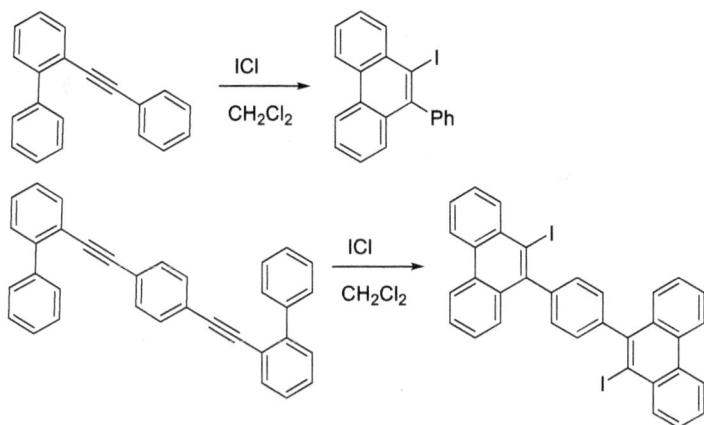

Figure 9.60: Electrophilic cyclization of alkynes using ICl [66].

Liu extended this approach to prepare dibenzo[g,p]chrysene derivatives by a two-step electrophilic cyclization with ICl, followed by a palladium-catalyzed intramolecular arylation (Figure 9.61) [67].

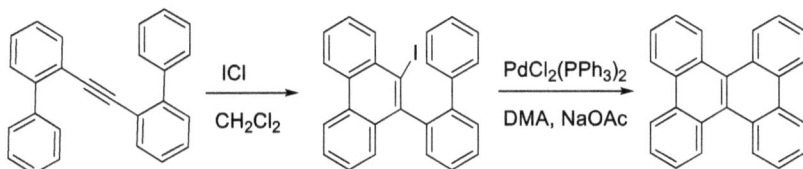

Figure 9.61: Synthesis of dibenzo[g,p]chrysenes via alkyne benzannulation/intramolecular arylation [67].

We saw above Brønsted acids such as TFA could be used for electrophilic alkyne benzannulations. Chalifoux and coworkers extended this approach to the preparation of pyrenes and peropyrenes [68]. For example, starting from the substituted 2,6-dialkynylbiphenyl, and initial cyclization was carried using TFA out to prepare the corresponding phenanthrene. The second cyclization was achieved using triflic acid (Figure 9.62). This approach has also been used for the synthesis of graphene nanoribbons, which we will see in Chapter 10.

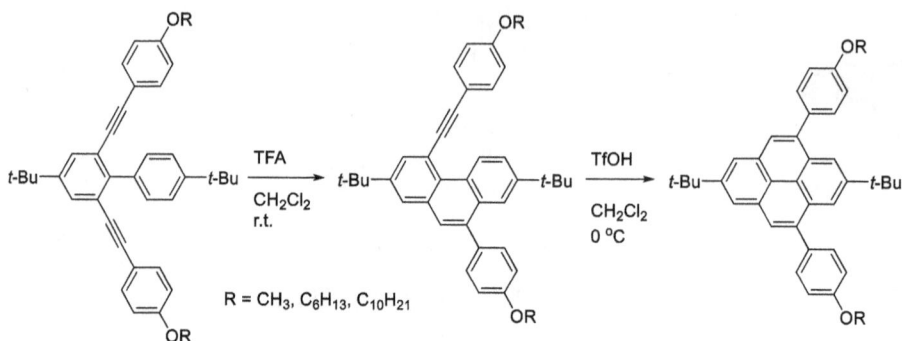

Figure 9.62: Synthesis of substituted pyrenes via twofold alkyne benzannulation [68].

9.3.12 Asao–Yamamoto benzannualation

In 2002, Asao, Yamamoto and coworkers reported the AuCl$_3$-catalyzed benzannulation of *ortho* arylethynyl benzaldehydes with alkynes in a formal [4+2] cycloaddition to form a substituted naphthalene (Figure 9.63) [69].

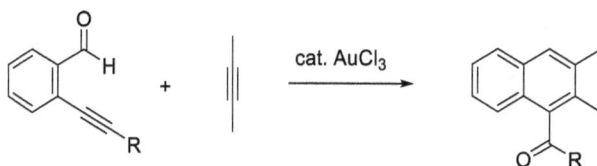

Figure 9.63: Gold-catalyzed Asao–Yamamoto benzannulation to prepare substituted naphthalenes [69].

Mechanistically, it is thought that the Lewis acid coordinates to the aryl alkyne, which allows an intramolecular cyclization of the carbonyl onto the alkyne (Figure 9.64). The resulting complex undergoes a Diels–Alder reaction with the alkyne, which is followed by cleavage of the C–Au bond and a rearrangement to form the naphthyl ketone.

Subsequent studies showed that the reaction could be carried out using Cu (OTf)$_2$ in the presence of a protic acid such as CF$_3$CO$_2$H [70]. In the presence of a Brønsted acid, the reaction outcome changes: the cycloadduct undergoes protonolysis of the C–Cu bond, which is followed by a retro-Diels–Alder reaction to give a substituted naphthalene and a carboxylic acid as a byproduct (Figure 9.65) [70]. In the case where the reactant is *o*-(phenylethynyl)-benzaldehyde, the byproduct is benzoic acid.

The scope of this reaction has been expanded to the preparation of a variety of fused aromatic systems by Dichtel and coworkers. For example, they demonstrated a benzannulation on bis(phenylethynyl)terphenyl with *o*-(phenylethynyl)benzaldehyde

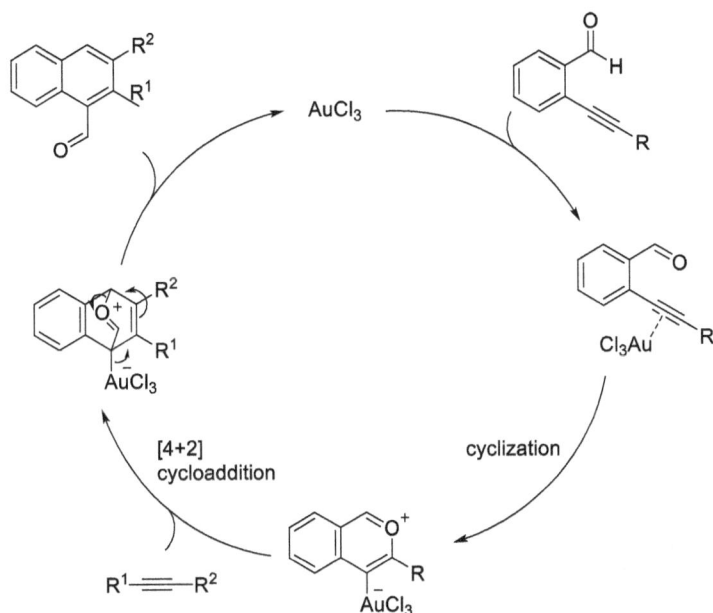

Figure 9.64: Proposed mechanism of the Aso–Yamamoto benzannulation with AuCl₃ [69].

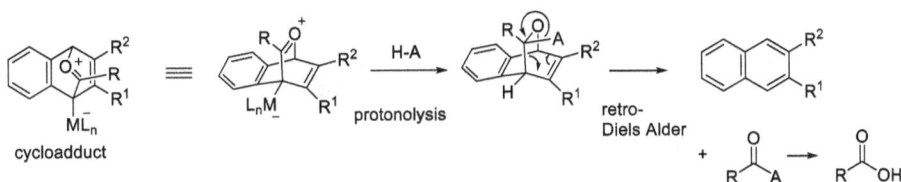

Figure 9.65: Proposed mechanism of benzannulation in the presence of Cu(OTf)₂ and acid.

to produce the corresponding doubly benzannulated product (Figure 9.66) [71]. The same method was used for the benzannulation of poly(phenylene ethynylene)s to produce poly(*ortho*-arylene)s.

When functionalized benzaldehyde derivatives are used with unsymmetrically substituted alkynes, two regioisomeric products can be produced (Figure 9.67). Dichtel and coworkers explored the factors controlling the regiochemistry of the benzannulation reaction. They showed that the regiochemistry of the reaction is determined by the step where cycloaddition of the alkyne occurs. The regioselectivity of the step is largely governed by electronic effects and not sterics, where an aryl group bearing electron-donating substituent is able to stabilize the partial positive charge on the alkyne carbon in the transition state [72].

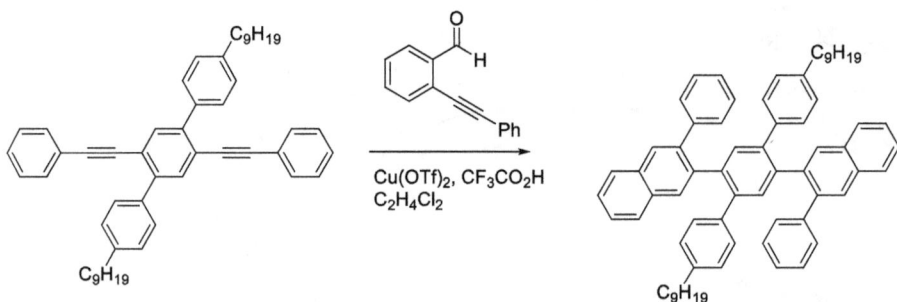

Figure 9.66: Synthesis of oligo(*ortho*-arylenes) via Asao–Yamamoto benzannulation.

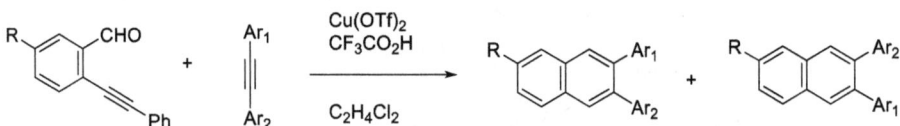

Figure 9.67: Different regioisomers formed during the Asao–Yamamoto benzannulation.

References

[1]   Harvey RG. Polycyclic Aromatic Hydrocarbons. New York: Wiley-VCH; 1997.

[2]   Clar E. Polycyclic Hydrocarbons. London: Academic Press; 1964.

[3]   Armit JW, Robinson R. Polynuclear heterocyclic aromatic types. Part II some anhydronium bases. J Chem Soc Trans. 1925;1604–18.

[4]   Clar E. The Aromatic Sextet. New York: Wiley; 1972.

[5]   Solà M. Forty years of Clar's aromatic π-sextet rule. Front Chem. 2013;1:22.

[6]   Walsh JC, Williams KLM, Lungerich D, Bodwell GJ. Synthesis of pyrene-4,5-dione on a 15 g scale. Eur J Org Chem. 2016;5933–36.

[7]   Fu PP, Lee HM, Harvey RG. Regioselective catalytic hydrogenation of polycyclic aromatic hydrocarbons under mild conditions. J Org Chem. 1980;45(14):2797–803.

[8]   Bukowska M, Harvey RG. Synthesis of 2-hydroxybenzo[a]pyrene, a tumorigenic phenol derivative of Benzo[a]pyrene. Polycycl Arom C. 1992;2(4):223–28.

[9]   Lee H, Harvey RG. Synthesis of 2, 7-dibromopyrene. J Org Chem. 1986;51:2847–48.

[10]   Clar E, Zander M. Syntheses of coronene and 1 : 2-7 : 8-Dibenzocoronene. J Chem Soc. 1957;4616–19.

[11]   Biermann D, Schmidt W. Diels-alder reactivity of polycyclic aromatic hydrocarbons. 1.Acenes and benzologs. J Am Chem Soc. 1980;102(9):3163–73.

[12]   Bouas-Laurent H, Castellan A, Desvergne JP, Lapouyade R. Photodimerization of anthracenes in fluid solution: Structural aspects. Chem Soc Rev. 2000;29(1):43–55.

[13]   Sugiyama N, Iwata M, Yoshioka M, Yamada K, Aoyama H. Photoxidation of anthracene. Chem Commun. 1968;1563.

[14]   Haworth RD. Syntheses of alkylphenanthrenes. Part 1. J Chem Soc. 1932;1125–33.

[15]   Feiser LF, Johnson WS. Syntheses of 1,2-Benzanthracene and chrysene. J Am Chem Soc. 1939;61:1647–54.

[16] Bachmann WE, Struve WS. The synthesis of derivatives of chrysene. J Org Chem. 1939;4 (4):456–63.

[17] Clar E. Die synthese des 1.2.3.4 Dibenz-tetracens. 44. Aromatische kohlenwasserstoffe. Chem Ber. 1948;81:68–71.

[18] Waldmann H, Mathiowetz H. On a new structure in the linear benzanthraquinone sequence. Ber Dtsch Chem Ges. 1931;64:1713–24.

[19] Weizmann C, Bergmann E, Bergmann F. Grignard reactiom with phthalic anhydrides. J Chem Soc. 1935;1367–70.

[20] Harvey RG, Leyba C, Konieczny M, Fu PP, Sukumaran KB, Novel A. Convenient synthesis of dibenz[a,c]anthracene. J Org Chem. 1978;43:3423–25.

[21] Leake PH. The pschorr synthesis. Chem Rev. 1956;56(1):27–48.

[22] Pschorr R. Neue synthese des phenanthrens und seiner derivate. Ber Dtsch Chem Ges. 1896;29(1):496–501.

[23] Wassmundt FW, Kiesman WF. Soluble catalysts for improved pschorr cyclizations. J Org Chem. 1995;60(1):196–201.

[24] Elbs K, Larsen E. Ueber Paraxylylphenylketon. Ber Dtsch Chem Ges. 1884;17(2):2847–2839.

[25] Cook JW. Polycyclic aromatic hydrocarbons. Part IV. Condensed derivatives of 1:2-benzanthracene. J Chem Soc. 1931;499–507.

[26] Davies W, Porter QN. The synthesis of polycyclic aromatic compounds. Part I. The reaction of quinones with vinylnaphthalenes and related dienes. J Chem Soc. 1957;4967–70.

[27] Corbett TG, Porter QN. Reactions of benzyne and 1,2-naphthyne with some dienes. Aust J Chem. 1965;18(11):1781–85.

[28] LeHoullier CS, Gribble GW. Twin annulation of naphthalene via a 1,5-Naphthodiyene synthon. New syntheses of chrysene and benzo[b,k]chrysene. J Org Chem. 1983;48(10):1682–85.

[29] Smith JG, Dibble PW. 2-(Dimethoxymethyl)benzyl Alcohol: A convenient isobenzofuran precursor. J Org Chem. 1983;48:5361–62.

[30] Naito K, Rickborn B. Experimental sectionIsobenzofuran: New approaches from 1,3-Dihydro-1-methoxyisobenzofuran. J Org Chem. 1980;45:4061–62.

[31] Crump SL, Netka J, Rickborn B. Preparation of isobenzofuran-aryne cycloadducts. J Org Chem. 1985;50:2746–50.

[32] Cava MP, Napier DR. Condensed cyclobutane aromatic systems. II. Dihalo derivatives of benzocyclobutene and benzocyclobutadiene dimer. J Am Chem Soc. 1957;79:1701–05.

[33] Cava MP, Shirley RL. Condensed cyclobutane aromatic compounds. X. Naphtho[b]cyclobutene. J Am Chem Soc. 1960;82(3):654–56.

[34] Psutka KM, Bozek KJA, Maly KE. Synthesis and mesomorphic properties of novel dibenz[a,c]anthracenedicarboximides. Org Lett. 2014;16(20):5442–45.

[35] Dini D, Calvete MJF, Hanack M, Pong RGS, Flom SR, Shirk JS. Nonlinear transmission of a tetrabrominated naphthalocyaninato indium chloride. J Phys Chem B. 2006;110 (25):12230–39.

[36] Swartz CR, Parkin SR, Bullock JE, Anthony JE, Mayer AC, Malliaras GG. Synthesis and characterization of electron-deficient pentacenes. Org Lett. 2005;7(15):3163–66.

[37] Minsky A, Rabinovitz M, Facile A. Synthesis of some polycyclic hydrocarbons – application of phase-transfer catalysis in the bis-wittig reaction. Synthesis (Stuttg). 1983;497–98.

[38] Yang CX, Yang DTC, Harvey RG. Application of ultrasound tothe synthesis of polycyclic arenes via the bis-wittig reaction. Synlett. 1992;10:799–800.

[39] Xu D, Penning TM, Blair IA, Harvey RG. Synthesis of phenol and quinone metabolites of benzo[a]pyrene, a carcinogenic component of tobacco smoke implicated in lung cancer. J Org Chem. 2009;74(2):597–604.

[40] Blum J, Zimmerman M. Photocyclization of substituted 1,4-distyrylbenzenes to dibenz [a,h] anthracenes. Tetrahedron. 1972;28(2):275–80.

[41] Karasch N, Alston TG, Lewis HB, Wolf W. The photochemical conversion of o-terphenyl into triphenylene. J Chem Soc Chem Commun. 1965;242–43.

[42] Liu L, Yang B, Katz TJ, Poindexter MK. Improved methodology for photocyclization reactions. J Org Chem. 1991;56(12):3769–75.

[43] Grzybowski M, Skonieczny K, Butenschön H, Gryko DT. Comparison of oxidative aromatic coupling and the scholl reaction. Angew Chem Int Ed. 2013;52(38):9900–30.

[44] Scholl R, Mansfeld J. Meso-Benzdianthron(helianthron), meso-Naphthodianthron, and a new way of receiving flavanthren. Ber Dtsch Chem Ges. 1910;43:1734–46.

[45] Scholl R, Seer C, R W. Perylen, a high condensated aromatic hydrocarbon C20H12. Ber Dtsch Chem Ges. 1910;43:2202–09.

[46] Scholl R, Seer C. Elimination of aromatically combined hydrogen and condensation of aromatic nuclei by means of aluminium chloride. Justus Liebigs Ann Chem. 1913;394:111–77.

[47] Scholl R, Seer C. Synthesis of violanthrone. Justus Liebigs Ann Chem. 1913;398:82–96.

[48] Brown SP, Schnell I, Brand JD, Müllen K, Spiess HW. An investigation of π-π packing in a columnar hexabenzocoronene by fast magic-angle spinning and double-quantum 1H solid-state NMR spectroscopy. J Am Chem Soc. 1999;121(28):6712–18.

[49] Borner RC, Bushby RJ, Cammidge AN. Ferric chloride/methanol in the preparation of triphenylene-based discotic liquid crystals. Liq Cryst. 2006;33(11–12):1439–48.

[50] Mohr B, Enkelmann V, Wegner G. Synthesis of alkyl- and alkoxy-substituted benzils and oxidative coupling to tetraalkoxyphenanthrene-9,10-diones. J Org Chem. 1994;59(3):635–38.

[51] Simpson CD, Brand JD, Berresheim AJ, Przybilla L, Raeder HJ, Mullen K. Synthesis of a giant 222 carbon graphite sheet. Chem -Eur J. 2002;8(6):1424–29.

[52] Ormsby JL, Black TD, Hilton CL, Bharat, King BT. Rearrangements in the scholl oxidation: Implications for molecular architectures. Tetrahedron. 2008;64(50):11370–78.

[53] Dou X, Yang X, Bodwell GJ, Wagner M, Enkelmann V, Mullen K. Unexpected phenyl group rearrangement during an intramolecular scholl reaction leading to an alkoxy-substituted hexa-peri-hexabenzocoronene. Org Lett. 2007;9(13):2485–88.

[54] Sato T, Shimada S, Hata K. A new route to polycondensed aromatics: Photolytic formation of dibenzofg,opnaphthacene. J Chem Soc Chem Commun. 1970;(12):766–67.

[55] Sato T, Shimada S, Hata K, New A. Route to polycondensed aromatics: Photolytic formation of triphenylene and dibenzo[fg, op]naphthacene ring systems. Bull Chem Soc Jpn. 1971;44:2484–90.

[56] He J, Mathew S, Kinney ZJ, Warrell RM, Molina JS, Hartley CS. Tetrabenzanthanthrenes by mitigation of rearrangements in the planarization of ortho-phenylene hexamers. Chem Commun. 2015;51(33):7245–48.

[57] Daigle M, Picard-Lafond A, Soligo E, Morin JF. Regioselective synthesis of nanographenes by photochemical cyclodehydrochlorination. Angew Chem Int Ed. 2016;55(6):2042–47.

[58] Peña D, Escudero S, Pérez D, Guitián E, Castedo L. Efficient palladium-catalyzed cyclotrimerization of arynes: Synthesis of triphenylenes. Angew Chem Int Ed. 1998;37(19):2659–61.

[59] Iuliano A, Piccioli P, Fabbri D. Ring-closing olefin metathesis of 2,2′-divinylbiphenyls: A novel and general approach to phenanthrenes. Org Lett. 2004;6(21):3711–14.

[60] Bonifacio MC, Robertson CR, Jung JY, King BT. Polycyclic aromatic hydrocarbons by ring-closing metathesis. J Org Chem. 2005;70(21):8522–26.

[61] Lee J, Li H, Kalin AJ, Yuan T, Wang C, Olson T, et al. Extended ladder-type Benzo[k]tetraphene-derived oligomers. Angew Chem Int Ed. 2017;56(44):13727–31.

[62] Senese AD, Chalifoux WA. Nanographene and graphene nanoribbon synthesis via alkyne benzannulations. Molecules. 2019;24(1):118.

[63] Goldfinger MB, Swager TM. Fused polycyclic aromatics via electrophile-induced cyclization reactions: Application to the synthesis of graphite ribbons. J Am Chem Soc. 1994;116:7895–96.

[64] Goldfinger MB, Crawford KB, Swager TM. Directed electrophilic cyclizations: Efficient methodology for the synthesis of fused polycyclic aromatics. J Am Chem Soc. 1997;119 (20):4578–93.

[65] Yao T, Campo MA, Larock RC. Synthesis of polycyclic aromatic iodides via ICl-induced intramolecular cyclization. Org Lett. 2004;6(16):2677–80.

[66] Yao T, Campo MA, Larock RC. Synthesis of polycyclic aromatics and heteroaromatics via electrophilic cyclization. J Org Chem. 2005;70(9):3511–17.

[67] Li CW, Wang CI, Liao HY, Chaudhuri R, Liu RS. Synthesis of dibenzo[g,p]chrysenes from bis (biaryl)acetylenes via sequential ICl-induced cyclization and mizoroki-heck coupling. J Org Chem. 2007;72(24):9203–07.

[68] Yang W, Monteiro JHSK, de Bettencourt-dias A, Catalano VJ, Chalifoux WA. Pyrenes, peropyrenes, and teropyrenes: Synthesis, structures, and photophysical properties. Angew Chem Int Ed. 2016;55(35):10427–30.

[69] Asao N, Takahashi K, Lee S, Kasahara T, Yamamoto Y. AuCl3-catalyzed benzannulation: Synthesis of naphthyl ketone derivatives from o-alkynylbenzaldehydes with alkynes. J Am Chem Soc. 2002;124(43):12650–51.

[70] Asao N, Nogami T, Lee S, Yamamoto Y. Lewis acid-catalyzed benzannulation via unprecedented [4+2] cycloaddition of o-alkynyl(oxo)benzenes and enynals with alkynes. J AmChem Soc. 2003;125(36):10921–25.

[71] Arslan H, Saathoff JD, Bunck DN, Clancy P, Dichtel WR. Highly efficient benzannulation of poly (phenylene ethynylene)s. Angew Chem Int Ed. 2012;51(48):12051–54.

[72] Arslan H, Walker KL, Dichtel WR. Regioselective Asao-Yamamoto benzannulations of diaryl acetylenes. Org Lett. 2014;16(22):5926–29.

10 Synthesis of select classes of polycyclic aromatic hydrocarbons

10.1 Introduction

The previous chapter outlined some of the key reactions used to form polycyclic aromatic hydrocarbons. These reactions and others have been used to access a wide variety of polycyclic aromatic structures. In this chapter, we will consider a selection of classes of polycyclic aromatic hydrocarbons and outline different approaches to synthesize them. Given the variety of polycyclic aromatic structures, this will not be a comprehensive set of compounds. Instead, selected types of PAHs will be considered which illustrate some of the fundamental reactions discussed in previous chapters, as well as compounds that have been the focus of considerable research interest because of their interesting properties such as self-assembly, photophysical, or electronic properties.

10.2 Synthesis of linear fused aromatic structures – the acenes

Acenes consist of linear fused benzenoid rings (Figure 10.1). Regardless of the number of fused aromatic rings, they possess only one aromatic sextet, which results in lower aromatic stabilization as the acene length is increased. As a consequence, longer acenes are more reactive and prone to photodimerization and photooxidation reactions. Pentacenes and longer acenes also exhibit low HOMO–LUMO gaps, making them attractive candidates for applications in organic electronics as semiconducting materials [1, 2]. Here, we will focus on synthetic approaches to pentacenes and longer acenes because they have been the focus of considerable research efforts, in part because of their potential electronic properties, but also because of the challenge of preparing long acenes with limited stability.

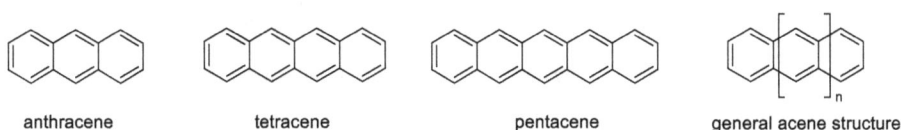

anthracene   tetracene   pentacene   general acene structure

Figure 10.1: Structures of acenes.

https://doi.org/10.1515/9783110562682-010

10.2.1 Synthetic approaches to pentacenes

The first synthesis of pentacene was reported by Clar and John in 1929 (Figure 10.2) [3]. They used the Elbs reaction (see Section 9.3.3) to prepare the pentacyclic framework. This reaction took place at high temperature (400 °C) in the presence of copper. The resulting 6,13-dihydropentacene could be dehydrogenated either by sublimation over copper at 380 °C or by refluxing in xylenes in the presence of chloranil .

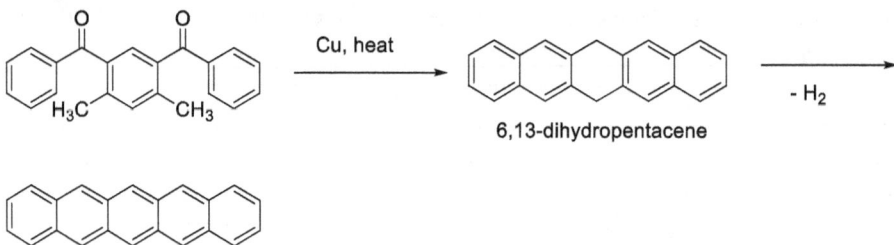

6,13-dihydropentacene

Figure 10.2: The first synthesis of pentacene by Clar and John [3].

Subsequently, Bailey and Madoff reported an alternative synthesis using a Diels–Alder reaction to prepare the hydrocarbon framework (Figure 10.3) [4]. Their synthesis involved the reaction of 1,2-dimethylenecyclohexane with benzoquinone in a twofold Diels–Alder reaction. The oxygen atoms were removed by thioacetal formation and desulfurization using Raney nickel. Dehydrogenation to give pentacene was achieved using palladium on carbon at ca. 250 °C.

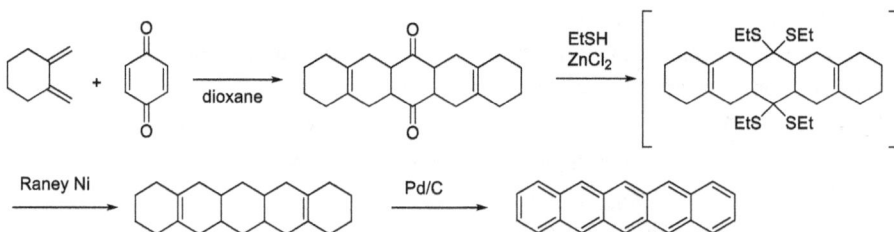

Figure 10.3: Synthesis of pentacene by Bailey and Madoff via a Diels–Alder reaction [4].

One of the most widely used approaches to pentacene derivatives involves base-mediated aldol condensation of phthalaldehyde with 1,4-cyclohexanedione to produce 6,13-pentacenequinone (Figure 10.4) [5]. The pentacenequinone can be reduced using methods such as aluminum in cyclohexanol [6]. Alternatively, the quinone units can be used as a functional handle for the introduction of other substituents. For example, treating a protected acetylene such as triisopropylsilyl acetylene with a strong base such as butyllithium or ethylmagnesium bromide generates the acetylide anion, which can undergo nucleophilic addition at the carbonyls to generate a diol. The crude diol can then be reduced using $SnCl_2$ to give the corresponding diethynyl pentacene (Figure 10.4) [7]. These ethynyl-substituted pentacenes are important because they are more resistant to photodimerization and photooxidation than the parent pentacene. Furthermore, the bulky ethynyl substituents impact packing in the solid state, which has important implications for their semiconducting properties [7].

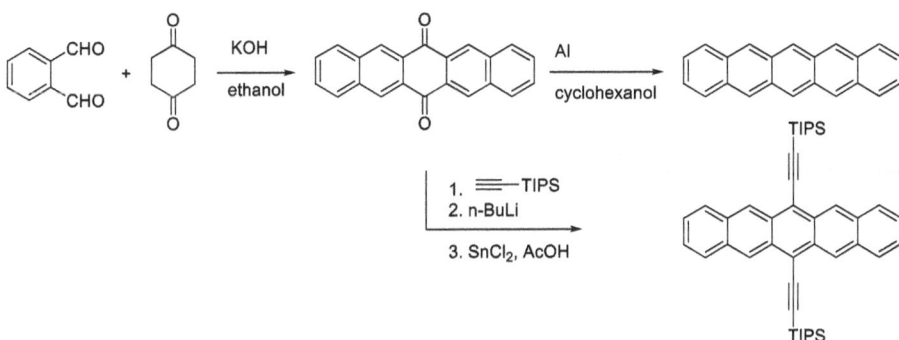

Figure 10.4: Synthesis of pentacene derivatives via condensation.

The Diels–Alder approach to pentacenes by Bailey and Madoff shown in Figure 10.3 requires extensive dehydrogenation to aromatize the rings. An alternative is to use an *o*-quinodimethane (*o*-xylylene) as the diene, which can be generated in situ, with benzoquinone as a dienophile to prepare pentacenequinone. For example, 1,4-dihydro-2,3-benzoxathiin-3-oxide eliminates SO_2 upon heating to generate the *o*-quinodimethane, which undergoes a twofold Diels–Alder reaction with benzoquinone to form the octahydropentacenequinone (Figure 10.5). This compound is then oxidized using bromine and pyridine in DMF to form pentacenequinone, which can be reduced to pentacene [8].

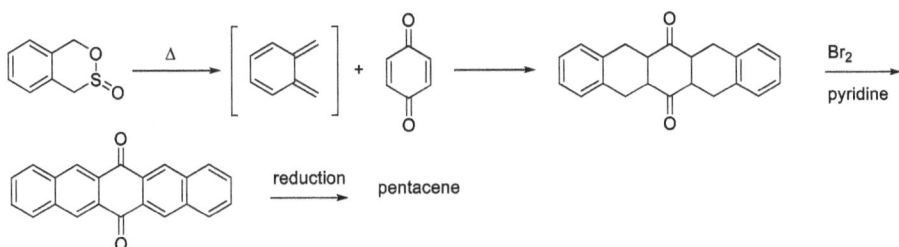

Figure 10.5: Synthesis of pentacene via a Diels–Alder reaction of *o*-quinodimethane [8].

In a related approach, Luo and Hart used the *o*-quinodimethane generated from benzocyclobutene to prepare pentacene (Figure 10.6) [9]. Specifically, benzocyclobutene was heated in the presence of the naphthalyne-furan adduct where it underwent a Diels–Alder reaction to give the pentacyclic framework. Acid-catalyzed dehydration gave the dihydropentacene, which was dehydrogenated using Pd/C.

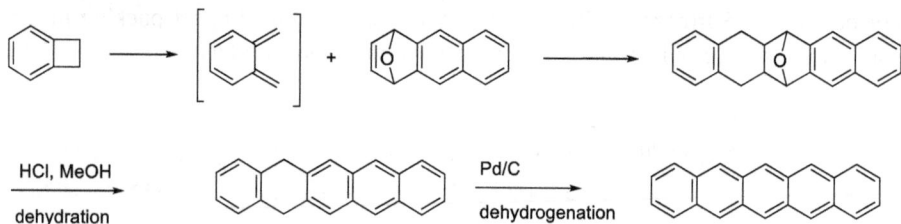

Figure 10.6: An alternative pentacene synthesis from benzocyclobutene [9].

Another variation on the Diels–Alder reaction using an o-quinodimethane is the Cava reaction, where the o-quinodimethane is generated from α,α,α′,α′-tetrabromo o-xylene via 1,4-elimination of Br_2 using sodium or potassium iodide (see Section 9.3.4) [10, 11]. Anthony and coworkers used this approach to access pentacenequinones and ethynyl pentacenes bearing electron-withdrawing groups (e.g., see Figure 10.7) [12].

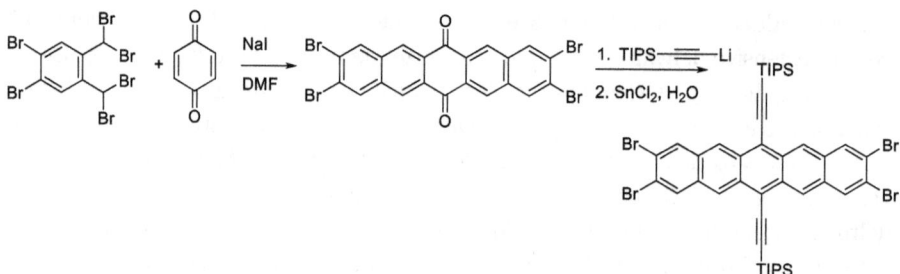

Figure 10.7: Synthesis of pentacenes using the Cava reaction [12].

Pentacene has also been prepared by a Diels–Alder reaction with a TMS-substituted isobenzofuran and a benzyne (Figure 10.8) [13]. Reaction of 1,4-dibromobenzene with lithium tetramethylpiperide (LTMP) generated a bromo-substituted benzyne, which underwent a cycloaddition reaction with the isobenzofuran to give the endoxide. This sequence was repeated with a second equivalent of the isobenzofuran and LTMP to give the pentacyclic framework. Treatment with trifluoroacetic acid leads to desilylation and formation of the diketone, which could be converted to pentacene upon reduction with $LiAlH_4$ and dehydration.

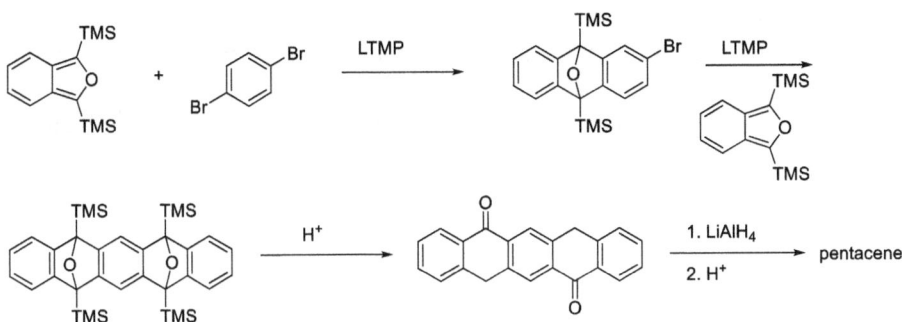

Figure 10.8: Synthesis of pentacene via Diels–Alder reactions of benzynes with isobenzofurans [13].

10.2.2 Synthesis of larger acenes

Some of the synthetic approaches used to prepare pentacene derivatives can also be used for the preparation of larger acenes. A challenge of preparing larger acenes is that they are more prone to photooxidation and photodimerization reactions. One of the early syntheses of hexacene makes use of a condensation approach (Figure 10.9) [14]. Condensation of 2,3-naphthalenedicarboxaldehyde and quinizarin gave the dihydroxy-hexacenequinone. This compound was reduced using hydrogen and zinc to give dihydrohexacene, which was dehydrogenated to yield hexacene.

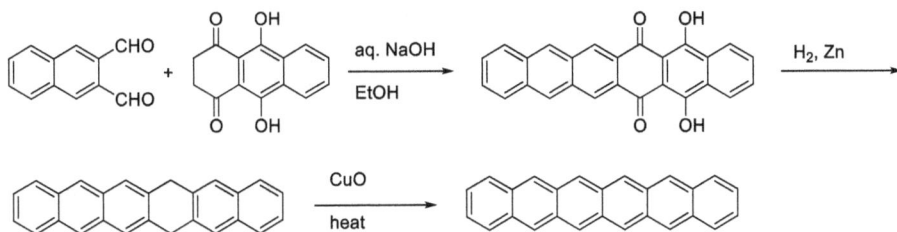

Figure 10.9: Synthesis of hexacene via condensation [14].

Strategies for the synthesis of hexacenes and higher acenes generally involve preparing acenes bearing stabilizing groups that prevent photodimerization and photooxidation, or by constructing the carbon skeleton as a masked and stable acene precursor, where the final acene is generated under inert conditions in the solid state or in a stabilizing matrix. An example of the former approach was reported by Anthony and coworkers, who showed that ethynyl-substituted hexacenes could be prepared from the corresponding hexacene- and heptacene-quinones (Figure 10.10) [15]. While the ethynyl triisopropylsilyl group was effective at stabilizing pentacene, the corresponding hexacenes and heptacenes were not stable in solution or in the solid state. However, using bulkier *t*-butyl groups on the silane lead to improvements in stability, such that the hexacene was stable as a solid, or in solution in the dark in air-free conditions. The heptacene was stable enough for spectroscopic characterization, but decomposed in solution, even under oxygen-free conditions.

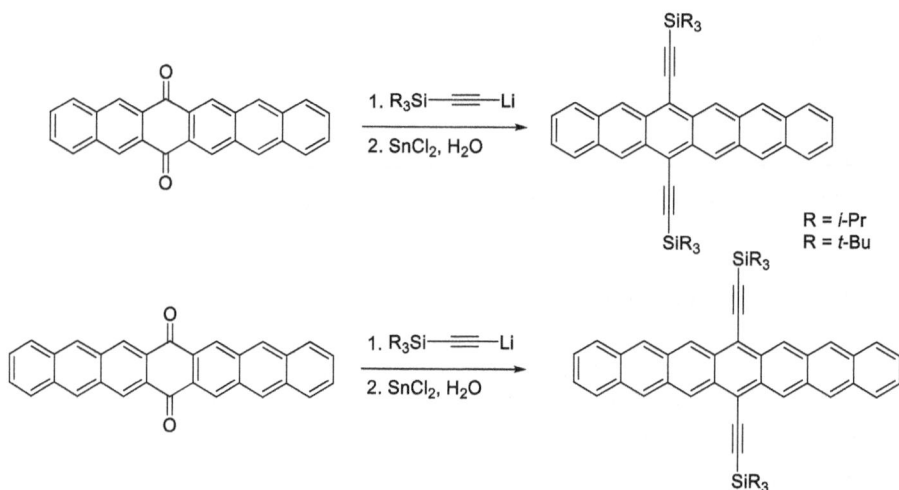

Figure 10.10: Synthesis of ethynyl-substituted hexacenes and pentacenes [15].

Neckers and coworkers showed that unsubstituted acenes such as hexacene and heptacene could be prepared by a photodecarbonylation as the last step in the synthesis (Figure 10.11) [16]. The synthesis began by reaction of 2,3-dibromonaphthalene with butyllithium to form the naphthalyne which underwent a Diels–Alder reaction with bicyclo[2,2,2]oct-2,3,5,6,7-pentaene. The product was dehydrogenated using chloranil in toluene, and the bridging alkene was dihydroxylated using OsO₄ to give the corresponding diol. The diol was then converted to 7,16-dihydro-7,16-ethanoheptacene-19,20-dione using a modified Swern oxidation with DMSO and trifluoroacetic acid. The dione was then used in a photodecarbonylation (a Strating–Zwanenburg reaction) to give the corresponding heptacene. This reaction was carried out both under inert conditions in solution and in an solid poly(methyl methacrylate) (PMMA) matrix. While the product could be characterized by UV-visible spectroscopy, mass spectrometry, and ¹H NMR, heptacene degraded even in the PMMA matrix. A similar approach was also reported for the preparation of hexacene [17], as well as larger acenes such as octacene, nonacene [18], and even undecacene [19].

The synthesis of higher acenes in a solid polymer matrix has the disadvantage of only being amenable to small-scale synthesis. More recently, Jancarik et al. reported a related approach for the bulk synthesis of higher acenes such as heptacene via a thermal decarbonylation reaction (Figure 10.12) [20]. The synthesis uses 7,7-dimethoxy-2,3,5,6-tetramethylenebicyclo[2.2.1]heptane as a diene in a Diels–Alder

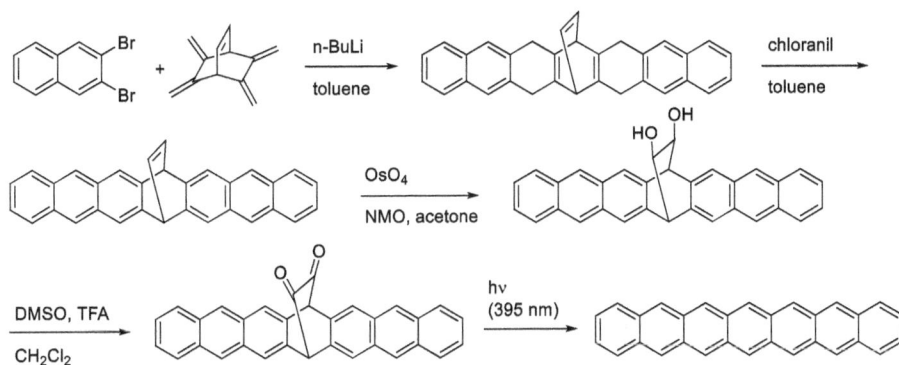

Figure 10.11: Synthesis of heptacene via photodecarbonylation [16].

reaction with naphthalyne, which was generated from the corresponding trimethyl-silyl naphthyl triflate. Then, the dimethyl acetal was cleaved to give the bicyclic ke-tone, which underwent thermal cheletropic decarbonylation at elevated temperatures to give heptacene. Interestingly, heptacene prepared under these conditions was ther-mally stable provide it is kept under inert conditions. When stored as a solid, it also did not undergo dimerization, perhaps because the solid state packing does not favor dimerization [20].

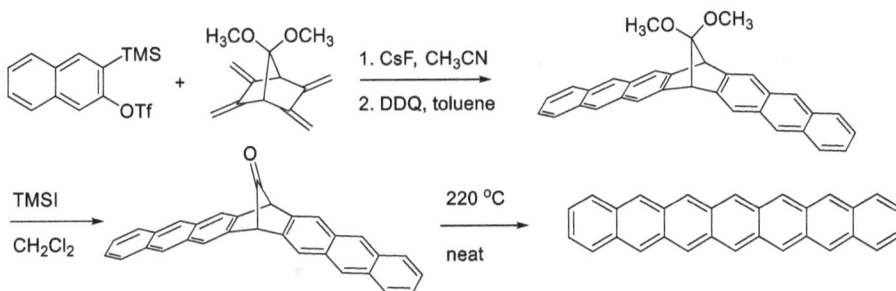

Figure 10.12: Synthesis of heptacene via thermal decarbonylation [20].

10.3 Synthesis of phenacenes

Phenacenes consist of angular fused benzene rings, as exemplified by phenan-threne, chrysene, and picene (Figure 10.13). Unlike the acenes, every other ring in in a phenacene possesses a Clar aromatic sextet, meaning they are much more sta-ble. Phenacenes are of interest as organic semiconducting materials such as field-effect transistors [21].

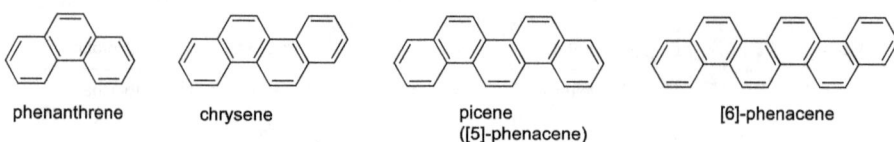

Figure 10.13: Some representative structures of phenacenes.

To illustrate some of the early approaches to the synthesis of phenacenes, we will consider the synthesis of picene using the Pschorr reaction (Section 9.3.2) as a key step (Figure 10.14) [22]. This synthesis involved a condensation of benzene-1,2-diacetonitrile with o-nitrobenzaldehyde to produce the stilbene derivative. The nitriles were then hydrolyzed to the corresponding carboxylic acid and the nitro groups were reduced to produce the anilines. At this stage, the Pschorr reaction was carried out: diazotization of the amines in the presence of copper gave the fused aromatic system. Finally, the carboxylic acid groups were removed via decarboxylation with soda lime to give picene.

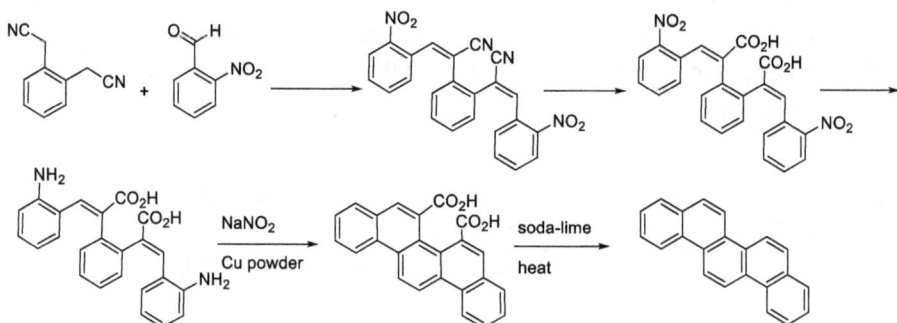

Figure 10.14: Synthesis of picene using the Pschorr reaction [22].

Another early approach for the synthesis of phenacenes is the Haworth synthesis, which we saw for the preparation of phenanthrene in Section 9.3.1. This general approach was used to prepare picene, as outlined in Figure 10.15 [23]. This synthesis starts with the Friedel–Crafts acylation of dihydrophenanthrene with succinic anhydride. This was followed by Wolff–Kishner reduction of the ketone and esterification of the carboxylic acid. The same sequence of Friedel–Crafts acylation, reduction, and esterification was repeated to add the four-carbon chain to the other aromatic ring. The dihydrophenanthrene unit was then converted to the corresponding phenanthrene using palladium on carbon. Finally, the esters were hydrolyzed to yield the dicarboxylic acid. Treatment with polyphosphoric acid led to the twofold electrophilic cyclization to give the desired regioisomer as the major product, along with a significant portion of one of the other regioisomeric products. Fortunately, these two

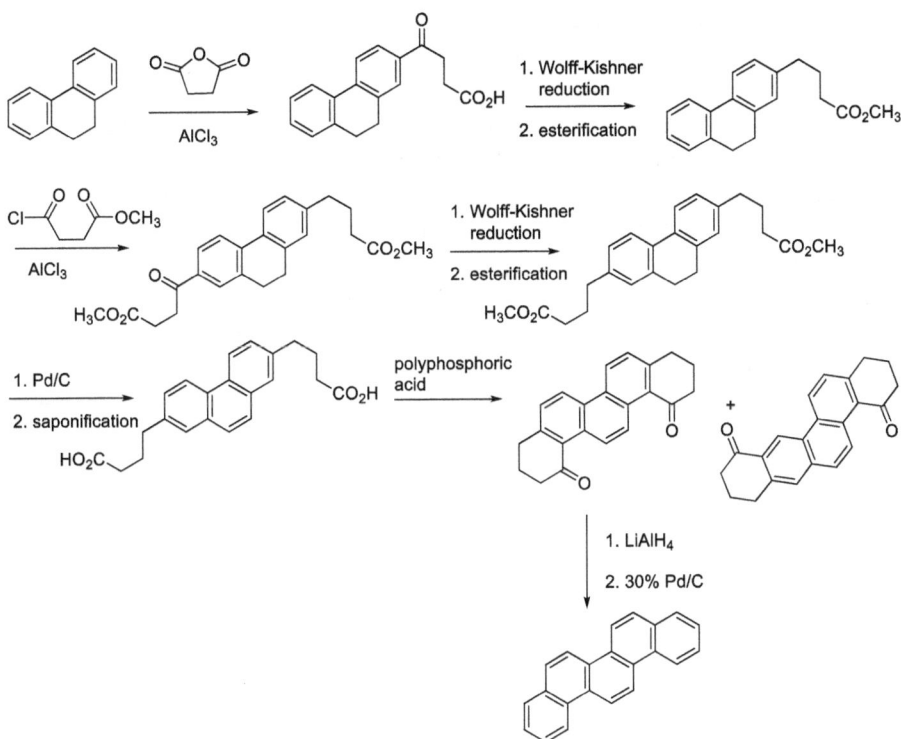

Figure 10.15: Preparation of picene via a Haworth synthesis [23].

products could be separated chromatographically. The desired regioisomer was then reduced using LiAlH$_4$ and aromatized using palladium on carbon to give picene.

Harvey and coworkers reported an approach to several PAHs by reacting enamine salts with aryl-substituted alkyl halides, followed by cyclization and oxidation. They applied this approach to the synthesis of picene, as outlined in Figure 10.16 [24]. The cyclohexyl enamine salt reacted with the alkyl iodides in a nucleophilic substitution to give the bis cyclohexanone derivative. In the presence of a strong acid such as methane sulfonic acid, electrophilic cyclization at the ketones, followed by dehydration occurred. Oxidation using DDQ led to dehydrogenation to yield the fully aromatic ring system.

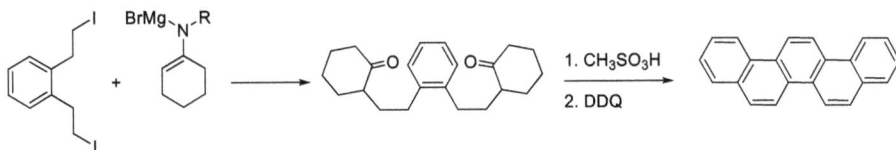

Figure 10.16: Synthesis of picene via enamine chemistry [24].

Harvey and coworkers also extended this approach to the synthesis of [6]phenacene (Figure 10.17) [24]. The synthesis started with electrophilic bromination of 3,7-di(*t*-butyl)naphthalene to give the dibromonaphthalene. Electrophilic removal of the *t*-butyl groups using AlCl₃ gave 1,5-dibromonaphthalene. This sequence highlights the use of bulky *t*-butyl groups to sterically control the regiochemistry of bromination, and also exploits the reversibility of electrophilic aromatic substitution using *t*-butyl groups. The dibromonaphthalene underwent twofold lithium–halogen exchange using butyllithium, and the resulting dianion was trapped with ethylene oxide to give the diol. The diol was converted to the dibromide using PBr₃ and then reacted with the enamine salt as described above. Acid-catalyzed cyclization and dehydration was followed by dehydrogenation using palladium on carbon gave [6]phenacene.

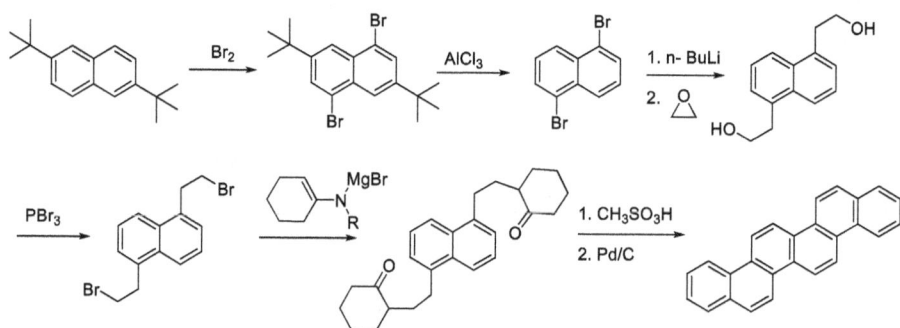

Figure 10.17: Synthesis of [6]phenacene [24].

The most common approach to the synthesis of phenacenes involves the formation of stilbene units – often by a Wittig reaction, followed by photocyclization. This approach was pioneered by Mallory and coworkers, with an example of the synthesis of [7]phenacene shown in Figure 10.18 [25].

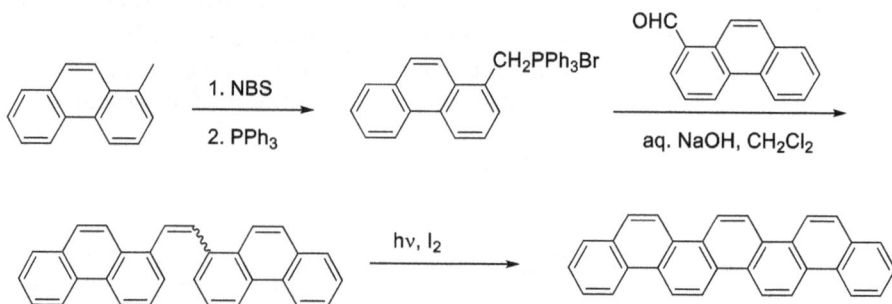

Figure 10.18: Mallory's synthesis of [7]phenacene via photocyclization [25].

10.4 Perylenes and rylenes

Fusing naphthalene units in a peri-condensed manner gives a class of compounds known as rylenes. The simplest compound in this class of is perylene, where two naphthalenes are peri-condensed. Extending the number of naphthalene units gives terrylene, quaterrylene, pentarylene, and so on. (Figure 10.19). Rylenes and their derivatives are of interest as dyes and as electron-accepting materials.

perylene   terrylene   quaterrylene

Figure 10.19: Representative structures of rylenes.

The first synthesis of perylene was reported by Scholl, who showed that 1,1′-binaphthyl could be converted to perylene using the classical Scholl reaction conditions of AlCl₃ at 140 °C (Figure 10.20) [26]. Subsequently, Clar used an intermolecular Scholl reaction using AlCl₃/NaCl melt with perylene and 1-bromonaphthalene to produce the corresponding brominated terrylene as a mixture of isomers, which was reductively debrominated to give terrylene (Figure 10.20) [27].

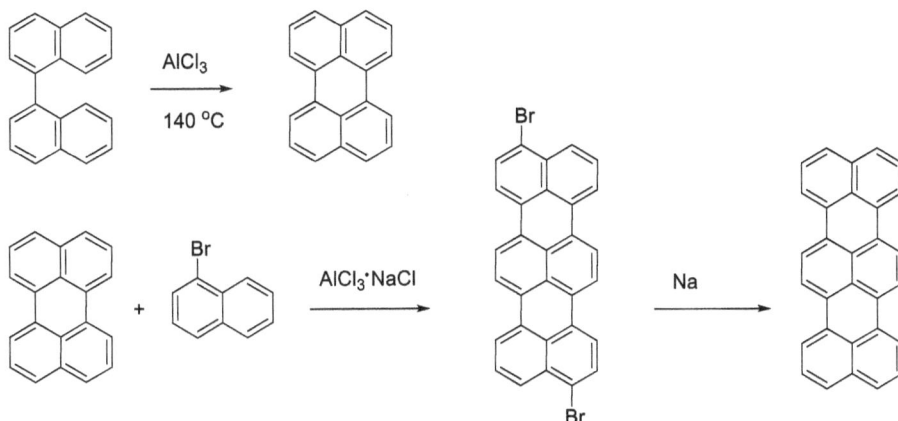

Figure 10.20: Synthesis of perylene and terrylene via Scholl reaction.

Another classical approach to the synthesis of perlyene involves the Haworth synthesis (Figure 10.21) [28]. Starting from anthracene, a twofold Friedel–Crafts acylation with succinic anhydride was followed by reduction (Clemmensen or Wolff–Kishner).

An acid-catalyzed electrophilic cyclization of the dicarboxylic acid gave the dione, which was followed by reduction and dehydrogenation to give perylene.

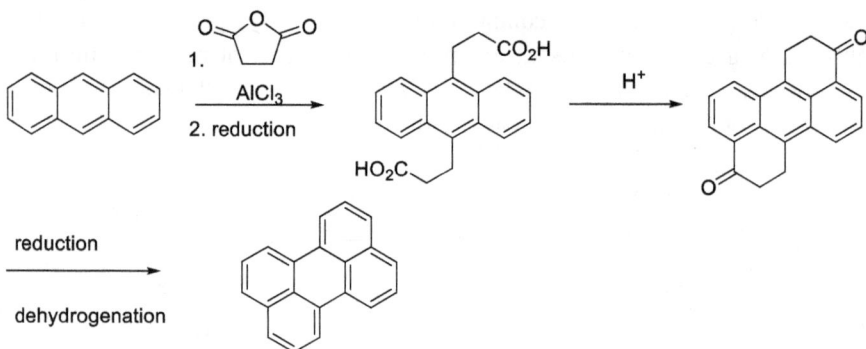

Figure 10.21: Preparation of perylene via a Haworth synthesis [28].

One of the challenges of preparing larger rylenes is their low solubility. To avoid this issue, solubilizing alkyl substituents are added. By adding *t*-butyl groups to naphthalene precursors, Müllen and coworkers were able to prepare quaterrylene (Figure 10.22) [29]. The synthesis began with a Suzuki–Miyaura cross-coupling with 4,4′dibromo-1,1′-binaphthyl and the *t*-butyl-substituted naphthalene boronic acid. The first set of ring-closing reactions was achieved using potassium metal in DME, while the final ring-closing to form the quaterrylene was achieved with a modified Scholl reaction using $AlCl_3$ and $CuCl_2$.

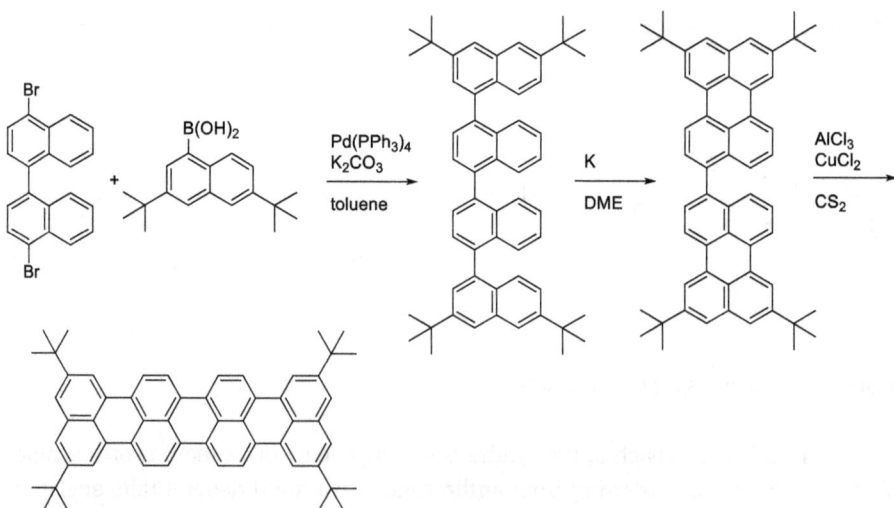

Figure 10.22: Synthesis of a substituted quaterrylene [29].

10.5 Synthesis of triphenylenes and related compounds

10.5.1 Triphenylenes

Triphenylenes consist of three benzene rings fused to give a compound with three-fold symmetry (Figure 10.23). By virtue of the three Clar sextets, triphenylene is stable. Substituted triphenylenes such as the hexaalkoxytriphenylenes, as shown in Figure 10.23, are of interest mostly because they can often exhibit columnar liquid crystal phases and exhibit potential as organic semiconductors [30].

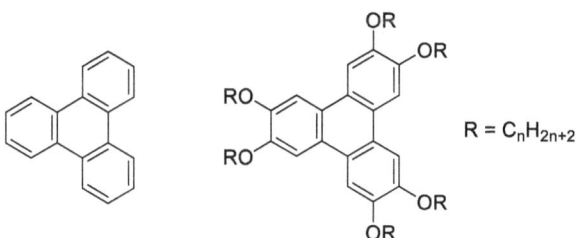

Figure 10.23: Triphenylene and a hexaalkoxytriphenylene.

An overview of some of the early triphenylene syntheses and the reactivity of triphenylenes was reviewed by Buess and Lawson [31]. Here we will consider a few representative examples. One of the classical approach to triphenylenes uses a Haworth synthesis (Figure 10.24), consisting of a Friedel–Crafts acylation of naphthalene using succinic anhydride, followed by reduction to produce the dicarboxylic acid [32]. Conversion to the acid chloride, followed by intramolecular Friedel–Crafts acylation gave the corresponding dione with the requisite carbon framework in place. The dione was converted to the triphenylene by reduction, dehydration, and dehydrogenation.

A related approach for the preparation used a Grignard reaction with succinic anhydride instead of electrophilic substitution. In this synthesis, 9-bromophenanthrene was converted to the corresponding Grignard reagent and reacted with succinic anhydride (Figure 10.25) [33]. The ketone was then reduced via the semicarbazide, and electrophilic cyclization gave cyclohexanone derivative. Reduction of the carbonyl with zinc gave tetrahydrotriphenylene, which could be dehydrogenated by heating in the presence of selenium.

The classical approaches described above are not readily extended to substituted triphenylenes, nor does do they tolerate many functional groups. Consequently, several modern synthetic approaches have been developed that make use of milder conditions and are amenable to the synthesis of triphenylenes with a variety of substitution patterns [34]. These methods can be classified in the following ways: (i) trimerization of appropriately substituted benzenes, (ii) coupling

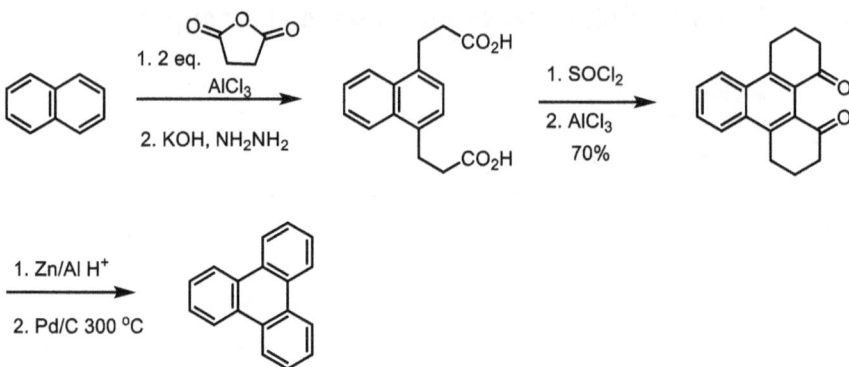

Figure 10.24: Synthesis of triphenylene using a Haworth synthesis [32].

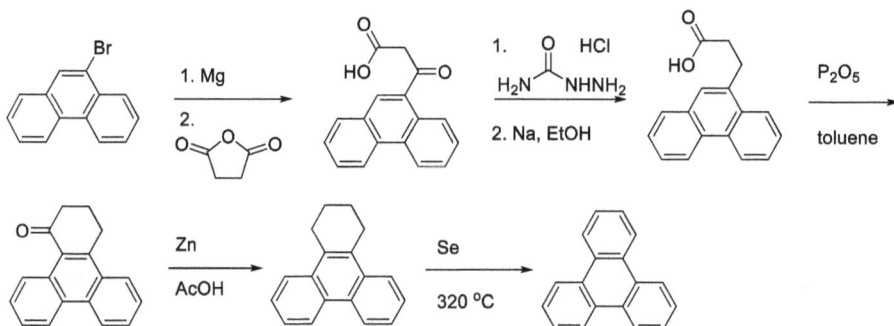

Figure 10.25: Synthesis of triphenylene via Grignard reaction with succinic anhydride [33].

of biphenyls with substituted benzenes, or (iii) a stepwise approach where cross coupling methods are used to prepare *ortho*-terphenyls, which are then subjected to oxidative cyclization to form the corresponding triphenylene (Figure 10.26). Each of these approaches has its advantages and limitations. The first approach is most suited to symmetrical triphenylenes, the second approach allows access to C_2-symmetric triphenylenes, while the third approach allows access to either C_2-symmetric triphenylenes, or if the cross-coupling is carried out sequentially, dissymmetric triphenylenes where each of the outer rings has different substituents.

As we saw in Chapter 9, oxidative cyclodehydrogenation is a viable approach for the formation of polycyclic aromatic systems by forming aryl–aryl bonds directly unsubstituted positions on a benzene ring (Section 9.3.7) [35]. This approach has been widely used for the preparation of electron-rich triphenylenes (Figure 10.27) [36]. The most common reagent used for this reaction is $FeCl_3$, but other reagents such as DDQ/CH_3SO_3H [37] or $MoCl_5$ [38] are also used. As an example, 1,2-dialkoxybenzenes undergo an oxidative cyclotrimerization in the presence of $FeCl_3$ to form the corresponding triphenylenes in high yields [39]. Oxidative cyclodehydrogenation can be

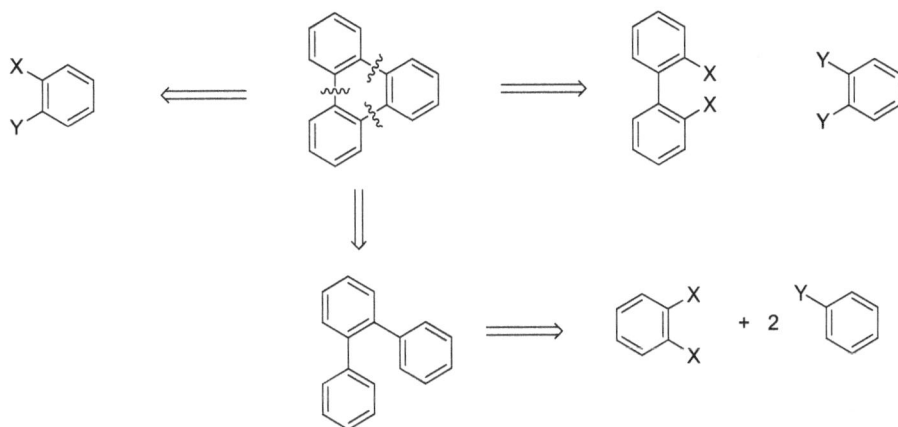

Figure 10.26: General approaches for the synthesis of triphenylenes.

used to access triphenylenes with lower symmetry as well. For example, biphenyls can react with substituted benzenes directly to form a triphenylene (Figure 10.27) [36, 40]. Depending on the substitution on the biphenyl, either C_2-symmetric or dissymmetric triphenylenes can be accessed. The substituted biphenyls are typically prepared either by cross-coupling (e.g., Suzuki–Miyaura) or homocoupling reactions. Oxidative cyclization can also be used to prepare triphenylenes from the corresponding o-terphenyl derivative (Figure 10.27). Again, the terphenyl is typically prepared by transition metal cross-coupling methods. The oxidative cyclization reaction is limited to electron-rich systems and does not work effectively for electron-deficient aromatic rings.

Oxidative cyclization of terphenyl derivatives has also been extended to the preparation of dibenz[a,c]anthracenes from the corresponding diaryl naphthalenes. For example, Williams and coworkers prepared a hexaalkoxy dibenz[a,c]anthracene by Suzuki coupling of a dibromonaphthalene with 3,4-dimethoxyphenylboronic acid to yield the diaryl naphthalene. The diaryl naphthalene was then treated with $FeCl_3$ to give the corresponding dibenz[a,c]anthracene (Figure 10.28) [41]. A similar strategy has been used to prepare a variety of substituted dibenz[a,c]anthracenes [42–45].

An alternative to the formation of triphenylenes via oxidative cyclodehydrogenation is a photocyclization (Mallory reaction – see Section 9.36) of the terphenyl. For example, triphenylene itself can be formed from o-terphenyl by irradiation in the presence of iodine (Figure 10.29) [46]. This method has also been applied to the preparation of substituted triphenylenes [47].

Triphenylenes can also be prepared by a sequence of cross-coupling reactions. For example, the tetraalkoxy biphenyl (prepared by Suzuki cross-coupling) was brominated and then converted into the corresponding diboronic acid via lithium–halogen exchange (Figure 10.30) [40]. The diboronic was then reacted with an

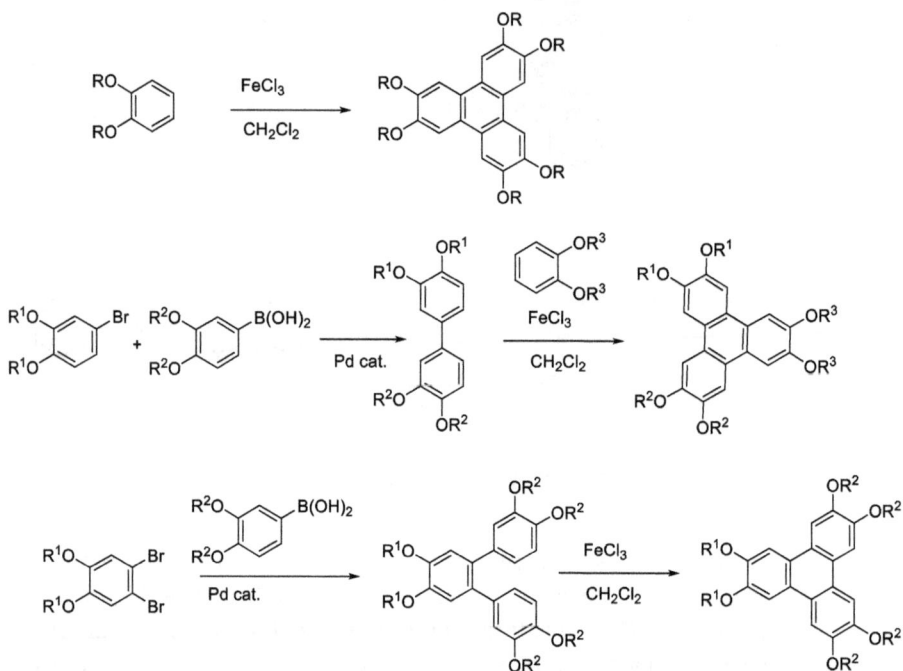

Figure 10.27: Representative triphenylene syntheses via oxidative cyclodehydrogenation.

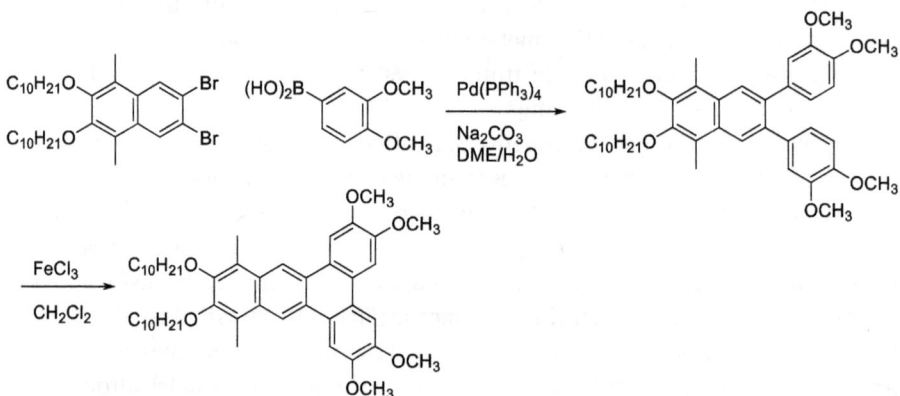

Figure 10.28: Synthesis of a substituted dibenz[a,c]anthracene [41].

ortho-dibromobenzene to yield the triphenylene. This approach can be used for the preparation of triphenylenes with low symmetry, bearing different substituents on each ring. Furthermore, unlike oxidative cyclization methods, this method is not limited electron-rich systems.

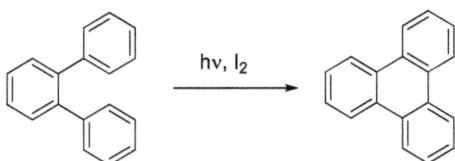

Figure 10.29: Synthesis of triphenylene via photocyclization.

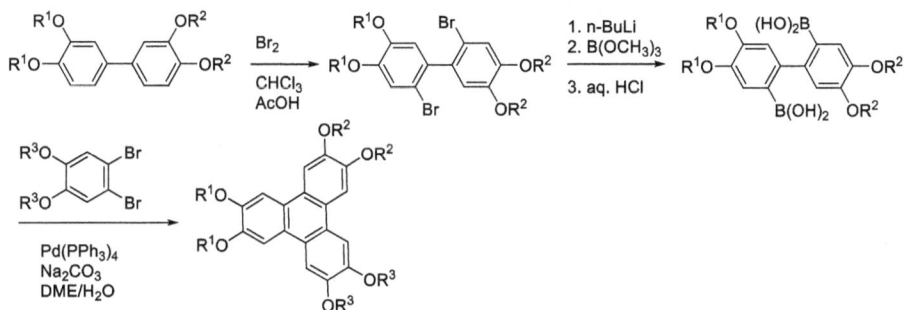

Figure 10.30: Synthesis of triphenylenes by Suzuki coupling [40].

Palladium-catalyzed aryne cyclotrimerization (Chapters 5 and 9) can be used to prepare triphenylenes. Peña and coworkers showed that in the presence of a Pd(0) catalyst such as Pd(PPh$_3$)$_4$, benzynes generated either from 1,2-dibromobenzene or the appropriate trimethylsilyl aryl triflate formed the corresponding triphenylene in moderate yield (Figure 10.31) [48].

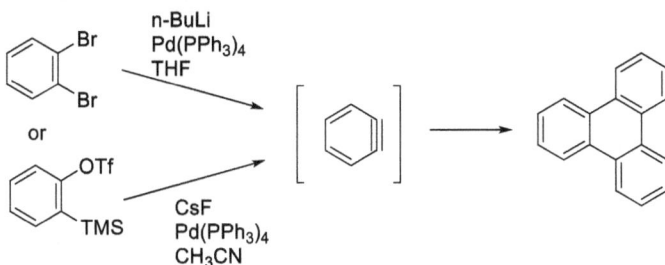

Figure 10.31: Synthesis of triphenylene via benzyne cyclotrimerization [48].

As discussed in Chapter 6, aryl–aryl bonds can be forms by the nickel-mediated homo-coupling of aryl halides. When *ortho*-dibromobenzenes are used, triphenylenes can be prepared – a reaction often referred to as the Yamamoto coupling (see 6.2.2) [49]. The Yamamoto coupling has the disadvantage that it requires a stoichiometric amount of an air-sensitive nickel complex such as Ni(COD)$_2$ that requires the use of a glove box. However, the reaction conditions allow the preparation of electron-deficient triphenylenes that are not readily prepared by oxidative cyclization reactions such as the Scholl reaction. For example, as shown in Figure 10.32, the nickel-mediated coupling of the electron-deficient ester and imide give the corresponding triphenylenes in good yields [50].

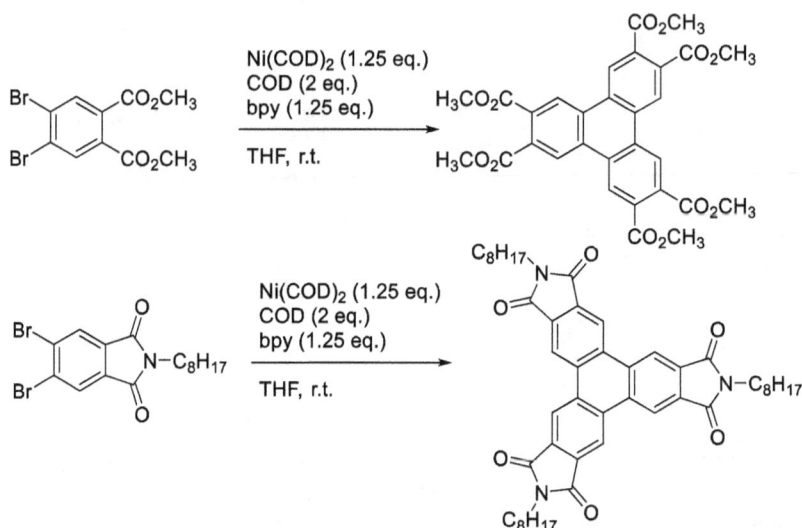

Figure 10.32: Synthesis of electron-deficient triphenylenes via Yamamoto coupling [50].

The Yamamoto coupling can also sometimes offer a more concise synthetic approach than palladium-catalyzed aryne cyclotrimerization methods [50]. However, for electron-rich substrates, the well-established oxidative cyclization methods such as the Scholl reaction are usually more convenient.

10.5.2 Trinaphthylenes and larger starphenes

Trinaphthylenes and starphenes are extended versions of triphenylene. However, the methods for the synthesis of these compounds are not well developed. Oxidative cyclization methods such as the Scholl reaction that are widely used for the preparation of triphenylenes are not effective for the preparation of trinaphthylenes and lead to complex mixtures [51]. The first synthesis of trinaphthylene consisted of a trimerization of naphthoquinone by heating at reflux in pyridine, followed by reduction using zinc dust or HI (Figure 10.33) [52]. These harsh reaction conditions are not necessarily amenable to the preparation of substituted trinaphthylenes.

Substituted trinaphthylenes have been prepared by annulation of the corresponding triphenylenes using a Cava reaction. For example, trinaphthylenes bearing imide groups were prepared from hexamethyltriphenylene by benzylic bromination,

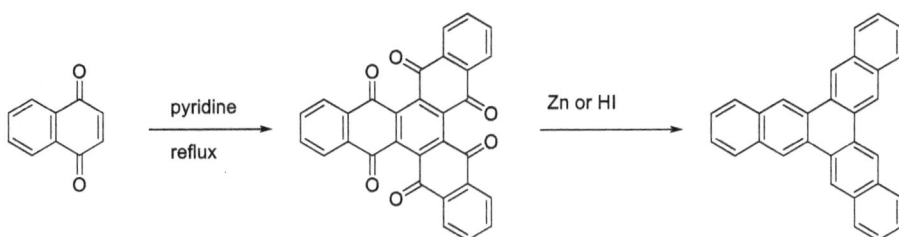

Figure 10.33: Synthesis of trinaphthylene [52].

followed by a Cava reaction with an *N*-alkyl maleimide (Figure 10.34) [53]. The hexamethyltriphenylene starting material was prepared from the corresponding aryl trimethylsilyl triflate via a palladium-catalyzed aryne cyclotrimerization. The disadvantages of this approach are the need to prepare the trimethylsilyl aryl triflate and the relatively low yield of both the aryne cyclotrimerization (34%) and the Cava reaction (35–40%).

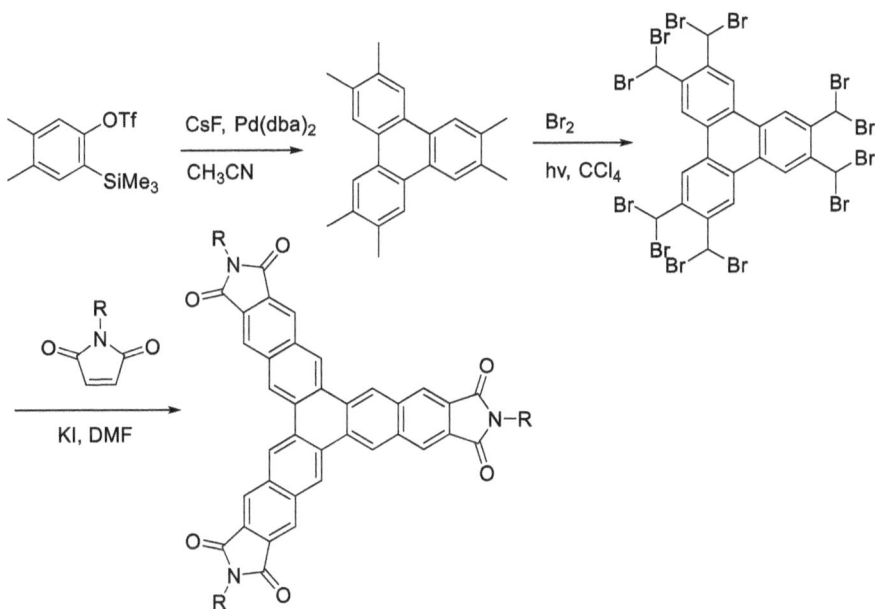

Figure 10.34: Synthesis of trinaphthylenes from triphenylenes via the Cava reaction [53].

Another approach toward the synthesis of substituted trinaphthylenes focused on an aryne cyclotrimerization of the appropriate naphthalene precursor (Figure 10.35) [51]. Starting from the appropriate 6,7-dibromo-2,3-dialkoxynaphthalene (which was prepared in three steps from 2,3-dihydroxynaphthalene), a lithium–halogen exchange was carried out and the aryllithium species was trapped with trimethyl borate and oxidized to give the bromo-naphthol. Silylation of the naphthol was achieved using hexamethyl-disilazane. A second lithium–halogen exchange resulted in migration of the silyl group and trapping the aryloxy anion with triflic anhydride to yield the aryne precursor. This approach for preparing aryne precursors was discussed in Section 5.2.4. Finally, treatment of the aryne precursor with CsF and catalytic Pd(PPh₃)₄ gave the desired trinaphthylene in modest yield.

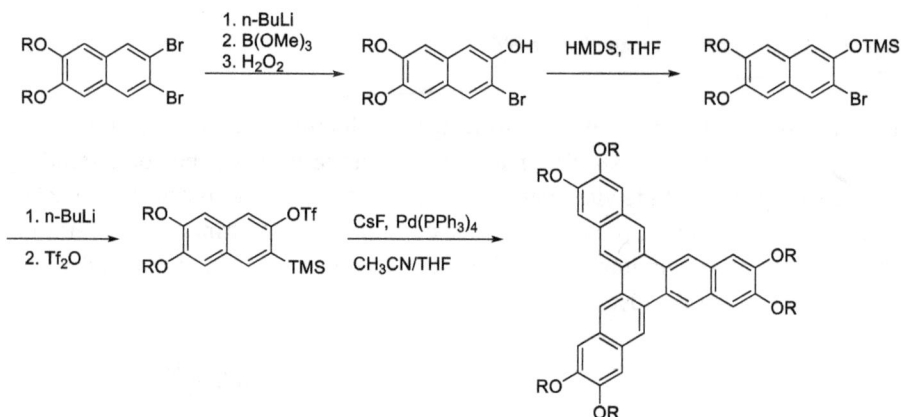

Figure 10.35: Synthesis of trinaphthylenes via aryne cyclotrimerization [51].

Subsequently, Bunz and coworkers showed that the same alkoxy-substituted tri-naphthylenes could be accessed in a more concise fashion by Yamamoto coupling of the 6,7-dibromo-2,3-dialkoxynaphthalenes (Figure 10.36) [54].

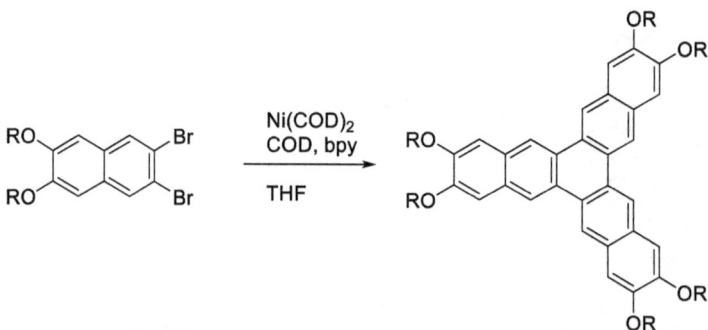

Figure 10.36: Synthesis of trinaphthylenes via Yamamoto coupling [54].

Larger starphenes have not been extensively studied, in part because of their low solubility and the challenges in their synthesis. Clar and Mullen reported the synthesis of "decastarphene," so named because it contains 10 aromatic rings (Figure 10.37)

[55]. The synthesis begins with a two-fold Friedel–Crafts acylation of phenanthrene using phthalic anhydride [56]. The carbonyl groups were reduced using Zn in aqueous KOH, and the product was subjected to another Friedel–Crafts acylation/reduction sequence using naphthalene 2,3-dicarboxylic anhydride. Treatment with zinc dust at high temperature leads to electrophilic cyclization and reduction. Finally, treatment with soda-lime gave the starphene. The synthesis requires high temperatures and harsh conditions and is not suitable for substituted starphenes. Furthermore, purification of the final product was challenging due to its low solubility.

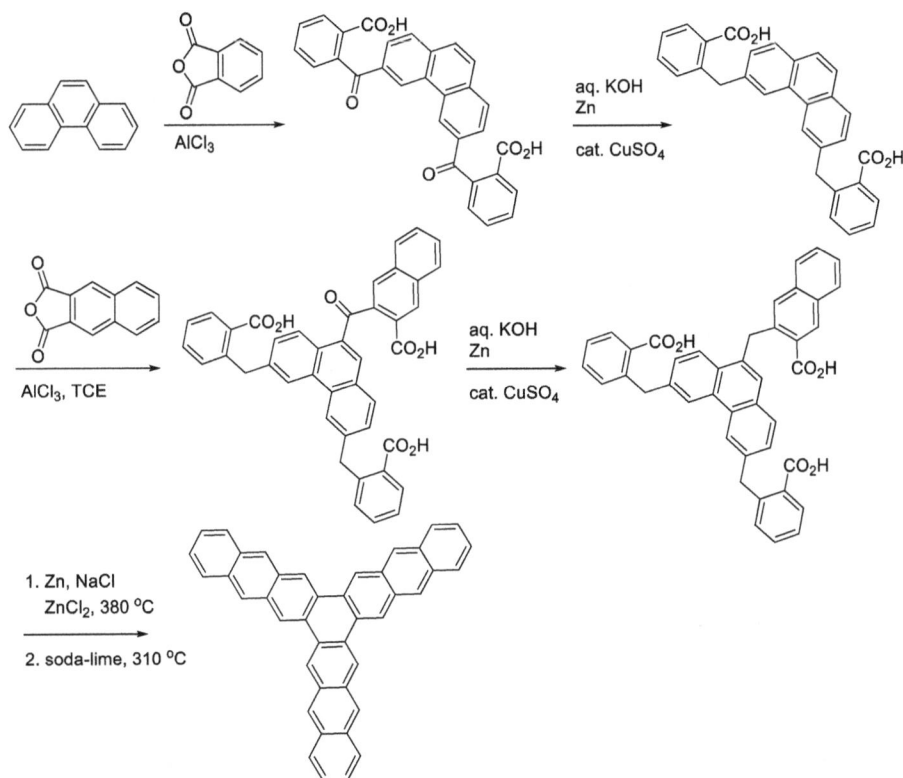

Figure 10.37: Synthesis of "decasterphene" [55].

Much more recently, Bunz and coworkers described the synthesis of soluble starphenes using the Yamamoto coupling (Figure 10.38) [57]. The synthesis began with a Cava reaction between 1,2-dibromo-4,5-bis(dibromomethyl)benzene and naphthoquinone using sodium iodide to give the dibromo-tetracenequinone. Ethynylation using ethynyltriisopropylsilane and butyllithium gave the diol, which was reduced to the corresponding tetracene using NaH_2PO_2. Finally, the dibromotetracene was trimerized using standard Yamamoto coupling conditions to give the desired starphene derivative. Unlike the parent starphenes, these compounds were much more soluble and amenable to characterization in solution and by X-ray crystallography.

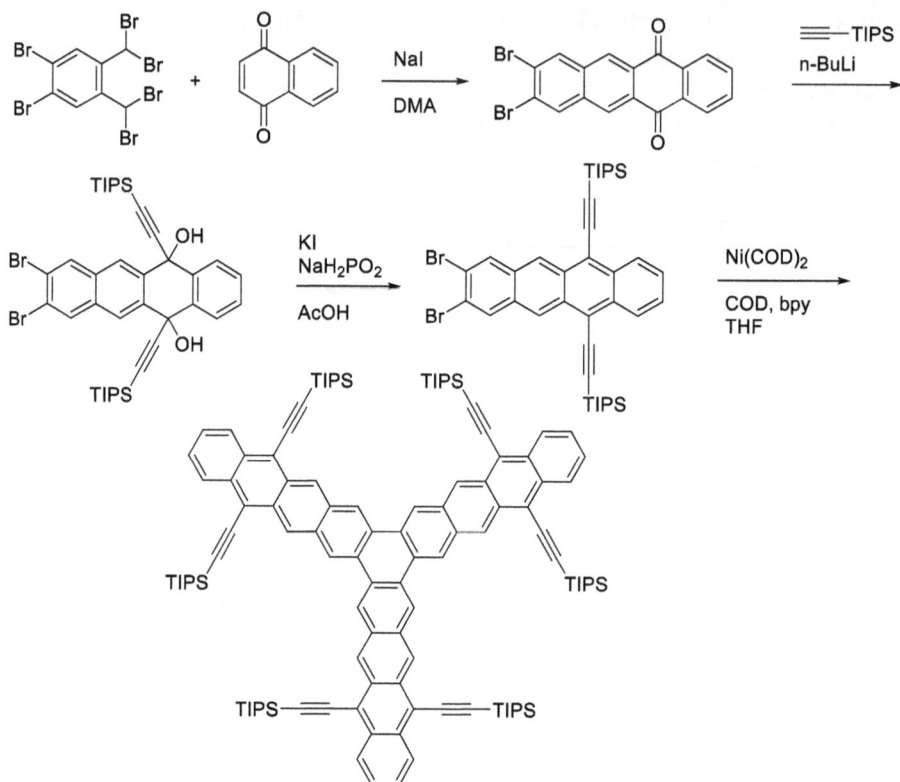

Figure 10.38: Synthesis of soluble starphenes via Yamamoto coupling [57].

A large starphene consisting of 16 fused benzene rings was prepared using methods inspired from recent synthesis of higher acenes along with a palladium-catalyzed aryne cyclotrimerization (Figure 10.39) [58]. As we saw for a synthesis of heptacene, the synthesis involved the Diels–Alder reaction of 7,7-dimethoxy -2,3,5,6-tetramethylenebicyclo[2.2.1]heptane with one equivalent of benzyne generated from 2-(trimethylsilyl)phenyl triflate in the presence of fluoride. The cyclo-adduct was then aromatized using DDQ. A second sequence of benzyne addition and oxidation was carried out with the diaryne precursor 1,4-bis(trimethylsilyl) phenyl-2,5-bis(triflate) to give an extended trimethylsilyl aryl triflate. This compound was treated with catalytic Pd$_2$(dba)$_3$ to effect the aryne cyclotrimerization. Finally, deprotection of the acetal using TMSI and cheletropic thermal decarbonylation gave the starphene [58].

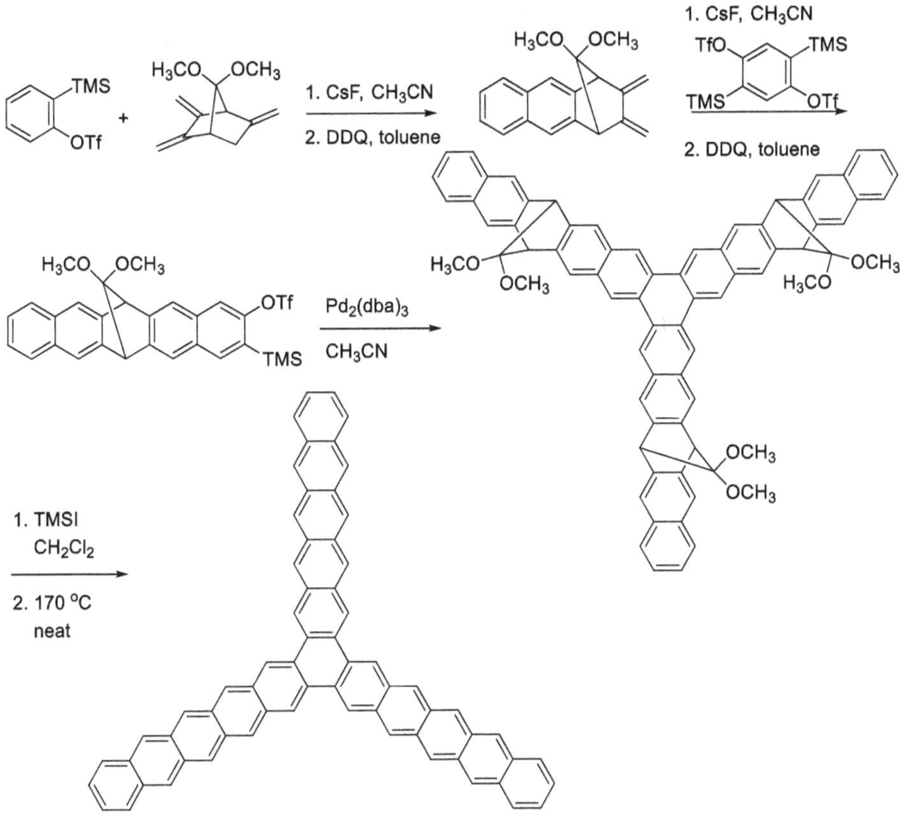

Figure 10.39: Synthesis of a large starphene with pentacene "branches" [58].

10.6 From coronenes to larger PAHs

Coronene is a planar, disk-shaped polycyclic aromatic compound consisting of six benzene rings that are *ortho*-fused in a cyclic fashion, creating a central six-membered ring. It is part of a class of compounds called circulenes, which we will discuss in more detail in Chapter 11.

The first synthesis of coronene was reported by Scholl and Meyer in 1932 (Figure 10.40) [59]. The synthesis begins with a Friedel–Crafts acylation of *m*-xylene with the diacid chloride to give the bis lactone, which was hydrolyzed and oxidized to give the hexa-carboxylic acid. This compound was reduced and underwent electrophilic cyclization to give the dibenzocoronene-tetraone derivative. The keto groups were removed by another reduction using HI and phosphorus and the product was then decarboxylated using soda lime in the presence of copper to give the di-*peri*-dibenzocoronene. This compound was then oxidized to the coronene-tetracarboxylic acid by heating in nitric acid, and finally decarboxylated using soda lime at 500 °C to give coronene.

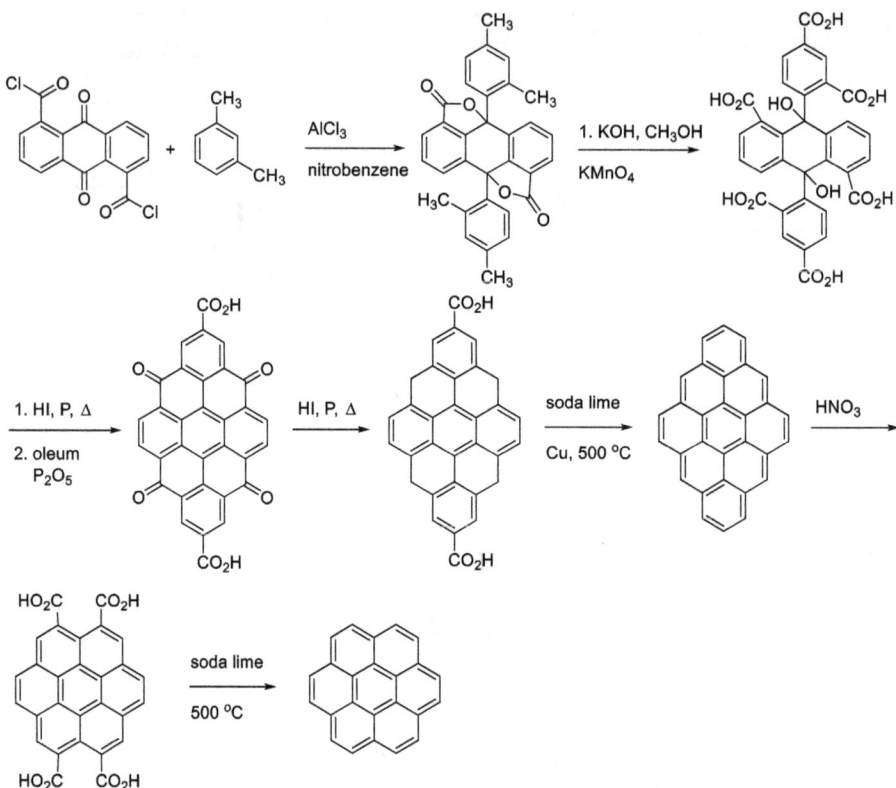

Figure 10.40: The first synthesis of coronene [59].

The synthesis by Scholl and Meyer long, used harsh conditions, and suffered from low yields. Clar and Zander developed a more concise synthesis of coronene starting from perylene (Figure 10.41) [60]. They reacted perylene with maleic anhydride in a Diels–Alder reaction in the presence of chloranil as an oxidizing reagent. The anhydride moiety was then removed by heating the Diels–Alder adduct in the presence of soda lime to produce the corresponding benzoperylene. This sequence of cycloaddition followed by decarboxylation and decarbonylation was then repeated to give coronene. A much more recent modification to Clar and Zander's approach showed that the benzoperylene could be converted to coronene by two-electron reduction of the benzoperylene using sodium metal and ultrasound, and reacting the resulting dianion with bromoacetaldehyde diethyl acetal (Figure 10.42) [61]. The acetal product was then cyclized by treatment with concentrated sulfuric acid under ultrasound conditions. This latter approach avoided the high temperature conditions for the decarboxylation/decarbonylation step and could more readily be performed on a larger scale.

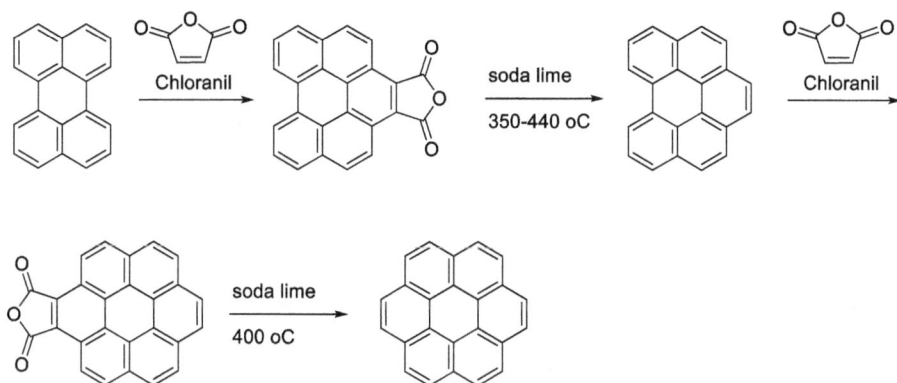

Figure 10.41: Synthesis of coronene via Diels–Alder reactions [60].

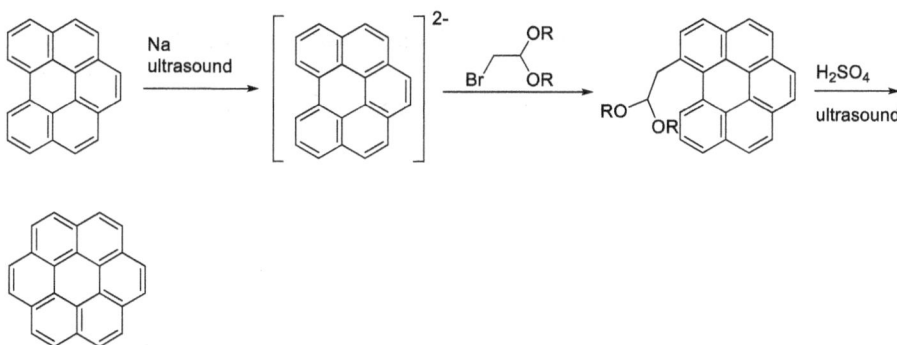

Figure 10.42: Alternative approach to coronene [61].

A distinctly different synthetic approach developed by Davy and Reiss made use of a photocyclization of a naphthalenophane diene as a key step (Figure 10.43) [62]. The synthesis began by treating 2,7-bis(bromomethyl)naphthalene with sodium sulfide to produce the bisthioether. The bisthioether was then methylated with $(CH_3)_3OBF_4$ to give the corresponding sulfonium salt. Treating the sulfonium salt with sodium hydride in THF lead to a Stevens rearrangement to give the substituted [2.2]naphthalenophane [63]. Another methylation with $(CH_3)_3OBF_4$ gave the sulfonium salt, which underwent elimination under basic conditions to give [2.2](2,7)-naphthalenophane-1,11-diene. Finally, UV irradiation in the presence of iodine gave coronene [62].

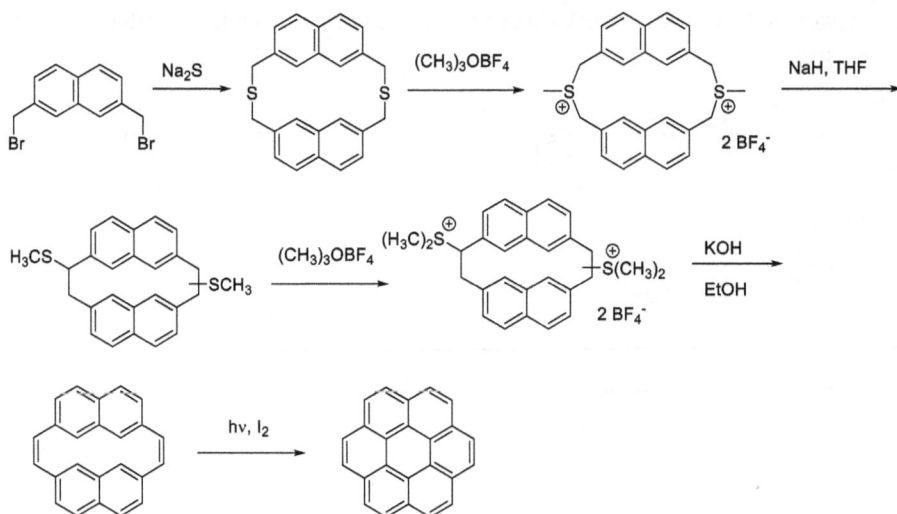

Figure 10.43: Synthesis of coronene from a cyclophane by photocyclization [62].

When the coronene system extended by peri-fusion of six additional benzene rings, the result is hexa-*peri*-hexabenzocoronene (Figure 10.44). This compound is stable, possessing seven Clar sextets, but is not surprisingly insoluble.

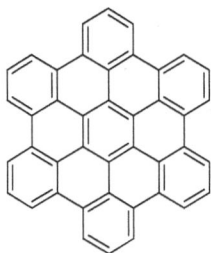

Figure 10.44: Hexa-*peri*-hexabenzocoronene.

One of the early syntheses of hexa-*peri*-hexabenzocoronene was reported by Clar and coworkers (Figure 10.45) [64]. The reaction involved bromination and ring-fusion at high temperatures.

An alternative approach to hexa-*peri*-hexabenzocoronene reported by Schmidt and coworkers started with the substituted anthraquinone, which was treated with phenyllithium to give the corresponding diol (Figure 10.46) [65]. Subsequent reaction with AlCl$_3$ in NaCl resulted in cyclodehydrogenation, which was followed by aromatization at high temperature in the presence of copper.

As mentioned above, hexa-*peri*-hexabenzocoronene is quite insoluble and challenging to characterize. The addition of solubilizing groups can be used to improve solubility, but the harsh conditions of the synthetic approaches described above are not likely

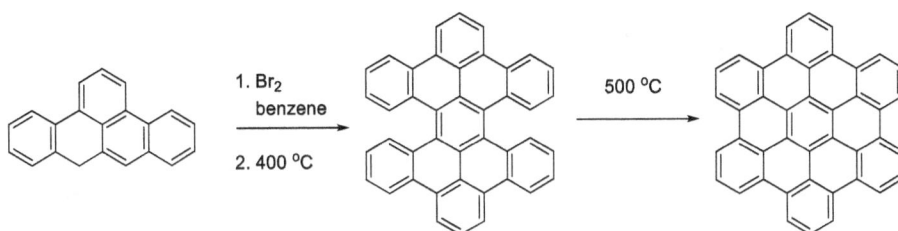

Figure 10.45: Clar's synthesis of hexa-*peri*-hexabenzocoronene [64].

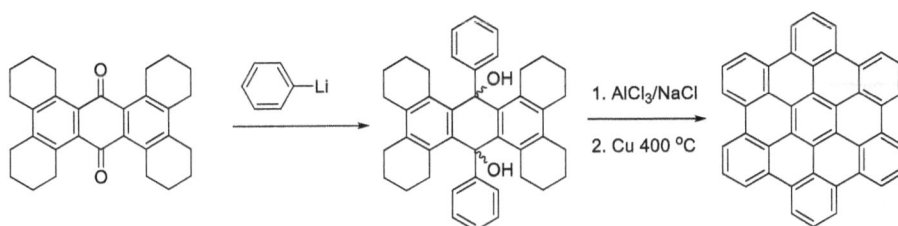

Figure 10.46: Synthesis of hexa-*peri*-hexabenzocoronene by Schmidt and coworkers [65].

suitable for the preparation of substituted hexa-*peri*-hexabenzocoronenes. Müllen and coworkers showed that the Scholl reaction or oxidative cyclodehydrogenation could be used effectively to prepare substituted hexa-*peri*-hexabenzocoronenes [66, 67]. These substituted hexa-*peri*-hexabenzocoronenes have been explored for their self-assembly to form columnar mesophases when substituted with flexible alkyl chains [67]. They are typically prepared by cyclodehydrogenation of the corresponding hexaphenylbenzene derivative, which is prepared either by transition metal-catalyzed alkyne cyclotrimerization (Section 8.2.2), or by a Diels–Alder reaction of the tetraphenylcyclopentadienone with a diphenyl acetylene (Section 8.2.1). One of the representative syntheses of these substituted hexa-*peri*-hexabenzocoronenes is shown in Figure 10.47 [67]. The synthesis began with the alkyl-substituted aniline, which was converted into the corresponding aryl iodide via diazotization and reaction with potassium iodide. A Sonogashira coupling reaction with trimethylsilylacetylene gave the ethynyl benzene. The TMS group was cleaved in the presence of fluoride, and then the terminal alkyne was subjected to another Sonogashira coupling to obtain the substituted diphenylacetylene. This compound underwent cobalt-catalyzed cyclotrimerization in the presence of $Co_2(CO)_8$. Finally, cyclodehydrogenation using $AlCl_3$ and $Cu(OTf)_2$ in carbon disulfide gave the hexa-*peri*-hexabenzocoronene.

This approach, reported by Müllen and coworkers, has been used to prepare larger PAHs. For example, reaction of 1,4-bis(phenylethynyl)benzene with tetraphenylcyclopentadienone gave octaphenylquinquephenyl (Figure 10.48) [68]. Oxidative cyclization

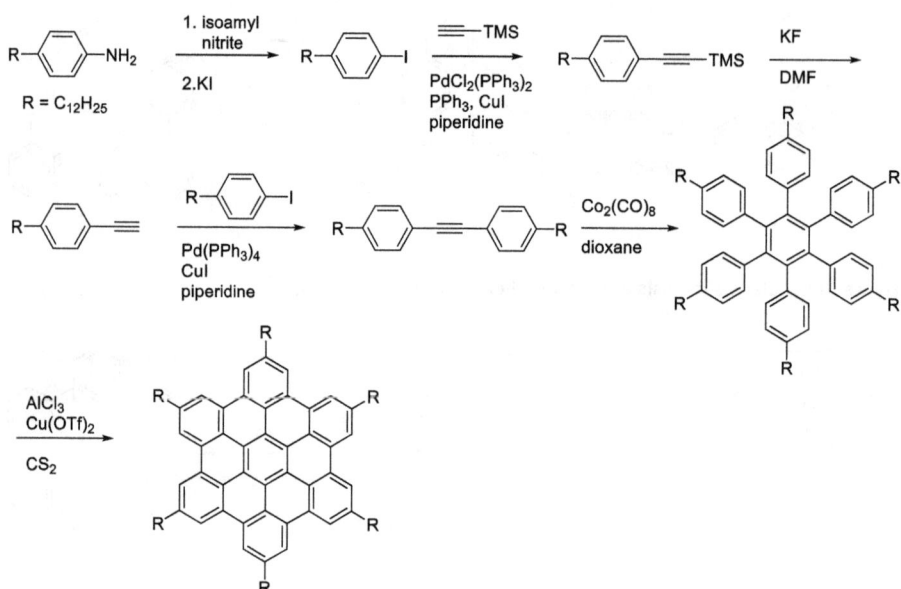

Figure 10.47: Synthesis of substituted hexa-*peri*-hexabenzocoronenes from hexaphenylbenzenes [67].

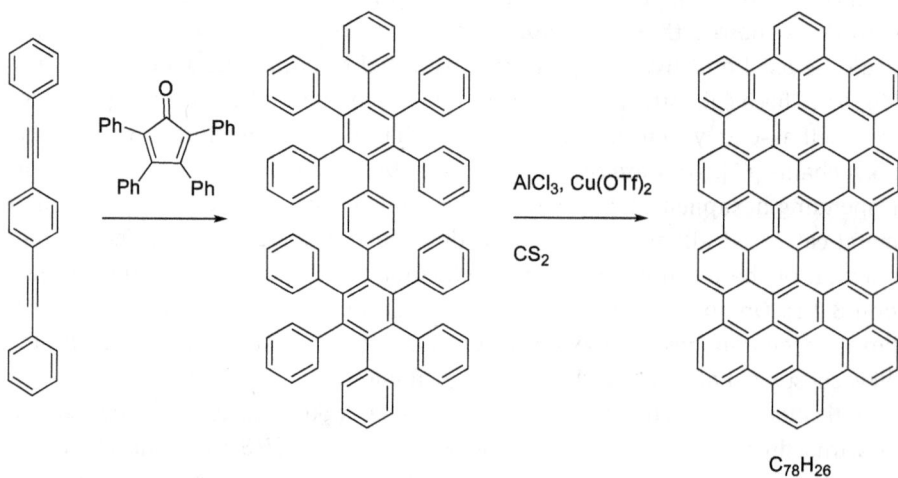

Figure 10.48: Synthesis of an elongated nanographene [68].

using AlCl$_3$ and Cu(OTf)$_2$ gave the elliptical PAH compound with the formula C$_{78}$H$_{26}$. This highly insoluble compound was characterized by mass spectrometry.

Using the same synthetic approach, Müllen and coworkers have extended constructed even larger graphitic systems. While these compounds are insoluble, they represent an approach to structurally defined nanographenes. For example, as outlined in

Figure 10.49, Müllen and coworkers were able to prepare a nanographene with the formula with 222 carbon atoms [69, 70]. The synthesis involves a tri-alkyne (prepared by Sonogashira cross-coupling) which underwent a two-fold Diels–Alder reaction with tetraphenylcyclopentadienone to give the bis(pentaphenylbenzene)-substituted acetylene, which was subjected to alkyne cyclotrimerization to give the highly branched oligophenylene with 37 benzene rings. Finally, cyclodehydrogenation gave the target nanographene. This latter step is impressive because it involved the formation of 54 carbon–carbon bonds in a single step.

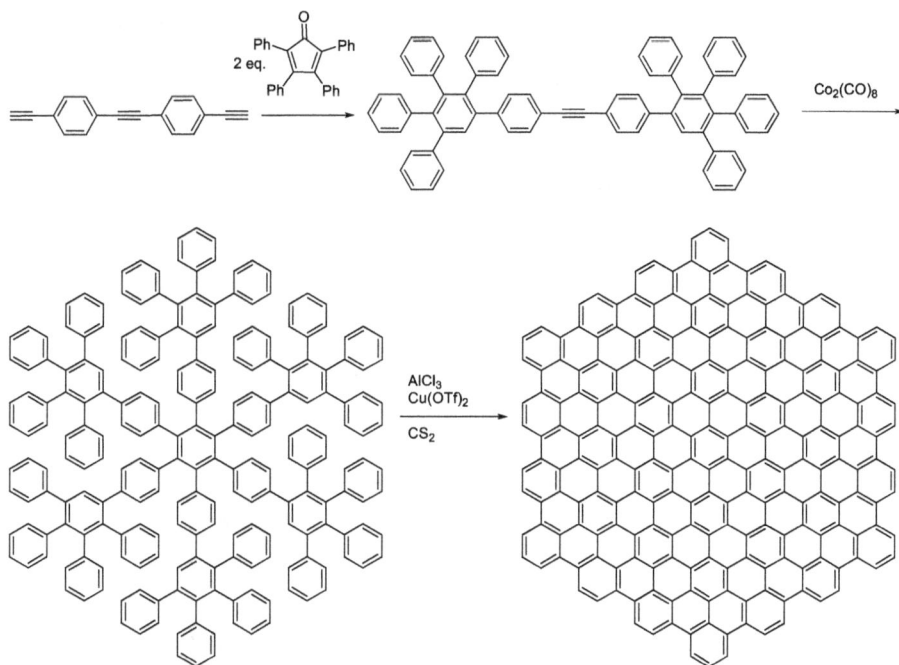

Figure 10.49: Synthesis of a larger nanographene [69, 70].

10.7 Graphene nanoribbons

Thus far, we have considered the preparation of structurally defined PAHs. If we extend the aromatic rings further, we can achieve graphene-like structures. One area of recent interest is the preparation of graphene nanoribbons (GNRs), which are narrow strips of graphene that have promising properties for organic electronics. These GNRs While there are "top-down" approaches for the preparation of GNRs from graphene including lithographic methods [71], sonochemical methods [72], and unzipping carbon nanotubes [73]. The bottom-up synthesis from well-defined molecular precursors has the potential advantage of preparing atomically precise GNRs [74, 75]. Below we will consider some select examples of synthetic approaches for the preparation of

GNRs, focusing primarily on solution-based methods. We will see that some of these methods for preparing GNRs draw on the synthetic methods for other PAHs.

For example, Müllen and coworkers used a Suzuki polymerization and Scholl reaction for the preparation of a GNR (Figure 10.50) [76]. Their synthesis began with a Suzuki coupling using a 2,3,5,6-tetraaryl-1,4-diiodobenzene with 4-bromo-phenylboronic acid to give the dibromohexaphenylbenzene. The bromo groups were then converted into the corresponding boronate esters by lithium–halogen exchange and trapping with 2-isopropoxy-4,4,5,5-tetramethyl-1,3,2-dioxaborolane. This monomer was then reacted with the diiodide from the first step under Suzuki cross-coupling conditions to yield an oligophenylene polymer. Finally, an oxidative cyclization using $FeCl_3$ have the fused GNR.

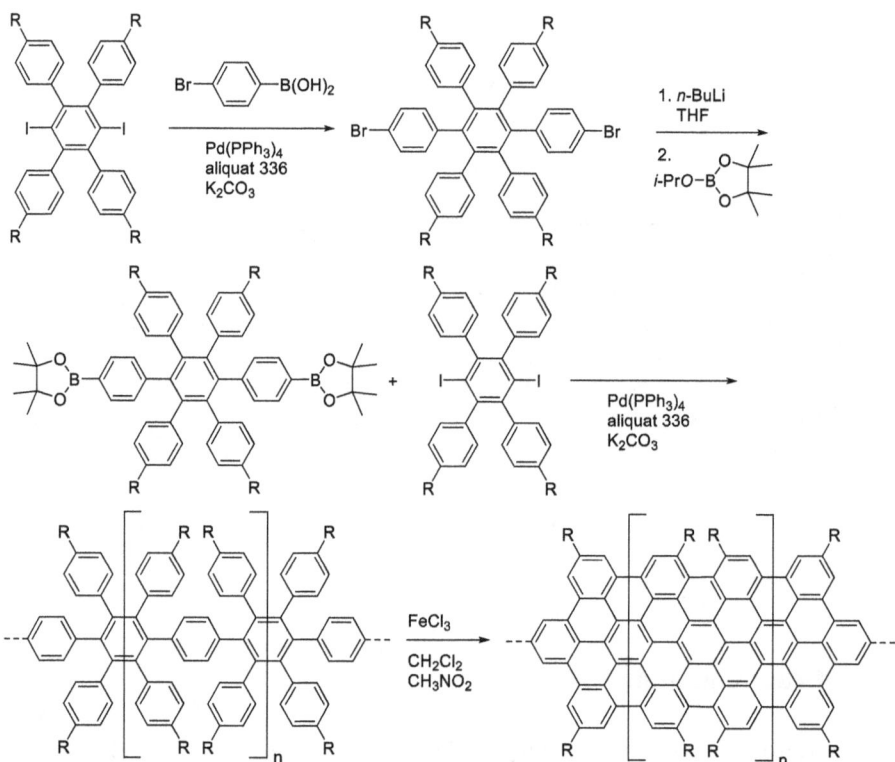

Figure 10.50: Synthesis of a GNR via Suzuki polymerization and Scholl reaction [76].

An alternative polymerization approach is to use a Yamamoto coupling. It the advantage of using one monomer, in contrast to the Suzuki coupling which typically has two different monomers to carry out the polymerization. For example, Sinitskii and coworkers used a Yamamoto coupling to prepare an arylated poly(triphenylene), which could then be converted to the GNR via a Scholl reaction (Figure 10.51) [77]. The synthesis began with electrophilic bromination of phenanthrenequinone, which then reacted with 1,3-diphenylacetone to prepare the cyclopentadienone. A Diels–Alder reaction with diphenylacetylene gave tetraphenyltriphenylene, which was subjected to a Yamamoto polymerization. Finally, cyclodehydrogenation was achieved using $FeCl_3$, giving a zigzag GNR.

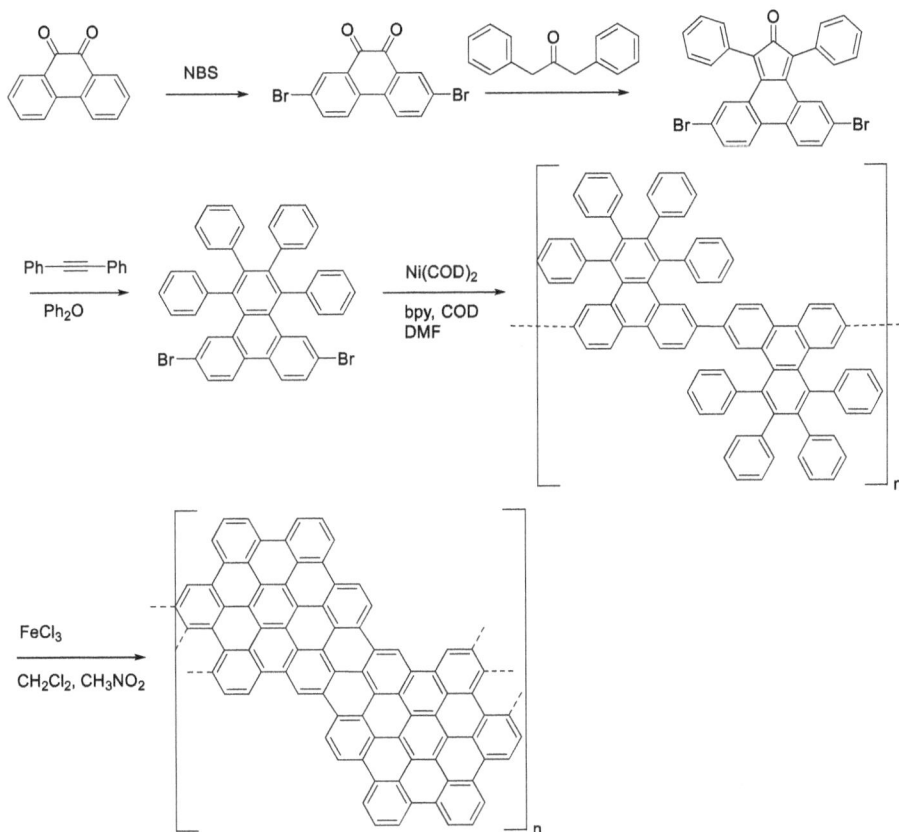

Figure 10.51: GNR synthesis involving a Yamamoto polymerization [77].

Müllen and coworkers also applied their method for preparing hexa-peri-hexaben-zocoronenes and other discrete nanographenes for the synthesis of GNRs. Specifically, they used a cyclopentadienone with a pendant acetylene unit to carry out a Diels–Alder polymerization, followed by a Scholl reaction to prepare the GNR (Figure 10.52) [78].

In the previous chapter we saw the Asao–Yamamoto benzannulation used to prepare polycyclic aromatic systems by reaction of diarylacetylenes with o-(phenyle-thynyl)benzaldehyde (Section 9.3.12). Dichtel and coworkers used this method to prepare a polyarene which could be converted into the corresponding GNR by

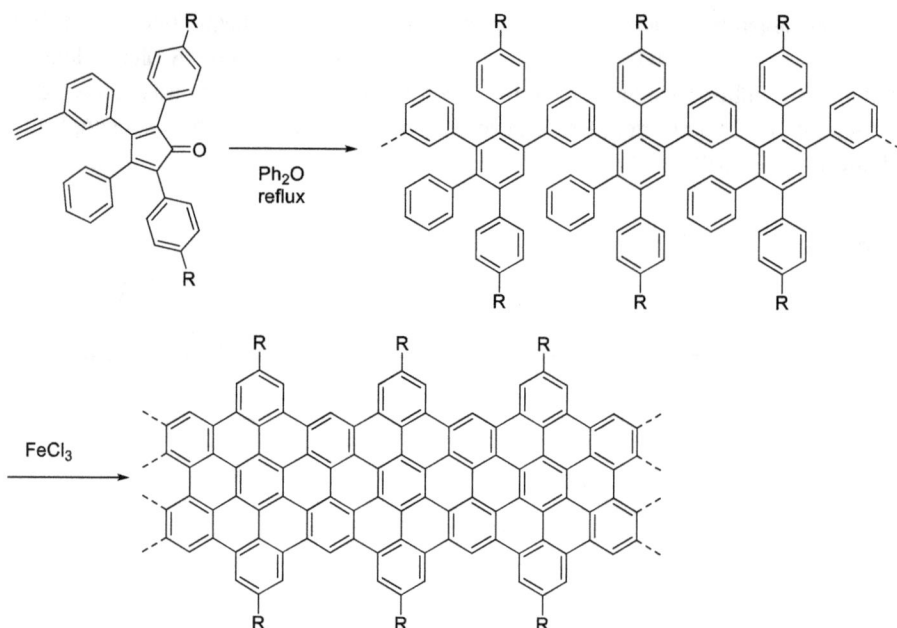

Figure 10.52: Preparation of a GNR by a Diels–Alder polymerization/Scholl reaction sequence [78].

oxidative cyclization (Figure 10.53) [79]. The synthesis began with conversion of the alkoxy-substituted bromonaphthalene to the corresponding boronic acid via lithium–halogen exchange and trapping with trimethyl borate. A twofold Suzuki coupling with the diethynyl dibromobenzene, gave the dinaphthylbenzene derivative. Cleavage of the trimethylsilyl groups using K_2CO_3 gave the dialkyne monomer. This monomer was reacted with 1,4-diiodobenzene as a co-monomer in a Sonogashira to give the poly(p-phenylene ethynylene). The poly(p-phenylene ethynylene) was then treated with o-(phenylethynyl)benzaldehyde under Asao–Yamamoto benzannulation conditions to give the corresponding polyarylene, which was treated with DDQ and methanesulfonic acid gave the GNR.

In one of the earliest reports of the synthesis of GNRs, Goldfinger and Swager used a combination of Suzuki coupling and an acid-catalyzed benzannulation (Figure 10.54) [80]. The synthesis of one of the monomers began with a twofold Sonogashira coupling of an alkoxyphenyl acetylene with 1,4-dibromo-2,5-diiodobenzene to yield the bis(arylethynyl)benzene derivative. The bromides were then converted to the corresponding iodides by lithium–halogen exchange, followed by trapping with iodine. This monomer was then reacted with the 2,5-dialkyl-benzene-1,3-diboronic acid in a Suzuki cross-coupling to produce a polyphenylene. Finally, benzannulation using trifluoroacetic acid produced the GNR.

In a related approach toward GNRs, Chalifoux and coworkers reported the synthesis of substituted GNRs based on a rylene core via Suzuki polymerization and acid-

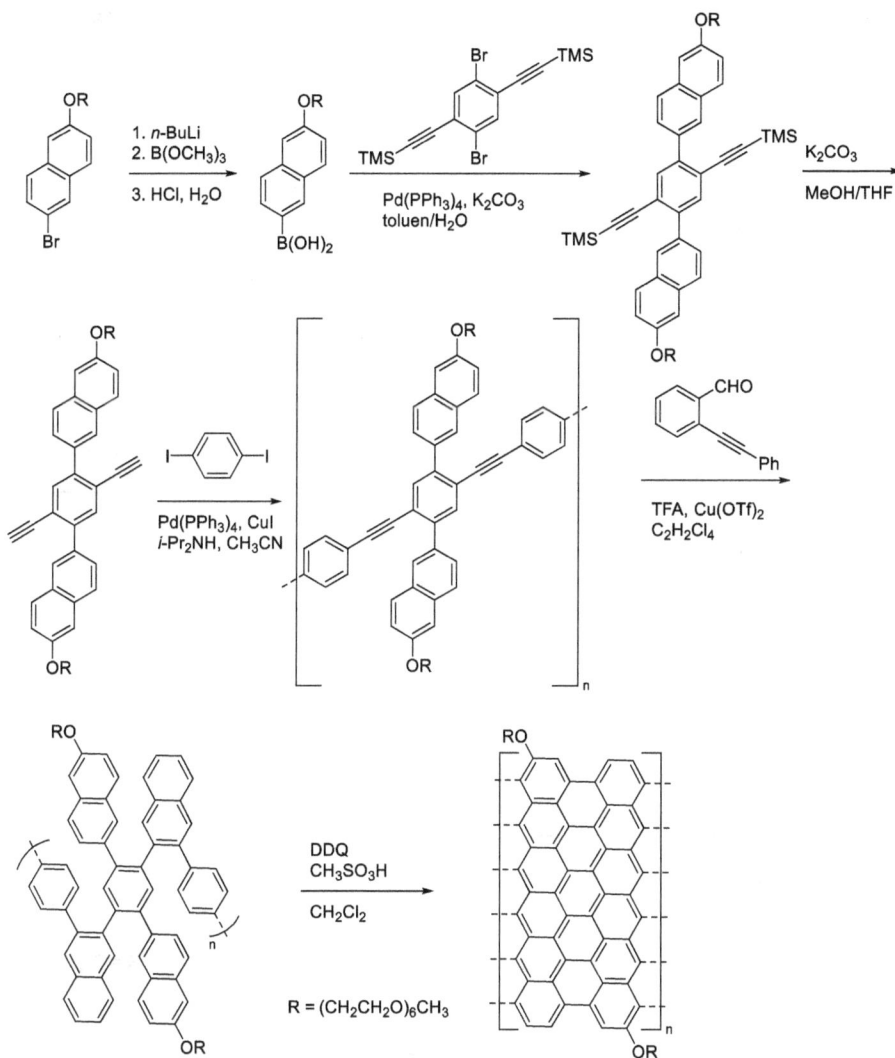

Figure 10.53: Synthesis of GNRs via Asao–Yamamoto benzannulation [79].

mediated benzannulation (Figure 10.55) [81]. The synthesis begins with the conversion of 4-bromo-2,6-diiodoaniline to the corresponding triazene using sodium nitrite and diethylamine. In Section 3.6.3, we saw that aryl triazenes can be used as masked aryl iodides. This was followed by a Sonogashira reaction using the (4-alkoxyphenyl)acetylene at the aryl iodides to give the bis(arylethynyl)benzene. The triazene was converted into the corresponding aryl iodide, which was converted into the corresponding boronate ester by lithium–halogen exchange, trapping with trimethyl borate, and treatment with pinacol. The bromo-substituted boronate ester was then polymerized under Suzuki cross-coupling conditions. Finally, treatment with trifluoroacetic acid followed by

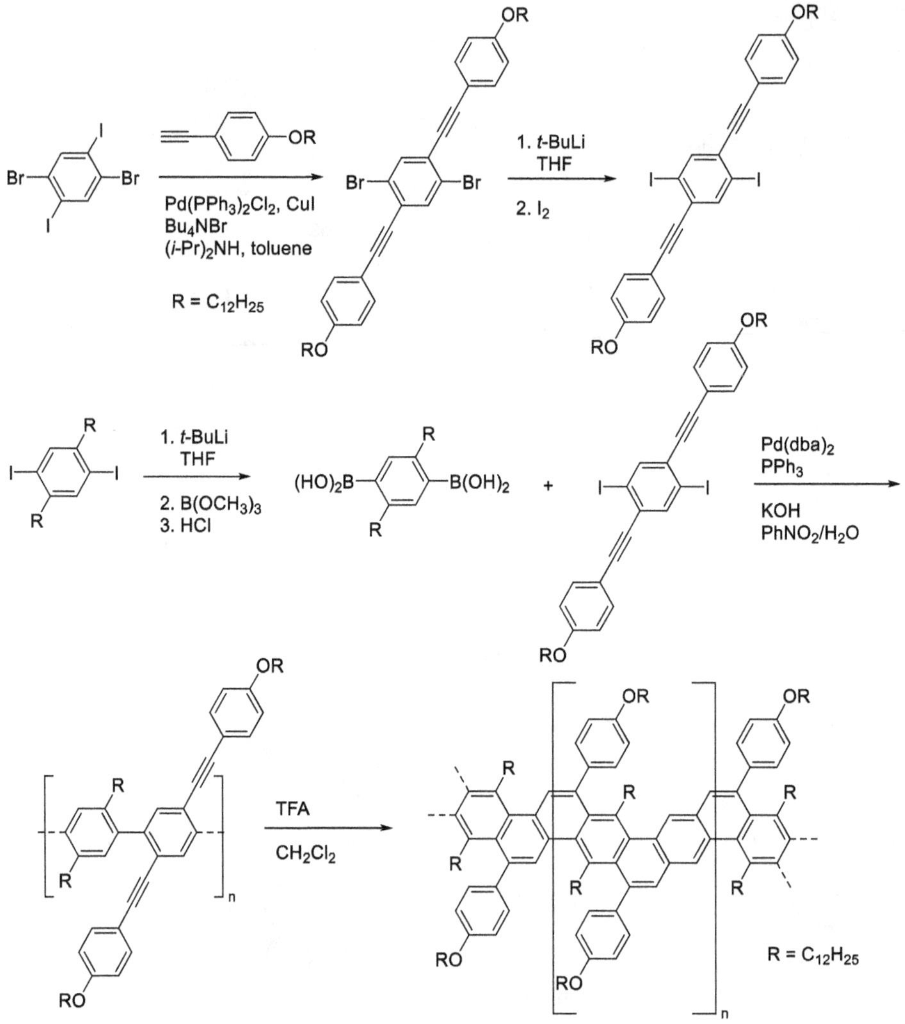

Figure 10.54: Swager's GNR synthesis via Suzuki polymerization and benzannulation [80].

triflic acid lead to benzannulation, giving the desired GNR with pendant alkoxyphenyl units as solubilizing groups.

In contrast to the solution-based synthetic approaches described above, Müllen, Fasel, and coworkers demonstrated the synthesis of GNRs from molecular precursors on metal surfaces at high temperatures [74]. For example, the monomer 10,10′-dibromo-9,9′bianthryl was thermally sublimed onto a gold surface resulting in dehalogenation to give a surface-stabilized biradical, which coupled to form a linear polymer (Figure 10.56). Subsequent heating to 400 °C led to cyclodehydrogenation to give the GNR with well-defined width, as shown by scanning tunneling microscopy

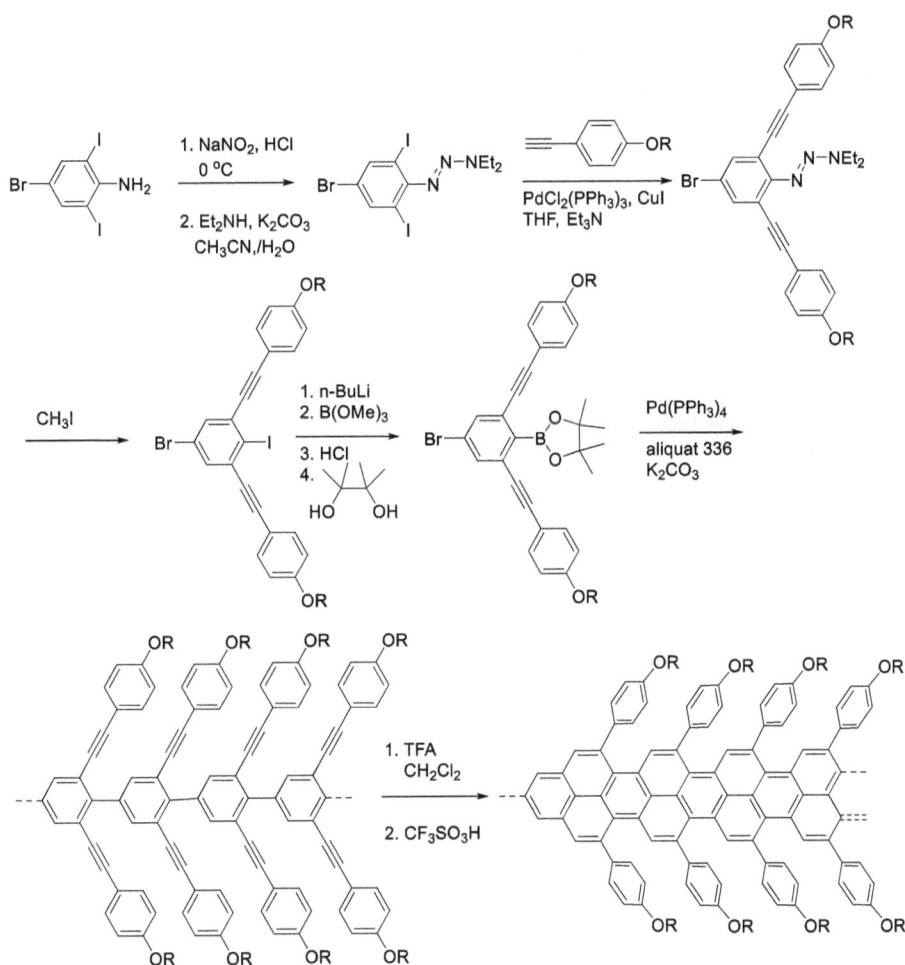

Figure 10.55: Synthesis of a rylene GNR via Suzuki polymerization and benzannulation [81].

and Raman spectroscopy [74]. Since this contribution, the on-surface synthesis method has been used to prepare a variety of different GNRs [75].

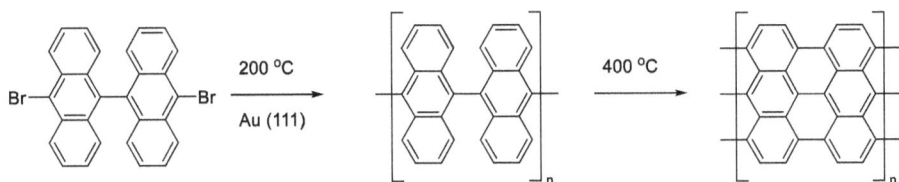

Figure 10.56: Synthesis of atomically precise GNR on a gold surface [74].

References

[1]   Anthony JE. The larger acenes: Versatile organic semiconductors. Angew Chem Int Ed. 2008;47(3):452–83.

[2]   Anthony JE. Functionalized acenes and heteroacenes for organic electronics. Chem Rev. 2006;106(12):5028–48.

[3]   Clar E, John F. Zur Kenntnis mehrkerniger aromatischer Kohlenwasserstoffe und ihrer Abkömmlinge, V. Mitteil.: Naphtho-anthracene, ihre Oxydationsprodukte und eine neue Klasse tiefgefärbter Kohlenwasserstoffe. Ber Dtsch Chem Ges. 1929;62:3021–29.

[4]   Bailey WJ, Madoff M. Cyclic dienes. II. A new synthesis of pentacene. J Am Chem Soc. 1953;75:5603–04.

[5]   Ried W, Anthofer F. Einfache Synthese für Pentacen-6,13-chinon. Angew Chem. 1953;65:601.

[6]   Bruckner V, Karczag (Wilhelms) A, Körmendy K, Meszaros M, Tomasz J. Einfache synthese des pentacens. Tetrahedron Lett. 1960;1(22):5–6.

[7]   Anthony JE, Brooks JS, Eaton DL, Parkin SR. Functionalized pentacene: Improved electronic properties from control of solid-state order. J Am Chem Soc. 2001 Sep 26;123(38):9482–83.

[8]   Martin N, Behnisch R, Hanack M. Syntheses and electrochemical properties of tetracyano-p-quinodimethane derivatives containing fused aromatic rings. J Org Chem. 1989;54 (11):2563–68.

[9]   Luo J, Hart H. Linear acene derivatives. New routes to pentacene and naphthacene and the first synthesis of a triptycene with two anthracene moieties.. J Org Chem. 1987;52 (22):4833–36.

[10]   Cava MP, Napier DR. Condensed cyclobutane aromatic systems. II. Dihalo derivatives of benzocyclobutene and benzocyclobutadiene dimer. J Am Chem Soc. 1957;79:1701–05.

[11]   Cava MP, Shirley RL. Condensed cyclobutane aromatic compounds. X. Naphtho[b] cyclobutene. J Am Chem Soc. 1960;82(3):654–56.

[12]   Swartz CR, Parkin SR, Bullock JE, Anthony JE, Mayer AC, Malliaras GG. Synthesis and characterization of electron-deficient pentacenes. Org Lett. 2005;7(15):3163–66.

[13]   Netka J, Crump SL, Rickborn B. Isobenzofuran-aryne cycloadducts: Formation and regioselective conversion to anthrones and substituted polycyclic aromatics. J Org Chem. 1986;51(8):1189–99.

[14]   Satchell MP, Stacey BE. New synthesis of hexacene. J Chem Soc C. 1971;468–69.

[15]   Payne MM, Parkin SR, Anthony JE. Functionalized higher acenes: Hexacene and heptacene. J Am Chem Soc. 2005;127(22):8028–29.

[16]   Mondai R, Shah BK, Neckers DC. Photogeneration of heptacene in a polymer matrix. J Am Chem Soc. 2006;128(30):9612–13.

[17]   Mondal R, Adhikari RM, Shah BK, Neckers DC. Revisiting the stability of hexacenes. Org Lett. 2007;9(13):2505–08.

[18]   Tönshoff C, Bettinger HF. Photogeneration of octacene and nonacene. Angew Chem Int Ed. 2010;49(24):4125–28.

[19]   Shen B, Tatchen J, Sanchez-Garcia E, Bettinger HF. Evolution of the optical gap in the acene series: Undecacene. Angew Chem Int Ed. 2018;57(33):10506–09.

[20]   Jancarik A, Levet G, Gourdon A. A practical general method for the preparation of long acenes. Chem-Eur J. 2019;25(9):2366–74.

[21]   Okamoto H, Hamao S, Eguchi R, Goto H, Takabayashi Y, Yen PYH, et al. Synthesis of the extended phenacene molecules, [10]phenacene and [11]phenacene, and their performance in a field-effect transistor. Sci Rep. 2019;9:4009.

[22]   Waldmann H, Pitschak G. A new picene synthesis. Justus Liebigs Ann Chem. 1937;527:183–89.

[23] Phillips DD. Polynuclear aromatic hydrocarbons. I. A new synthesis of picene. J Am Chem Soc. 1953;75:3223–26.

[24] Harvey RG, Pataki J, Cortex C, Di Raddo P, Yang C. A new general synthesis of polycyclic aromatic compounds based on enamine chemistry. J Org Chem. 1991;56:1210–17.

[25] Mallory FB, Butler KE, Evans AC. Phenacenes: A family of graphite ribbons. 1.Syntheses of some [7]phenacenes by stilbene-like photocyclization. Tetrahedron Lett. 1996;37 (40):7173–76.

[26] Scholl R, Seer C, Weitzenböck R. Perylen, a high condensated aromatic hydrocarbon C20H12. Ber Dtsch Chem Ges. 1910;43:2202–09.

[27] Clar E. Das Kondensationsprinzip, Ein Einfaches Neues Prinzip Im AufBau Der Aromatischen Kohlenwasserstoffe (Aromatische Kohlenwasserstoffe) 42. Chem Ber. 1948;81:52.

[28] Postovskii IY, Bednyagina NP. Synthesis of Perylene from Anthracene. Zhur Obs Chem USSR. 1937;7:2919–25.

[29] Bohnen A, Koch KH, Luettke W, Mullen K. Polyarylenes and polyarylenevinylenes. 2. Oligorylenes as models for poly-peri-naphthalene. Angew Chem. 1990;102(5):548–50.

[30] Bengs H, Closs F, Frey T, Funhoff D, Ringsdorf H, Siemensmeyer K. Highly photoconductive discotic liquid crystals structure-property relations in the homologous series of hexa-alkoxytriphenylenes. Liq Cryst. 1993;15(5):565–74.

[31] Buess CM, Lawson DD. The preparation, reactions, and properties of triphenylenes. Chem Rev. 1960;60(4):313–30.

[32] Rhaman AU, Tombesi OL. Double acylation of aromatic compounds. III. Synthesis of triphenylene by double succinylation of naphthalene. Chem Ber. 1966;99(6):1805–09.

[33] Bergmann E, Blum-Bergmann O. Synthesis of Triphenylene. J Am Chem Soc. 1937;59 (8):1441–42.

[34] Pérez D, Guitián E. Selected strategies for the synthesis of triphenylenes. Chem Soc Rev. 2004;33(5):274–83.

[35] Grzybowski M, Skonieczny K, Butenschön H, Gryko DT. Comparison of oxidative aromatic coupling and the scholl reaction. Angew Chem Int Ed. 2013;52(38):9900–30.

[36] Boden N, Bushby RJ, Cammidge AN. A quick-and-easy route to unsymmetrically substituted derivatives of triphenylene: Preparation of polymeric discotic liquid crystals. J Chem Soc Chem Commun. 1994;465–66.

[37] Zhai L, Shukla R, Wadumethrige SH, Rathore R. Probing the arenium-ion (protonTransfer) versus the cation-radical (electron transfer) mechanism of scholl reaction using DDQ as oxidant. J Org Chem. 2010;75(14):4748–60.

[38] Kumar S, Manickam M. Oxidative trimerization of o-dialkoxybenzenes to hexaalkoxytriphenylenes: Molybdenum(v) chloride as a novel reagent. Chem Commun. 1997;1615–16.

[39] Borner RC, Bushby RJ, Cammidge AN. Ferric chloride/methanol in the preparation of triphenylene-based discotic liquid crystals. Liq Cryst. 2006;33(11–12):1439–48.

[40] Goodby JW, Hird M, Toyne KJ, Watson T. A novel, efficient and general synthetic route to unsymmetrical triphenylene mesogens using palladium-catalysed cross-coupling reactions. J Chem Soc Chem Commun. 1994;1701–02.

[41] Lau K, Foster J, Williams V. Synthesis of a hexaalkoxybenzo[b]triphenylene mesogen. Chem Commun. 2003 Sep ;7(17):2172–73.

[42] Paquette JA, Yardley CJ, Psutka KM, Cochran MA, Calderon O, Williams VE, et al. Dibenz[a,c] anthracene derivatives exhibiting columnar mesophases over broad temperature ranges. Chem Commun. 2012 Jul ;11(48):8210–12.

[43] Psutka KM, Williams J, Paquette JA, Calderon O, Bozek KJA, Williams VE, et al. Synthesis of substituted dibenz[a,c]anthracenes and an investigation of their liquid-crystalline properties. Eur J Org Chem. 2015;(7):1456–63.

[44] Psutka KM, Bozek KJA, Maly KE. Synthesis and mesomorphic properties of novel dibenz[a,c] anthracenedicarboximides. Org Lett. 2014;16(20):5442–45.

[45] Paquette JA, Psutka KM, Yardley CJ, Maly KE. Probing the structural features that influence the mesomorphic properties of substituted dibenz[a,c]anthracenes. Can J Chem. 2017;95 (4):399–409.

[46] Sato T, Shimada S, Hata K. A new route to polycondensed aromatics: Photolytic formation of triphenylene and dibenzo[fg, op]naphthacene ring systems. Bull Chem Soc Jpn. 1971;44:2484–90.

[47] Bushby RJ, Hardy C. Regiospecific synthesis of 2,3,6,7,10,11-hexasubstituted triphenylenes by oxidative photocyclisation of 3,3”,4,4′,4”,5′-hexasubstituted 1,I ′ : 2′1 “-terphenyls. J Chem Soc Perkin Trans 1. 1986;721–23.

[48] Peña D, Escudero S, Pérez D, Guitián E, Castedo L. Efficient palladium-catalyzed cyclotrimerization of arynes: Synthesis of triphenylenes. Angew Chem Int Ed. 1998;37 (19):2659–61.

[49] Zhou ZH, Yamamoto T. Research on carbon-carbon coupling reactions of haloaromatic compounds mediated by zerovalent nickel complexes. Preparation of cyclic oligomers of thiophene and benzene and stable anthrylnickel(II) complexes. J Organomet Chem. 1991;414 (1):119–27.

[50] Schroeder ZW, Ledrew J, Selmani VM, Maly KE. Preparation of substituted triphenylenes via nickel-mediated Yamamoto coupling. RSC Adv. 2021;11:39564–69.

[51] Lynett PT, Maly KE. Synthesis of substituted trinaphthylenes via aryne cyclotrimerization. Org Lett. 2009;11(16):3726–29.

[52] Pummerer R, Luttringhaus A, Fick R, Pfaff A, Riegelbauer G, Rosenhauer E. Polymerization processes. Condensation of 1,4-naphthoquinone to triphthaloylbenzene with pyridine. Ber Dtsch Chem Ges. 1938;71B:2569–83.

[53] Yin J, Qu H, Zhang K, Luo J, Zhang X, Chi C, et al. Electron-deficient triphenylene and trinaphthylene carboximides. Org Lett. 2009;11(14):3028–31.

[54] Rüdiger EC, Rominger F, Steuer L, Bunz UHF. Synthesis of Substituted Trinaphthylenes. J Org Chem. 2016;81(1):193–96.

[55] Clar E, Mullen A. The non-existence of a threefold aromatic conjugation in linear benzologues of triphenylene (starphenes). Tetrahedron. 1968;24(23):6719–24.

[56] Clar E, Kelly W. Aromatic hydrocarbons. LXVII. Heptaphene and 2,3,8,9-dibenzopicene. J Am Chem Soc. 1953;76:3502–04.

[57] Rüdiger EC, Porz M, Schaffroth M, Rominger F, Bunz UHF. Synthesis of soluble, alkyne-substituted trideca- and hexadeca-starphenes. Chem-Eur J. 2014;20(40):12725–28.

[58] Holec J, Cogliati B, Lawrence J, Berdonces-Layunta A, Herrero P, Nagata Y, et al. A large starphene comprising pentacene branches. Angew Chem Int Ed. 2021;60(14):7752–58.

[59] Scholl R, Meyer K. Synthese des anti-diperi-Dibenz-coronens und dessen Abbau zum Coronen (Hexabenzo-benzol). (Mitbearbeitet von Horst v. Hoeßle und Solon Brissimdji). Ber Dtsch Chem Ges. 1932;65:902–15.

[60] Clar E, Zander M. Syntheses of Coronene and 1 : 2-7 : 8-Dibenzocoronene. J Chem Soc. 1957;4616–19.

[61] Van Dijk JTM, Hartwijk A, Bleeker AC, Lugtenburg J, Cornelisse J. Gram scale synthesis of benzo[ghi]perylene and coronene. J Org Chem. 1996;61(3):1136–39.

[62] Davy JR, Reiss JA. Synthesis of [2,2] (2,7)naphthalenophane-1,11-diene. J Chem Soc Chem Commun. 1973;806–07.

[63] Davy JR, Reiss JA. Stevens rearrangements in a napthalene cyclophane. Tetrahedron Lett. 1972;13(35):3639–42.

[64] Clar E, Ironside CT, Zander M. The electronic interaction between benzenoid rings in condensed aromatic hydrocarbons. 1: 12–2:3-4:5-6:7-8:9-10:11-hexabenzocoronene, 1:2-3:4-5:6-10:11-tetrabenzoanthanthrene, and 4:5-6:7-11:12-13:14-tetrabenzoperopyrene. J Chem Soc. 1959;142–47.

[65] Hendel W, Khan ZH, Schmidt W. Hexa-peri-benzocoronene, a candidate for the origin of the diffuse interstellar visible absorption bands ?. Tetrahedron. 1986;42(4):1127–34.

[66] Stabel A, Herwig P, Mullen K, Rabe JP. Diode-like current-voltage curves fo a single molecule-tunneling sepctroscopy with submolecular resolution of an alkylated, pericondensed hexabenzocoronene. Angew Chem Int Ed. 1995;34(15):1609–11.

[67] Herwig P, Kayser CW, Mullen K, Spiess HW. Columnar mesophases of alkylated hexa-perihexabenzocoronenes with remarkably large phase widths. AdvMater. 1996;8(6):510–13.

[68] Müller M, Iyer VS, Kübel C, Enkelmann V, Müllen K. Polycyclic aromatic hydrocarbons by cyclodehydrogenation and skeletal rearrangement of oligophenylenes. Angew Chem Int Ed. 1997;36(15):1607–10.

[69] Iyer VS, Wehmeier M, Diedrich J, Keegstra MA, Mullen K. From hexa-peri-hexabenzocoronene to "superacenes.". Angew Chem Int Ed. 1997;36(15):1603–07.

[70] Simpson CD, Brand JD, Berresheim AJ, Przybilla L, Raeder HJ, Mullen K. Synthesis of a giant 222 carbon graphite sheet. Chem-Eur J. 2002;8(6):1424–29.

[71] Chen Z, Lin Y-M, Rooks MJ, Avouris P. Graphene nano-ribbon electronics. Phs E. 2007;40 (2):228–32.

[72] Li X, Wang X, Zhang L, Lee S, Dai H. Chemically derived, ultrasmooth graphene nanoribbon semiconductors. Science. 2008;319:1229–32.

[73] Kosynkin DV, Higginbotham AL, Sinitskii A, Lomeda JR, Dimiev A, Price BK, et al. Longitudinal unzipping of carbon nanotubes to form graphene nanoribbons. Nature. 2009;458:872–76.

[74] Cai J, Ruffieux P, Jaafar R, Bieri M, Braun T, Blankenburg S, et al. Atomically precise bottom-up fabrication of graphene nanoribbons. Nature. 2010;466(7305):470–73.

[75] Houtsma RSK, De La Rie J, Stöhr M. Atomically precise graphene nanoribbons: Interplay of structural and electronic properties. Chem Soc Rev. 2021;50(11):6541–68.

[76] Yang X, Dou X, Rouhanipour A, Zhi L, Räder HJ, Müllen K. Two-dimensional graphene nanoribbons. J Am Chem Soc. 2008;130(13):4216–17.

[77] Vo TH, Shekhirev M, Kunkel DA, Morton MD, Berglund E, Kong L, et al. Large-scale solution synthesis of narrow graphene nanoribbons. Nat Commun. 2014;5:1–8.

[78] Narita A, Feng X, Hernandez Y, Jensen SA, Bonn M, Yang H, et al. Synthesis of structurally well-defined and liquid-phase-processable graphene nanoribbons. Nat Chem. 2014;6:126–32.

[79] Gao J, Uribe-Romo FJ, Saathoff JD, Arslan H, Crick CR, Hein SJ, et al. Ambipolar transport in solution-synthesized graphene nanoribbons. ACS Nano. 2016;10(4):4847–56.

[80] Goldfinger MB, Swager TM. Fused polycyclic aromatics via electrophile-induced cyclization reactions: Application to the synthesis of graphite ribbons. J Am Chem Soc. 1994;116:7895–96.

[81] Yang W, Lucotti A, Tommasini M, Chalifoux WA. Bottom-up synthesis of soluble and narrow graphene nanoribbons using alkyne benzannulations. J Am Chem Soc. 2016;138 (29):9137–44.

11 Nonplanar aromatic compounds

11.1 Introduction

Most aromatic compounds are planar – indeed one of the accepted criteria for aromaticity is planarity, in part because delocalization of the π-electrons requires an essentially parallel alignment of the atomic p-orbitals on adjacent atoms. However, it is easy to imagine that small deviations from planarity would not disrupt aromaticity. The question then becomes: at what point does deviation from planarity disrupt aromaticity? The answer to this may depend in part on what measure of aromaticity is being used. The deliberate synthesis of nonplanar aromatic systems has been motivated in part to probe the limits of aromaticity, and in part to explore the synthesis of challenging molecular structures. Additionally, however, nonplanar aromatic compounds often exhibit interesting properties that make them functional materials, and can potentially serve as synthetic precursors for the bottom-up synthesis of fullerenes and carbon nanotubes. For all of these reasons, the synthesis of nonplanar aromatic compounds is an area of active research.

One of the challenges of preparing nonplanar aromatic compounds is of course overcoming the normal tendency of aromatic compounds to be planar. Contorting aromatic rings out of plane may disrupt aromaticity and introduce strain, so it comes with an energetic cost. In general, there are two approaches for the preparation of nonplanar aromatic systems: i) aromatic compounds with steric crowding that forces the molecules to adopt a nonplanar geometry, and ii) systems that adopt a nonplanar geometry due to the constraints of covalent bonds. The first approach is exemplified by compounds such as helicenes and "twistacenes," while examples of the second approach includes fullerene fragments, cyclophenylenes, and some cyclophanes. In this chapter, we will explore some of the synthetic approaches to select classes of nonplanar aromatic systems. While many of the synthetic methods are based on reactions seen in previous chapters, the synthetic strategies also necessarily include ways to overcome or circumvent the strain in the nonplanar systems.

11.2 Helicenes and related contorted polycyclic compounds

11.2.1 Helicenes

Helicenes are polycyclic aromatic hydrocarbons consisting of *ortho*-fused benzene rings [1, 2]. To avoid steric congestion, these compounds adopt helical geometry. Helicenes are named according to the number of aromatic rings that comprise them (Figure 11.1).

https://doi.org/10.1515/9783110562682-011

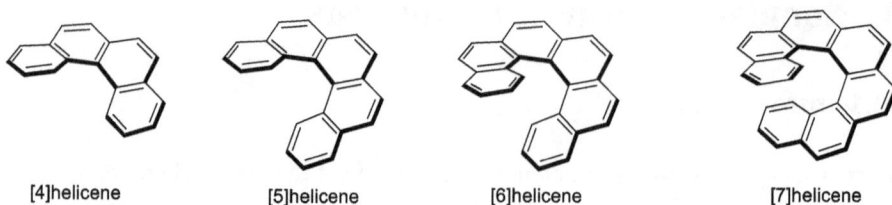

[4]helicene   [5]helicene   [6]helicene   [7]helicene

Figure 11.1: Representative helicene structures.

One of the interesting features of these compounds is that their helical configuration means that they are chiral, existing as either a right-handed or left-handed helix [3]. The right-handed helix is denoted P, while the left-handed helix is denoted M (Figure 11.2). The smaller helicenes are not configurationally stable and can racemize easily. The larger helicenes are configurationally stable and exhibit interesting chiroptical properties.

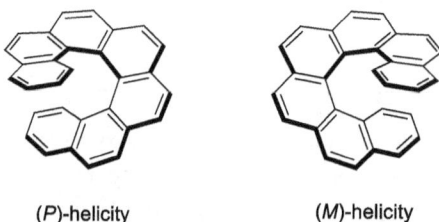

(P)-helicity   (M)-helicity

Figure 11.2: Helical chirality in helicenes.

The earliest syntheses of helicenes focused on [5]helicene, and involved the Pschorr cyclization of diazonium salts as a key step (see Section 3.6.4). As shown in Figure 11.3, reduction of nitro-substituted bis-stilbene was followed by a Pschorr cyclization, which

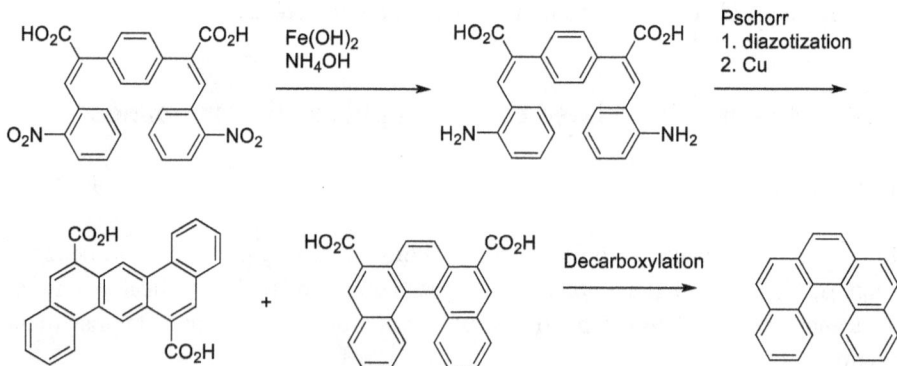

Figure 11.3: Synthesis of [5]helicene via Pschorr reaction [4].

gave the helicene precursor along with the non-helical isomer as a side product [4]. This reaction highlights one of the challenges associated with the preparation with helicenes – cyclization reactions such as this can lead to poor selectivity for the helicene product. Subsequent decarboxylation gave [5]helicene.

Another early synthetic approach toward helicenes made use of intramolecular Friedel–Crafts cyclization reactions, similar to strategies used to prepare polycyclic aromatic hydrocarbons seen in Chapter 9. For example, in 1956, Newman and coworkers used the dicarboxylic acid derivative and conducted sequential Friedel–Crafts acylation reactions, followed by Wolff–Kishner reductions to remove the carbonyl groups (Figure 11.4) [5]. Finally, aromatization was achieved at high temperature in the presence of rhodium and alumina to give [6]helicene.

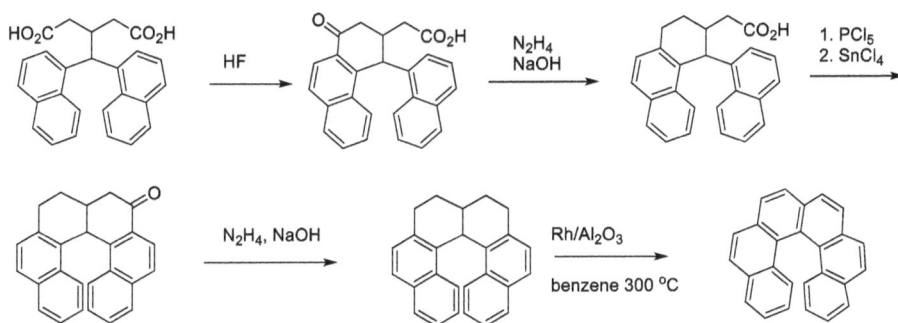

Figure 11.4: Synthesis of [6]helicene via Friedel–Crafts cyclizations [5].

A widely used strategy for the preparation of helicenes made use of photocyclization and oxidation (see Section 9.3.6). For example, in 1967, photocyclization and oxidation was used to prepare [4]helicene (Figure 11.5) [6]. However, when the same reaction was used to prepare [5]helicene, 1,12-benzoperylene was isolated as the major product [7]. In this case, presumably [5]helicene is formed and undergoes a subsequent photocyclization reaction (Figure 11.5).

In the same year, Martin reported the synthesis of [7]helicene via photocyclization (Figure 11.6) [8]. This synthesis was significant because it was the first report of a [7]helicene. The synthesis begins with a Friedel–Crafts acylation of phenanthrene with the acid chloride, which was followed by reduction and dehydration to give the stilbene analog. Photocyclization using a mercury lamp in the presence of iodine gave [7]helicene in good yield. This photocyclization has been extended to the preparation of helicenes as large as [13]helicene [9].

One of the challenges of photocyclization methods and indeed other cyclization reactions such as the Pschorr reaction is control of the regiochemistry of cyclization to form a helical product. This challenge is illustrated in Figure 11.7, which shows the two potential photocyclization products that can be produced during the preparation

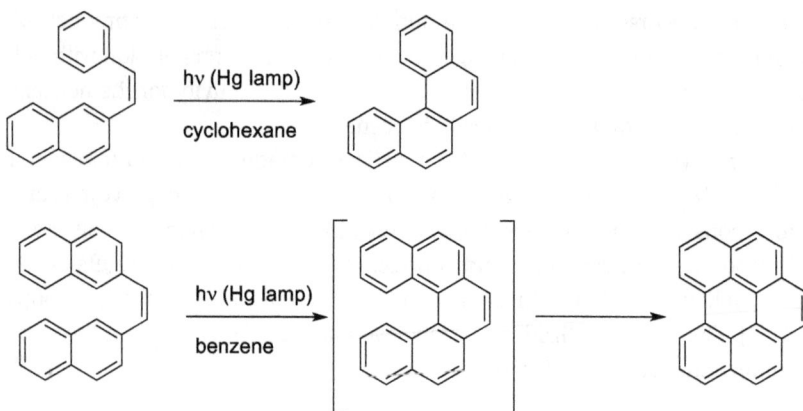

Figure 11.5: Photocyclization to form [4]helicene and 1,12-benzoperylene [6, 7].

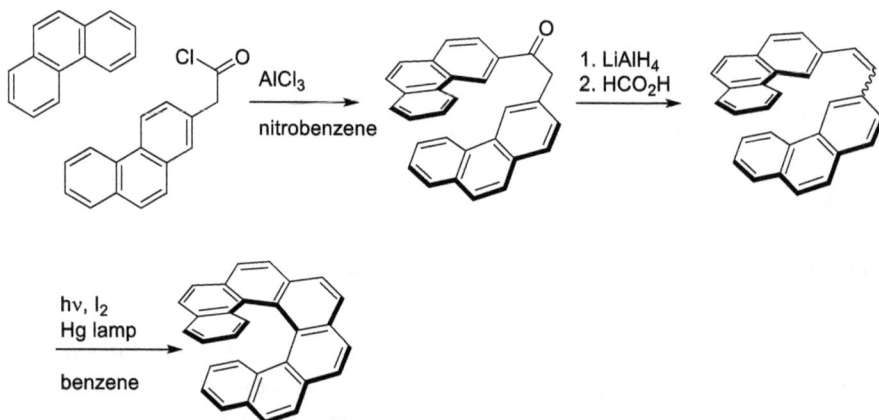

Figure 11.6: Synthesis of [7]helicene via photocyclization [8].

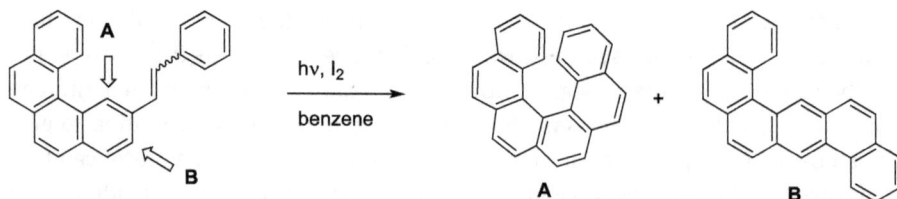

Figure 11.7: Possible regioisomers during photocyclization reactions.

of [6]helicene. If photocyclization occurs at position A, the desired helicene is obtained. However, if photocyclization occurs at position B, benzo[a]naphtha[1,2-h]anthracene is obtained.

To prevent the undesired regioisomer, both the Martin and Katz groups used a bromo substituent at position "B" as a protective group to block the undesired photocyclization (Figure 11.8) [10, 11]. The bromo substituent can then be removed by lithium–halogen exchange and trapping the aryllithium with water.

Figure 11.8: Controlling regiochemistry of photocyclization using a "bromo auxiliary" [10, 12].

Another strategy for the preparation of helicenes involves Diels-Alder reactions [13, 14]. Among the earliest examples of this approach involved starting with 3,3′,4,4′-tetrahydro-1,1′-binaphthalene as a diene and reacting it with maleic anhydride, followed by removal of the anhydride and aromatization to give [5]helicene (Figure 11.9) [14].

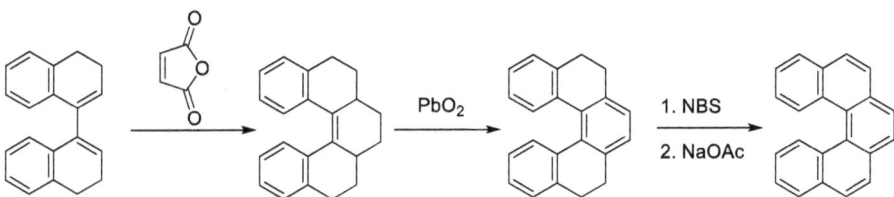

Figure 11.9: Synthesis of [5]helicene via Diels–Alder reaction [14].

A similar approach has been used for the preparation of substituted helicenes by using similar dienes and dienophiles such as *N*-phenylmaleimide, quinones, and benzynes (Figure 11.10) [15, 16].

A noteworthy advance in the synthesis of helicene derivatives via Diels–Alder reactions reported by Katz and coworkers involve the Diels–Alder reaction of divinylarenes with benzoquinone to prepare the corresponding helicenebisquinones (Figure 11.11) [17, 18]. This synthesis demonstrated the viability of Diels–Alder reactions for preparing functionalized helicenes and allow them to be prepared on a gram scale.

Yet another approach for preparing helicenes was inspired by Kharash's method for the preparation of phenanthrene by treating 2,2′-bis(bromomethyl)-1,1′-biphenyl with potassium amide in liquid ammonia (Figure 11.12) [19]. The reaction has been

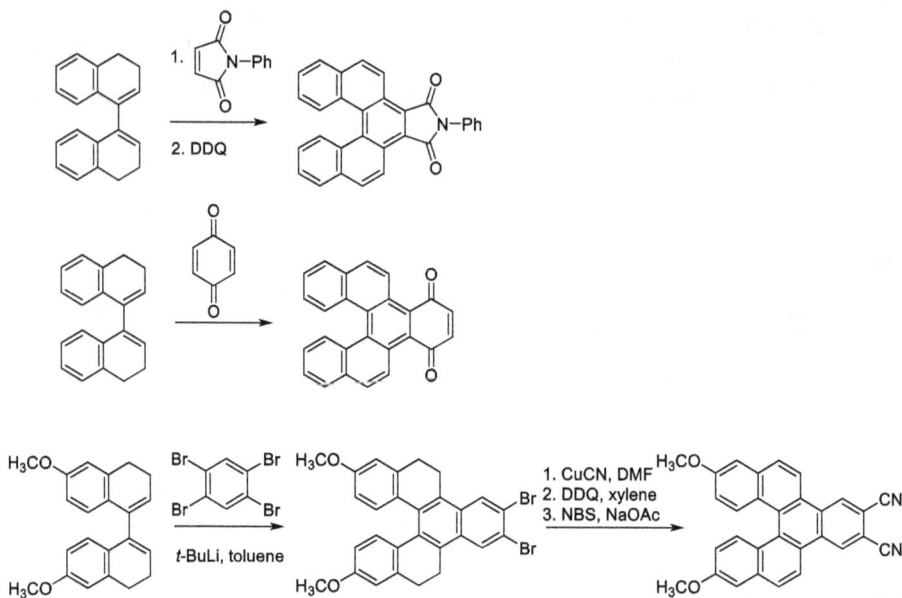

Figure 11.10: More examples of helicene syntheses via Diels–Alder reactions [15, 16].

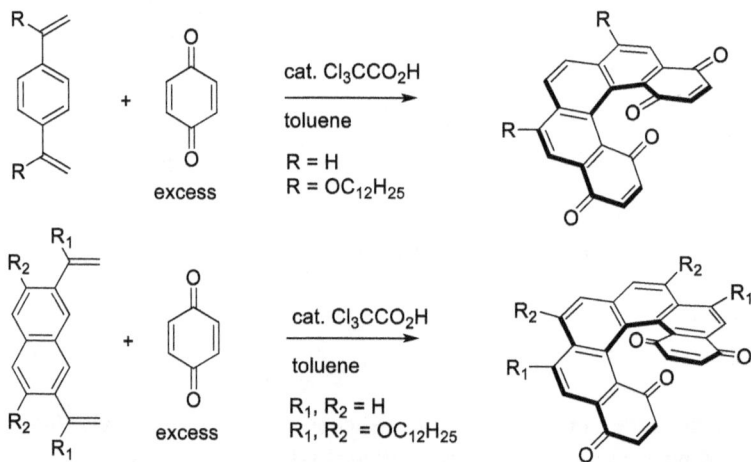

Figure 11.11: Synthesis of helicenebisquinones via Diels–Alder reactions [17, 18].

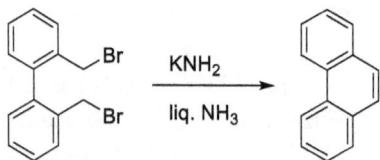

Figure 11.12: Preparation of phenanthrene [19].

referred to as a carbenoid coupling or a "benzylic-type" coupling, although studies by Gingras and coworkers suggest that the reaction does not proceed through a carbene intermediate [20].

Although the mechanism of the reaction is not fully understood, the reaction has been used successfully to prepare helicenes [21, 22]. For example, the synthesis of [5]helicene is outlined in Figure 11.13 [22]. The triflate starting material, which was derived from 2,2′-binol, was subjected to Kumada coupling to introduce the methyl groups. Benzylic bromination gave the bis-bromomethyl binaphthyl derivative, which was treated with LiHMDS in THF and HMPA to give [5]helicene in good yield. This approach has also been used for the preparation of [7]helicene [20].

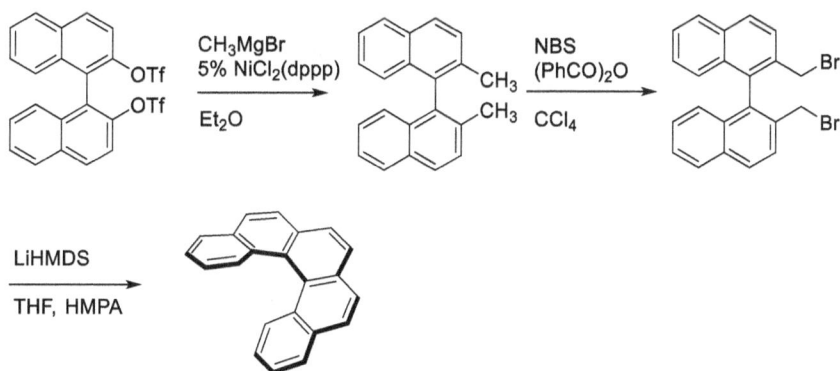

Figure 11.13: Synthesis of [5]helicene via a carbenoid coupling [22].

Two conceptually similar approaches to helicenes consist of a McMurry coupling [22], as well as ring-closing metathesis (Figure 11.14) [23]. While the yields of the McMurry coupling were quite low, the olefin metathesis reaction produced helicenes in high yield.

Another approach to helicenes consists of intramolecular [2 + 2 + 2] cyclotrimerization of alkynes (see Section 8.2.2). An example of this approach is shown in Figure 11.15 [24]. This reaction begins with a selective Sonogashira coupling of 2-bromoiodobenzene with trimethylsilylacetylene, followed by an in situ removal of the silyl group and addition of 1-iodo-2-bromonaphthalene, which gave a second selective Sonogashira coupling. This sequence enabled the one-pot preparation of the dissymmetric alkyne, and highlights selectivity for aryl iodides over aryl bromides in Sonogashira couplings. The resulting dibromo-substituted diarylacetylene was then subjected to a twofold palladium-catalyzed Suzuki cross-coupling to give the trialkyne. Deprotection of the alkynes was followed by a nickel-mediated cyclotrimerization reaction to give the helicene derivative in good yield.

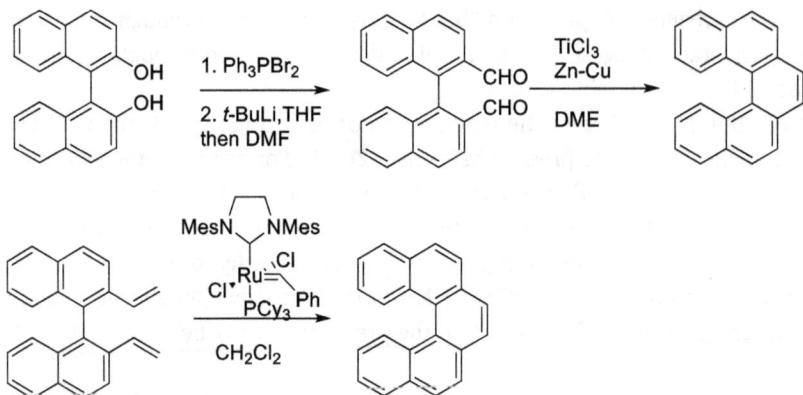

Figure 11.14: Helicenes via McMurry coupling and ring-closing olefin metathesis [22, 23].

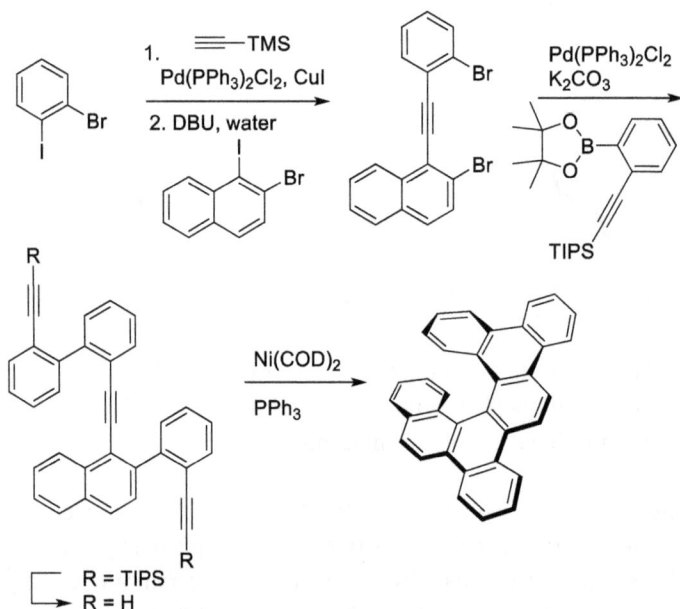

Figure 11.15: Synthesis of helicenes via alkyne cyclotrimerization [24].

11.2.2 Stereoselective helicene syntheses

One of the most interesting structural features of helicenes is their chirality. As such, there is an interest in preparing enantiomerically pure helicenes [3]. One way to achieve enantiomerically enriched helicenes is by separation of enantiomers. As an example, resolution of [6]helicene was achieved by forming a diastereomeric charge transfer complex with 2-(2,4,5,7-tetranitro-9-fluorenylidene-aminooxy)propionic acid (TAPA), shown in Figure 11.16 [5, 25]. Another approach involved the manual separation of enantiomorphic crystals – an approach that relies on crystallization of the compounds as chiral conglomerates [26, 27]. A more common approach for separating helicene enantiomers is by chiral stationary phase HPLC. At the same time, several approaches to stereoselective synthesis of helicenes have been explored. During the preparation of helicenes via photocyclization, attempts have been made to control chirality using circularly polarized light to achieve the photocyclization [28], and by carrying the photocyclization in a chiral nematic medium as the solvent [29]. While both of these approaches are conceptually interesting and elegant, the enantiomeric excess achieved for both was very low.

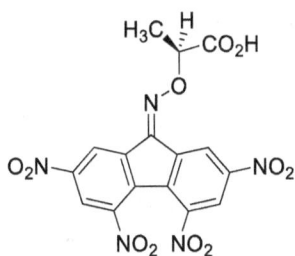

Figure 11.16: "(−)-TAPA," a chiral compound, used for the resolution of [6]helicene.

More conventional approaches for achieving stereochemical control such as chiral auxiliaries or chiral catalysts have been explored and are in some cases effective for the preparation of enantio-enriched helicenes. Enantiomerically pure chiral groups or auxiliaries have been attached to helicene precursors to induce a preferred helical configuration via a diastereoselective reaction. For example, the enantiopure paracyclophane was attached to a [4]helicene via a Wittig reaction, producing a stilbene derivative that could undergo photocyclization with high diastereoselectivity (Figure 11.17) [30].

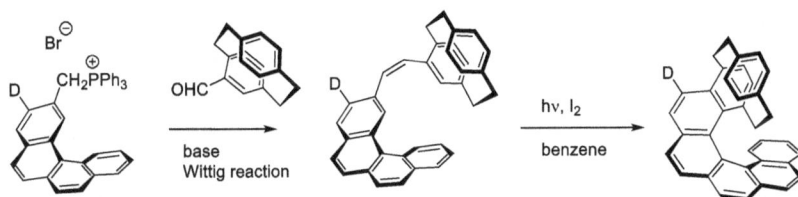

Figure 11.17: Diastereoselective helicene formation incorporating a chiral paracyclophane [30].

Chiral menthyl groups have been used to induce a diastereoselective photocyclization as outlined in Figure 11.18 [31]. In the first example, a substituted 5-helicene bearing two menthyl groups was prepared in a 7:3 mixture of diastereomers. When

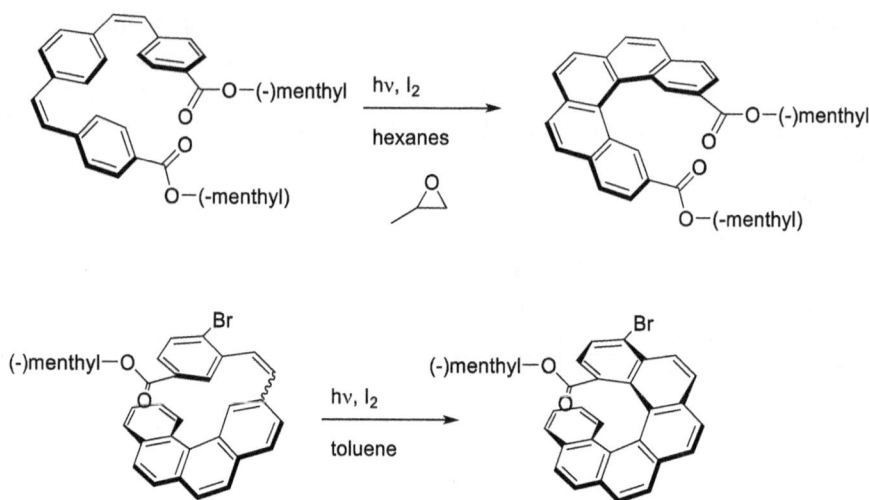

Figure 11.18: Menthyl groups as chiral auxiliaries for diastereoselective helicene synthesis [31, 32].

this strategy was combined with a bromo auxiliary, a substituted [6]helicene was prepared in good yield and diastereomeric ratios as high as 98:2 [32]. Unlike the previous examples, where the chiral component remains in the structure, attaching the menthyl groups as esters means that these can function as true chiral auxiliaries that can be removed to give the enantiomerically enriched helicene.

As another example, a chiral cyclopentane was incorporated into a stilbene precursor and used to direct the helicity upon photocyclization. Hydrolysis and dehydration of the chiral cyclopentanol derivatives gave the corresponding helicene terminated with cyclopentadienyl rings in enantiomeric excess of greater than 89% (Figure 11.19) [33]. This functionality was exploited to prepare helical metallocenes. When two chiral cyclopentane groups were used in conjunction with a bromo group to prevent formation of the linear photocyclization product, a diastereomeric ratio of more than 99:1 was achieved for the photocyclization [34]. The bromo group was then removed by lithium–halogen exchange and treatment with aqueous ammonium chloride, and the chiral groups were removed as in the previous example by hydrolysis and dehydration.

The examples of diastereoselective helicene syntheses presented thus far have all focused photocyclization as the key step to form the helicene. However, diastereoselective syntheses of helicenes has also been achieved using other approaches. For example, Diels–Alder reactions of vinyl aromatic compounds with benzoquinone bearing a chiral sulfoxide group has been used to prepare helicenebisquinones with enantiomeric excesses of up to 88% (Figure 11.20) [35].

An alternative to a diastereoselective helicene synthesis using covalently linked chiral moieties is a true asymmetric synthesis using a chiral catalyst. For example,

R=TBDMS

R=TBDMS

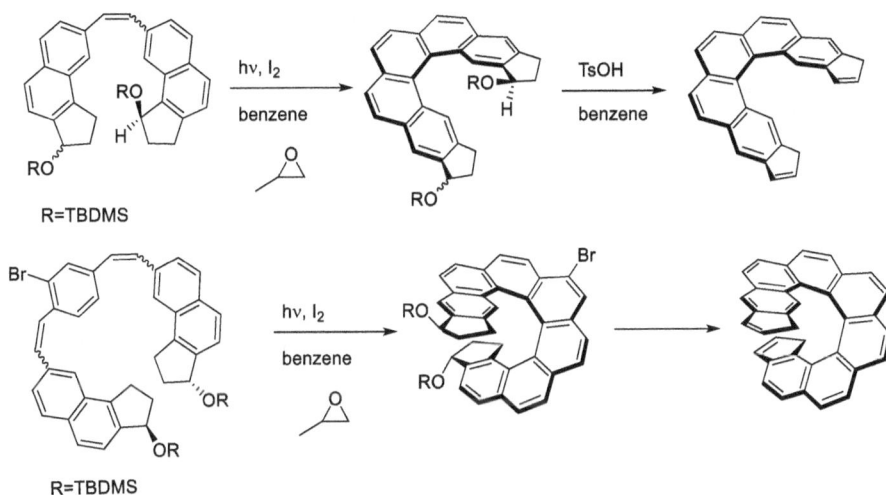

Figure 11.19: Diastereoselective helicene synthesis using a chiral cyclopentanol auxiliary [33, 34].

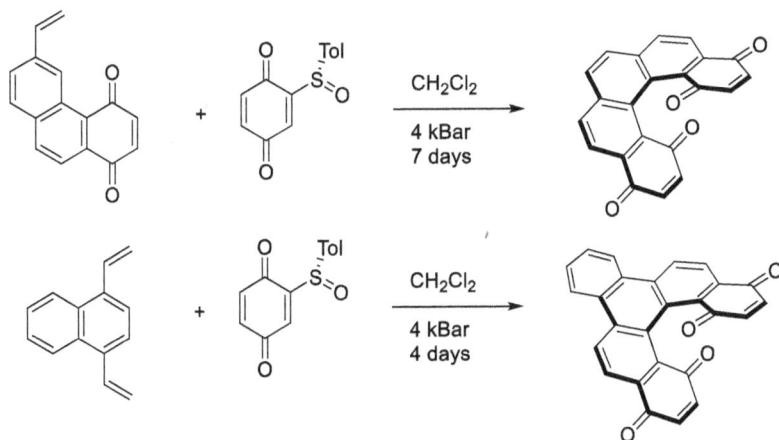

Figure 11.20: Diastereoselective Diels–Alder reactions with a chiral benzoquinone [35].

olefin metathesis using a chiral ruthenium catalyst has been used to prepare enantio-enriched [7]helicenes (Figure 11.21) [36]. Depending on the catalyst used, either the (M) or (P) helical forms could be favored. It should be noted that the bianthryl precursor is actually racemic, so this enantioselective reaction is essentially a kinetic resolution, where the catalyst reacts preferentially with one enantiomer. Consequently, yields are modest.

R = H, CH$_3$

(P)

(M)

Chiral catalyst:

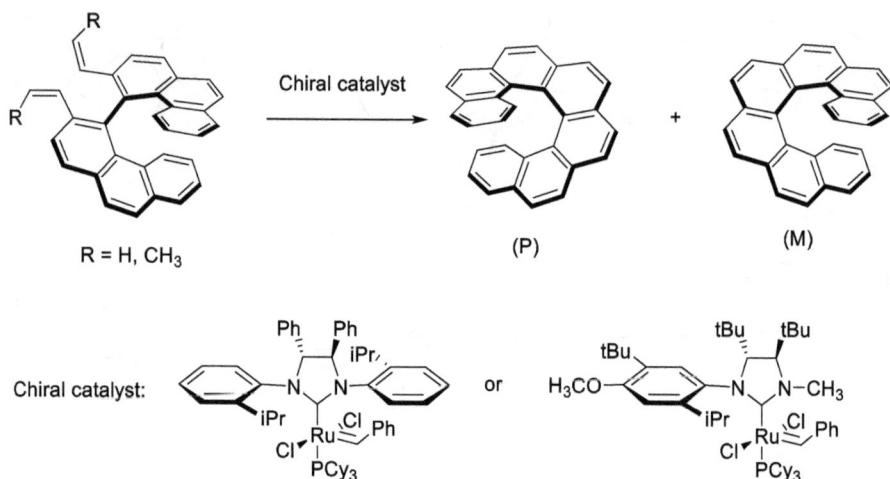

Figure 11.21: Helicene synthesis by asymmetric olefin metathesis [36].

As a final example, enantioselective synthesis of a substituted [5]helicenes was reported using palladium-catalyzed aryne and alkyne [2 + 2 + 2] cocyclotrimeriza-tion (see Chapters 5 and 9) in the presence of a chiral ligand (Figure 11.22) [37]. Spe-cifically, the trimethylsilyl naphthyl triflate and dimethyl acetylene dicarboxylate was treated with CsF in the presence of catalytic Pd$_2$(dba)$_3$ and a chiral ligand such as BINAP. Under these conditions, modest enantiomeric excesses of up to 60% were observed.

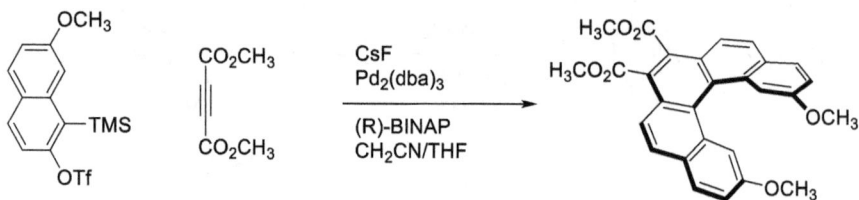

Figure 11.22: [5]Helicenes via palladium-catalyzed [2 + 2 + 2] cocyclotrimerization [37].

11.2.3 Other benzannulated PAHs

There are many nonplanar polycyclic aromatic hydrocarbons that feature steric crowd-ing similar to helicenes and often have structural features similar to helicenes, but which are more complex. For example, hexabenzotriphenylene (Figure 11.23) is nonpla-nar and can be thought of as a set of three [5]helicenes fused at the central ring. The compounds can adopt either a D$_3$-symmetric propeller conformation or a C$_2$-symmetric conformation.

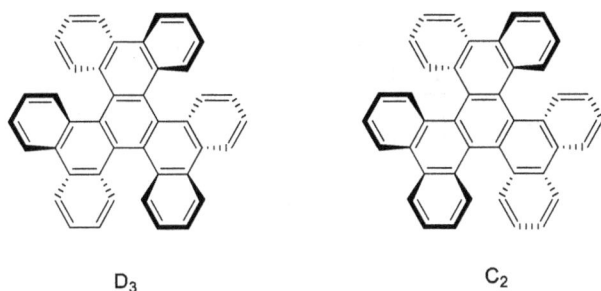

Figure 11.23: Conformations of hexabenzotriphenylene.

The first definitive synthesis of hexabenzotriphenylene was reported by Pascal and coworkers, who used flash vacuum pyrolysis of phenanthrene-9,10-dicarboxylic anhydride (Figure 11.24), which was prepared by photocyclization of diphenyl maleic anhydride [38]. Using this approach, the flash vacuum pyrolysis at 700 °C gave hexabenzotriphenylene in a 5% yield. X-ray crystallographic analysis revealed that hexabenzotriphenylene adopted a D_3-symmetric, propeller-like conformation.

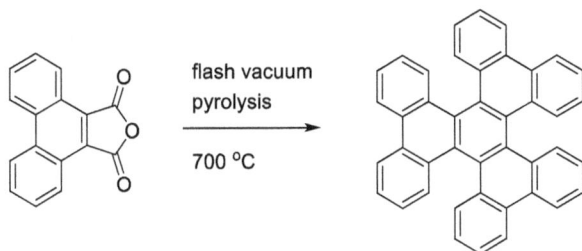

Figure 11.24: Synthesis of hexabenzotriphenylene by flash vacuum pyrolysis.

Subsequently, Peña et al. showed that hexabenzotriphenylene could be prepared via a palladium-catalyzed aryne cyclotrimerization (see Chapter 9) as outlined in Figure 11.25 [39]. The synthesis began with bromination of 9-phenanthrol, which was followed silylation both at the bromo substituent and the alcohol. Following that, the trimethylsilyl ether was cleaved and the resulting phenoxide was trapped with triflic anhydride to give the trimethyl silyl triflate as the aryne precursor. Treatment of this compound with CsF in the presence of $Pd_2(dba)_3$ led to aryne formation and cyclotrimerization to give hexabenzotriphenylene. Interestingly, the compound prepared by aryne cyclotrimerization yielded the C_2-symmetric conformation, which slowly converted to the more stable D_3-symmetric conformation in solution [40].

This methodology was extended to other contorted polycyclic aromatic hydrocarbons containing helicene units embedded in the structure. For example, the aryne cyclotrimerization of 1,2-naphthalyne gave a mixture of products – the major product

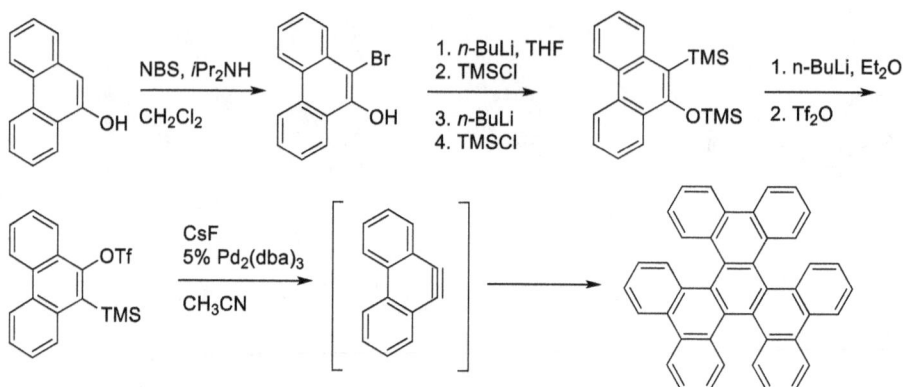

Figure 11.25: Synthesis of hexabenzotriphenylene via aryne cyclotrimerization [39].

consisting of a [5]helicene unit and a [4]helicene unit, and the minor product consisting of three fused [4]helicenes (Figure 11.26) [39]. When 3,4-didehydrophenanthrene was used as the aryne precursor, the product consisted of a [7]helicene fused with a [5]helicene [41].

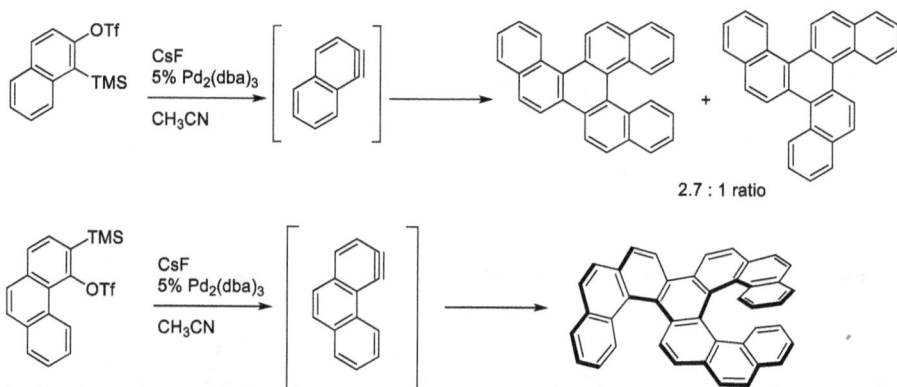

Figure 11.26: Preparation of other contorted PAHs via aryne cyclotrimerization [39, 41].

In Chapter 9, we saw the synthesis of hexa-*peri*-hexabenzocoronenes, which are planar PAH systems. In contrast, when benzene rings are cata-fused to coronene, a nonplanar PAH with [4]helicene is obtained. Despite their nonplanar structures, these hexa-*cata*-hexabenzocoronenes can still stack effectively and have been investigated for their potential semiconducting properties [42–44]. One of the first syntheses of these compounds was reported by Nuckolls and coworkers (Figure 11.27) [42]. The synthesis began by converting pentacenone to the corresponding thioketone using Lawesson's reagent. The thione was then reacted with diphenyl diazomethane in a

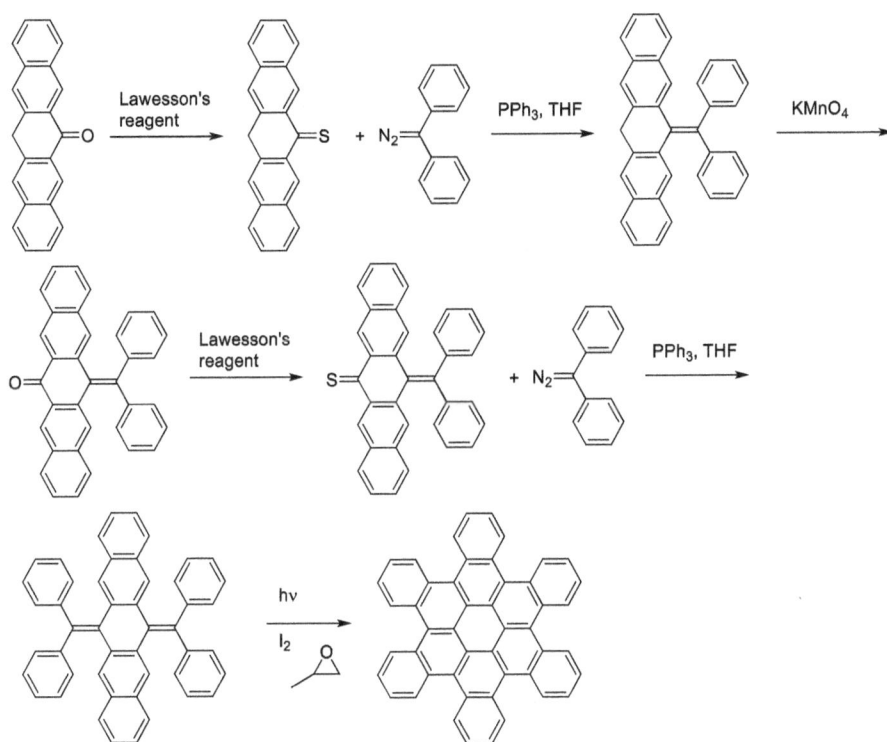

Figure 11.27: Synthesis of hexa-*cata*-hexabenzocoronene via olefination and photocyclization [42].

Barton–Kellogg olefination to produce the tetrasubstituted alkene. A benzylic oxidation of these compounds using KMnO$_4$ gave another carbonyl, which was subjected to the same sequence thionation and olefination to give the bisolefin. This was photocyclized in the presence of iodine and propylene oxide to give hexa-*cata*-hexabenzocoronene. The same approach could be applied to the synthesis of substituted systems.

An alternative approach to access hexa-*cata*-hexabenzocoronenes involves uses Suzuki coupling and Scholl reactions as the key steps (Figure 11.28) [43, 45]. The synthesis begins by a reaction of 6,13-pentacenequinone with CBr$_4$ and PPh$_3$ (a Ramirez reaction) to give the bis(dibromoolefin). This compound was then reacted with phenylboronic acid in a Suzuki–Miyaura cross-coupling to give the tetraphenyl bisolefin, which was cyclized using FeCl$_3$. Again, this synthetic approach could also be applied to the synthesis of substituted hexa-*cata*-hexabenzocoronenes.

In Chapter 9, we saw the nickel-mediated Yamamoto coupling as an alternative to palladium-catalyzed aryne cyclotrimerization for the preparation of threefold symmetric polycyclic aromatic hydrocarbons such as triphenylenes and trinaphthylenes. This reaction has not been widely explored for the preparation of nonplanar polycyclic aromatic hydrocarbons. In one notable exception, Gingras and coworkers reported the

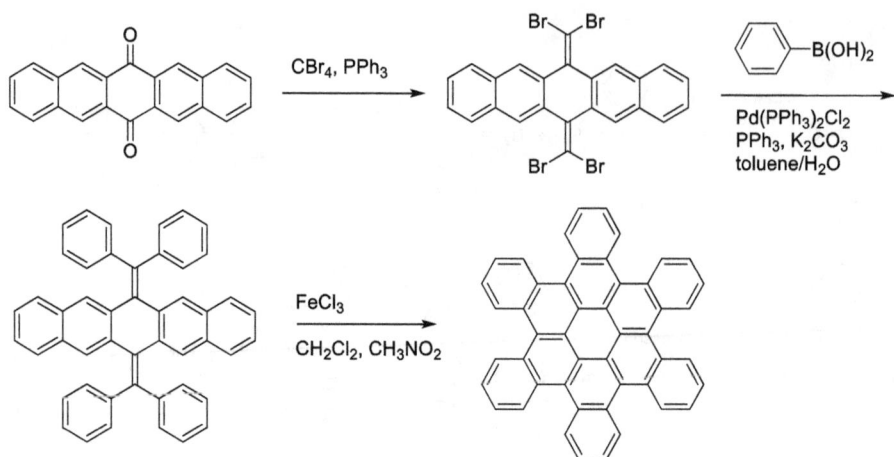

Figure 11.28: Synthesis of hexa-*cata*-hexabenzocoronenes via Suzuki coupling and Scholl reaction [45].

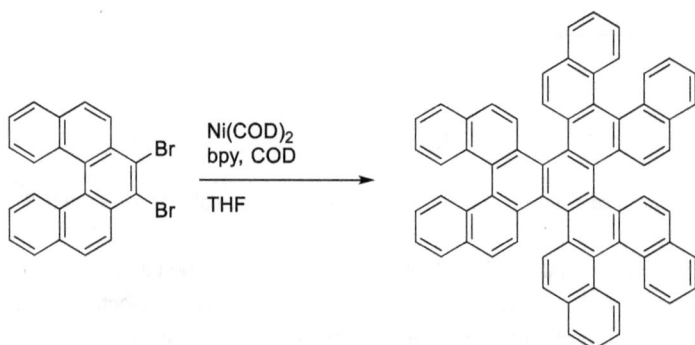

Figure 11.29: Synthesis of PAHs containing helicene units via Yamamoto coupling [46].

Yamamoto cyclotrimerization of a dibromo-substituted [5]helicene to prepare a large PAH with six helicene units embedded within its structure (Figure 11.29) [46].

11.3 Twisted acenes

We have already explored the synthesis of acenes (such as tetracene and pentacene) in Chapter 10. While we generally think of acenes as being planar, the crystal structures of many acenes do show minor deviations from planarity and the energetic cost of deformation from planarity is relatively modest. Because of this, it is possible to induce a twist from planarity in acenes by attaching several bulky substituents to the periphery [47]. To relieve the strain imposed by the steric interactions of the bulky substituents, the acene shows a twist from planarity. For example, decaphenylanthracene shows a twist of 63° from end-to-end (Figure 11.30) [48]. This strategy can be used in conjunction with benzannulation, combining some of the features seen in the contorted aromatics in the previous example with bulky substituents to induce a twist. An example of this approach is seen in diphenyltetrabenz[a,c,h,j]anthracene which features an end to end twist of 66° (Figure 11.30) [49].

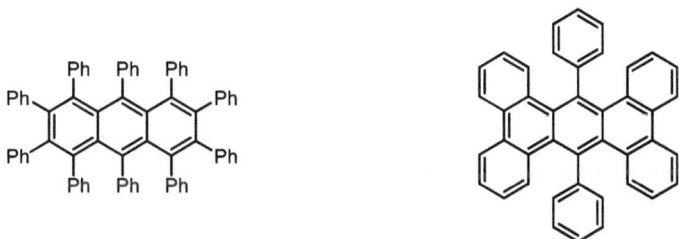

Figure 11.30: Examples of twisted acenes (a.k.a. "twistacenes").

The challenge of synthesizing these compounds is overcoming the steric crowding. Consequently, many of the syntheses of twisted acenes employ high temperatures and are often low yielding. The synthesis of larger twisted acenes often employs Diels–Alder reactions with arynes as key steps. For example, as shown in Figure 11.31, decaphenylanthracene was prepared by reacting the tetraphenylanthranilic acid with isoamyl nitrite to prepare the diazonium carboxylate [48]. The diazonium carboxylate underwent in situ elimination to generate the benzyne, which reacts with the isobenzofuran derivative in a Diels–Alder reaction. The Diels–Alder adduct was then reduced using Zn to give the target decaphenylanthracene in low yield.

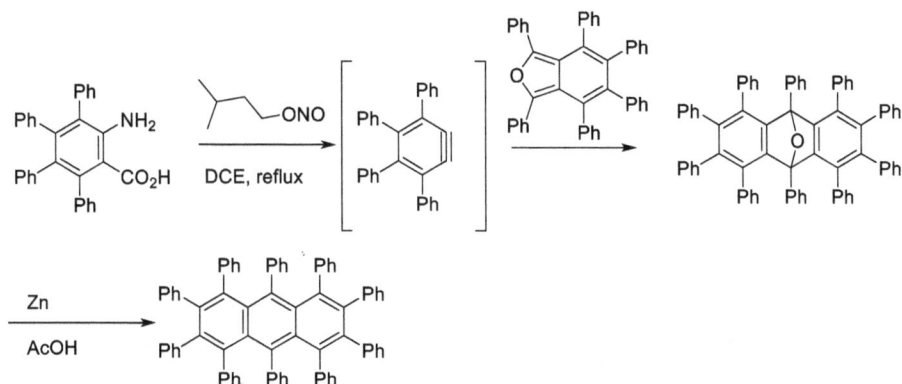

Figure 11.31: Synthesis of decaphenylanthracene [48].

A similar approach was used for the preparation of dibenzotetracene (Figure 11.32) [50]. The resulting tetracene derivative featured an impressive end-to-end twist of 105°.

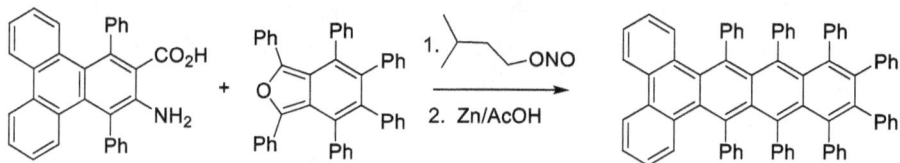

Figure 11.32: Synthesis of a twisted dibenzotetracene [50].

Diels–Alder reactions of arynes has also been used to prepare twisted benzannulated pentacenes and even heptacenes. For example, in the preparation of tetrabenzopentacene outlined in Figure 11.33, the isobenzofuran reacted with the diaryne generated from tetrabromoterphenyl in the presence of n-BuLi [51]. It is likely that the elimination to form the arynes happens sequentially such that the second benzyne formation occurs after the first benzyne has reacted. Again, the use of the isobenzofuran requires a final reduction step to remove the oxygen and generate the fully aromatized system. In this case that reduction is achieved using n-BuLi in the presence of TiCl$_3$. The tetrabenzopentacene product is particularly interesting because the twisted geometry makes it chiral and the barrier to inversion is high enough that it is configurationally stable. A similar approach was used to prepare the benzannulated heptacene (Figure 11.33) [52]. However, in this case, the arynes were generated from trimethylsilyl aryl triflate, which is reacted with the cyclopentadienone.

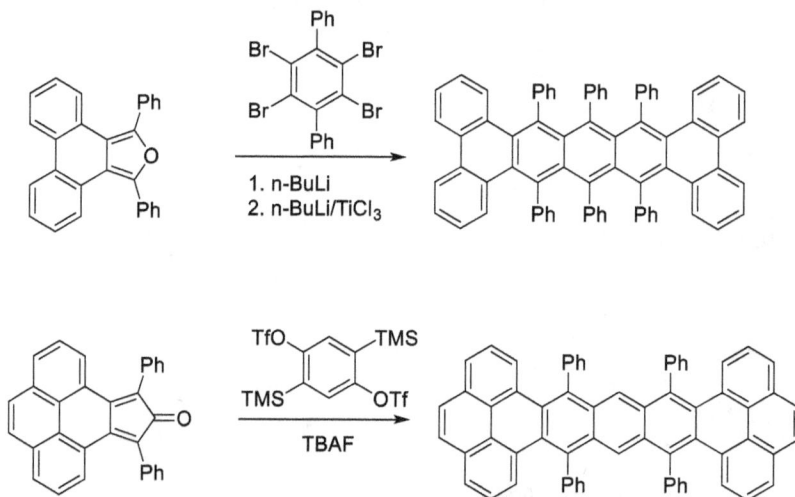

Figure 11.33: Synthesis of twisted pentacenes and heptacenes via aryne chemistry [51, 52].

11.4 Circulenes and related curved polycyclic aromatic hydrocarbons

Circulenes are a class of compounds where benzene rings are *ortho*-fused to produce polycyclic system with central rings of varying sizes and benzene rings on the periphery (Figure 11.34). They are named according to the number of benzenes joined to form a ring, which also defines the size of the central ring. [6]-Circulene, which is also referred to as coronene, is a planar polycyclic aromatic system (see Chapter 10). When the central ring of coronene is replaced with either a smaller ring (such as the five-membered ring of corannulene), or a larger ring (such as in [7]- or [8]-circulene), a nonplanar structure results. Corannulenes adopt a bowl-shaped curved structure, while the [7]- and [8]-circulenes adopt a saddle-shaped curved structure [53].

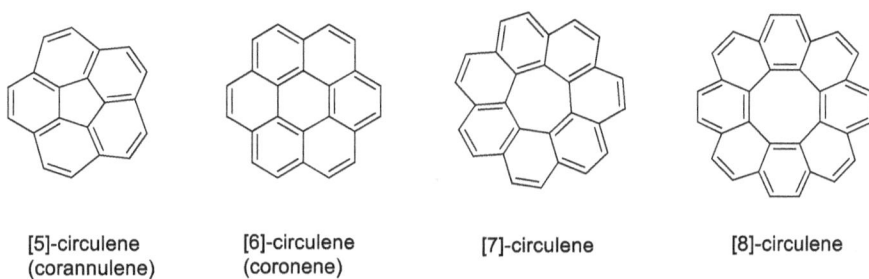

| [5]-circulene (corannulene) | [6]-circulene (coronene) | [7]-circulene | [8]-circulene |

Figure 11.34: Structures of the circulene family.

11.4.1 [5]-Circulenes – corannulenes

Corannulene is a bowl-shaped compound with five fused benzene rings. It has been the subject of interest because it is a fragment of fullerenes such as C_{60} and could potentially be used for the bottom-up synthesis of fullerenes [54]. It is also of interest because its curved structure makes it suitable for molecular recognition with fullerenes [55–57].

The first preparation of corannulene was reported in 1966 by Lawton and Barth using a 16-step synthesis [58]. In the early 1990s, Scott and coworkers developed a much more concise synthesis using flash vacuum pyrolysis as a key step (Figure 11.35) [59, 60]. In Scott's synthesis, acenaphthenone underwent a one-pot condensation with the triketone, followed by a Diels–Alder reaction with norbornadiene to give the disubstituted fluoranthrene derivative. This dione was treated with PCl_5 to yield the bis(chloroethylene) derivative. This compound was subjected to flash vacuum pyrolysis at high temperature to give corannulene as the major product in 35–40% yield.

Soon after Scott's concise synthesis, Siegel reported an alternative synthesis of corannulene (Figure 11.36) [61]. The synthesis began with 2,7-dimethylnaphthalene,

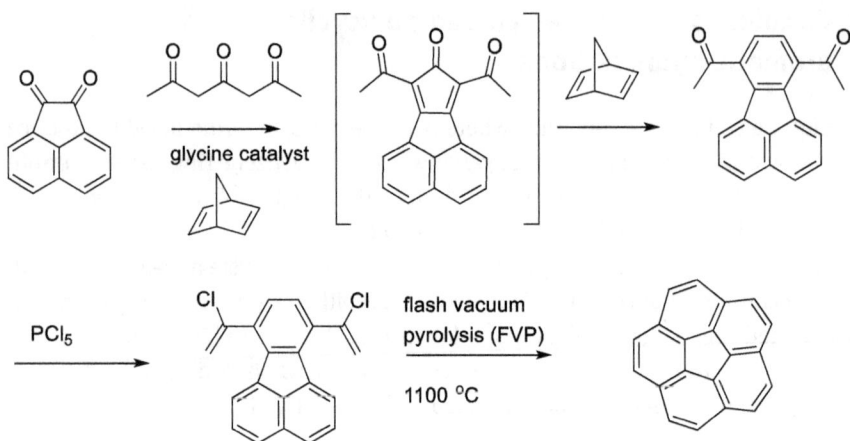

Figure 11.35: Scott's synthesis of corannulene via flash vacuum pyrolysis [60].

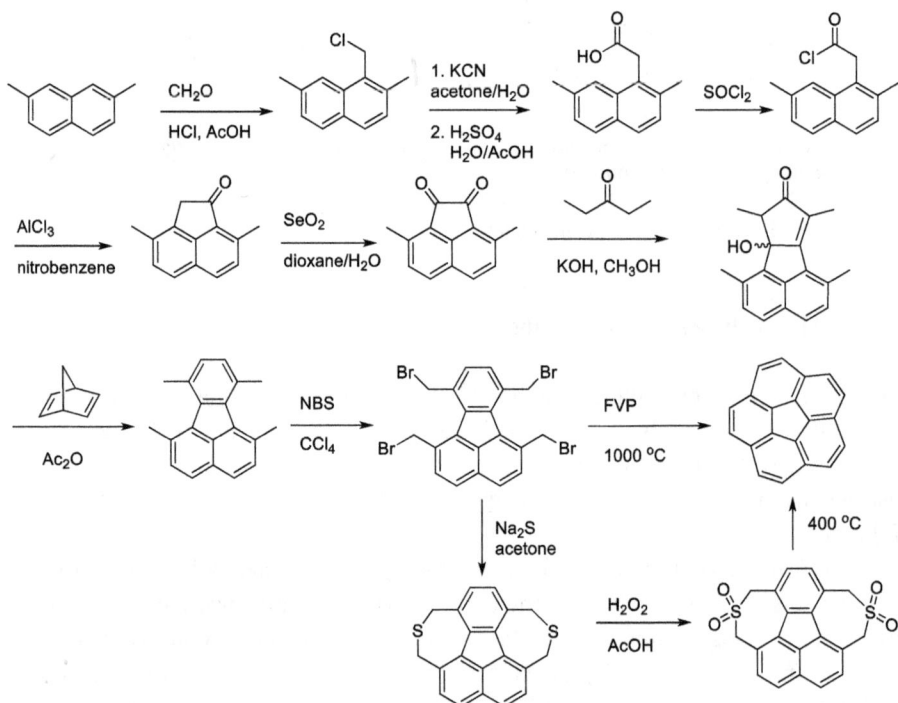

Figure 11.36: Siegel's corannulene synthesis [61].

which was subjected to chloromethylation using paraformaldehyde and HCl, followed by cyanation, and hydrolysis to produce the carboxylic acid. The carboxylic acid was converted to the acid chloride using SOCl₂ and then underwent an intramolecular

Friedel–Crafts acylation to produce the ketone. Subsequent oxidation gave 2,7-dimethylacenaphthenone, which was reacted with pentanone under basic conditions, followed by a Diels–Alder reaction with norbornadiene to give 1,6,7,10-tetramethylfluoranthene [62]. The tetramethylfluoranthene was then reacted with NBS to achieve benzylic bromination. The brominated compound could be used to generate corannulene directly by flash vacuum pyrolysis. Alternatively, treatment with sodium sulfide gave the bis-sulfide, which could be oxidized to the sulfone using hydrogen peroxide in acetic acid. Heating to 400 °C led to extrusion of SO_2 gave corannulene in low yield along with side products such as dihydro- and tetrahydrocorannulenes, which could be converted into corannulene using palladium on carbon.

Rabideau reported an alternative synthesis of corannulene from 1,6,7,10-tetramethylfluoranthene [62], which was a key intermediate in Siegel's synthesis (Figure 11.37) [63]. Instead of benzylic bromination to introduce one bromine on each of the methyl groups, Rabideau and coworkers conducted the benzylic bromination to introduce two bromines on each methyl, yielding 1,6,7,10-tetrakis(dibromomethyl)fluoranthene. Under basic conditions, this compound underwent a benzylic coupling, similar to that seen in some of the helicene syntheses. The resulting tetrabromocorannulene could be debrominated using LiAlH₄ and DDQ, or used to prepare functionalized corannulene derivatives. Modifications of this synthetic approach have been used to prepare corannulene in kilogram quantities [64].

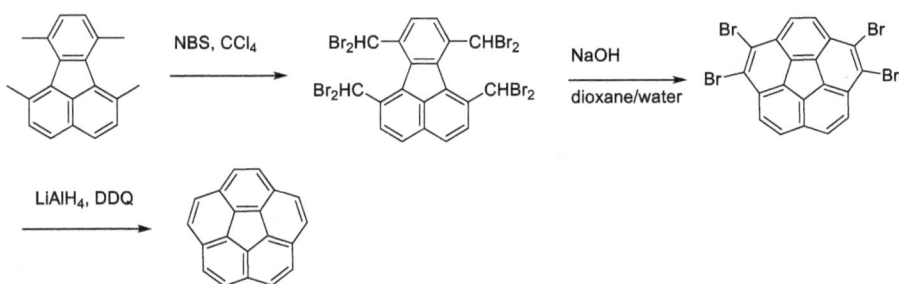

Figure 11.37: Synthesis of corannulene via a benzylic coupling [63].

Another bowl-shaped fullerene fragment is sumanene, which consists of four benzenoid rings and three five-membered rings (Figure 11.38).

Figure 11.38: Structure of sumanene.

In 2003, Sakurai developed a concise synthesis of sumanene from norbornadiene and features a ring-opening metathesis/ring-closing metathesis sequence as the key steps (Figure 11.39) [65]. The synthesis begins by lithiation of norbornadiene and trapping with 1,2-dibromoethane, followed by treatment with tributyltin chloride to give the unsaturated bromo-substituted stannane. This was subjected to a copper-mediated trimerization to give benzotris(norbornadiene) as a mixture of *syn* and *anti* isomers. The *syn* isomer was treated with ethylene in the presence of Grubbs catalyst, which lead to ring-opening metathesis of the norbornadiene units, followed by a ring-closing metathesis to give the cyclized product. Oxidation using DDQ gave sumanene.

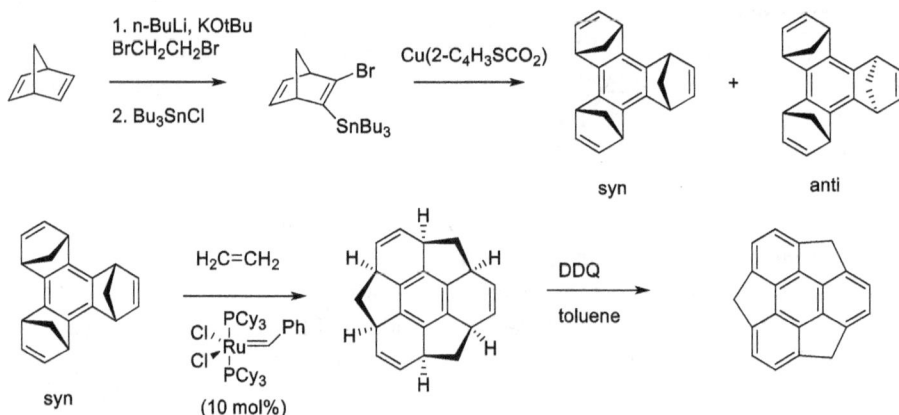

Figure 11.39: Synthesis of sumanene [65].

11.4.2 [7]-Circulenes and related systems

Introducing seven-membered rings into PAHs, such as in [7]-circulene, induces a curvature. The first synthesis of [7]-circulene was reported by Yamamoto and coworkers as outlined in Figure 11.40 [66, 67]. The synthesis begins by reduction of 5,5'-dimethyl-2,2'-dinitrobiphenyl using tin in HCl. The diamino biphenyl was converted into the dibromide via diazonium salt formation and reaction with mercuric bromide. Following this, a benzylic bromination gave the bromomethyl groups, which were converted into the thiols. At this stage, a reaction of the dithiol with 2,7-bis(bromomethyl)naphthalene gave the dithiacyclophane. The sulfides were methylated using $(MeO)_2CHBF_4$, and the product underwent a Stevens rearrangement with sodium hydride to give the bis(sulfide). Oxidation with mCPBA, followed by pyrolysis gave the unsaturated cyclophane. A photochemical cyclization of the unsaturated cyclophane yielded the fused system with 6 aromatic rings. Following this, a twofold lithium–halogen exchange on the dibromide and trapping with DMF gave the diformyl compound. This compound was subjected to McMurry coupling conditions to give [7]-circulene.

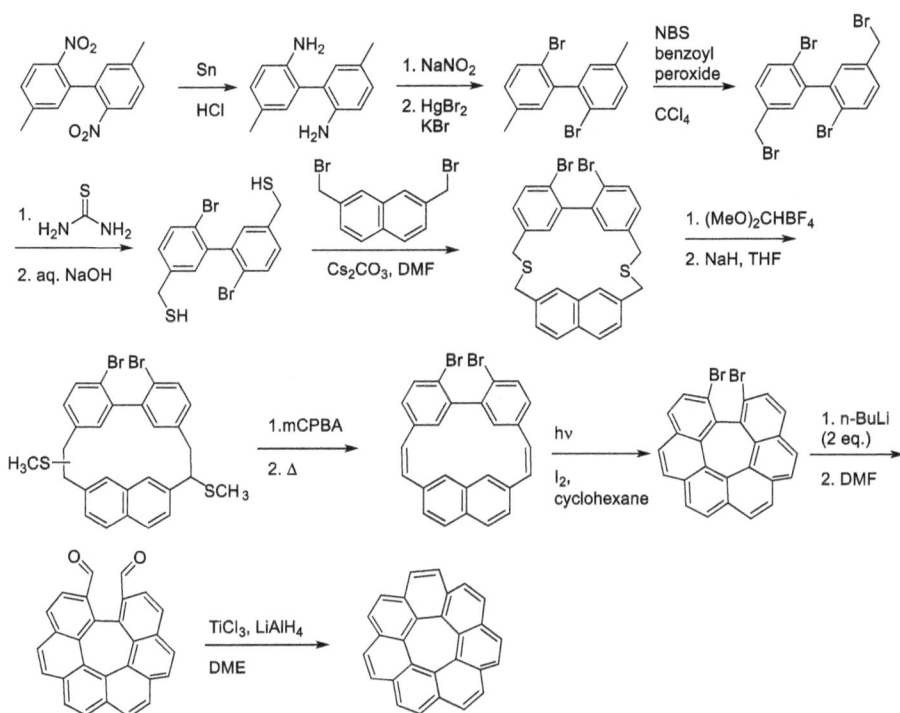

Figure 11.40: Synthesis of [7]-circulene [66].

Exploiting the idea that the introduction of seven- and eight-membered rings into polycyclic aromatic systems can introduce curvature, several other approaches for the synthesis of these compounds have been explored.

For example, Scott, Itami, and coworkers have prepared a propeller-shaped PAH containing seven-membered rings (Figure 11.41) [68]. This synthesis of this compound started from the brominated PAH, which reacted with the 2-boryl-2′chlorobiphenyl in a threefold Suzuki coupling. The product was then subjected to threefold direct arylation to give the final propeller-shaped molecule bearing three seven-membered rings.

11.5 Cyclophanes and cyclophenylenes

Another way to distort aromatic systems from planarity is to constrain them to a ring. Here we will explore some of the synthetic approaches to nonplanar cyclophanes and paracyclophenylenes, where the incorporation of aromatic structures into a ring causes curvature. These systems are of fundamental interest for the unusual photophysical and electronic properties that arise from curvature, for probing

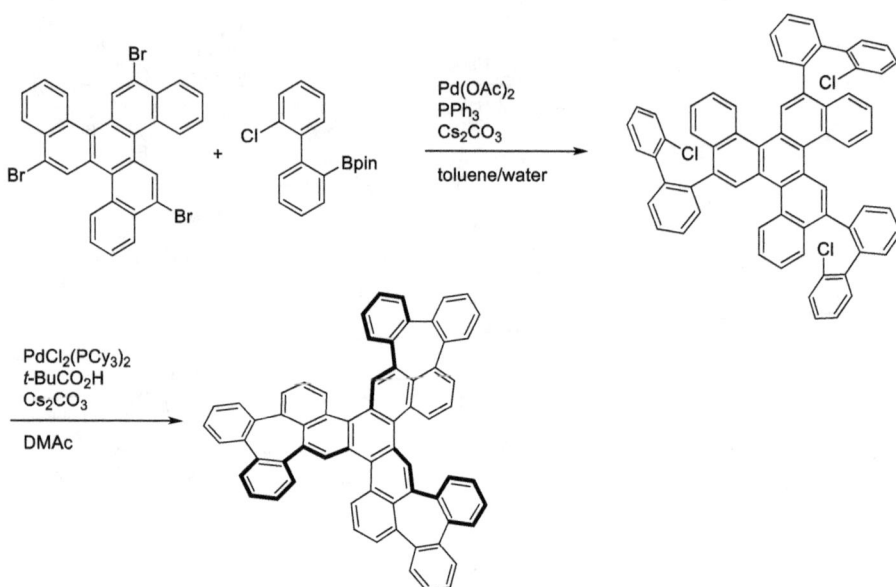

Figure 11.41: Synthesis of a contorted PAH containing three cyclooctatetraene units [68].

the limits of aromaticity, and as potential molecular precursors for the bottom-up synthesis of carbon nanotubes.

11.5.1 Cyclophanes

Cyclophanes consist of an aromatic ring with a chain bridging two non-adjacent positions on the ring. Some representative general cyclophanes are shown in Figure 11.42. Cyclophanes can also include more than one aromatic unit, as in [2.2]paracyclophane, as well as more than one bridging unit. Cyclophanes have also been prepared that contain polycyclic aromatic units [69].

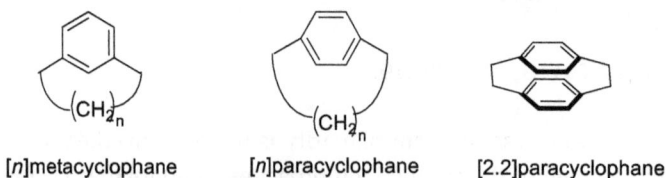

[*n*]metacyclophane   [*n*]paracyclophane   [2.2]paracyclophane

Figure 11.42: Representative cyclophane structures.

Cyclophanes present an attractive approach for accessing nonplanar aromatic systems by controlling the length of the bridging chain. Specifically, shortening the length of the bridging chain, imposes strain that causes the aromatic ring to distort from planarity. In simple [*n*]cyclophanes containing one benzene ring, small distortions from planarity can be observed. More dramatic distortions from planarity are observed in some polycyclic systems. For example, Bodwell and coworkers have shown that compounds such as 2,7-pyrenophanes can show significant deviations from planarity (Figure 11.43) [70].

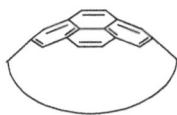

Figure 11.43: The general structure of a 2,7-pyrenophane.

To prepare these compounds, they relied on a valence isomerization/dehydrogenation (VID) reaction, which involved 6π-electron electrocyclic ring closure of the corresponding [2.2]metacyclophane diene, followed by oxidation, as depicted in Figure 11.44 [70].

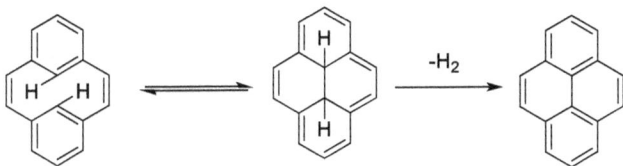

Figure 11.44: Pyrene via a valence isomerization/dehydrogenation (VID) reaction.

This VID reaction had been used to successfully prepare a number of different nonplanar pyrenophanes with varied linking groups. One example is outlined in Figure 11.45 [70]. The synthesis began by alkylating the phenol using the appropriate dibromoalkane and was followed by reduction of the esters and conversion to the benzyl bromides. The sulfide-containing cyclophane was then prepared using sodium sulfide with alumina. The sulfides were then methylated using $(MeO)_2CHBF_4$, which underwent a Stevens rearrangement to give the desired product as a mixture of isomers. A subsequent methylation and elimination gave the cyclophane diene, which underwent the VID reaction to yield the pyrenophanes. In this particular example, the cyclophane diene was not isolated and underwent electrocyclic ring-closing followed by a sequence of [1,5]-hydride shifts prior to oxidation.

To achieve a more highly curved aromatic system, an extended aromatic system, teropyrene, was incorporated into a cyclophane (Figure 11.46) [71].

This synthesis also made use of the VID reaction as the key step for forming the fully conjugated teropyrene unit, but followed a different approach to access the cyclophane diene precursor. The synthesis began with a two-fold Friedel–Crafts alkylation of

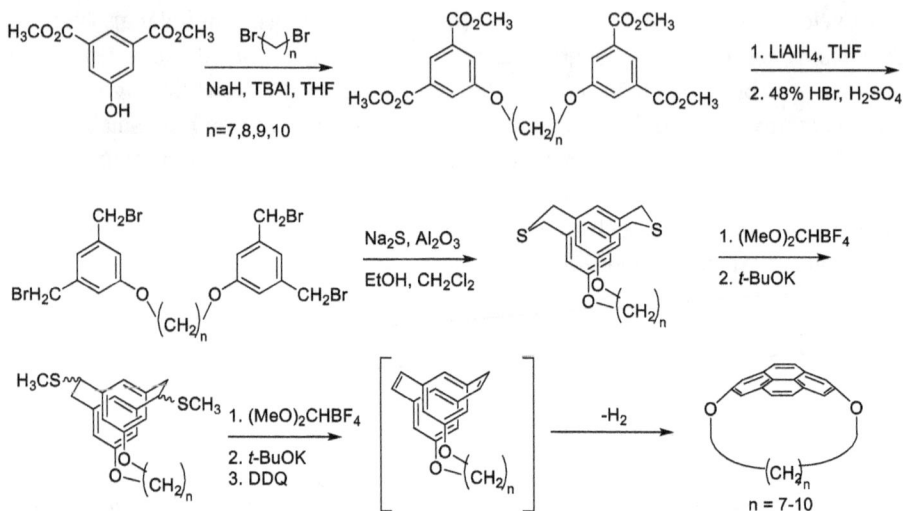

Figure 11.45: Synthesis of 2,7-pyrenophanes [70].

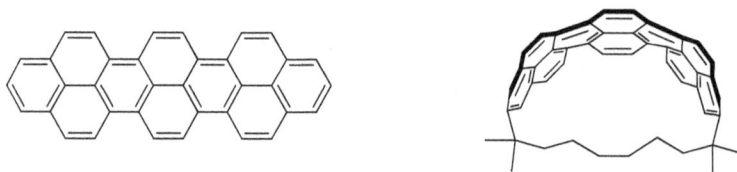

Figure 11.46: Terepyrene and a (2,11)tereopyrenophane [71].

pyrene with 2,9-dichloro-2,9-dimethyldecane using $AlCl_3$ as the Lewis acid (Figure 11.47). Electrophilic substitution of pyrene usually takes place at the 1-, 3-, 6-, and 8-positions. However, Friedel–Crafts alkylation with hindered electrophiles such as *t*-butyl groups takes place at the less hindered 2- and 7-positions. This regioselectivity was exploited to achieve substitution at the 2-position of the pyrenes. The compound was then formylated using Cl_2CHOCH_3 and $TiCl_4$ (Rieche formylation – see Section 2.3.4) at the 6-positions of the pyrene rings. The formyl groups were then bridged using a McMurry coupling. The sequence of formylation and McMurry coupling was repeated to yield the cyclophane diene, which underwent VID using DDQ in *m*-xylene in very good yield, giving a highly curved cyclophane.

11.5.2 Cycloparaphenylenes

Cycloparaphenylenes (CPPs) consist of phenyl rings linked via the *para*-positions to form a ring. These compounds have been a synthetic target for some time because

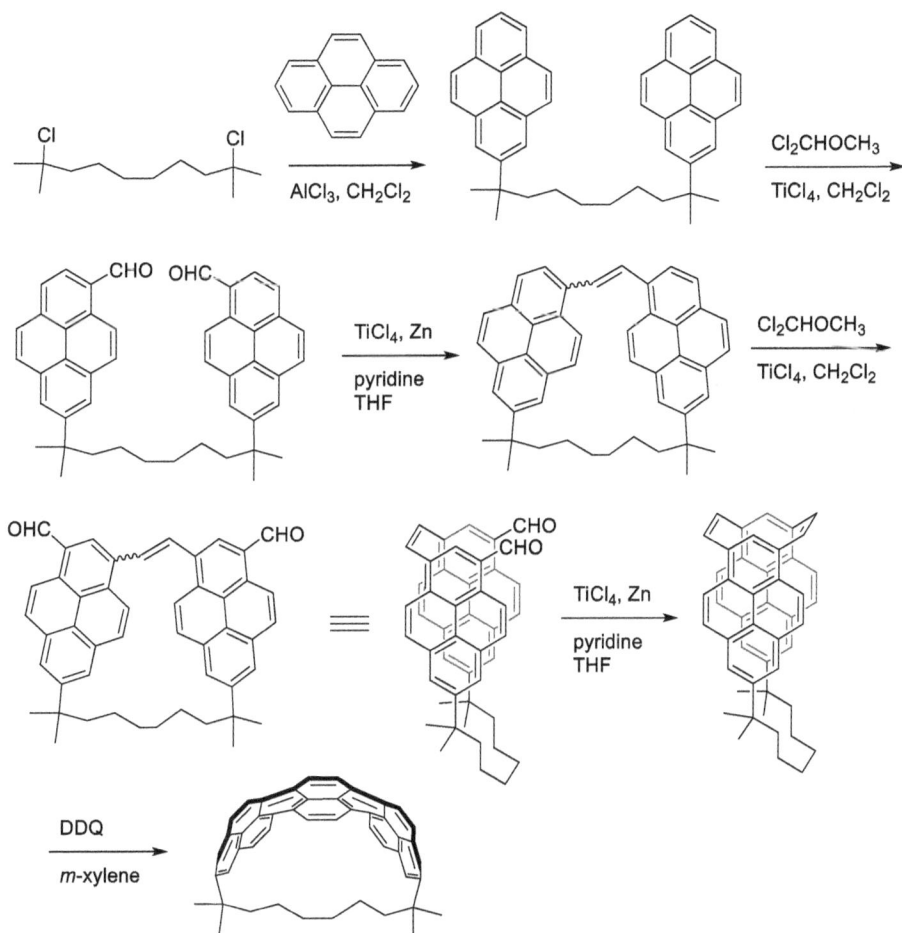

Figure 11.47: Synthesis of a (2,11)teropyrenophane [71].

of the inherent synthetic challenge. They can also be thought of as the basic repeat unit of an armchair carbon nanotube. As such, the controlled synthesis of CPPs could serve as a platform for the bottom-up synthesis of carbon nanotubes with uniform diameter.

The first successful synthesis of a CPP was reported by Jasti and Bertozzi [72]. Their approach consisted of preparing macrocycles containing cyclohexadienes, which served as precursors to aryl rings. Their synthesis is outlined in Figure 11.48. It began by treating 1,4-diiodobenzene with *n*-butyllithium to effect a lithium–halogen exchange. The aryllithium was then reacted with benzoquinone to undergo nucleophilic addition to the quinone carbonyls. The resulting adduct was then protected using methyl iodide. The iodides were then converted into the boronate esters by lithium–halogen exchange and treatment with isopropyl pinacol borate. The resulting

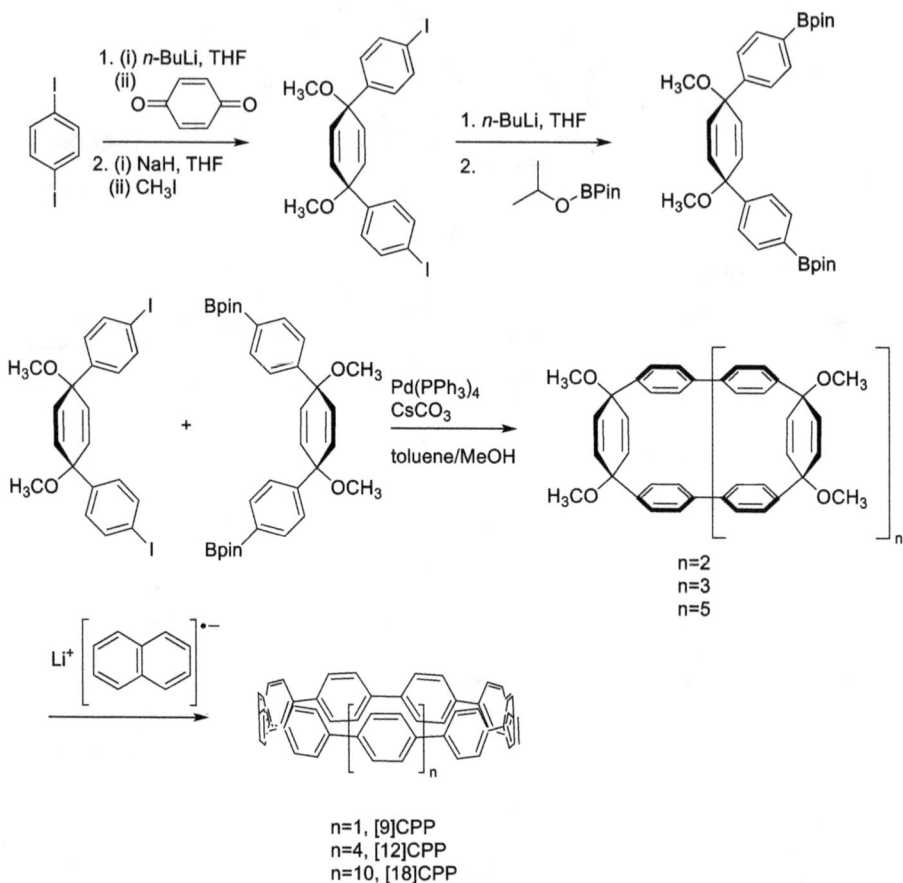

Figure 11.48: The first synthesis of [9]CPP, [12]CPP, and 18[CPP] [72].

boronate ester was reacted with the iodide in a Suzuki cross-coupling to give a mixture of macrocycles containing 9, 12, and 18 rings in low yields. The macrocycle containing nine rings was formed in the lowest yield (ca. 2%), but this product was unexpected given the starting materials and likely results from a homocoupling of the diboronate ester as one of the steps toward macrocycle formation. These different macrocycles were readily separated using column chromatography and then converted to the aromatic CPPs by a reductive aromatization method using lithium naphthalenide as the reducing agent. This reductive aromatization is thought to proceed by a one-electron reduction, which results in C–O bond cleavage to generate lithium methoxide and a stabilized radical intermediate (Figure 11.49). A second one-electron reduction gave an alkyllithium intermediate, which was aromatized by a second equivalent of lithium methoxide. Despite the low temperatures, the reductive aromatization still provides the strained CPPs in reasonable yield. This synthetic approach had the drawback that the

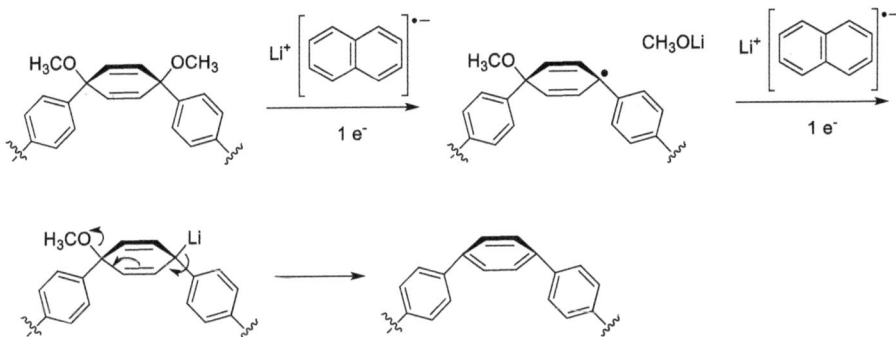

Figure 11.49: Proposed mechanism for reductive aromatization.

macrocyclization step was low yielding and nonselective, requiring separation of the different macrocycles. Nonetheless, the synthesis of these CPPs was an impressive achievement because it demonstrated that these strained CPPs could be accessed synthetically and opened up a new field of study into cyclophenylenes.

The reductive aromatization was later used by Xia and Jasti to prepare the highly strained [6]CPP [73]. This synthesis is outlined in Figure 11.50. Unlike the initial synthesis of CPPs, which resulted in a mixture of macrocycles that required separation, this synthesis was designed to prepare [6]CPP selectively. The synthesis started with oxidation of 4-bromo-4′[(trimethylsilyl)oxy]biphenyl using phenyliodine(III) diacetate (PIDA) and water to give the unsaturated ketone. The alcohol was deprotonated and the compound was treated with the biphenyl aryllithium, which performed a nucleophilic addition to the ketone. The resulting diol was then protected as the methyl ethers, similar to the first CPP synthesis. The silyl protecting group was removed and the resulting phenol was subjected to another oxidation at the using PIDA. The resulting unsaturated ketone underwent nucleophilic addition with 4-bromophenyllithium, and the resulting diol was again protected as the methyl ethers as in the previous steps. This gave a pentacyclic precursor with terminal bromo substituents. This compound was reacted with the boronate ester of 1,4-benzenediboronic acid in a Suzuki coupling to give the desired macrocycle in a low yield of 12%. The low yield of this macrocyclization step is not surprising given the strain of the product. With this compound in hand, the reductive aromatization using lithium napthalenide, followed by quenching with iodine, proceeded in reasonable yield to give the target [6]CPP.

Following Jasti and Bertozzi's synthesis of CPPs, Itami and coworkers reported a selective synthesis of [12]CPP (Figure 11.51) [74]. As in Jasti and Bertozzi's synthesis, 4-iodophenyllithium was generated by lithium–halogen exchange, but was reacted with 1,4-cyclohexanedione to give the saturated cyclohexanediol bearing iodophenyl groups. This common building block was protected with methoxymethyl (MOM) groups and converted to the corresponding diboronate via a Miyaura coupling. The cyclic tetramer, bearing the desired 12 rings, was prepared by a sequence of two Suzuki cross-coupling

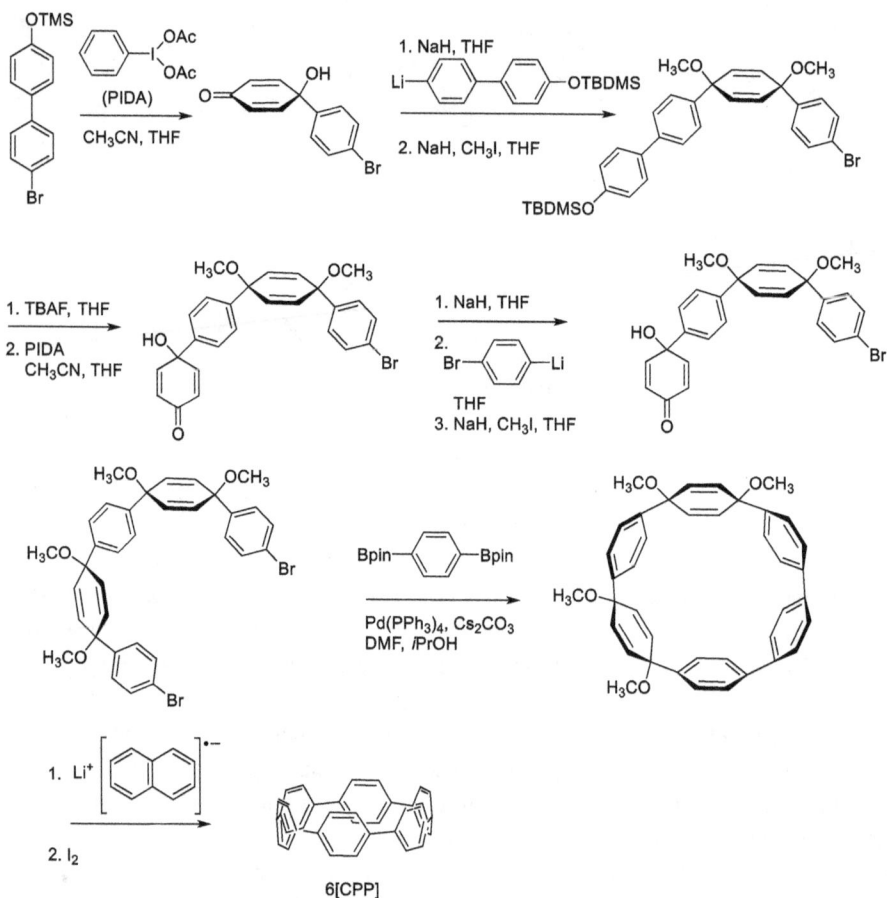

Figure 11.50: Synthesis of [6]CPP [73].

reactions. The tetrameric macrocycle was converted into [12]CPP in good yield by treatment with *p*-toluenesulfonic acid under microwave conditions. This last step achieved the deprotection of the methoxymethyl protecting groups, eightfold dehydration, and oxidative aromatization in a single step. Subsequently, Itami and coworkers demonstrated that this approach could be used for the size-selective preparation of larger CPPs [75].

The CPP syntheses discussed thus far all make use of palladium-catalyzed Suzuki cross-coupling as the key step for preparing the macrocycles that serve as the precursors for the CPPs. Itami subsequently explored a "shotgun" approach to CPP formation using a single tricyclic monomer using nickel-mediated Yamamoto coupling conditions (Figure 11.52) [76]. The synthesis began with 1,4-dibromobenzene, which was converted to the corresponding 4-bromophenylcerium reagent by lithium–halogen exchange and treatment with CeCl₃. The arylcerium reagent was reacted with 1,4-cyclohexanedione

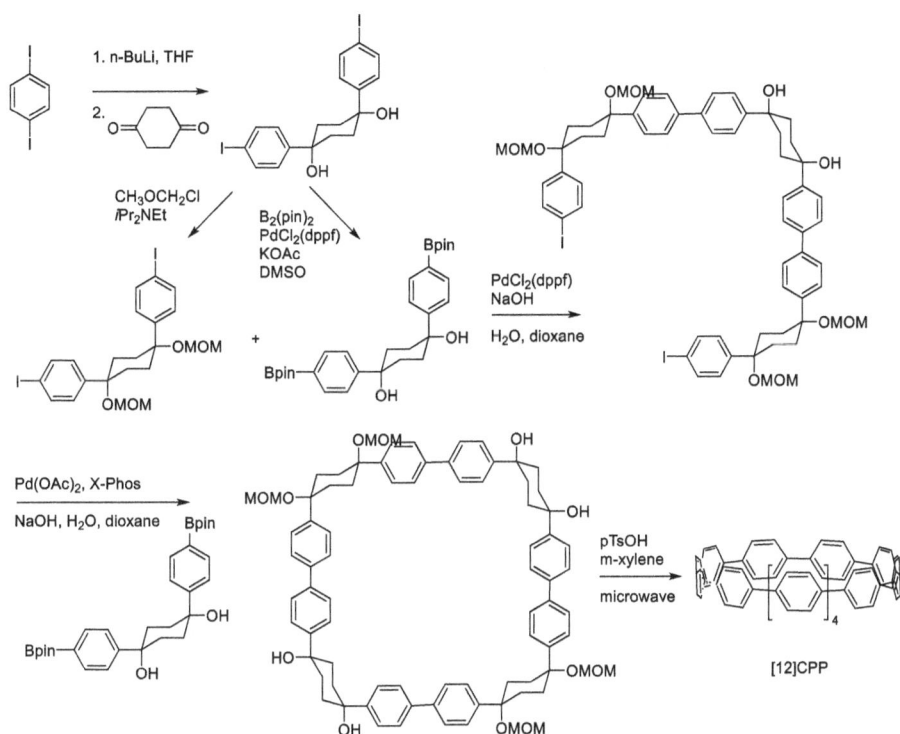

Figure 11.51: Itami's selective synthesis of [12]CPP [74].

in a twofold nucleophilic addition to give the *cis*-diol, which was protected with MOM groups. The phenylcerium reagent was chosen to avoid monoaddition and give the desired *cis*-diol in high selectivity. The resulting dibromo monomer was subjected to Yamamoto coupling conditions and gave dodecacyclic macrocycle in a reasonable yield of 22%. The yield could be further improved to 27% when the corresponding iodides were used. With this macrocycle in hand, deprotection and aromatization proceeded to give [12]CPP good yield using $NaHSO_4 \cdot H_2O$ in *m*-xylene/DMSO at reflux in air – a modification of the microwave conditions described previously. Subsequently, it was shown that one of the side products of the macrocyclization gave the [9]CPP precursor. This synthetic approach was significant because is concise and simple, making it amenable to larger scale preparation of [12]CPP.

In another ground-breaking study, Yamago and coworkers reported the synthesis of highly strained [8]CPP using a distinctly different approach. Rather than using six-membered rings as "masked" benzene rings with less strain, they prepared an organometallic macrocycle with relatively little strain (Figure 11.53) [77]. Specifically, they reacted 4,4′-bis(trimethylstannyl)biphenyl with dichloro(cycloocta-1,5-diene)platinum(II) to give the square planar macrocycle. After a ligand exchange with 1,1′-bis(diphenylphosphino)ferrocene, treatment with bromine led to reductive elimination

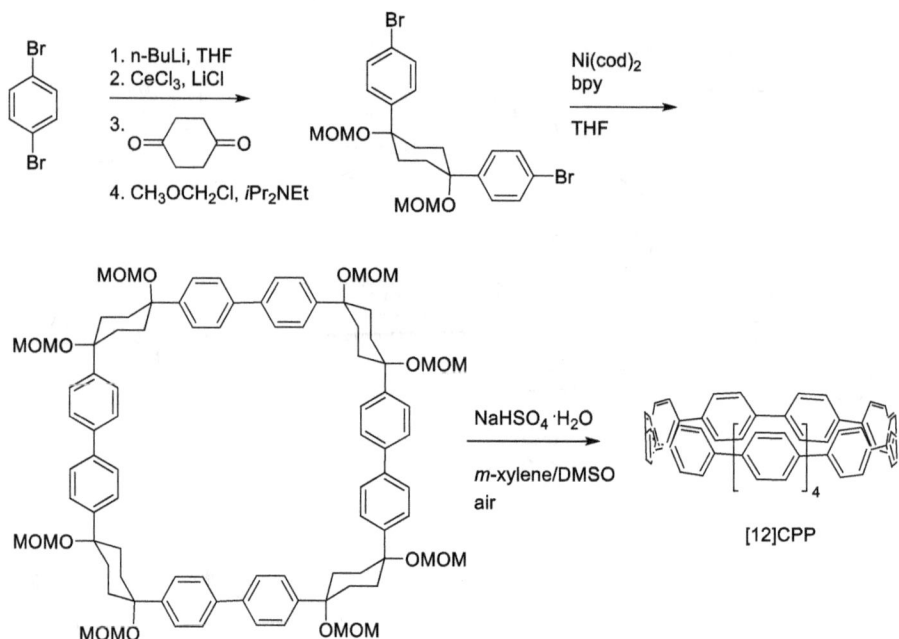

Figure 11.52: Concise synthesis of [12]CPP using a nickel-mediated Yamamoto coupling [76].

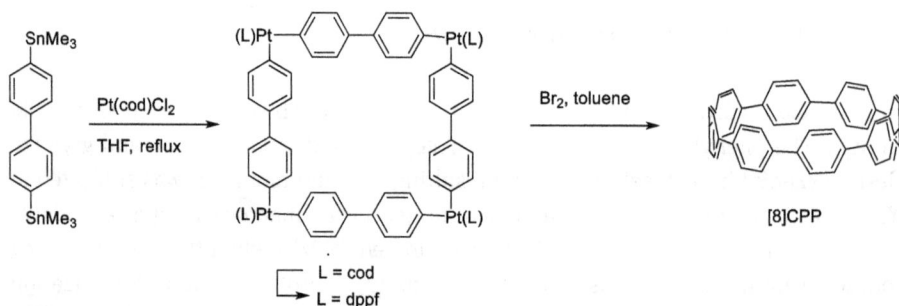

Figure 11.53: Yamago's synthesis of [8]CPP from a platinum macrocycle [77].

to form the aryl–aryl bonds of [8]CPP. This synthesis is impressive because it produced the highly strained CPP in only three steps.

The synthesis of paracyclophenylenes and other carbon rings is a rapidly evolving field and has been extended to the synthesis of carbon nanocages [78, 79] and other topologically interesting structures [80]. In the coming years, we are certain to see new methods developed for the preparation of aromatic compounds and the application of these methods to the preparation of elaborate carbon nanostructures that capture the imagination, and whose properties are waiting to be explored.

References

[1] Gingras M. One hundred years of helicene chemistry. Part 1: Non-stereoselective syntheses of carbohelicenes. Chem Soc Rev. 2013;42(3):968–1006.

[2] Shen Y, Chen C, Reactions DÀA, Reactions FÀC, Cyclizations M. Helicenes : Synthesis and Applications. Chem Rev. 2012;112:1463–535.

[3] Gingras M, Félix G, Peresutti R. One hundred years of helicene chemistry. Part 2: Stereoselective syntheses and chiral separations of carbohelicenes. Chem Soc Rev. 2013;42 (3):1007–50.

[4] Cook JW. Polycyclic aromatic hydrocarbons. Part XII. The orientation of derivatives of 1 : 2-Benzanthracene, with notes on the preparation of some new homologues, and on the isolation of 3:4:5:6-dibenxphenanthrene. J Chem Soc. 1933;1592–97.

[5] Newman MS, Lednicer D. The synthesis and resolution of hexahelicene. J Am Chem Soc. 1956;78:4765–70.

[6] Scholz M, Muhlstadt M, Dietz F. Chemie angeregter zustände. I. Mitt. Die richtung der photocyclisierung naphthalinsubstituierter äthylene. Tetrahedron Lett. 1967;8(7):665–68.

[7] Dietz F, Scholz M. Chemie angeregter zustände – IV : Die photocyclisierung der drei isomeren distyrylbenzole. Tetrahedron. 1968;24(24):6845–49.

[8] Flammang-Barbieux M, Nasielski J, Martin RH. Synthesis of heptahelicene (1) benzo [c] phenanthro [4, 3-g]phenanthrene. Tetrahedron Lett. 1967;8(8):743–44.

[9] Martin RH, Morren G, Schurter JJ. [13]Helicene and [13]helicene-10,21-d2. Tetrahedron Lett. 1969;10:3683–88.

[10] Martin RH, Schurter JJ. The photocyclisation of 1,2-diarylethylenes: Determination of the chemical structure of the cyclised products. Tetrahedron Lett. 1969;10(42):3679–82.

[11] Liu L, Katz TJ. Bromine auxiliaries in photosyntheses of [5]helicenes. Tetrahedron Lett. 1991;32(47):6831–34.

[12] Liu L, Yang B, Katz TJ, Poindexter MK. Improved methodology for photocyclization reactions. J Org Chem. 1991;56(12):3769–75.

[13] Weidlich HA. Synthese kondensierter Ringsysteme (I. Mitteil.). Ber Dtsch Chem Ges. 1938;71:1203–04.

[14] Altman Y, Ginsburg D. Alicyclic studies. XIV. Improved synthesis of dibenzo[c,g]phenanthrene and benzo[g,i]perylene. J Chem Soc. 1959;468–69.

[15] Minuti L, Taticchi A, Marrocchi A, Gacs-Baitz E. Diels-Alder reaction of 3,3',4,4'-Tetrahydro -1,1'-binaphthalene. One-pot synthesis of pentahelicenebenzoquinone. Tetrahedron. 1997;53 (20):6873–78.

[16] Sooksimuang T, Mandal BK. [5]Helicene-fused phthalocyanine derivatives. New members of the phthalocyanine family. J Org Chem. 2003;68(2):652–55.

[17] Willmore ND, Liu L, Katz TJ. A Diels–Alder route to [5]- and [6]-helicenes. Angew Chem Int Ed. 1992;31:1093–95.

[18] Katz TJ, Liu L, Willmore ND, Fox JM, Rheingold AL, Shi S, et al. An efficient synthesis of functionalized helicenes. J Am Chem Soc. 1997;119(42):10054–63.

[19] Kharasch MS, Nudenberg W, Fields EK. Synthesis of polyenes. IV. J Am Chem Soc. 1944;66 (8):1276–79.

[20] Gingras M, Dubois F. Synthesis of carbohelicenes and derivatives by "carbenoid couplings.". Tetrahedron Lett. 1999;40(7):1309–12.

[21] Maigrot N, Mazaleyrat J-P. New and improved synthesis of optically pure (r)-and (s)-2,2'-dimethyl-1, 1'-binaphthyl and related compounds. Synthesis (Stuttg). 1985;317–20.

[22] Dubois F, Gingras M. Syntheses of [5]-helicene by McMurry or carbenoid couplings. Tetrahedron Lett. 1998;39(28):5039–40.

[23] Collins SK, Grandbois A, Vachon MP, Côté J. Preparation of helicenes through olefin metathesis. Angew Chem Int Ed. 2006;45(18):2923–26.

[24] Jančařík A, Rybáček J, Cocq K, Chocholoušová JV, Vacek J, Pohl R, et al. Rapid access to dibenzohelicenes and their functionalized derivatives. Angew Chem Int Ed. 2013;52 (38):9970–75.

[25] Newman MS, Lutz WB, Lednicer D. A new reagent for resolution by complex formation; The resolution of phenanthro-[3,4-c]phenanthrene. J Am Chem Soc. 1955;77:3420–21.

[26] Martin RH, Flammang-Barbieux M, Cosyn JP, Gelbcke M. 1-Synthesis of octa- and nonahelicenes. 2-New syntheses of hexa- and heptahelicenes. 3-Optical rotation and O.R.D. of heptahelicene. Tetrahedron Lett. 1968;9(31):3507–10.

[27] Martin RH, Marchant MJ. Resolution and optical properties ([α]max, ord and cd) of hepta-, octa- and nonahelicene. Tetrahedron. 1974;30(2):343–45.

[28] Moradpour A, Nicoud JF, Balavoine G, Kagan H, Tsoucaris G. Photochemistry with circularly polarized light. The synthesis of optically active hexahelicene. J Am Chem Soc. 1971;93 (9):2353–54.

[29] Nakazaki M, Yamamoto K, Fujiwara K, Maeda M. Mechanically directed absolute asymmetric syntheses of helicenes in a twisted nematic mesophase. J Chem Soc Chem Commun. 1979;1086–87.

[30] Tribout J, Martin RH, Doyle M, Wynberg H. Chemical assignment of absolute configuration in the helicene and heterohelicene series Part XXXIV. Tetrahedron Lett. 1972;13(28):2839–42.

[31] Pearson MSM, Carbery DR. Studies toward the photochemical synthesis of functionalized [5]- and [6] carbohelicenes. J Org Chem. 2009;74(15):5320–25.

[32] Vanest JM, Martin RH. Helicenes: A striking temperature dependence in a chemically induced asymmetric photosynthesis. Recl Trav Chim Pays-Bas. 1979;98:113.

[33] Sudhakar A, Katz TJ. Asymmetric synthesis of helical metallocenes. J Am Chem Soc. 1986;108 (1):179–81.

[34] Gilbert AM, Katz TJ, Geiger WE, Robben MP, Rheingold AL. Synthesis and properties of an optically active helical bis-cobaltocenium ion. J Am Chem Soc. 1993;115(8):3199–211.

[35] Carren MC, Herna R, Urbano A, Alder D. Enantioselective approach to both enantiomers of helical bisquinones. J Org Chem. 1999;64(4):1387–90.

[36] Grandbois A, Collins SK. Enantioselective synthesis of [7]helicene: Dramatic effects of olefin additives and aromatic solvents in asymmetric olefin metathesis. Chem-Eur J. 2008;14 (30):9323–29.

[37] Caeiro J, Peña D, Cobas A, Pérez D, Guitián E. Asymmetric catalysis in the [2 + 2 + 2] cycloaddition of arynes and alkynes: Enantioselective synthesis of a pentahelicene. Adv Synth Catal. 2006;348(16–17):2466–74.

[38] Barnett L, Ho DM, Baldridge KK, Pascal RA. The structure of hexabenzotriphenylene and the problem of overcrowded "D(3h)" polycyclic aromatic compounds. J Am Chem Soc. 1999;121 (4):727–33.

[39] Peña D, Pérez D, Guitián E, Castedo L. Synthesis of hexabenzotriphenylene and other strained polycyclic aromatic hydrocarbons by palladium-catalyzed cyclotrimerization of arynes. Org Lett. 1999;1(10):1555–57.

[40] Peña D, Cobas A, Pérez D, Guitián E, Castedo L. Kinetic control in the palladium-catalyzed synthesis of c2-symmetric hexabenzotriphenylene. A conformational study. Org Lett. 2000;2 (11):1629–32.

[41] Peña D, Cobas A, Pérez D, Guitián E, Castedo L. Dibenzo[a,o]phenanthro[3,4-s]pycene, a configurationally stable double helicene: Synthesis and determination of its conformation by NMR and GIAO calculations. Org Lett. 2003;5(11):1863–66.

[42] Xiao S, Myers M, Miao Q, Sanaur S, Pang K, Steigerwald ML, et al. Molecular wires from contorted aromatic compounds. Angew Chem Int Ed. 2005;44(45):7390–94.

[43] Gorodetsky AA, Chiu CY, Schiros T, Palma M, Cox M, Jia Z, et al. Reticulated heterojunctions for photovoltaic devices. Angew Chem Int Ed. 2010;49(43):7909–12.

[44] Ball M, Zhong Y, Wu Y, Schenck C, Ng F, Steigerwald M, et al. Contorted polycyclic aromatics. Acc Chem Res. 2015;48(2):267–76.

[45] Kumar S, Pola S, Huang CW, Islam MM, Venkateswarlu S, Tao YT. Polysubstituted hexa-cata-hexabenzocoronenes: Syntheses, characterization, and their potential as semiconducting materials in transistor applications. J Org Chem. 2019;84(13):8562–70.

[46] Berezhnaia V, Roy M, Vanthuyne N, Villa M, Naubron JV, Rodriguez J, et al. Chiral nanographene propeller embedding six enantiomerically stable [5]helicene units. J Am Chem Soc. 2017;139(51):18508–11.

[47] Pascal RA. Twisted acenes. Chem Rev. 2006;106(12):4809–19.

[48] Qiao X, Padula MA, Ho DM, Vogelaar NJ, Schutt CE, Pascal RA. Octaphenylnaphthalene and decaphenylanthracene. J Am Chem Soc. 1996;118(4):741–45.

[49] Pascal RA, Mcmillan WD, Van Engen D. Remarkably twisted polycyclic aromatic hydrocarbon. J Am Chem Soc. 1986;108(18):5652–53.

[50] Qiao X, Ho DM, Pascal RA. An extraordinarily twisted polycyclic aromatic hydrocarbon. Angew Chem Int Ed. 1997;36(13–14):1531–32.

[51] Lu J, Ho DM, Vogelaar NJ, Kraml CM, Pascal RA. A pentacene with a 144° twist. J Am Chem Soc. 2004;126(36):11168–69.

[52] Duong HM, Bendikov M, Steiger D, Zhang Q, Sonmez G, Yamada J, et al. Efficient synthesis of a novel, twisted and stable, electroluminescent "twistacene.". Org Lett. 2003;5(23):4433–36.

[53] Rickhaus M, Mayor M, Juríček M. Chirality in curved polyaromatic systems. Chem Soc Rev. 2017;46(6):1643–60.

[54] Scott LT. Fragments of fullerenes: Novel syntheses, structures and reactions. Pure Appl Chem. 1996;68(2):291–300.

[55] Mizyed S, Georghiou PE, Bancu M, Cuadra B, Rai AK, Cheng P, et al. Embracing C60 with multiarmed geodesic partners. J Am Chem Soc. 2001;123(51):12770–74.

[56] Georghiou PE, Tran AH, Mizyed S, Bancu M, Scott LT. Concave polyarenes with sulfide-linked flaps and tentacles: New electron-rich hosts for fullerenes. J Org Chem. 2005;70(16):6158–63.

[57] Sygula A, Fronczek FR, Sygula R, Rabideau PW, Olmstead MM. A double concave hydrocarbon buckycatcher. J Am Chem Soc. 2007;129(13):3842–43.

[58] Barth WE, Lawton RG. The corannulene. J Am Chem Soc. 1971;93(1967):1730–45.

[59] Scott LT, Hashemi MM, Meyer DT, Warren HB. Corannulene. A convenient new synthesis. J Am Chem Soc. 1991;113(18):7082–84.

[60] Scott LT, Cheng P, Hashemi MM, Bratcher MS, Meyer DT, Warren HB. Corannulene. A three-step synthesis. J Am Chem Soc. 1997;119(45):10963–68.

[61] Borchardt A, Fuchicello A, Kilway KV, Baldridge KK, Siegel JS. Synthesis and dynamics of the corannulene nucleus. J Am Chem Soc. 1992;114(5):1921–23.

[62] Borchard A, Hardcastle K, Gantzel P, Siegel JS. 1,6,7,10-Tetramethylfluoranthene: Synthesis and structure of a twisted polynuclear aromatic hydrocarbon. Tetrahedron Lett. 1993;34 (2):273–76.

[63] Sygula A, Rabideau PW, Practical A. Large scale synthesis of the corannulene system to represent the polar cap of buckminsterfullerene C 60, was first most recently we have been working on the development of solution-phase syntheses for these systems since, if they could be. J Am Chem Soc. 2000;122:6323–24.

[64] Butterfield AM, Gilomen B, Siegel JS. Kilogram-scale production of corannulene. Org Proc. 2012;16(4):664–76.

[65] Sakurai H, Daiko T, Hirao T. A synthesis of sumanene, a fullerene fragment. Science. 2003;301:1878.

[66] Yamamoto K, Harada T, Okamoto Y, Chikamatsu H, Nakazaki M, Kai Y, et al. Synthesis and molecular structure of [7]circulene. J Am Chem Soc. 1988;110(11):3578–84.

[67] Yamamoto K. Extended systems of closed helicene. Synthesis and characterization of [7] and [7.7]-circulene. Pure Appl Chem. 1993;65(1):157–63.

[68] Kawai K, Kato K, Peng L, Segawa Y, Scott LT, Itami K. Synthesis and structure of a propeller-shaped polycyclic aromatic hydrocarbon containing seven-membered rings. Org Lett. 2018;20(7):1932–35.

[69] Ghasemabadi PG, Yao T, Bodwell GJ. Cyclophanes containing large polycyclic aromatic hydrocarbons. Chem Soc Rev. 2015;44(18):6494–518.

[70] Bodwell GJ, Bridson JN, Cyrański MK, Kennedy JWJ, Krygowski TM, Mannion MR, et al. Nonplanar aromatic compounds. 8.1 synthesis, crystal structures, and aromaticity investigations of the 1,n-dioxa[n](2,7)pyrenophanes. How does bending affect the cyclic π-electron delocalization of the pyrene system?. J Org Chem. 2003;68(6):2089–98.

[71] Merner BL, Dawe LN, Bodwell GJ. 1,1,8,8-tetramethyl[8](2,11)teropyrenophane: Half of an aromatic belt and a segment of an (8,8) single-walled carbon nanotube. Angew Chem Int Ed. 2009;48(30):5487–91.

[72] Jasti R, Bhattacharjee J, Neaton JB, Bertozzi CR. Carbon nanohoop structures. J Am Chem Soc. 2008;130:17646–47.

[73] Xia J, Jasti R. Synthesis, characterization, and crystal structure of [6]cycloparaphenylene. Angew Chem Int Ed. 2012;51(10):2474–76.

[74] Takaba H, Omachi H, Yamamoto Y, Bouffard J, Itami K. Selective synthesis of [12] cycloparaphenylene. Angew Chem Int Ed. 2009;48(33):6112–16.

[75] Omachi H, Matsuura S, Segawa Y, Itami K. A modular and size-selective synthesis of [n]cycloparaphenylenes: A step toward the selective synthesis of [n,n] single-walled carbon nanotubes. Angew Chem Int Ed. 2010;49(52):10202–05.

[76] Segawa Y, Miyamoto S, Omachi H, Matsuura S, Šenel P, Sasamori T, et al. Concise synthesis and crystal structure of [12]cycloparaphenylene. Angew Chem Int Ed. 2011;50(14):3244–48.

[77] Yamago S, Watanabe Y, Iwamoto T. Synthesis of [8]cycloparaphenylene from a square-shaped tetranuclear platinum complex. Angew Chem Int Ed. 2010;49(4):757–59.

[78] Matsui K, Segawa Y, Namikawa T, Kamada K, Itami K. Synthesis and properties of all-benzene carbon nanocages: A junction unit of branched carbon nanotubes. Chem Sci. 2013;4 (1):84–88.

[79] Matsui K, Segawa Y, Itami K. All-benzene carbon nanocages: Size-selective synthesis, photophysical properties, and crystal structure. J Am Chem Soc. 2014;136:16452–58.

[80] Segawa Y, Levine DR, Itami K. Topologically unique molecular nanocarbons. Acc Chem Res. 2019;52(10):2760–67.

Index

(diacetoxyiodo)benzene 103
(MeO)$_2$CHBF$_4$ 350, 353
(phenyl)[*o*-(trimethylsilyl)phenyl]iodonium
    triflate 103
[18] annulene 9
[2+2] cycloadditions 109
[4+2] cycloaddition 106
[4 + 2] cycloaddition 106
[4+4] photodimerization 255
[6]phenacene 298
[7]phenacene 298
[Ir(OMe)(cod)]$_2$ 204
[n]phenylenes 220, 224
1,10-phenanthroline 194, 209
1,1-dimethoxyethylene 110
1,3-dipolar cycloadditions 111
1,4-cyclohexadiene 125–126
1,4-cyclohexanedione 290
1,4-diradical 125–126
18-crown-6 ether 104, 192
1-aminobenzotriazole 101–102
2,2′bipyridine 135
2-methoxyfuran 106
2-methyl-3-butyn-2-ol 155, 157
3-hydroxyindole 3
4,4′-di-*tert*-butyl-2,2′-dipyridyl 204–205, 209

abemaciclib 203, 205
acenes 251–252, 255, 261, 289, 293–295, 310
acetamide 61–62
acetyl hypofluorite 16
acetylides 57
acridones 42
activating groups 20, 24
active pharmaceutical ingredients 1, 3–4, 10
acylium ion 18
addition–elimination mechanism 39
AlCl$_3$ 78, 259, 261, 268–270, 298–300,
    314–316
aldehydes 40, 46
alizarin 2
alkyllithium 51
alkyllithiums 74–75, 85, 88, 100
alkyne benzannulation 279–282
alkyne cyclotrimerization 121–122, 218–221,
    224, 315, 317
alkynes 153, 156–157

altretamine 56
amides 75–76
amidine 203
aminobenzoic acid 101
amphetamine 179
anilines 16, 18, 20, 26, 29, 42, 51, 58,
    61, 64
anionic *N* Fries rearrangement 82
anionic *ortho*-Fries rearrangement 80–83
anisidine 98
anisole 74–75
anisotropy of induced current density 10
anthracene 2, 249, 251–256, 259–260, 264,
    266, 277
anthranilic acid 112
anthraquinone 2, 118–120
anthraquinones 77–79, 217
antiaromatic 5, 7, 9
arenium ion 13
arm-chair periphery 247
aromatic sextet *See* Clar sextet
aromatic stabilization energy 6–7
aryl amines 191–192
aryl bromides 51
aryl cation 59, 61, 67
aryl halide 132–133, 139–140, 142–143,
    145–147, 149–154, 159–160, 164, 168
aryl iodides 64–65
aryl radical 59–60, 65–67
aryl zinc 145–146
arylboronic acids 73, 200
aryllithium 143, 145–146
aryllithium species 73–74, 76–79, 81, 83–86,
    88, 95, 114, 191, 200–201
arylsilanes 73
arylstannanes 73
aryne cyclotrimerization 121–122, 277–278,
    305–308, 310, 341–343
arynes 95, 105, 107–111, 113, 115, 121,
    262–264, 276
Asao–Yamamoto benzannulation 282, 284,
    319, 321
atrazine 56–57
atropisomerism 174–176
azides 112
azobenzenes 63–64, 66
azobis(isobutyronitrile) 28

https://doi.org/10.1515/9783110562682-012

B(OCH$_3$)$_3$ 143
B(OiPr)$_3$ 143
Baeyer–Villiger 78
Baeyer–Villiger oxidation 259
Balz–Schiemann reaction 61–62
Barton–Kellogg olefination 343
BaryPhos 181
bay region 247
BBr$_3$ 62, 78, 179, 192
benzenedithiol 48
benzimidazole 204
benzocyclobutene 291–292
benzocyclobutenes 109, 219–220
benzofuran 78
benzoin condensation 46
benzophenone imine 198
benzophenones 42, 46
benzoquinone 290–291, 333, 338–339, 355
benzothiadiazole-1-1-dioxide 102
benzotriazoles 112
benzylic bromination 349–350
benzyne 52, 95–104, 106–110, 112–114, 116–118,
    120–121, 125, 163, 292, 305, 310, 333
Bergman cyclization 125–126
BHT 24
bianthraquinone 134
biaryl 65, 131–134, 136, 138, 145–147, 150, 161,
    163, 171, 174–177, 179–181
Bidentate nucleophiles 45
BINAP 175, 177, 198, 340
BINOL 175, 177, 179
biphenyl 65
biphenylenes 109
biphenyls 138, 176
biphyscion 134–135
bis(pinacolato)diboron 201–202, 204–205
bis(trimethylsilyl)acetylene 219–221, 223–224
Bischler-Napieralski isoquinoline
    synthesis 242
boronic acid 143–145, 147, 170–172
Breslow intermediate 47
bromoanisole 98–99, 117
bromobenzene 51
bromofluorobenzene 99–100
bromophenols 103
bromopyridines 54
bromoxynil 24
Buchwald–Hartwig reaction 196
butyllithium 290, 294, 298, 309

cacalol 78, 80
caerulomycin C 90–91
calicheamicins 126
carbamates 51, 76, 81–82
carbazole 43
carbazoles 162, 172
carbenoid coupling 335
carbon nanotubes 329, 352, 355
carbonyl groups 42
Castro–Stephens reaction 153
cata-condensed PAHs 247
catechols 45
catenarin 78
Cava reaction 265–266, 292, 306–307, 309
Celecoxib 63
cesium carbonate 195
cesium fluoride 53, 104, 111, 123
C–H insertion 161, 163–164, 167, 170–171
cheletropic reaction 107, 216–217, 295, 310
chiral auxiliaries 177
chirality 174–175, 330, 337–340, 346
chloranil 290, 294, 312
chlorobenzene 40
chloromethylation 348
chloropyridine 55
chlorosulfonation 15
chrysene 252, 261–263, 267, 281, 295
circulenes 347, 350
citalopram 160
Clar sextet 250–252, 254, 256
Clayden rearrangement 51
Clemmensen reduction 27, 259
closed-shell configuration 5
Co$_2$(CO)$_8$ 224
Combes quinolone synthesis 239
complex-induced proximity effect 75
concerted nucleophilic aromatic
    substitution 52
copper metal 132–133, 159, 192
copper(I) salts 59–60, 192
copper(I) thiophene-2-carboxylate 133
corannulene 210, 347–349
coronene 210, 311–314, 342, 347
covalent organic frameworks 45, 192
CpCo(CO)$_2$ 219–220, 224
cross-coupling reaction 131, 133–134, 139–147,
    149–150, 154, 160, 168–169, 173, 176, 179–180
cryptaustoline 113–114
cryptowoline 113–114

Cs$_2$CO$_3$ 196
CsF 308, 340–341
Cu(OAc)$_2$ 194
CuI 153, 157
cuprous chloride 195
cuprous cyanide 159
cuprous iodide 194–195
cyanamid 232
cyanuric acid 56
cyanuric chloride 56–57
cyanuric fluoride 57
cycloadditions 95, 105, 109, 111
cyclobutadiene 5, 7–8
cyclodehydrochlorination 275
cyclodehydrogenation 268, 270–276,
    302–304, 314–315, 317, 319, 322
cycloparaphenylenes 354–358
cyclopentadienones 216
cyclophanes 329, 351–353

Dacarbazine 64
damnacathol 120
DavePhos 166
DDQ 255, 270, 297, 302, 310, 320,
    349–350, 354
deactivating groups 18, 20–21, 23–24, 26
decaphenylanthracene 345
defucogilvocarin V 147–148
dehalogenation 133
dehydrobenzannulenes 155
dehydrogenative coupling 138, 173, 198, 268
dengibsin 84
dengibsinin 84, 145–146
diamagnetic anisotropy 8
diaminocyclohexane 194
diaryl ethers 191
diastereomeric charge transfer complex 337
diazonium carboxylate 95, 101, 107, 345
diazonium salt 39, 58–67, 101, 191, 142, 157,
    159, 225, 330, 350
dibenz[a,c]anthracene 303–304
dibromonitrobenzene 132–133
dichloromethyl methyl ether 19
dichlorotetrazine 47
didehydrobenzene 95
Diels–Alder reaction 106–108, 215–220, 255,
    261–264, 290, 310, 312, 317, 333–334,
    345–347, 349
diethyl malonate 101, 117

difluoroaniline 192, 203
dihaloarenes 100
dihydroxylation 254
diiodoethane 77
dimethyl acetylenedicarboxylate 122
dimethylamine 56, 64
direct arylation 160–163, 165–168,
    170–173, 351
directed metalation groups 75–77, 84
directed *ortho* metalation 74–78, 81, 83, 85,
    92, 103–104, 145, 163
directed remote metalation 82–85
directing effect 41
disproportionation 132, 136
dithianthrene 48
duloxetine 43
dyes 1–2, 4
dynamic covalent chemistry 47
dynemicin 107–108
dynemicins 126

Elbs reaction 260–261, 290
electron-withdrawing group 41
electrophilic aromatic substitution 13, 162
elimination–addition mechanism 52, 97
ene reaction 110
enediyne 125
enolate 114, 117
ent-clavilactone B 115
erythroglaucin 78
esperamicins 126
estrone 220
ethylmagnesium bromide 290
eupolauramine 84

FeCl$_3$ 138–139, 176, 268, 270, 302–303,
    318–319, 343
felodipine 231
Fischer indole synthesis 234–236
fjord region 247
flash vacuum pyrolysis 341, 347–349
fluorenones 84, 123–124, 145, 162
fluoride 41, 43, 47, 51, 54, 61–62
fluorobenzenes 45
fluoronaphthalene 43
fluoronitrobenzene 51
fluorosulfonate 53
fluoxetine 43
flupirtine 55–56

formyl group 77–78, 168
formylation 78
fredericamycin A 78
Friedel–Crafts acylation 17–18, 27–28,
    256–259, 261, 296, 299, 301, 309, 311,
    331, 349
Friedel–Crafts alkylation 17–18, 27–28, 30
Friedländer quinoline synthesis 241–242
Fries rearrangement 32
furans 34, 57, 167, 207

gilvocarins 106, 147, 163, 165
Goldberg reaction 191
Gomberg–Bachmann reaction 65–66, 131
gossypol 174, 178, 181–182
graphene nanoribbons 281, 317–323
Grignard 107, 115
Grignard reaction 301–302
Grignard reagents 73, 82, 95, 139, 149, 153,
    177, 200, 259, 262
Grubbs catalyst 277, 350
gymnopusin 84

halodesilylation 31
halogen dance reaction 86–92
halogenation 16
halomethylation 18
halomethylium ion 18
Hantzsch pyridine synthesis 231
hasubanonine 225
Hauser annulation 118
Haworth synthesis 256–258, 296–297,
    299–302
Heck reaction 150–153, 238–239
helicenebisquinones 333–334
helicenes 329, 336, 340
heptacene 251–252, 293–295, 310
heteropolymolybdovanadic acid 173
hexabenzotriphenylene 340–342
hexacene 293–294
hexamethyldisilazane 103
hexa-peri-hexabenzocoronene 271, 314–315
hexaphenylbenzene 216, 315
hexaphenylbenzenes 89–90
Hinsberg thiophene synthesis 229
Hiyama coupling 148–149
Hiyama–Denmark reaction 149
homocoupling 131–137, 139, 172
Hückel's rule 5

hydrazine 40, 229, 232
hydrazone 101
hydrogenation 6, 254–255
hypophosphorous acid 60

ICl 280–281
imidazoles 229
imidazolium salt 46, 53
indigo 2–3, 62
indoles 234–239
indolyne 114–115
inductive effects 95, 97–98, 118
iodine monochloride 16
iodomethane 64, 77, 79, 81, 84
isoamyl nitrite 58, 101, 345
isobenzofurans 263–264, 292, 345–346
isodesmic reactions 7
isoquinolines 234, 242–243

Japp–Klingemann reaction 236–237

ketorolac 57
kinetic isotope effects 53–54
kinetic versus thermodynamic control 33
Knorr pyrrole synthesis 228
Knorr quinolone synthesis 239–240
Kolbe–Schmitt process 23
K-region 247, 252–255, 277
Kumada–Corriu coupling 149–150, 335

lactonization 82
Larock heteroannulation 237–238
Lawesson's reagent 342
leaving group 39, 41, 49–50, 53, 55
leaving groups 41–42, 55
ligands 141–142, 180, 192, 196
linezolid 194
liquid crystal 301
liquid crystal display 4
liquid crystals 62, 84, 131
lithium diisopropyl amide 75, 79, 82–83, 88,
    90, 92
lithium diisopropylamide 134, 145, 147
lithium mercury amalgam 100
lithium naphthalenide 356
lithium tetramethylpiperide 79, 88, 292
lithium–halogen exchange 73, 85–86, 88–89,
    100, 103, 114, 143, 157, 298, 303, 307, 318,
    320–321, 333, 338, 350, 355, 357–358

losartan 144–145
luminescence 43–45, 155, 199–200
lycorines 108
lysergic acid 108–109

maleic anhydride 255, 262, 333, 341
malonic esters 45
malononitrile 168
martinellic acid 193
mauveine 2
McMurry coupling 335–336, 350, 354
mCPBA 217, 350
Meisenheimer complex 39–41, 49–50
melleine 117
menthyl groups 337
methoxymethyl ether 76–77
methyl thioglycolate 62
methylpiperazine 43
Michael acceptor 118
migratory insertion 123, 150–151, 220, 237
Mitsunobu reaction 62
Miyaura borylation 181–182, 200–204, 209
Miyaura coupling 225, 357
MoCl₅ 270, 302
multicomponent reaction 113, 115–116
mumbaistatin 120

N,N'-difluoro-1,4-diazoniabicyclo[2.2.2]octane
    salts 16
N,N-dimethylglycine 195
nanographenes 316, 319
Nanokid 157–158
naphthalyne 207, 262, 292, 294–295
naphthoquinone 306, 309
naphthoquinones 217
naproxen (TM) 153, 175
N-bromosuccinimide 16, 28
n-Bu₃SnEt₂ 196
n-BuLi See n-butyllithium
n-butyllithium 74–75
N-chlorosuccinimide 16
Negishi coupling 145–146
N-fluoro-bis(trifluoromethansulfonyl)amine 16
NHCs See N-heterocyclic carbenes
N-heterocyclic carbenes 46
Ni(COD)₂ 135, 305
Ni(PPh₃)₄ 135
nitration 14, 191
nitrene 101

nitric acid 14–15, 29
nitronium tetrafluoroborate 15
N-methylcrinasiadine 114
NO₂⁺ 14
nomenclature of PAHs 247
nonacene 294
norbornadiene 347, 349–350
N-phenylmaleimide 333
nuclear magnetic resonance 8
nucleophiles 42–43, 46, 49–52, 58–59, 95, 97,
    113
nucleophilic addition 39, 41, 53, 55, 59
nucleophilic aromatic substitution 39–47, 50,
    52–55, 57, 62, 192, 198, 203, 209
nucleus-independent chemical shift 9

o-bis(trimethylsilyl)benzene 103
ochratoxin 81
octacene 294
ofloxacin 43–44
olefin metathesis 224–225, 277–279, 335–336,
    339–340
open-shell configuration 5
o-quinodimethanes 219–220, 265, 291–292
organic light-emitting diodes 4
organosilane 149
organotin reagents 147
OsO₄ 254, 294
o-trimethylsilyl aryl triflate 82, 95, 102–104,
    112, 121
oxazoles 229–230
oxazolines 76
oxidation of PAHs 253
oxidation of the benzylic positions 28
oxidative addition 60, 132–133, 136, 140,
    142–143, 153–154, 159, 161–162, 164, 168,
    192, 196, 202, 205, 237
o-xylylenes. See o-quinodimethanes

Paal–Knorr furan synthesis 227
Paal–Knorr pyrrole synthesis 228
palladium 62, 134, 139–143, 146, 148–151,
    153–154, 159–162, 166–168, 170–173, 180
palladium-catalyzed aryl amination 195
papaverine 243
Pb(OAc)₄ 101
PCl₅ 55
Pd(OAc)₂ 124, 141, 145, 150, 162, 164, 166, 170,
    172

Pd(PPh₃)₂Cl₂ 141
Pd(PPh₃)₄ 121–122, 141, 147, 305, 308
Pd₂(dba)₃ 122, 141, 160, 180–181, 310,
      340–341
pentacene 252–253, 256, 258, 290–293, 311
pentacenequinone 290–291, 343
pentacenone 342
pentarylene 299
peracetic acid 217
peri-condensed PAHs 247
periodic acid 16
Perkin condensation 260
perylene 206, 299–300, 312
pharmaceutical 43, 61
pharmaceuticals 3, 131, 147
phenacenes 295–296, 298
phenanthrene 122, 249, 251, 257, 260,
      267, 278, 281, 295–296, 309, 331,
      333–334, 341
phenanthrenequinone 266
phenanthridine 124
phenanthrols 84
PhenoFluor 53
phenols 16, 20, 29, 32, 47, 53, 58–59, 76
phenoxazines 45
phenylhydrazine 234, 236
phenylhydrazines 63
phenyliodine(III) diacetate 357
phenyllithium 96, 314
PhNTf₂ 103
photocyclization 267–268, 298, 303, 305,
      313–314, 331–333, 337–338, 341, 343
photodecarbonylation 294–295
photodimerization 289, 291, 293
photooxidation 289, 291, 293
phthalaldehyde 290
phthalic anhydride 258–259
picene 252, 262, 295–297
Pictet–Spengler reaction 242
pinacol boronate ester 144
pivalic acid 166–167, 172
pKₐ
 – of benzene 74
POCl₃ 19, 55, 203
polycyclic aromatic hydrocarbons 95, 107, 122
polymers 45
polyphosphoric acid 296
Pomeranz–Fritsch reaction 243
potassium amide 87–88, 97

potassium anilide 87
potassium aryl trifluoroborates 144
potassium carbonate 192, 194–195
potassium ferricyanide 62, 139
potassium hydride 52–53
potassium hydrogen fluoride 144
potassium hydroxide 42
potassium iodide 292, 315
potassium t-butoxide 96, 203, 209
propylene oxide 267
protiodesilylation 31, 220, 223
protodeborylation 200
Pschorr cyclization 67, 131, 260, 296, 330
Pschorr reaction 67
pyrazine 36
pyrazines 56, 230–232
pyrazoles 230
pyrene 206, 210, 252–254, 258–259,
      266–267
pyridazine 36
pyridazines 56, 230–233
pyridine 36, 195, 208
pyridine N-oxides 36
pyridine-N-oxide 169
pyridines 55–57, 164, 168, 230–231
pyridone 79
pyridones 55
pyridyl ethers 55
pyrimidine 36
pyrimidines 56, 230–231, 233
pyrrole 34–35, 207–208
pyrroles 57, 167

quaterrylene 299–300
quinizarin 293
quinoline 208
quinolines 234, 239–241
quinones 253, 258, 265, 333

Ramirez reaction 343
Raney nickel 290
reduction of acyl groups 27
reduction of nitro groups 26
reductive elimination 60, 123, 132, 136,
      140–141, 143, 150, 153–154, 159, 161, 164,
      168, 170–171, 192, 196, 202, 205, 237
regiochemistry 21, 97–99, 160, 162–163,
      167–168, 172, 206, 209, 253–255,
      257–258, 283, 331

regiochemistry of electrophilic aromatic
    substitution 20, 41, 97–99, 117–118
retro-Brook reaction 103
reversibility 47
Rieche formylation 19, 354
rizatriptan 237–238
Rosendmund–von Braun reaction 159–160
rubiadin 120
RuO$_3$ 254
rylenes 299–300

Sandmeyer reaction 59–60
Scholl reaction 268–270, 299–300, 306, 315,
    318–320
S$_E$Ar mechanism 13
sec-butyllithium 75
selenium 301
semiconductors 301
sildenafil 229
silyl migration 103–104
Singulair™ 151–152
Skraup reaction 240
Smiles rearrangement 48–49, 116, 120–121
S$_N$1 reaction 59
SnCl$_2$ 63, 291
sodium amide 51–52, 87–88, 95–97, 99, 109,
    113–114, 117
sodium borohydride 60
sodium hydride 43, 46, 50, 313, 350
sodium iodide 65
sodium nitrite 58, 61–62, 64, 101
sodium sulfide 42, 313
sodium t-butoxide 196
Sonogashira coupling 153–159, 221–223, 315,
    320, 335
sparteine 179
stability of aromatic compounds 6
stannous chloride 26
starphenes 306, 308–310
stereoselective synthesis 337
Stevens rearrangement 313, 350, 353
stilbenes 260, 267
Stille coupling 146–148, 168
Strating-Zwanenburg reaction 294
strychnine 235
succinic anhydride 256, 258, 296, 299,
    301–302
sulfonamides 76

sulfonanilide 195
sulfonation 15, 30–31
sulfones 49, 76
sulfuric acid 61, 78
sulfuryl fluoride 53
sumanene 349–350
supramolecular chemistry 47
Suzuki–Miyaura coupling 143–145, 147, 163,
    170, 179–182, 200, 202–204, 266, 278,
    300, 303, 318, 320, 335, 343–344, 351,
    356–358
Swern oxidation 179

TADF See thermally activated delayed
    fluorescence
taiwanins 122
taranabant 160
t-butyllithium 75, 114
teropyrene 353
terrylene 299
tetraalkylammonium tribromides 16
tetrabutylammonium fluoride 104, 111, 264
tetracene 252, 256
tetrafluoroboric acid 60, 62
tetrafluorophthalonitrile 48
tetrafluoroterephthalonitrile 45
tetrahydropyrene 255, 258
tetramethylammonium fluoride 53
tetramethylethylenediamine 75
tetramethylethylynediamine 75, 81
tetraphenylanthracene 107
tetraphenylcyclopentadienone 106–107, 216
the harmonic oscillator model of aromaticity 8
thermally activated delayed fluorescence 44,
    199, 203
thianthrene 48
thiazoles 229
thiocarbamates 51
thioindigo 62–63
thiophene 34–35, 207
thiophene-1,1-dioxides 217
thiophenes 57, 139, 147, 167
thioxanthones 42
thymol 61
thymoxamine 61
TiCl$_4$ 19, 78
tin(II) chloride 157
transmetallation 140, 143, 149, 154, 170

triarylamines 194, 198–199
triazenes 63–65, 155, 321
triazines 56–57
tribromobenzene 61
triflic acid 103
triflic anhydride 103
trifluorobenzonitrile 41
trifluoroperacetic acid 217
triisopropylsilyl acetylene 290
trimethylsilyl acetylene 155, 157, 335
trimethylsilyl chloride 77, 103
trimethylsilyl iodide 65
trinaphthylenes 306–308
trioxaazatriangulene 192–193
triphenylene 121–122, 252, 258, 275
triphenylenes 301–307
triptycene 107
trisphaeridine 114
Truce–Smiles rearrangement 50–51
twisted acenes 345

Ullmann coupling 132–135, 176–178, 191–192
Ullmann reaction 132–135, 176–178
umpolung 46
undecacene 294
ureas 51
UV-visible absorption spectra 252

valence isomerization/dehydrogenation 353
Vilsmeier–Haack reaction 19, 35, 242
vinylnaphthalene 261–262
vinyltributyltin 148
viriditoxin 180

welwitindolinone alkaloids 114
Wheland intermediate 13
Wittig reaction 266, 278, 298, 337
Wolff–Kishner reduction 27, 259, 296, 331

xanthate 42
xanthones 42, 47
xantphos 198
xerography 194

Yamamoto coupling 137, 305–306, 308–310, 318, 343–344, 358, 360

zigzag periphery 247
zinc 293, 301, 306, 309
Zn(CN)$_2$ 159–160
ZnCl$_2$ 146

β-hydride elimination 123, 141, 149–151
σ-bond insertion 117–119

www.ingramcontent.com/pod-product-compliance
Lightning Source LLC
Chambersburg PA
CBHW080707220326
41598CB00033B/5332